Advanced Structured Materials

Volume 2

For further volumes:
http://www.springer.com/series/8611

Andreas Öchsner • Graeme E. Murch
Editors

Heat Transfer in Multi-Phase Materials

Springer

Editors
Andreas Öchsner
University of Technology Malaysia
Faculty of Mechanical Engineering
Department of Solid Mechanics
and Design
Johor
Malaysia

and

The University of Newcastle
Centre for Mass and Thermal Transport
in Engineering Materials
School of Engineering
Australia

Prof. Dr. Graeme E. Murch
University of Newcastle
Centre for Mass and Thermal Transport
in Engineering Materials
School of Engineering
Callaghan New South Wales
Australia

ISSN 1869-8433
ISBN 978-3-642-27039-0 ISBN 978-3-642-04403-8 (eBook)
DOI 10.1007/978-3-642-04403-8
Springer Heidelberg Dordrecht London New York

Cover design: SPi Publisher Services

Printed on acid-free paper

Springer is part of Springer Science+Business Media (www.springer.com)

Preface

In many demanding applications such as automotive or aerospace, common engineering materials reaching their limits and new developments are required to fulfill the increasing demands on performance, characteristics, and functions. The properties of materials can be increased, for example, by combining different materials to achieve better properties than a single constituent or by shaping the material or constituents into a specific structure. Many of these new materials reveal a much more complex behavior than traditional engineering materials due to their advanced structure or composition. The purpose of this book is to cover one of the important physical characteristics, that is thermal properties, in detail from different points of view. This book aims to provide readers not only with a good understanding of the fundamentals but also with an awareness of recent advances in properties determination and applications of multiphase materials. The book contains 14 chapters written by experts in the relevant fields from academia and from major national laboratories/research institutes.

The first part of the book covers materials where two or more solid phases form the composite. The second part is related to porous and cellular materials where two or more solid phases form certain shapes of cells with an empty or air-filled space. Typical representatives of this group are foamed polymers or metals, which have a significant potential in multifunctional applications. The last part of the book covers problem where fluids in a solid structure fulfill technical functions – such as in the case of combustion – or significantly determining the overall characteristics of the material.

The editors wish to thank all the chapter authors for their participation and cooperation, which made this text possible.

Finally, we would like to thank the team at Springer, especially Dr. Christoph Baumann, for their excellent cooperation during the whole phase of the project.

January 2011

Andreas Öchsner
Graeme E. Murch

Contents

Contributors

Khairul Alam Department of Mechanical Engineering, Ohio University, Athens, OH 45701, USA, alam@ohio.edu

Mihnea S. Anghelescu Department of Mechanical Engineering, Ohio University, Athens, OH 45701, USA, ma138903@ohio.edu

Marcelo Antunes Centre Català del Plàstic, Departament de Ciència dels Materials i Enginyeria Metallúrgica, Universitat Politècnica de Catalunya, C/Colom 114, 08222 Terrassa, Barcelona, Spain, marcelo.antunes@upc.edu

Dominique Baillis Université de Lyon, CNRS, INSA-Lyon, CETHIL, UMR5008, 69621 Villeurbanne, France and Université Lyon 1, 69622 Villeurbanne, France, dominique.baillis@insa-lyon.fr

A. Barletta Dipartimento di Ingegneria Energetica, Nucleare e del Controllo Ambientale (DIENCA), Alma Mater Studiorum, Università di Bologna, Viale Risorgimento 2, 40136 Bologna, Italy, antonio.barletta@unibo.it

I.V. Belova The University of Newcastle, Callaghan, NSW 2308 Australia, Irina.Belova@newcastle.edu.au

Helmut J. Böhm Institute of Lightweight Design and Structural Biomechanics, Vienna University of Technology, Gußhausstr. 27–29, 1040, Vienna, Austria, hjb@ilsb.tuwien.ac.at

Adrian Bradu Department of Mechanical Engineering, Ohio University, Athens, OH 45701, USA, ab315905@ohio.edu

Thiago A. Carniel Department of Mechanical Engineering, Center for Technological Sciences, Santa Catarina State University-UDESC, Campus Avelino Marcante, Joinville SC, 89223-100, Brazil, rokkyy@gmail.com

Rémi Coquard Société Etude Conseils Calcul en Mécanique des Structures (ECMS), 66, boulevard Niels, Bohr, 69603 Villeurbanne Cedex, France, remi.coquard@ec2-ms.fr

Manuel Ernani Cruz Federal University of Rio de Janeiro, Politécnica/COPPE/UFRJ, Av. Horácio Macedo 2030, 68503, 21941-972, Rio de Janeiro, RJ, Brazil, manuel@mecanica.coppe.ufrj.br

Marcelo J.S. de Lemos Instituto Tecnológico de Aeronáutica – ITA, 12228-900, São José dos Campos, São Paulo, Brazil, delemos@ita.br

T. Fiedler The University of Newcastle, Callaghan, NSW 2308 Australia, Thomas.Fiedler@newcastle.edu.au

José L. Lage Mechanical Engineering Department, Bobby B. Lyle School of Engineering, Southern Methodist University, Dallas, TX 75275-0337, USA, JLL@SMU.EDU

J.Z. Liang Research Division of Green Function Materials and Equipment, College of Industry Equipment and Control Engineering, South China University of Technology, Guangzhou 510640, P.R. China, liangjz@scut.edu.cn

Carlos Frederico Matt Electric Power Research Center, Av. Horácio Macedo, 354 CP 68007, Rio de Janeiro 21941-590, RJ, Brazil, cfmatt@cepel.br

L.F. Menezes Department of Mechanical Engineering, Universidade de Coimbra, 3030 Coimbra, Portugal, luis.menezes@dem.uc.pt

Pablo A. Muñoz-Rojas Department of Mechanical Engineering, Center for Technological Sciences, Santa Catarina State University-UDESC, Campus Avelino Marcante, Joinville SC, 89223-100, Brazil, pablo@joinville.udesc.br

Sergio Nogales Institute of Lightweight Design and Structural Biomechanics, Vienna University of Technology, Gußhausstr. 27–29, 1040, Vienna, Austria, snogales@ilsb.tuwien.ac.at

Andreas Öchsner Faculty of Mechanical Engineering, University of Technology Malaysia, 81310, UTM Skudai Johor, Malaysia and The University of Newcastle, Callaghan, NSW 2308, Australia, Andreas.Oechsner@gmail.com

Heinz E. Pettermann Institute of Lightweight Design and Structural Biomechanics, Vienna University of Technology, Gußhausstr. 27–29, 1040, Vienna, Austria, pettermann@ilsb.tuwien.ac.at

Jaona Randrianalisoa Université de Lyon, CNRS, INSA-Lyon, CETHIL, UMR5008, 69621 Villeurbanne, France and Université Lyon 1, 69622 Villeurbanne, France, jaona.randrianalisoa@insa-lyon.fr

Miguel Ángel Rodríguez-Pérez Cellular Materials Laboratory (CellMat), Condensed Matter Physics Department, University of Valladolid, 47011 Valladolid, Spain, marrod@fmc.uva.es

Emilio C.N. Silva Department of Mechatronics and Mechanical Systems Engineering, Mechanical Engineering Building, University of São Paulo, São Paulo SP, 05508-900, Brazil, ecnsilva@usp.br

Ramvir Singh Department of Physics, University of Rajasthan, Jaipur 302 004, India, singhrvs@rediffmail.com, rvs2020@gmail.com

Eusebio Solórzano Cellular Materials Laboratory (CellMat), Condensed Matter Physics Department, University of Valladolid, 47011 Valladolid, Spain, esolo@fmc.uva.es

Anthony G. Straatman Department of Mechanical and Materials Engineering, The University of Western Ontario, London, ON, Canada N6A 5B9, astraatman@eng.uwo.ca

F. Teixeira-Dias GRIDS–DAPS Division of Armour & Protection Systems, Department of Mechanical Engineering, Universidade de Aveiro, Campus Universitário de Santiago, 3810-193 Aveiro, Portugal, ftd@ua.pt

José Ignacio Velasco Centre Català del Plàstic, Departament de Ciència dels Materials i Enginyeria Metallúrgica, Universitat Politècnica de Catalunya, C/Colom 114, 08222 Terrassa, Barcelona, Spain, jose.ignacio.velasco@upc.edu

Part I
Composite Materials
(two or more solid phases)

Continuum Modeling of Diffusive Transport in Inhomogeneous Solids

Helmut J. Böhm, Heinz E. Pettermann, and Sergio Nogales

Abstract General features of homogenization and localization in studying the conduction behavior of inhomogeneous materials are introduced and two groups of methods for solving such problems are presented. First, mean field and bounding approaches are discussed and comparisons between the predictions of relevant methods are given. Next, modeling approaches to studying discrete microstructures are covered, the main emphasis being put on periodic homogenization and windowing procedures. Finally, an application of the methods to diamond particle reinforced aluminum is presented, in which interfacial effects play an important role.

1 Introduction

The mathematical description of diffusive transport rests, on the one hand, on Fick's first law, which postulates that the diffusive flux goes from regions of high values of some conserved quantity, Φ, to regions of low values. It can be written formally as

$$\mathbf{j}(\mathbf{x}) = -\mathscr{D}(\mathbf{x})\,\mathbf{g}(\mathbf{x}), \tag{1}$$

where $\mathbf{j}(\mathbf{x})$ is the vector of the diffusive flux, $\mathbf{g}(\mathbf{x}) = \nabla\Phi(\mathbf{x})$ is the gradient of the conserved quantity (also referred to as intensity), and $\mathscr{D}(\mathbf{x})$ is a scalar diffusivity or a diffusivity tensor of order two describing a position dependent material behavior. In the absence of sources and sinks, the flux is divergence-free, $\nabla\mathbf{j}(\mathbf{x}) = 0$, which, in combination with (1) leads to the description of diffusive phenomena by Laplace equations of the type

$$\nabla(\mathscr{D}(\mathbf{x})\,\nabla\Phi(\mathbf{x})) = 0. \tag{2}$$

H.J. Böhm (✉), H.E. Pettermann, and S. Nogales
Institute of Lightweight Design and Structural Biomechanics, Vienna University of Technology, Gußhausstr. 27–29, A-1040, Vienna, Austria
email: hjb@ilsb.tuwien.ac.at, pettermann@ilsb.tuwien.ac.at, snogales@ilsb.tuwien.ac.at

A. Öchsner and G.E. Murch (eds.), *Heat Transfer in Multi-Phase Materials,*
Adv Struct Mater 2, DOI 10.1007/8611_2010_43,

In combination with appropriate boundary conditions and, where appropriate, interfacial conditions, the partial differential equation (2) pertains to steady-state diffusive transport in both homogeneous and inhomogeneous media.

A number of phenomena in inhomogeneous solids, among them thermal and electrical conduction, electrostatic and magnetostatic behaviors, species diffusion (including moisture diffusion giving rise to hygral strains and stresses in polymeric constituents of composites), and antiplane elasticity can be described by formalisms of the above type, see, e.g., [22, 42, 59]. In what follows the emphasis is put on the thermal conduction behavior, i.e., the conserved quantity, Φ, is enthalpy and the equations are typically written in terms of the temperature, T. The flux vector then is the heat flux, $\mathbf{q}$, the gradient vector is the temperature gradient, $\mathbf{d} = \nabla T$, and the material property tensor is the conductivity tensor, $\mathscr{K}$, its inverse being referred to as the resistivity tensor, $\mathscr{R} = \mathscr{K}^{-1}$. All results pertaining to thermal conduction can be directly applied to the other diffusion phenomena listed above by switching in the appropriate variables.

The present chapter provides a basic discussion of models for studying thermal conduction in inhomogeneous materials showing at least two, well separated, characteristic length scales, viz., a macroscale pertinent to samples, components and structures, at which the heterogeneity of the media is not evident, and a microscale, where regions occupied by the different constituents (or phases) are clearly distinguishable. Pertinent materials are, e.g., composite materials, which typically display a matrix–inclusion topology on the microscale, polycrystals, all grains of which are topologically equivalent, and porous as well as cellular materials, the void phase of which is usually treated as being non-conducting.

On the one hand, the thermal conductivity of inhomogeneous media that consist of dissimilar constituents with conductivities $\mathscr{K}^{(\mathrm{p})}$ is to be described in terms of the behavior of an energetically equivalent, homogeneous "comparison material" that shows an effective conductivity $\mathscr{K}^*$. The task of deducing $\mathscr{K}^*$ from the spatial arrangement as well as the material properties of the phases (and, where appropriate, from the behavior of the interfaces between them) is referred to as homogenization. On the other hand the fluxes and gradients on the microscale (referred to as the microfields) are to be obtained for given conditions on the macroscale, a task referred to as localization. These two main tasks in analyzing the behavior inhomogeneous solids are accomplished by applying continuum-level models.

Estimates of the macroscopic conduction properties of inhomogeneous materials have elicited scientific interest for some 135 years, and a fair number of theoretical and semi-empirical models have been devised [22, 52, 59]. In modeling work published over the past 50 years two strands of development have played important roles. On the one hand, analytical and semi-analytical estimates and bounds have been developed that explicitly or implicitly use statistical information on the phase geometry within some representative volume element. Some of the resulting mean field estimates and bounding methods will be discussed in Sect. 2. On the other hand, the thermal fields in specific phase arrangements may be evaluated at high accuracy, in most cases by using numerical engineering methods. Such "discrete microfield approaches" are presented in Sect. 3 and are discussed in detail in other chapters of the present work.

All methods discussed in the following have close analogues in continuum micromechanics of materials, see, e.g., [5, 7], a research field that focuses on modeling the mechanical behavior of inhomogeneous materials. Due to the lower order of the tensors involved in describing diffusion phenomena the latter tend, however, to be somewhat simpler to model than (thermo)mechanical behaviors. In addition, the effects of material symmetries are less complex in conduction problems, with, e.g., geometries of cubic symmetry giving rise to macroscopically isotropic conduction behavior [46].

Throughout the present chapter scalars are denoted by variables typeset in italic fonts or by Greek letters. For vectors and tensors (of order 2) minor case letters in bold and upper case letters in calligraphic fonts, respectively, are used. Volume averages are denoted as

$$\langle f \rangle = \frac{1}{\Omega_\mathrm{s}} \int_{\Omega_\mathrm{s}} f(\mathbf{x})\, \mathrm{d}\Omega, \tag{3}$$

where Ω_s is a suitable volume element. Provided this volume element is sufficiently large and shows no significant macroscopic gradients of composition or of the heat flux and intensity fields, the latter can be split into "slow" and "fast" contributions,

$$\mathbf{q}(\mathbf{x}) = \langle \mathbf{q} \rangle + \mathbf{q}'(\mathbf{x}) \quad \text{and} \quad \mathbf{d}(\mathbf{x}) = \langle \mathbf{d} \rangle + \mathbf{d}'(\mathbf{x}), \tag{4}$$

respectively. Here $\langle \mathbf{q} \rangle$ and $\langle \mathbf{d} \rangle$ are the macroscopic (slow) fields, whereas $\mathbf{q}'(\mathbf{x})$ and $\mathbf{d}'(\mathbf{x})$ stand for the microscopic fluctuations, for which the relation $\langle \mathbf{q}' \rangle = \langle \mathbf{d}' \rangle = \mathbf{0}$ holds.

In the present chapter special emphasis is put on interactions between effective material properties and microstructure that go beyond phase volume fraction effects; in the case of composites they involve the topology of the phase arrangement, the shapes and positions of inhomogeneities as well as their orientation and size distributions. The discussion focuses mainly on engineering-relevant issues and is neither intended to provide a formal review of the state of the art in conduction modeling nor to go into mathematical details of the methods. Readers interested in the latter aspects are referred to the books by Torquato [59] and Milton [42] as well as to [19] and [35].

2 Mean Field Estimates and Bounds for Conduction Properties

Mean field estimates and many bounding methods for studying the thermal conduction behavior of inhomogeneous materials are formulated in terms of phase averaged gradients and fluxes,

$$\langle \mathbf{d} \rangle^{(\mathrm{p})} = \frac{1}{\Omega^{(\mathrm{p})}} \int_{\Omega^{(\mathrm{p})}} \mathbf{d}(\mathbf{x}) \,\mathrm{d}\Omega \quad \text{and} \quad \langle \mathbf{q} \rangle^{(\mathrm{p})} = \frac{1}{\Omega^{(\mathrm{p})}} \int_{\Omega^{(\mathrm{p})}} \mathbf{q}(\mathbf{x}) \,\mathrm{d}\Omega, \tag{5}$$

respectively, where the superscript (p) denotes a constituent and $\Omega^{(\mathrm{p})}$ the volume occupied by it. From the definition of phase averaging the relations between the macroscopic fields and the phase averaged (or mean) microscopic fields follow immediately as

$$\langle \mathbf{d} \rangle = \sum_{(\mathrm{p})} \xi^{(\mathrm{p})} \langle \mathbf{d} \rangle^{(\mathrm{p})} \quad \text{and} \quad \langle \mathbf{q} \rangle = \sum_{(\mathrm{p})} \xi^{(\mathrm{p})} \langle \mathbf{q} \rangle^{(\mathrm{p})}, \tag{6}$$

provided the interfaces between the phases are perfect. Here $\xi^{(\mathrm{p})} = \Omega^{(\mathrm{p})} \big/ \sum_{(\mathrm{k})} \Omega^{(\mathrm{k})}$ stands for the volume fraction of constituent (p) and the sums run over all phases.

The phase averaged fields are linked to the macroscopic fields by the (phase averaged) gradient and flux concentration tensors, $\overline{\mathscr{A}}^{(\mathrm{p})}$ and $\overline{\mathscr{B}}^{(\mathrm{p})}$, defined via the relations

$$\langle \mathbf{d} \rangle^{(\mathrm{p})} = \overline{\mathscr{A}}^{(\mathrm{p})} \langle \mathbf{d} \rangle \quad \text{and} \quad \langle \mathbf{q} \rangle^{(\mathrm{p})} = \overline{\mathscr{B}}^{(\mathrm{p})} \langle \mathbf{q} \rangle, \tag{7}$$

respectively. Obviously, the concentration tensors depend on both the phase arrangement and the phase properties. From (6) and (7) the concentration tensors can be seen to fulfill the relationships

$$\sum_{(\mathrm{p})} \xi^{(\mathrm{p})} \overline{\mathscr{A}}^{(\mathrm{p})} = \mathscr{U} \quad \text{and} \quad \sum_{(\mathrm{p})} \xi^{(\mathrm{p})} \overline{\mathscr{B}}^{(\mathrm{p})} = \mathscr{U}, \tag{8}$$

where $\mathscr{U}$ denotes the unit tensor of rank two.

In the present section linear conduction behavior is assumed at the phase and macroscopic levels, i.e.,

$$\begin{aligned} \langle \mathbf{q} \rangle^{(\mathrm{p})} &= -\mathscr{K}^{(\mathrm{p})} \langle \mathbf{d} \rangle^{(\mathrm{p})} \qquad & \langle \mathbf{d} \rangle^{(\mathrm{p})} &= -\mathscr{R}^{(\mathrm{p})} \langle \mathbf{q} \rangle^{(\mathrm{p})} \\ \langle \mathbf{q} \rangle &= -\mathscr{K}^{*} \langle \mathbf{d} \rangle & \langle \mathbf{d} \rangle &= -\mathscr{R}^{*} \langle \mathbf{q} \rangle \end{aligned} \tag{9}$$

in accordance with (1). The superscript asterisk is used to denote effective properties.

With the exception of Sects. 2.6 and 4 the interfaces between phases are assumed to be perfect, i.e., there are no temperature jumps across interfaces.

2.1 General Relations

By using (6)–(9) the effective conductivity and resistivity tensors of an inhomogeneous material can be obtained from the phase properties and concentration tensors as

$$\mathscr{K}^{*} = \sum_{(\mathrm{p})} \xi^{(\mathrm{p})} \mathscr{K}^{(\mathrm{p})} \overline{\mathscr{A}}^{(\mathrm{p})} \quad \text{and} \quad \mathscr{R}^{*} = \sum_{(\mathrm{p})} \xi^{(\mathrm{p})} \mathscr{R}^{(\mathrm{p})} \overline{\mathscr{B}}^{(\mathrm{p})}. \tag{10}$$

Accordingly, once the phase concentration tensors are known both the homogenization and localization problems are solved within the mean field framework, which explains the central role of the concentration tensors in the present section.

The gradient and flux concentration tensors of a given phase (p) are linked by the equations

$$\overline{\mathscr{A}}^{(\mathrm{p})} = \mathscr{R}^{(\mathrm{p})}\overline{\mathscr{B}}^{(\mathrm{p})}\mathscr{K}^{*} \quad \text{and} \quad \overline{\mathscr{B}}^{(\mathrm{p})} = \mathscr{K}^{(\mathrm{p})}\overline{\mathscr{A}}^{(\mathrm{p})}\mathscr{R}^{*}. \tag{11}$$

If the effective and phase conductivity tensors of a two-phase composite are known, they can be used to generate phase concentration tensors via relations of the type

$$\begin{aligned} \xi^{(\mathrm{m})}\overline{\mathscr{A}}^{(\mathrm{m})} &= (\mathscr{K}^{(\mathrm{m})} - \mathscr{K}^{(\mathrm{i})})^{-1}(\mathscr{K}^{*} - \mathscr{K}^{(\mathrm{i})})\xi^{(\mathrm{i})}\overline{\mathscr{A}}^{(\mathrm{i})} \\ &= (\mathscr{K}^{(\mathrm{i})} - \mathscr{K}^{(\mathrm{m})})^{-1}(\mathscr{K}^{*} - \mathscr{K}^{(\mathrm{m})}), \end{aligned} \tag{12}$$

where (m) and (i) denote the matrix and reinforcement (fibers, particles) phases.

2.2 *Dilute Inhomogeneities*

In order to generate expressions for the gradient and flux concentration tensors required in mean field methods, recourse is typically made to inclusions or inhomogeneities of ellipsoidal shape. When a uniform "free" gradient $\mathbf{d}_\mathrm{f}$ is first induced in an inclusion and the latter is then placed into a gradient-free matrix of equal conductivity, a uniform in-situ inclusion gradient, $\mathbf{d}_\mathrm{c}$, results that can be described by the expression

$$\mathbf{d}_\mathrm{c} = \mathscr{S}\mathbf{d}_\mathrm{f}, \tag{13}$$

by analogy to Eshelby's [15] relation in elasticity. The tensor $\mathscr{S}$ is referred to as the depolarization tensor, diffusion Eshelby tensor or conduction Eshelby tensor. For spheroidal inclusions that are embedded in an isotropic matrix or a transversally isotropic matrix (the out-of-plane axis of which corresponds to the spheroid's axis of rotation), the non-diagonal terms of the depolarization tensor vanish in the material frame of reference and the diagonal components can be expressed as

$$S(1,1) = 1 - g(t) \quad \text{and} \quad S(2,2) = S(3,3) = \frac{g(t)}{2}, \tag{14}$$

where $g(t)$ takes the form

$$g(t) = 1 + \frac{1}{t^2 - 1}\left[1 - \frac{t}{2(t^2-1)^{1/2}}\ln\left(\frac{t + (t^2-1)^{1/2}}{t - (t^2-1)^{1/2}}\right)\right] \tag{15}$$

for $t \geq 1$ and

$$g(t) = 1 + \frac{1}{t^2 - 1}\left[1 - \frac{t}{(1-t^2)^{1/2}} \arctan\left(\frac{1-t^2}{t}\right)\right] \tag{16}$$

for $t \leq 1$ [20]. Here the parameter t is defined as

$$t = a\frac{K_{\mathrm{T}}^{(\mathrm{m})}}{K_{\mathrm{A}}^{(\mathrm{m})}}, \tag{17}$$

where $K_{\mathrm{A}}^{(\mathrm{m})}$ and $K_{\mathrm{T}}^{(\mathrm{m})}$ are the axial and transverse conductivities of the matrix material, respectively. a is the aspect ratio of the inclusions, with $a > 1$ denoting prolate (or fiber-like) and $a < 1$ oblate (or platelet-like) ellipsoids of revolution. For isotropic matrix behavior (17) reduces to $t = a$, so that the depolarization tensor depends only on the shape of the inclusion. For the special case of a spherical inclusion, $a = 1$, in an isotropic matrix the non-zero terms of the depolarization tensor are

$$S(1,1) = S(2,2) = S(3,3) = \frac{1}{3}. \tag{18}$$

Proceeding by analogy to Eshelby's equivalent inclusion approach [15], the fields in inhomogeneous inclusions embedded in a matrix — referred to as inhomogeneities — can be evaluated on the basis of (13). The inhomogeneity gradient and flux concentration tensors pertaining to a single ellipsoidal inclusion (i) embedded in a matrix (m), known as dilute concentration tensors, are then obtained as

$$\begin{aligned}\mathscr{A}_{\mathrm{dil}}^{(\mathrm{i,m})} &= \{\mathscr{U} + \mathscr{S}^{(\mathrm{i,m})}\mathscr{R}^{(\mathrm{m})}[\mathscr{K}^{(\mathrm{i})} - \mathscr{K}^{(\mathrm{m})}]\}^{-1} \\ \mathscr{B}_{\mathrm{dil}}^{(\mathrm{i,m})} &= \{\mathscr{U} + \mathscr{K}^{(\mathrm{m})}[\mathscr{U} - \mathscr{S}^{(\mathrm{i,m})}][\mathscr{R}^{(\mathrm{i})} - \mathscr{R}^{(\mathrm{m})}]\}^{-1}\end{aligned} \tag{19}$$

by analogy to Hill [29] and Benveniste [4]. Here the designators $\mathscr{A}_{\mathrm{dil}}^{(\mathrm{i,m})}$, $\mathscr{B}_{\mathrm{dil}}^{(\mathrm{i,m})}$ and $\mathscr{S}^{(\mathrm{i,m})}$ are used to stress that the concentration and depolarization tensors pertain to an inhomogeneity (i) embedded in matrix (m). The corresponding matrix concentration tensors can be obtained via (8).

When inserted into (10) the above concentration tensors allow describing the thermal conduction behavior of inhomogeneous materials of matrix–inclusion topology, non-ellipsoidal inhomogeneities being approximated by suitable ellipsoidal or spheroidal shapes. However, because the Eshelby formalism pertains to single inhomogeneities that are not subject to perturbation effects due to the presence of neighbors, such models are restricted to dilute inhomogeneity volume fractions of a few percent.

2.3 Non-Dilute Inhomogeneities

In order to obtain methods for studying heterogeneous materials with freely selectable inhomogeneity volume fractions, the interactions between neighboring inhomogeneities must be accounted for. Within the mean field framework these interactions are typically dealt with in a collective way. This is done either by modifying the mean gradient and flux fields individual inhomogeneities are subjected to, which gives rise to effective field methods, or by adapting the behavior of the embedding medium, which leads to effective medium approaches. Because individual two- or many-particle interactions are not resolved in such methods, they are often referred to as non-interacting models.

2.3.1 Mori–Tanaka Methods

Mori–Tanaka methods [43] follow the effective field strategy, i.e., they subject dilute inhomogeneities to (a-priori unknown) matrix fields that differ from the macroscopic gradients and fluxes. In terms of concentration tensors this ansatz translates into the relationships

$$\overline{\mathcal{A}}^{(i)}_{\mathrm{MT}} = \mathcal{A}^{(i,m)}_{\mathrm{dil}} \overline{\mathcal{A}}^{(m)}_{\mathrm{MT}} \qquad \overline{\mathcal{B}}^{(i)}_{\mathrm{MT}} = \mathcal{B}^{(i,m)}_{\mathrm{dil}} \overline{\mathcal{B}}^{(m)}_{\mathrm{MT}} \tag{20}$$

linking the non-dilute (Mori–Tanaka) inhomogeneity concentration tensors, $\overline{\mathcal{A}}^{(i)}_{\mathrm{MT}}$ and $\overline{\mathcal{B}}^{(i)}_{\mathrm{MT}}$, the dilute inhomogeneity concentration tensors described in Sect. 2.2, $\mathcal{A}^{(i,m)}_{\mathrm{dil}}$ and $\mathcal{B}^{(i,m)}_{\mathrm{dil}}$, and the unknown Mori–Tanaka matrix concentration tensors, $\overline{\mathcal{A}}^{(m)}_{\mathrm{MT}}$ and $\overline{\mathcal{B}}^{(m)}_{\mathrm{MT}}$, respectively [4]. By plugging (20) into (8), the Mori–Tanaka concentration tensors are obtained as

$$\begin{aligned}
\overline{\mathcal{A}}^{(m)}_{\mathrm{MT}} &= [\xi^{(m)} \mathcal{U} + \sum_{(j)\neq(m)} \xi^{(j)} \mathcal{A}^{(j,m)}_{\mathrm{dil}}]^{-1} & \overline{\mathcal{A}}^{(i)}_{\mathrm{MT}} &= \mathcal{A}^{(i)}_{\mathrm{dil}} [\xi^{(m)} \mathcal{U} + \sum_{(j)\neq(m)} \xi^{(j)} \mathcal{A}^{(j,m)}_{\mathrm{dil}}]^{-1} \\
\overline{\mathcal{B}}^{(m)}_{\mathrm{MT}} &= [\xi^{(m)} \mathcal{U} + \sum_{(j)\neq(m)} \xi^{(j)} \mathcal{B}^{(j,m)}_{\mathrm{dil}}]^{-1} & \overline{\mathcal{B}}^{(i)}_{\mathrm{MT}} &= \mathcal{B}^{(i)}_{\mathrm{dil}} [\xi^{(m)} \mathcal{U} + \sum_{(j)\neq(m)} \xi^{(j)} \mathcal{B}^{(j,m)}_{\mathrm{dil}}]^{-1}
\end{aligned} \tag{21}$$

from which the effective conductivity and resistivity tensors can be evaluated via (10) as

$$\begin{aligned}
\mathcal{K}^{*}_{\mathrm{MT}} &= [\xi^{(m)} \mathcal{K}^{(m)} + \sum_{(i)\neq(m)} \xi^{(i)} \mathcal{K}^{(i)} \mathcal{A}^{(i,m)}_{\mathrm{dil}}] [\xi^{(m)} \mathcal{U} + \sum_{(i)\neq(m)} \xi^{(i)} \mathcal{A}^{(i,m)}_{\mathrm{dil}}]^{-1} \\
\mathcal{R}^{*}_{\mathrm{MT}} &= [\xi^{(m)} \mathcal{R}^{(m)} + \sum_{(i)\neq(m)} \xi^{(i)} \mathcal{R}^{(i)} \mathcal{B}^{(i,m)}_{\mathrm{dil}}] [\xi^{(m)} \mathcal{U} + \sum_{(i)\neq(m)} \xi^{(i)} \mathcal{B}^{(i,m)}_{\mathrm{dil}}]^{-1}.
\end{aligned} \tag{22}$$

Mori–Tanaka modeling schemes are explicit, fairly simple and obviously distinguish between a matrix phase and one or more inhomogeneity phases. Accordingly, they pertain to materials consisting of a contiguous matrix that is reinforced by (disconnected) aligned ellipsoidal inhomogeneities, typical examples being "classical" composites and closed-cell porous materials. Whereas in elasticity multi-phase Mori–Tanaka models of composites containing aligned reinforcements of different aspect ratios and different stiffnesses may give rise to invalid effective elastic tensors [16], no such difficulties are encountered for diffusive behavior due to the lower order of the tensors involved.

Mori–Tanaka methods were first applied to conduction problems by Hatta and Taya [27], who obtained scalar expressions of the type

$$K^*_{\rm MT} = K^{(\rm m)} + 3\xi^{(\rm i)} \frac{K^{(\rm m)}(K^{(\rm i)} - K^{(\rm m)})}{3K^{(\rm m)} + (1-\xi)\,(K^{(\rm i)} - K^{(\rm m)})} \tag{23}$$

for the effective coefficients of conductivity of macroscopically isotropic composites consisting of an isotropic matrix reinforced by isotropic spherical particles.

Mori–Tanaka methods can be extended in an ad-hoc way to studying composites reinforced by non-aligned inhomogeneities. For this purpose the dilute gradient concentration tensors pertaining to individual inhomogeneities, $\mathscr{A}^{(\rm i,m)}_{\rm dil}$, are rotated from their local coordinate system, which is described by the Euler angles φ, ψ and θ, to the global coordinate system, to give $\mathscr{A}^{(\rm i)\angle}_{\rm dil}(\varphi,\psi,\theta)$. A concentration tensor pertaining to dilute inhomogeneities of all orientations present in the material can then be evaluated by orientational averaging as

$$\langle\langle \mathscr{A}^{(\rm i)\angle}_{\rm dil} \rangle\rangle = \int_0^{2\pi}\int_0^{2\pi}\int_0^{\pi} \mathscr{A}^{(\rm i)\angle}_{\rm dil}(\varphi,\psi,\theta)\,\rho(\varphi,\psi,\theta)\,{\rm d}\varphi\,{\rm d}\psi\,{\rm d}\theta, \tag{24}$$

where the orientation distribution function ρ is normalized such that $\langle\rho\rangle = 1$. Applying this concept of orientational averaging to (22) gives the effective conductivity tensor as

$$\mathscr{K}^*_{\rm MT} = [\xi^{(\rm m)}\mathscr{K}^{(\rm m)} + \xi^{(\rm i)}\langle\langle \mathscr{K}^{(\rm i)}\mathscr{A}^{(\rm i)\angle}_{\rm dil}\rangle\rangle]\,[\xi^{(\rm m)}\,\mathscr{U} + \xi^{(\rm i)}\langle\langle \mathscr{A}^{(\rm i)\angle}_{\rm dil}\rangle\rangle]^{-1}, \tag{25}$$

all inhomogeneities being subsumed into a single phase (i). The pertinent phase gradient concentration tensors can be written as

$$\overline{\mathscr{A}}^{(\rm m)}_{\rm MT} = [\xi^{(\rm m)}\,\mathscr{U} + \xi^{(\rm i)}\langle\langle \mathscr{A}^{(\rm i)\angle}_{\rm dil}\rangle\rangle]^{-1} \quad \text{and} \quad \overline{\mathscr{A}}^{(\rm i)}_{\rm MT} = \langle\langle \mathscr{A}^{(\rm i)\angle}_{\rm dil}\rangle\rangle \overline{\mathscr{A}}^{(\rm m)}_{\rm MT}. \tag{26}$$

Being an orientational average over all inhomogeneities, the concentration tensor $\overline{\mathscr{A}}^{(\rm i)}_{\rm MT}$ is of limited practical use in localization. However, following [11], concentration tensors

$$\overline{\mathscr{A}}^{(\rm i)\angle}_{\rm MT}(\varphi,\psi,\theta) = \mathscr{A}^{(\rm i)\angle}_{\rm dil}\overline{\mathscr{A}}^{(\rm m)}_{\rm MT} \tag{27}$$

can be introduced to find the temperature gradient in individual inhomogeneities of any given orientation. Even though "extended Mori–Tanaka models" following (24)–(27) – like their equivalents formulated in terms of the flux concentration tensors $\mathscr{B}_{\text{dil}}^{(\text{i})\angle}(\varphi, \psi, \theta)$ – do not fulfill the condition of aligned inhomogeneities inherent in the Mori–Tanaka concept, they provide satisfactory results in most situations. More flexible but more complex modeling schemes were proposed by Ponte Castañeda and Willis [51] and further developed by other authors [10].

2.3.2 Classical Self-Consistent Schemes

The simplest effective medium approaches are classical self-consistent schemes, which are based on embedding inhomogeneities directly into the effective medium. In terms of conductivity, resistivity and concentration tensors the core statement of such models can be denoted as

$$\mathscr{K}_{\text{SC}}^{*} = \sum_{(\text{p})} \xi^{(\text{p})} \mathscr{K}^{(\text{p})} \mathscr{A}_{\text{dil}}^{(\text{p},*)} \quad \text{or} \quad \mathscr{R}_{\text{SC}}^{*} = \sum_{(\text{p})} \xi^{(\text{p})} \mathscr{R}^{(\text{p})} \mathscr{B}_{\text{dil}}^{(\text{p},*)} . \tag{28}$$

The above relationships are implicit because the dilute concentration tensors of an ellipsoidal inhomogeneity embedded in the effective medium, $\mathscr{A}_{\text{dil}}^{(\text{p},*)}$ and $\mathscr{B}_{\text{dil}}^{(\text{p},*)}$, as well as the depolarization tensor $\mathscr{S}^{(\text{p},*)}$ used in describing them via (19), must be evaluated with respect to the unknown effective tensors $\mathscr{K}_{\text{SC}}^{*}$ and $\mathscr{R}_{\text{SC}}^{*}$. Solutions can be obtained by self-consistent iteration, in which, e.g., steps $n-1$ and n for the first expression in (28) are linked by the equations

$$\begin{aligned} \mathscr{K}_{\text{SC},n}^{*} &= \sum_{(\text{p})} \mathscr{K}^{(\text{p})} [\, \mathscr{U} + \mathscr{S}_{n-1}^{(\text{p},*)} \mathscr{R}_{\text{SC,n}-1} (\mathscr{K}^{(\text{p})} - \mathscr{K}_{\text{SC,n}-1})]^{-1} \\ \mathscr{R}_{\text{SC},n}^{*} &= (\mathscr{K}_{\text{SC},n}^{*})^{-1} . \end{aligned} \tag{29}$$

For aligned spheroidal, but non-spherical inhomogeneities that show transversally isotropic or higher material symmetry, the $\mathscr{K}_{\text{SC},n}^{*}$ are typically transversally isotropic; this case can be handled by the expressions for $\mathscr{S}^{(\text{p},*)}$ given in (15) and (16).

In contrast to Mori–Tanaka methods (28) and (29) are symmetrical in terms of their constituents. Accordingly, classical self-consistent schemes do not generally describe a contiguous matrix phase into which inhomogeneities are embedded. As a consequence, such methods are best suited to modeling heterogeneous materials that do not show a matrix–inclusion microtopology over at least part of their range of phase volume fractions, typical examples being polycrystals and composites with interwoven phase topologies.

Classical self-consistent descriptions for the effective conductivity of inhomogeneous materials were introduced by Bruggemann [8]. For the special case of a macroscopically isotropic solid consisting of two isotropic constituents, (1) and (2), the classical self-consistent scheme leads to the nonlinear equation

$$\xi^{(1)} \frac{K^{(1)} - K^*_{\mathrm{SC}}}{K^{(1)} + 2K^*_{\mathrm{SC}}} = \xi^{(2)} \frac{K^*_{\mathrm{SC}} - K^{(2)}}{K^{(2)} + 2K^*_{\mathrm{SC}}} \tag{30}$$

in the effective coefficients of conductivity K^*_{SC}. In (30) either phase percolates if its volume fraction exceeds a value of $\frac{1}{3}$. This makes classical self-consistent methods unsuitable for describing porous materials with high pore volume fractions.

2.3.3 Differential Schemes

A further effective medium approach is based on successive steps of homogenizing dilute volume fractions of inhomogeneities in an effective medium consisting of the matrix plus previously processed, smaller inhomogeneities. In the resulting differential schemes the effective conductivity tensor can be described by analogy to [23] by the system of differential equations

$$\frac{\mathrm{d}\mathscr{K}^*_{\mathrm{D}}}{\mathrm{d}\xi^{(\mathrm{i})}} = \frac{1}{\xi^{(\mathrm{m})}} [\mathscr{K}^{(\mathrm{i})} - \mathscr{K}^*_{\mathrm{D}}] \mathscr{A}^{(\mathrm{i},*)}_{\mathrm{dil}} \tag{31}$$

using the initial conditions

$$\mathscr{K}^*(\xi^{(\mathrm{i})} = 0) = \mathscr{K}^{(\mathrm{m})}. \tag{32}$$

This initial value problem can be integrated numerically by algorithms such as Runge–Kutta methods and for many cases analytical solutions are available [50]. As the numerical solution proceeds the depolarization tensor $\mathscr{S}^{(\mathrm{i},*)}$ and the dilute concentration tensor $\mathscr{A}^{(\mathrm{i},*)}_{\mathrm{dil}}$ must be re-evaluated at each step to account for the current, in general anisotropic, effective medium by analogy to the classical self-consistent scheme.

Being based on multiple successive homogenization steps, differential schemes implicitly assume a very wide distribution of inhomogeneity sizes. They pertain to matrix–inclusion composites by construction.

The special case of isotropic spherical particles in an isotropic matrix results in a scalar initial value problem which can be integrated analytically to give the nonlinear equation

$$\frac{K^{(\mathrm{i})} - K^*_{\mathrm{D}}}{K^{(\mathrm{i})} - K^{(\mathrm{m})}} \left(\frac{K^{(\mathrm{m})}}{K^*_{\mathrm{D}}} \right)^{\frac{1}{3}} = \xi^{(\mathrm{m})} \tag{33}$$

for the effective coefficients of conductivity K^*_{D} [8].

2.3.4 Other Estimates

A number of models based on single-inclusion solutions have been proposed that do not fall into the above categories of mean field descriptions. Probably the best known of them is the Maxwell–Garnett approximation [39], the results of which coincide with the Mori–Tanaka solutions for the appropriate phase geometries. Kerner [33] proposed a generalized self-consistent scheme, in which spherical or cylindrical inhomogeneities are coated in a layer of matrix material of uniform thickness, the latter being chosen to obtain the proper phase volume fractions of the composite, and embedded in the effective medium. The estimates for the effective conductivity recovered by this approach again coincide with the Mori–Tanaka predictions.

Advanced effective field estimates, which can include pair interactions of inhomogeneities, are provided by multi-particle effective field methods [9].

Torquato [58] proposed estimates for the effective conductivity of macroscopically isotropic inhomogeneous materials on the base of expansions. Information on details of the phase arrangement is introduced into these estimates via the three-point microstructural parameter $\zeta(\xi^{(\mathrm{i})})$, which is also used in three-point bounds, compare the following Sect. 2.4.

2.4 Bounds

Bounds on the effective properties of inhomogeneous materials have been based on minimum energy principles. They introduce statistical information on the microstructure via n-point correlation functions, which describe, e.g., the probability of finding n randomly positioned points simultaneously in the same phase. The more microstructural information that is incorporated into the bounds, the tighter they become.

The simplest bounds on the effective conductivity are the Wiener [62] bounds,

$$\left(\sum_{(\mathrm{p})} \xi^{(\mathrm{p})} \mathscr{R}^{(\mathrm{p})}\right)^{-1} \leq \mathscr{K}^* \leq \sum_{(\mathrm{p})} \xi^{(\mathrm{p})} \mathscr{K}^{(\mathrm{p})}, \tag{34}$$

which use constant flux and gradient fields, respectively. Being one-point bounds, they contain microstructural information only in the form of the phase volume fractions $\xi^{(\mathrm{p})}$. They are too slack to be of practical use in most situations.

Considerably tighter bounds, which are sensitive to the macroscopic symmetry of inhomogeneous materials, can be obtained by combining the variational principle due to Hashin and Shtrikman [24] with phase-wise uniform trial fields. In their original form these bounds pertain to macroscopically isotropic materials, such as composites reinforced by particles or by randomly oriented fibers. They were extended to materials containing aligned ellipsoidal inhomogeneities [63] and, for the case of two-phase solids, were shown to coincide with Mori–Tanaka estimates [61].

Three-point bounds on the effective conductivity [3] can be expressed in terms of a three-point microstructural parameter $\zeta(\xi^{(i)})$, which depends on the shapes, size distributions and arrangement of inhomogeneities and has been evaluated for a number of configurations relevant to two-phase particle and fiber reinforced composites, see, e.g., [59]. Such "improved bounds" are considerably tighter than Hashin–Shtrikman-type bounds.

Most bounds on the effective conductivities of inhomogeneous materials pertain to perfect interfaces between the constituents. However, bounds are also available for the special cases of imperfect and superconducting interfaces [34, 60]. For porous materials the lower bounds are typically trivial and equal to zero.

2.5 Comparisons of Predictions

In this section selected results are presented on the overall thermal conduction behavior of two-phase composites in dependence on the reinforcement volume fraction $\xi^{(i)}$. For macroscopically isotropic particle reinforced composites the comparisons are based on spherical particulates of Al_2O_3 embedded in a matrix of pure silver, see Table 1 for the material parameters. For these constituents the thermal conductivity contrast, $c = K^{(i)}/K^{(m)}$, takes a value of approximately 0.088.

Figure 1 shows predictions for the effective conductivities obtained for the above composite. The one-point Wiener bounds can be seen to be very slack, whereas the two-point Hashin–Shtrikman (H/S) bounds are considerably tighter. Because the matrix shows a higher conductivity than the reinforcements the Mori–Tanaka method (MTM) corresponds to the upper Hashin–Shtrikman bound, as do the generalized self-consistent scheme and Maxwell–Garnet estimates. The three-point bounds (3PB) and Torquato's three-point estimates (3PE) were evaluated for randomly positioned, non-overlapping spherical particles of equal size using expressions for the statistical parameter $\zeta(\xi^{(i)})$ given by Miller and Torquato [41], which are available for $\xi^{(i)} < 0.6$. The three-point bounds are much tighter than the Hashin–Shtrikman bounds and the three-point estimates nearly coincide with the upper three-point bound. The classical self-consistent estimates (CSCS) show the typical behavior of closely approaching one Hashin–Shtrikman bound at low inhomogeneity volume fractions and the other for high values of $\xi^{(i)}$. These estimates do not fall within the three-point bounds for $\xi^{(i)} \gtrsim 0.3$ because they do not describe matrix–inclusion microtopologies for all volume fractions, as noted in

Table 1 Constituent conductivities of the silver matrix and the Al_2O_3 particles used in generating Fig. 1. Both constituents are taken to be isotropic

	K (W/mK)
matrix	427.0
particles	37.7

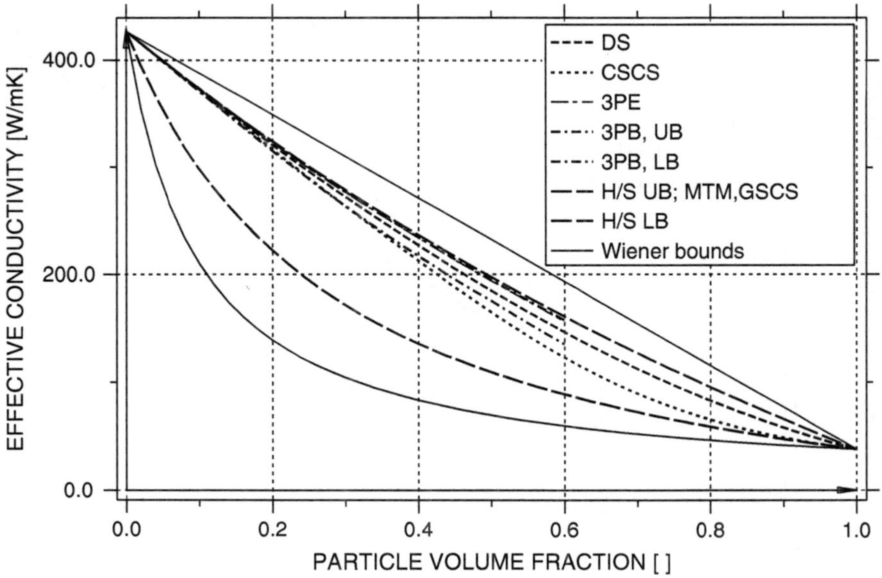

Fig. 1 Comparison of selected estimates and bounds for the effective conductivity of a silver matrix reinforced by equiaxed Al_2O_3 particles (see text for designators of models)

Table 2 Constituent conductivities of the polyetherimide matrix and the T-300 graphite fibers used in generating Figs. 2 and 3 [21]

	K_A (W/mK)	K_T (W/mK)
Matrix	0.22	0.22
Fibers	8.40	0.84

Sect. 2.3.2. Finally, the differential scheme (DS) leads to predictions that are markedly lower than the three-point estimates, but do not violate the three-point bounds for the composite considered. All estimates discussed in Sect. 2.3 can be seen to comply with the Hashin–Shtrikman bounds.

A second set of comparisons pertains to a polyetherimide matrix reinforced by aligned short T-300 graphite fibers of aspect ratio $a = 10$, compare Table 2. This composite shows a transversally isotropic macroscopic response. The thermal conduction behavior of the graphite fibers is strongly anisotropic, so that the axial and transverse thermal conductivity contrasts take values of $c_A = 38.2$ and $c_T = 3.82$, respectively.

Figures 2 and 3 compare selected estimates and bounds for the effective axial and transverse conductivities, respectively, of the above composite. Due to the macroscopic transverse isotropy of the material clear qualitative differences are evident between the two sets of curves. It is interesting to note that the upper Wiener bound, which is a rule-of-mixture expression, closely approaches the upper Hashin–Shtrikman bound for the axial conductivity, i.e., despite the rather

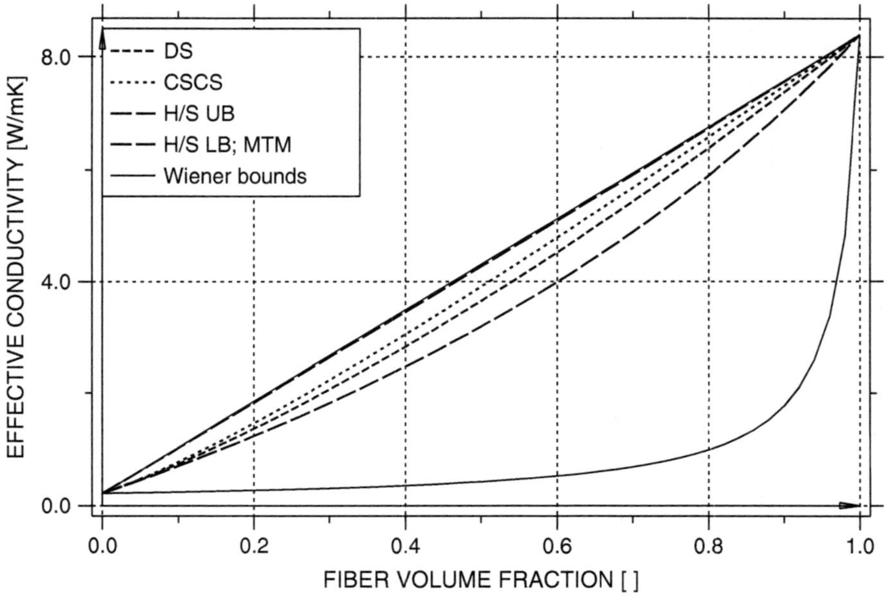

Fig. 2 Comparison of selected mean field estimates and bounds for the effective axial conductivity of a polyetherimide matrix reinforced by aligned short fibers ($a = 10$) (see text for designators of models)

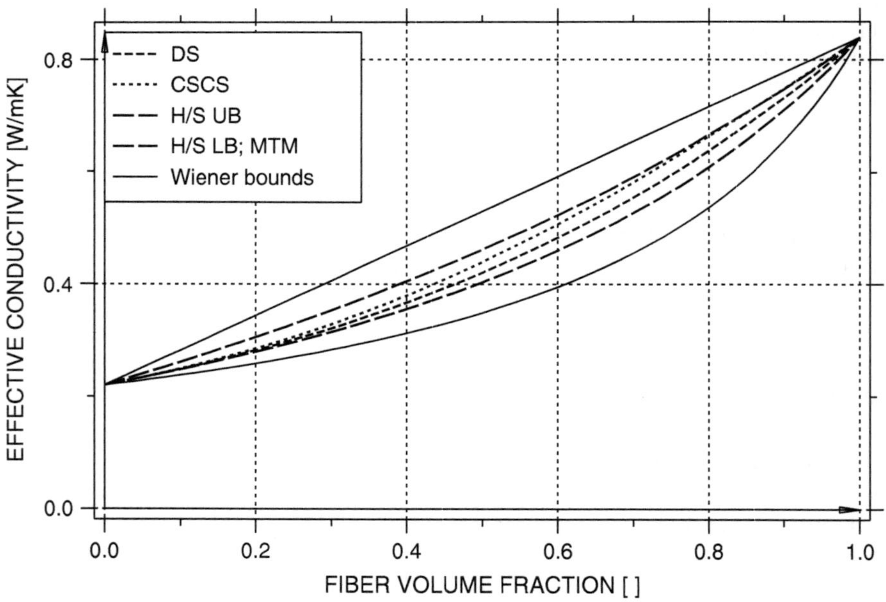

Fig. 3 Comparison of selected mean field estimates and bounds for the effective transverse conductivity of a polyetherimide matrix reinforced by aligned short fibers ($a = 10$) (see text for designators of models)

moderate aspect ratio of $a = 10$ the axial conduction behavior of the short fiber composite is very similar to that of a continuously reinforced material. Because the axial and transverse conductivities of the graphite fibers exceed that of the matrix, the Mori–Tanaka estimates coincide with the lower Hashin–Shtrikman bounds in axial and transverse conduction.

Expressions for the statistical parameter $\zeta(\xi^{(i)})$ are not available for configurations of the type studied in Figs. 2 and 3, so that neither three-point bounds nor three-point estimates are given. The behaviors predicted by the classical self-consistent and differential schemes are qualitatively similar to the ones shown in Fig. 1 and these two estimates comply with the Hashin–Shtrikman bounds.

2.6 Imperfect Interfaces and Non-Spherical Inhomogeneities

The mean field and bounding expressions presented in Sects. 2.2–2.4 pertain to composites with perfect interfaces and give predictions that are independent of the absolute size of the inhomogeneities. The presence of finite interfacial conductances, however, is well known to give rise to a marked size effect in the macroscopic conductivities [25].

Unless the distribution of the interfacial conductances is confocal with a given ellipsoidal inhomogeneity, they lead to the loss of the Eshelby property, i.e., the microfields are inhomogeneous. The same holds for inhomogeneities of non-ellipsoidal shapes. These two cases can be handled in a two-phase setting by the semi-analytical "replacement tensor" approach developed by Duschlbauer et al. [12, 13]. This modeling strategy is based on numerically evaluating the effective conductivity, $\mathscr{K}^*_{\rm dil}$, as well as the gradient and flux fields of a configuration consisting of a single inhomogeneity that is embedded in the matrix at a very low volume fraction of, say, $\xi^{(i)}_{\rm dil} \leq 10^{-3}$. The inhomogeneous fields in the inhomogeneity, which are due to a non-ellipsoidal shape and/or to finite interfacial conductances, are volume averaged and the resulting "replacement concentration tensor", $\mathscr{A}^{(\rm i,m;r)}_{\rm dil}$, is extracted. Furthermore, a "replacement inhomogeneity conductivity", $\mathscr{K}^{(\rm i;r)}$, must be evaluated from the relationship

$$\mathscr{K}^{(\rm i;r)} = \mathscr{K}^{(\rm m)} + \frac{1}{\xi^{(i)}_{\rm dil}}[\mathscr{K}^*_{\rm dil} - \mathscr{K}^{(\rm m)}](\mathscr{A}^{(\rm i,m;r)}_{\rm dil})^{-1}, \tag{35}$$

which ensures that (10) is fulfilled. By replacing $\mathscr{K}^{(\rm i)}$ and $\mathscr{A}^{(\rm i,m)}_{\rm dil}$ with $\mathscr{K}^{(\rm i;r)}$ and $\mathscr{A}^{(\rm i,m;r)}_{\rm dil}$, respectively, in mean field methods of the types presented in Sect. 2.3, estimates are obtained for the effective conductivity of materials containing non-ellipsoidal inhomogeneities and/or showing finite interfacial conductivities.

The replacement tensor approach is well suited to Mori–Tanaka methods. For spherical isotropic inhomogeneities this combination allows to recover the well-known results of Hasselman and Johnson [26],

$$K^* = K^{(m)} \frac{2\xi^{(i)}\left(\frac{K^{(i)}}{K^{(m)}} - \frac{2K^{(i)}}{dh} - 1\right) + \frac{K^{(i)}}{K^{(m)}} + \frac{4K^{(i)}}{dh} + 2}{\xi^{(i)}\left(1 - \frac{K^{(i)}}{K^{(m)}} + \frac{2K^{(i)}}{dh}\right) + \frac{K^{(i)}}{K^{(m)}} + \frac{4K^{(i)}}{dh} + 2}, \tag{36}$$

where d stands for the diameter of the particles and h for the interfacial conductance. Applications of the replacement tensor approach to effective medium models are feasible but not particularly attractive, because they require carrying out the numerical evaluation of the replacement tensors with respect to the new reference material after each iteration.

3 Discrete Microstructure Models for Conduction Properties

Discrete microstructure models of thermal conduction in inhomogeneous solids aim at precisely evaluating the temperature, gradient and flux fields in specific, well defined phase geometries. As sketched in Fig. 4, which pertains to a transverse section of a unidirectionally reinforced composite, there are three major types of discrete microstructure models, viz., periodic microfield models, embedding models and windowing models. These methods use different approximations for

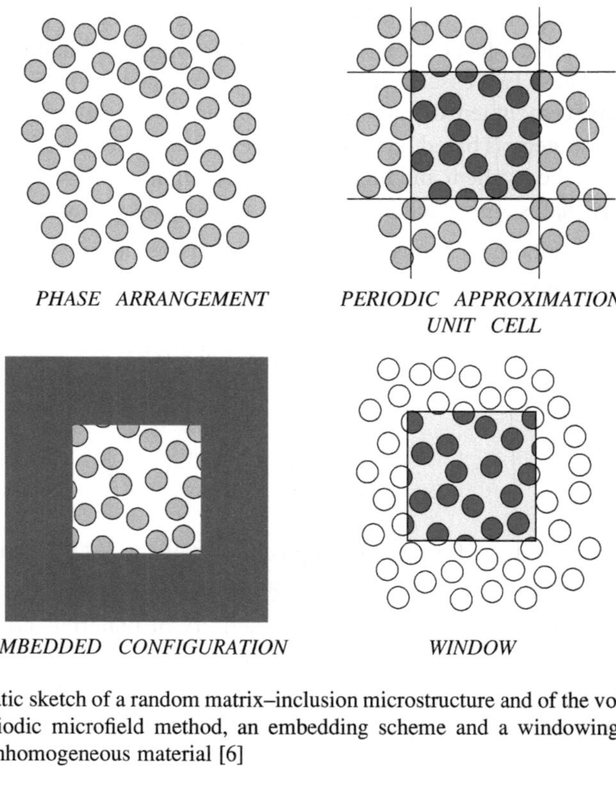

Fig. 4 Schematic sketch of a random matrix–inclusion microstructure and of the volume elements used by a periodic microfield method, an embedding scheme and a windowing approach for studying this inhomogeneous material [6]

studying volume elements of limited size and typically employ numerical engineering methods for evaluating the microfields. The following discussion mainly assumes the use of Finite Element (FE) methods for this purpose; for a more detailed discussion of its application to problems involving inhomogeneous materials see, e.g., [5, 7]. FE approaches are well suited to describing complex phase geometries and many implementations are capable of handling finite interfacial conductances. An alternative solution approach, the Lattice Monte Carlo method, is discussed in detail in [17].

In the ideal case, the volume elements used in discrete microstructure models are representative volume elements (RVEs), the size of which is sufficient for containing all relevant statistical information on the actual phase arrangement to be studied. In practice, however, limitations in computer power have tended to restrict simulations to smaller volume elements, which typically are only approximations to proper RVEs.

Volume elements used for studying composite materials range from very simple, such as cubic arrays of spherical reinforcements, to highly complex phase arrangements that are generated by computer codes or extracted from experimental data. Computer generated microstructures may, on the one hand, involve a considerable number of randomly positioned and, where applicable, randomly oriented inhomogeneities, Voronoi cells based on suitable "clouds" of points, or similar constructs. On the other hand, experimentally determined statistical descriptors of phase arrangements can be used to generate microstructures that are statistically equivalent to some target material, a process known as statistical reconstruction [53]. Experimental procedures for generating "real structure" phase arrangement have included serial sectioning, compare [57], and computed tomography, see, e.g., [32]. It is worth noting that identical geometrical models using identical discretizations can be used for both thermal and mechanical analysis.

3.1 Periodic Homogenization

Periodic microfield (or periodic homogenization) methods describe periodic model materials, the effective properties of which are used to approximate the behavior of actual, non-periodic inhomogeneous materials. Such approaches typically represent a tradeoff between basing the model on a relatively small volume element, the periodic unit cell, and being restricted to periodic phase arrangements. For unit cells that are sufficiently large to be RVEs no approximation in terms of phase arrangement statistics is involved.

Unit cells are volume elements with periodic phase arrangements that tile the computational space by translation and this way provide a complete description of infinitely extended periodic media. In n-dimensional space each unit cell is associated with n linearly independent translation vectors, $\mathbf{p}_i$. Whereas the volume of the smallest unit of periodicity is uniquely defined, the $\mathbf{p}_i$ describing such volumes

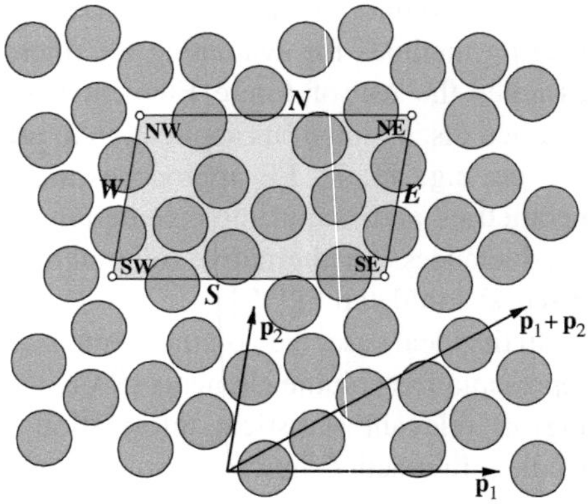

Fig. 5 Planar periodic phase arrangement with unit cell and periodicity vectors

and the shapes of unit cells of minimal size ("minimum unit cells") are non-unique. Figure 5 shows a planar periodic phase arrangement with a minimum unit cell and the corresponding translation vectors, $\mathbf{p}_1$ and $\mathbf{p}_2$. The vectors $\mathbf{p}_1 \pm \mathbf{p}_2$ are also valid translation vectors and may be used in building other unit cells of different shapes.

The surface of a unit cell can be split into at least n regions, Γ_k, each of which consists of two parallel surface elements, Γ_{k^-} and Γ_{k^+}, that are separated by a vector of the type $\sum_i m_i \mathbf{p}_i$, where the m_i are integer numbers. In the case of quadrilateral or hexahedral unit cells, each region Γ_k is uniquely associated with the translation vector $\mathbf{p}_k$. For example, in Fig. 5 there are two surface regions, Γ_1 consisting of faces W and E separated by the periodicity vector $\mathbf{p}_1$, as well as Γ_2 consisting of faces N and S separated by $\mathbf{p}_2$. The surface elements do not have to be planar.

Boundary conditions must be prescribed to the unit cells in such a way that the periodicity of gradients and fluxes is ensured. For quadrilateral or hexahedral unit cells this is done by coupling the temperatures on corresponding points of the surface elements Γ_{k^-} and Γ_{k^+}, T_{k^-} and T_{k^+}, according to the relationship

$$\Delta T_k = T_{k^+} - T_{k^-} = \langle \mathbf{d} \rangle \, \mathbf{p}_k, \tag{37}$$

i.e., the temperature difference in the direction of a periodicity vector depends directly on the macroscopic temperature gradient, $\langle \mathbf{d} \rangle$. The symmetry and antisymmetry boundary conditions often employed in mechanical analysis of inhomogeneous materials [5, 7] are of limited use in periodic homogenization of diffusive transport.

In typical Finite Element practice (37) can be implemented in terms of "multipoint constraints" that link four temperature degrees of freedom. For the configuration shown in Fig. 5 one node, say, SW, acts as an "anchor node" the temperature of which is fixed. The temperature differences according to (37), ΔT_1 and ΔT_2, are carried by the "master nodes", SE and NW, which are used to couple the temperatures of nodes on the "slave faces", E and N, to those of the corresponding nodes on the "master faces", W and S, respectively. In such a scheme the master nodes control the macroscopic gradients, the master faces handle the microscopic fluctuations, and the slaves are used to enforce periodicity. For example, when local coordinates $\tilde{s}$ are employed on the E and W faces, the constraint equations for surface region Γ_1 in Fig. 5 result as

$$T_{\mathrm{E}}(\tilde{s}) - T_{\mathrm{W}}(\tilde{s}) = T_{\mathrm{SE}} - T_{\mathrm{SW}} = \Delta T_1. \tag{38}$$

Obviously, the meshing of the two surface elements making up a surface region Γ_k must be compatible to allow the application of (38) in terms of nodal constraints. The above implementation of the periodicity boundary conditions can be easily extended to three-dimensional unit cells using, e.g., the naming scheme sketched in Fig. 6.

Once the unit cells have been defined and appropriate periodicity boundary conditions applied, the volume elements must be subjected to appropriate loads corresponding to macroscopic gradients or fluxes. This can be done, on the one hand, by using the mathematical framework known as asymptotic homogenization or "homogenization theory", which is based on explicitly introducing macroscopic and microscopic coordinates into the formulation of the problem, see, e.g., [1, 2, 36, 37, 49]. Such asymptotic homogenization approaches are discussed in detail in [38] of the present work and have typically required specialized analysis software.

Alternatively, the "method of macroscopic degrees of freedom" [40] can be adapted to prescribing appropriate thermal conduction load cases to unit cells. It makes use of the fact that in constraint equations such as (38) the macroscopic fields are carried by the master nodes, which can, accordingly, be employed to apply macroscopic gradients and fluxes to the unit cell. Prescribing macroscopic gradients is straightforward: An appropriate temperature difference ΔT_k is evaluated from

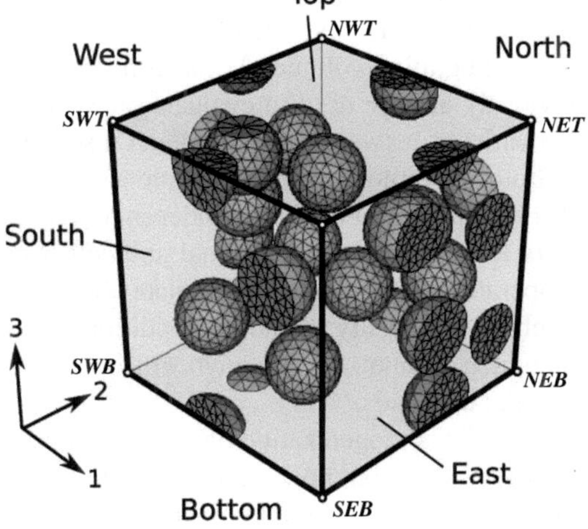

Fig. 6 Cube-shaped periodic unit cell containing 15 randomly positioned spherical particles of equal size at a volume fraction of $\xi^{(i)} = 0.15$. Designators of the six faces (East, West, North, South, Top, Bottom) and of the vertices are given [7]

(37) and applied via appropriate temperatures of the master and anchor nodes in an equivalent of (38). With analysis codes that can handle concentrated nodal fluxes, a method proposed by Smit et al. [56] for mechanical unit cell analysis can be adapted to specify a given far-field flux $\mathbf{q}^{\mathrm{a}}$ to the unit cell. This encompasses evaluating the thermal power passing through a slave surface, scaling it for the cell's cross sectional area, and applying it as a concentrated nodal flux to the associated master node. The direction of this nodal flux is normal to the orientation of the slave face. For example, the concentrated flux in vertical direction to be prescribed to master node SE of the unit cell sketched in Fig. 5 takes the form

$$q_{\mathrm{SE},2} = \frac{1}{\Gamma_{\mathrm{N}}} \int_{\Gamma_{\mathrm{N}}} (\mathbf{q}^{\mathrm{a}})^{\mathrm{T}} \, \mathbf{n}_{\Gamma}(\mathbf{x}) \, \mathrm{d}\Gamma . \tag{39}$$

The volume averaged temperature gradient vector, $\langle \mathbf{d} \rangle$, set up in a unit cell loaded by prescribed macroscopic fluxes can be be evaluated directly by inserting the temperatures of the anchor and master nodes into (37). If loading was done by prescribed macroscopic gradients, the resulting macroscopic fluxes can be obtained from the "reaction fluxes" at the master nodes, by surface integration over the nodal fluxes at the cell boundaries, or by numerically approximating the volume average of the components of the flux vector as specified by (3). Applications of the method of macroscopic degrees of freedom to modeling the conduction behavior of composites can be found, e.g., in [6, 14].

For evaluating volume averages of scalars or the components of vectors and tensors algorithms of the type

$$\langle f \rangle = \frac{1}{\Omega_{\mathrm{s}}} \int_{\Omega_{\mathrm{s}}} f(\mathbf{x}) \, \mathrm{d}\Omega \approx \frac{1}{\Omega_{\mathrm{s}}} \sum_{l=1}^{N} f_l \, \Omega_l \tag{40}$$

can be used to advantage with many Finite Element codes. Here f_l and Ω_l are the function value and the integration weight (in terms of the volume pertaining to the integration point), respectively, associated with the l-th integration point within a given integration volume Ω_{s} that contains N integration points. Equation (40) can obviously also be used to evaluate phase averages, which is of interest for direct comparisons between mean field and discrete microfield models.

Both asymptotic homogenization and the method of macroscopic degrees of freedom require n linearly independent load cases for obtaining the full conductivity tensor of an n-dimensional unit cell from (9). Because, as stated in (37), the temperature increments accumulate along the periodicity vectors, models that combine periodicity boundary conditions with temperature dependent phase conductivities cannot be unequivocally interpreted in terms of infinitely extended periodic materials.

Periodic homogenization has proved to be a very flexible modeling tool and has been the most commonly used discrete microstructure approach to studying the conduction behavior of inhomogeneous solids.

3.2 Embedding Methods

Embedding methods combine a geometrically highly resolved core with a surrounding outer region that is described by some smeared-out model, compare Fig. 4 (lower left). Such modeling approaches do not require any periodicity constraints but give rise to boundary layers of perturbed fields where core and embedding regions meet.

The conduction response of the outer region may, on the one hand, be approximated via some analytical method, such as one of the mean field schemes discussed in Sect. 2.3. The resulting models are well suited to studying the microfields in problems that involve macroscopic gradients of the microstructure or of the fields, e.g., the diffusion behavior in graded materials, or nonlinear phase conductivities. On the other hand, the conduction behavior of the embedding region can be set equal to the averaged response of the core, which gives rise to self-consistent schemes for evaluating the effective conductivity of inhomogeneous materials with linear phase behavior.

Both of the above approaches have been fairly widely employed in continuum micromechanics, especially for "zooming in" on regions of interest such as crack tips. However, little use appears to have been made of embedding schemes involving complex core regions in studying the conduction behavior of inhomogeneous materials.

3.3 Windowing Methods

Whereas periodic homogenization aims at describing the macroscopic constitutive behavior of inhomogeneous materials in the limit of very large samples, windowing methods concentrate on evaluating the responses of finite sized samples or "windows", the results being referred to as apparent (rather than effective) properties. Windows are volume elements of simple shape, are extracted from the inhomogeneous medium at random positions and with random orientations, as sketched in Fig. 7, and are typically too small to be proper representative volumes.

By analogy to [28] the evaluation of the macroscopic properties of a window may be based on an integral version of the Hill condition [30] in thermal conduction,

$$\int_{\Gamma} [[\mathbf{q}(\mathbf{x}) - \langle \mathbf{q} \rangle]^{\mathrm{T}} \mathbf{n}_{\Gamma}(\mathbf{x})][T(\mathbf{x}) - \langle \mathbf{d} \rangle^{\mathrm{T}} \mathbf{x}]\, \mathrm{d}\Gamma = 0, \tag{41}$$

which is a formal statement of the energetic equivalence of microscopic (resolved at the phase level) and macroscopic (homogenized) descriptions.

The above condition can be fulfilled by applying appropriate boundary conditions to a given window. Two ways of doing so consist of prescribing either uniform

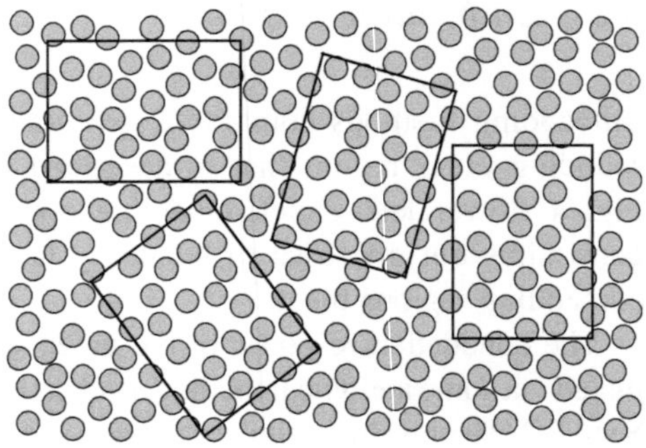

Fig. 7 Schematic depiction of a planar phase arrangement and four rectangular windows of equal size [6]

fluxes, $\mathbf{q}^{\mathrm{a}}$, or uniform gradients, $\mathbf{d}^{\mathrm{a}}$, to the whole surface of the window, so that the normal fluxes, q_{n}, or the temperatures, T, on the boundaries vary as

$$q_{\mathrm{n}}(\mathbf{x}) = (\mathbf{q}^{\mathrm{a}})^{\mathrm{T}}\mathbf{n}_{\Gamma}(\mathbf{x}) \quad \text{or} \quad T(\mathbf{x}) = (\mathbf{d}^{\mathrm{a}})^{\mathrm{T}}\mathbf{x}, \tag{42}$$

respectively. This way either the first or the second term of (41) is set identically to zero by the uniform Neumann (UNBC) and uniform Dirichlet (UDBC) boundary conditions, respectively. Macrohomogeneous boundary conditions following (42) have been shown to give rise to lower and upper estimates for the thermal conductivity of a given volume element, and ensemble averages over such estimates provide bounds on the apparent conductivity tensor [48]. Hierarchies of bounds can be generated from sets of windows of different sizes [47], bringing out effects of the size of the volume elements.

Alternatively, the Hill condition (41) can be enforced by making the product under the integral vanish separately for each face of the volume element. This can be achieved by prescribing either uniform normal flux components q_i^{a} or uniform tangential gradient components d_i^{a} to each face i of the window. There are a fair number of possible choices for such mixed uniform boundary conditions (MUBC). One of them provides estimates that have been found to agree with the results of periodic homogenization for periodic volume elements of orthotropic or higher symmetry and to tend to provide reasonable results for lower symmetries. This set of load cases, which was first proposed by Jiang et al. [31] and has been referred to as "periodicity compatible mixed uniform boundary conditions" [7], is listed in Table 3. It provides a convenient windowing procedure for obtaining estimates of the apparent conductivity tensor from regularly shaped samples of inhomogeneous solids with linear phase conductivities. Because it is formulated in terms of nonzero tangential gradients and zero normal fluxes, it is also suitable for handling porous materials, but cannot deal with phases of infinite conductivity. Furthermore, it was found to be compatible with models employing finite interfacial conductances [44]. The applicability of the scheme to nonlinear conduction behavior has not been investigated in depth yet.

Table 3 The three linearly independent uniform gradient load cases constituting the periodicity compatible mixed uniform boundary conditions for thermal conduction [7]

	Load case 1	Load case 2	Load case 3
East	$T = d_1^a l_1/2$	$q_1^a = 0$	$q_1^a = 0$
West	$T = -d_1^a l_1/2$	$q_1^a = 0$	$q_1^a = 0$
North	$q_2^a = 0$	$T = d_2^a l_2/2$	$q_2^a = 0$
South	$q_2^a = 0$	$T = -d_2^a l_2/2$	$q_2^a = 0$
Top	$q_3^a = 0$	$q_3^a = 0$	$T = d_3^a l_3/2$
Bottom	$q_3^a = 0$	$q_3^a = 0$	$T = -d_3^a l_3/2$

The nomenclature of the faces follows Fig. 6, the window is assumed to be of hexahedral shape and aligned with the coordinate directions, and the l_i are the window's side lengths

It may be noted that for the case of periodic phase arrangements there is a fourth approach to fulfilling (41). It consists of making the fluctuations of the non-uniform boundary fields cancel out on paired parallel faces that show identical fluctuations but face normals of opposite orientations. This strategy leads directly to the periodicity boundary conditions discussed in Sect. 3.1.

4 Sample Application

Metal matrix composites (MMCs) consisting of diamond particles embedded in aluminum or copper matrices are of considerable technological interest for use as heat sink materials, because they promise effective conductivities exceeding those of the metallic matrices. Due to the presence of finite conductances at the interfaces between matrix and reinforcements the macroscopic conductivities of such composites show a marked dependence on the particle size.

Because, on the one hand, the diamond particles are of cubo-octahedral shape and, on the other hand, there is experimental evidence that the diamonds show different interfacial conductances on their $\{100\}$ and $\{111\}$ faces [54], such composites are interesting targets for mean field and discrete microfield modeling. In the following, values of $h_{\{100\}} = 100$ MW/m^{2}K and $h_{\{111\}} = 20$ MW/m^{2}K are assumed for the interfacial conductances at the two sets of faces. The isotropic conductivities of particles and matrix are chosen as $k^{(i)} = 1800$ W/mK and $k^{(m)} = 237$ W/mK, respectively. For particles that are regular tetrakaidekahedra of 200 μm diameter an "equivalent" homogeneous interfacial conductance of $h_{\text{hom},200} \approx 27.7$ MW/m^{2}K was evaluated numerically [44].

Discrete microfield studies were carried out with both unit cell and windowing models. Periodic phase arrangements were generated by first setting up periodic arrays of 20 randomly positioned, equally sized spherical particles using the two-step algorithm of Segurado [55]. Randomly oriented regular tetrakaidekahedra were then inscribed into the spheres. Figure 8 shows three of the resulting unit cells, the particles being tetrakaidekahedra in two of them and spheres in the third. The particle centers are identical in the three phase arrangements and the particle sizes were adjusted to give an inhomogeneity volume fraction of $\xi^{(i)} = 0.34$ in each

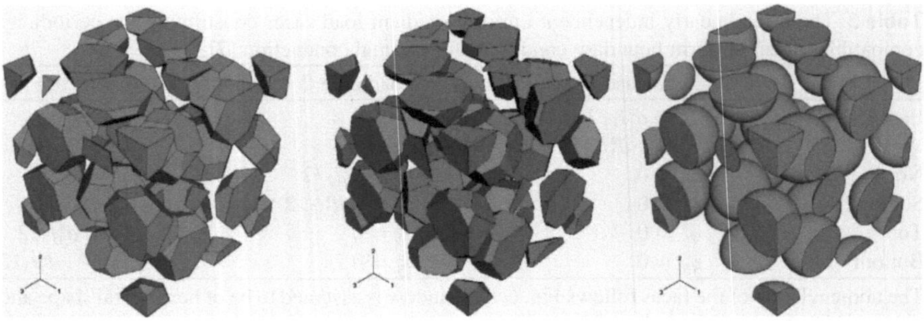

Fig. 8 Three unit cells of nominal particle volume fraction $\xi^{(i)} = 0.34$ used for modeling the thermal conduction behavior of diamond reinforced aluminum [45]

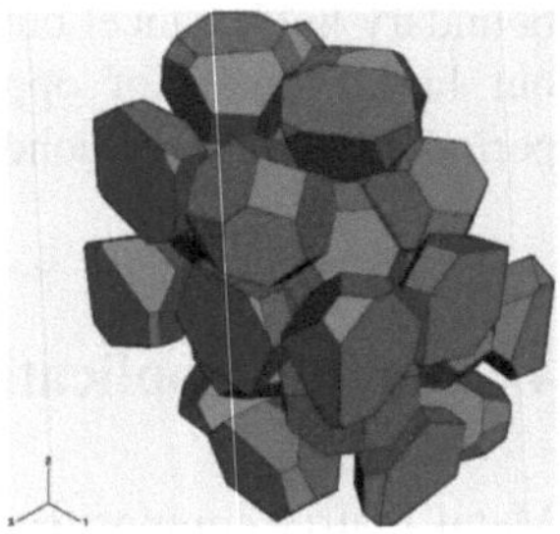

Fig. 9 Non-periodic volume element of particle volume fraction $\xi^{(i)} = 0.445$ [44]

case. Inhomogeneous interfacial conductances were specified for the tetrakaidekahedra, whereas the equivalent homogeneous conductance $h_{\mathrm{hom},200}$ was used for the spheres. The windowing studies were based on non-periodic volume elements containing some 26 particles. Such geometries are less prone to geometrical jamming than are periodic unit cells, so that phase arrangements with higher particle volume fractions can be generated [18]. A volume element of this type with a particle volume fraction of $\xi^{(i)} = 0.445$ is shown in Fig. 9.

Within the mean field framework, the Mori–Tanaka method in combination with the replacement tensor approach presented in Sect. 2.6 is suitable for handling the effects of the inhomogeneous interfacial conductances and the polyhedral particle shape. In order to properly handle the size effect, the replacement tensors must be evaluated separately for each size of inclusion studied. Because the tetrakaidekahedra with inhomogeneous interfacial conductances show cubic symmetry, the replacement tensors in conductivity are isotropic, so that no angular averaging is required for modeling randomly oriented particles.

For spherical inhomogeneities that show a uniform interfacial conductance h the components of the replacement tensors (which are isotropic diagonal tensors under these conditions) can be evaluated analytically, the replacement conductivity for particles of diameter d being given by

$$K^{(\mathrm{i;r})} = K^{(\mathrm{i})} \frac{dh}{dh + 2K^{(\mathrm{i})}} . \tag{43}$$

Furthermore, a critical diameter of the particles, d_c, at which the replacement conductivity of the inhomogeneity is equal to the matrix conductivity, can be evaluated as

$$d_c = \frac{2}{h} \frac{K^{(i)} K^{(m)}}{K^{(i)} - K^{(m)}}, \tag{44}$$

compare, e.g., [6]. Particles exceeding this critical radius improve the effective conductivity compared to the matrix, whereas smaller particles lead to a deterioration of the macroscopic conductivity. For the present material parameters this critical diameter evaluates as $d_c \approx 19.7\ \mu\text{m}$.

Figure 10 presents various predictions for the dependence of the macroscopic conductivity of diamond–aluminum composites on the particle diameter [45] at a particle volume fraction of $\xi^{(i)} = 0.34$. The Mori–Tanaka results for conductivities obtained for spheres with ideal interfaces (MTM, perfect) and for fully debonded reinforcements (MTM/voids), which behave like spherical voids, are independent of the particle diameter. The three-point bounds of Torquato and Rintoul [60], evaluated for spherical particles of equal size having a homogeneous interfacial conductance of $h_{\text{hom},200}$, are marked as 3PB, hom. The two bounds can be seen to "cross over" at $d \approx d_c$, and the lower bound approaches the Mori–Tanaka model

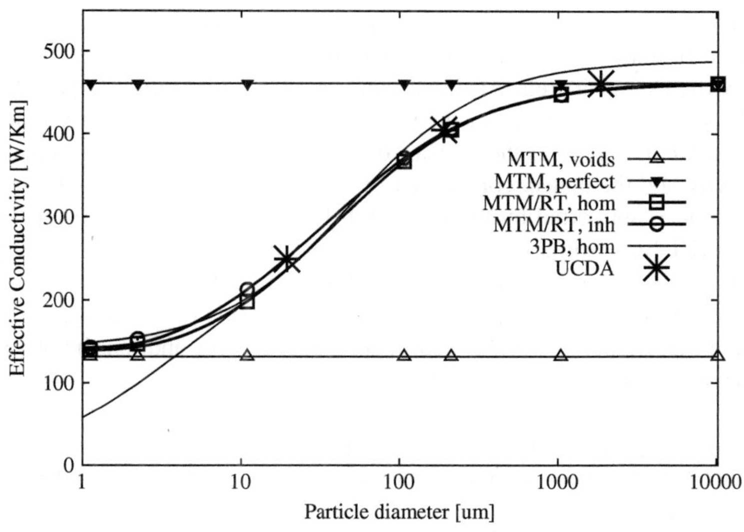

Fig. 10 Predictions of the effective conductivity of diamond–aluminum MMCs [45]. Mori–Tanaka results for fully debonded (MTM, voids) and perfectly bonded (MTM, perfect) spherical particles, three-point bounds (3PB, hom) for spherical particles with interfacial conductances, Mori–Tanaka predictions using the replacement tensor algorithm for particles with homogeneous (MTM/RT, hom) and inhomogeneous (MTM/RT, inh) interfaces, as well as unit cell results (UCDA) for tetrakaidekahedral particles are shown. The reinforcement volume fraction is $\xi^{(i)} = 0.34$

with a perfect interface for $d \gg d_c$. The Mori–Tanaka method using the replacement tensor scheme was evaluated both for the homogeneous interfacial conductance of $h_{hom,200}$ (MTM/RT, hom) and for the inhomogeneous conductances (MTM/RT, inh). The latter model can be seen to give slightly higher effective conductances for values of d of the same order of magnitude as d_c. The unit cell predictions, marked as UCDA, agree closely with the "MTM/RT, inh" results and tend to lie slightly above them. Taken together, the results shown in Fig. 10 indicate that the influence of the tetrakaidekahedral particle shape and of the inhomogeneity of the interfacial conductances on the macroscopic conduction behavior of the diamond–aluminum composites is rather limited.

The apparent conductivities predicted by windowing models and the effective conductivities obtained by combining the replacement tensor approach with the Mori–Tanaka method are compared in Table 4 for different particle sizes and a fixed particle volume fraction of $\xi^{(i)} = 0.445$. The windowing studies are based on the non-periodic volume element shown in Fig. 9 and employ the three types of boundary conditions discussed in Sect. 3.3, viz., uniform Neumann (UNBC), uniform Dirichlet (UDBC) and mixed uniform (MUBC) boundary conditions, the latter being prescribed according to Table 3. All models pertain to tetrakaidekahedral particles of equal size having inhomogeneous interfacial conductances and all windowing results are directional averages over three load cases.

As expected, the UNBC consistently provide lower and the UDBC upper estimates. Although among the windowing results the mixed uniform boundary conditions provide the best agreement with the RT/MTM estimates for all particle sizes, the difference between the predictions of the two models exceed 10% for the largest particles considered. An important reason for this rather large difference probably lies in the small size of the volume element used, the influence of which becomes more important as the contrast between the inhomogeneities' replacement conductivity and the matrix conductivity grows with increasing particle size. This interpretation is supported by the marked differences between the lower (UNBC) and upper (UDBC) estimates, which exceed 50% for all particle sizes — for proper representative volume elements these differences should approach zero. Considerably larger volume elements, which could mitigate this discrepancy, on the one hand have high requirements for computational resources and, on the other hand, are expensive in terms of analysts' time for preprocessing.

Table 4 Comparison of apparent and effective conductivities K (W/mK) of a diamond–aluminum MMC with tetrakaidekahedral particles of volume fraction $\xi^{(i)} = 0.445$ at four different particle sizes [45]

	UNBC	MUBC	UDBC	RT/MTM
d = 2 μm $< d_c$	87.2	128.7	277.5	129.2
d = 20 μm $\approx d_c$	189.9	266.7	378.7	253.1
d = 200 μm $> d_c$	337.0	492.0	571.9	466.8
d = 2,000 μm $\gg d_c$	397.2	602.4	669.2	546.2

Windowing results obtained with the volume element shown in Fig. 9 pertain to uniform Neumann (UNBC), uniform Dirichlet (UDBC) and mixed uniform (MUBC) boundary conditions. RT/MTM denotes effective conductivities evaluated with a Mori–Tanaka scheme using replacement tensors

Discrete microstructure approaches can provide detailed information on the macroscopic responses and microfields of composites provided the sizes of the reinforcements do not differ drastically. For polydisperse inhomogeneity sizes spanning more than, say, an order of magnitude the need for discretizations that are sufficiently fine to resolve the smallest reinforcements makes discrete microstructure models based on numerical engineering methods prohibitively expensive. In many cases, however, estimates of the influence of the size effect in composites reinforced by polydispersely sized particles with imperfect interfacial conductances can be obtained with mean field methods.

The different inhomogeneity phases in the Mori–Tanaka expressions for the effective conductivity, (22), may be interpreted as pertaining to particles of N different size classes, the inhomogeneity volume fractions $\xi^{(\mathrm{i})}$ being used to account for the particle size distribution. Judging from Fig. 10, no gross errors are introduced by approximating the particles of diamond–aluminum MMCs by spheres with suitable homogeneous interfacial conductances. This allows the use of simple expressions for the replacement tensors, compare (43), and for the depolarization tensor, compare (18), which can be plugged into (22) to obtain a Mori–Tanaka expression for the effective coefficient of conductivity in the form [6]

$$K^*_{\mathrm{MT}} = K^{(\mathrm{m})} + \frac{\sum_{i=1}^{N} \xi^{(\mathrm{i})} (K^{(\mathrm{i;r})} - K^{(\mathrm{m})}) A_{\mathrm{dil}}^{(\mathrm{i;r})}}{\xi^{(\mathrm{m})} + \sum_{i=1}^{N} \xi^{(\mathrm{i})} A_{\mathrm{dil}}^{(\mathrm{i;r})}}, \tag{45}$$

where the replacement dilute gradient concentration factor is obtained from (19) as

$$A_{\mathrm{dil}}^{(\mathrm{i;r})} = \frac{3K^{(\mathrm{m})}}{2K^{(\mathrm{m})} + K^{(\mathrm{i;r})}}. \tag{46}$$

Equation (45) is sufficiently simple to allow resolving particle size distributions by hundreds of size classes. It was used study the influence of monomodal and bimodal particle size distributions on the overall conduction response of diamond–aluminum MMCs [6, 44].

5 Closing Remarks

In the present chapter two groups of models were discussed that are capable of providing estimates of the macroscopic diffusive transport behavior of inhomogeneous solids in dependence of the microstructure, the constituent properties and, where applicable, the behavior of the interfaces between the phases. The different modeling schemes cover considerable ranges in terms of complexity, flexibility, spatial resolution, capability and computational cost, allowing a suitable modeling approach to be found for most problems.

The majority of the methods are "bidirectional" in terms of the length scales involved, i.e., they are capable of handling both homogenization and localization. The input parameters of the models can be neatly separated into geometrical descriptors and material parameters of the constituents, which is very helpful in carrying out "virtual experiments".

References

1. Auriault, J.: Effective macroscopic description for heat conduction in periodic composites. Int J Heat Mass Transf **26**, 861–869 (1983)
2. Auriault, J.: Upscaling heterogeneous media by asymptotic expansions. J Eng Mech ASCE **128**, 817–822 (2002)
3. Beasley, J., Torquato, S.: Bounds on the conductivity of a suspension of random impenetrable spheres. J Appl Phys **60**, 3576–3581 (1986)
4. Benveniste, Y.: A new approach to the application of Mori–Tanaka's theory in composite materials. Mech Mater **6**, 147–157 (1987)
5. Böhm, H.: A Short Introduction to Basic Aspects of Continuum Micromechanics. Tech. Rep. (ILSB Arbeitsbericht 206), TU Wien, Vienna, Austria (2009). http://www.ilsb.tuwien.ac.at/links/downloads/ilsbrep206.pdf
6. Böhm, H., Nogales, S.: Mori–Tanaka models for the thermal conductivity of composites with interfacial resistance and particle size distributions. Compos Sci Technol **68**, 1181–1187 (2008)
7. Böhm, H., Pahr, D., Daxner, T.: Analytical and numerical methods for modeling the thermomechanical and thermophysical behavior of microstructured materials. In: Silberschmidt, V. (ed.) Computational and Experimental Mechanics of Advanced Materials. CISM Courses and Lectures, vol. 514, pp. 167–223. Springer, Vienna, Austria (2009)
8. Bruggemann, D.: Berechnung verschiedener physikalischer Konstanten von heterogenen Substanzen. I. Dielektrizitätskonstanten und Leitfähigkeiten der Mischkörper aus isotropen Substanzen. Ann Phys **24**, 636–679 (1935)
9. Buryachenko, V.: Micromechanics of Heterogeneous Materials. Springer, New York, NY (2007)
10. Duan, H., Karihaloo, B., Wang, J., Yi, X.: Effective conductivities of heterogeneous media containing multiple inclusions with various spatial distributions. Phys Rev **B**, 174203 (2006)
11. Duschlbauer, D.: Computational Simulation of the Thermal Conductivity of MMCs under Consideration of the Inclusion–Matrix Interface. Reihe **5**, Nr.561, VDI-Verlag, Düsseldorf, Germany (2004)
12. Duschlbauer, D., Pettermann, H., Böhm, H.: Heat conduction of a spheroidal inhomogeneity with imperfectly bonded interface. J Appl Phys **94**, 1539–1549 (2003)
13. Duschlbauer, D., Pettermann, H., Böhm, H.: Numerical simulation of the thermal conductivity of MMCs – the effect of thermal interface resistance. Mater Sci Technol **19**, 1107–1114 (2003)
14. Duschlbauer, D., Böhm, H., Pettermann, H.: Computational simulation of composites reinforced by planar random fibers: homogenization and localization by unit cell and mean field approaches. J Compos Mater **40**, 2217–2234 (2006)
15. Eshelby, J.: The determination of the elastic field of an ellipsoidal inclusion and related problems. Proc R Soc Lond A Math Phys Sci **241**, 376–396 (1957)
16. Ferrari, M.: Asymmetry and the high concentration limit of the Mori–Tanaka effective medium theory. Mech Mater **11**, 251–256 (1991)

17. Fiedler, T, Belova, IV., Öchsner, A, Murch, GE.: Lattice Monte Carlo analysis of thermal diffusion in multi-phase materials. Springer, Heidelberg (2011). doi:10.1007/8611_2010_6
18. Flaquer, J., Ríos, A., Martín-Meizoso, A., Nogales, S., Böhm, H.: Effect of diamond shapes and associated thermal boundary resistance on thermal conductivity of diamond-based composites. Comput Mater Sci **41**, 156–163 (2007)
19. Furmański, P.: Heat conduction in composites: homogenization and macroscopic behavior. Appl Mech Rev **50**, 327–356 (1997)
20. Giraud, A., Gruescu, C., Do, D., Homand, F., Kondo, D.: Effective thermal conductivity of transversely isotropic media with arbitrary oriented ellipsoidal inhomogeneities. Int J Solids Struct **44**, 2627–2647 (2007)
21. Harte, A., McNamara, J.: Use of micromechanical modelling in the material characterisation of overinjected thermoplastic composites. J Mater Process Technol **173**, 376–383 (2006)
22. Hashin, Z.: Analysis of composite materials – a survey. J Appl Mech Trans ASME **50**, 481–505 (1983)
23. Hashin, Z.: The differential scheme and its application to cracked materials. J Mech Phys Solids **36**, 719–733 (1988)
24. Hashin, Z., Shtrikman, S.: A variational approach to the theory of the effective magnetic permeability of multiphase materials. J Appl Phys **33**, 3125–3131 (1962)
25. Hasselman, D., Donaldson, K.: Effect of reinforcement particle size on the thermal conductivity of a particulate-silicon carbide-reinforced aluminum matrix composite. J Am Ceram Soc **75**, 3137–3140 (1992)
26. Hasselman, D., Johnson, L.: Effective thermal conductivity of composites with interfacial thermal barrier resistance. J Compos Mater **21**, 508–515 (1987)
27. Hatta, H., Taya, M.: Effective thermal conductivity of a misoriented short fiber composite. J Appl Phys **58**, 2478–2486 (1985)
28. Hazanov, S.: Hill condition and overall properties of composites. Arch Appl Mech **68**, 385–394 (1998)
29. Hill, R.: A self-consistent mechanics of composite materials. J Mech Phys Solids **13**, 213–222 (1965)
30. Hill, R.: The essential structure of constitutive laws for metal composites and polycrystals. J Mech Phys Solids **15**, 79–95 (1967)
31. Jiang, M., Ostoja-Starzewski, M., Jasiuk, I.: Scale-dependent bounds on effective elastoplastic response of random composites. J Mech Phys Solids **49**, 655–673 (2001)
32. Kenesei, P., Borbély, A., Biermann, H.: Microstructure based three-dimensional finite element modeling of particulate reinforced metal matrix composites. Mater Sci Eng A Struct **387**, 852–856 (2004)
33. Kerner, E.: The electrical conductivity of composite media. Proc Phys Soc B **69**, 802–807 (1956)
34. Lipton, R., Talbot, D.: Bounds for the effective conductivity of a composite with an imperfect interface. Proc R Soc Lond A Math Phys Sci **457**, 1501–1517 (2001)
35. Markov, K.: Elementary micromechanics of heterogeneous media. In: Markov, K., Preziosi, L. (eds.) Heterogeneous Media: Micromechanics Modeling Methods and Simulations, pp. 1–162. Birkhäuser, Boston, MA (2000)
36. Matt, C., Cruz, M.: Application of a multiscale finite-element approach to calculate the effective thermal conductivity of particulate media. Comput Appl Math **21**, 429–460 (2002)
37. Matt, C., Cruz, M.: Effective thermal conductivity of composite materials with 3-D microstructures and interfacial thermal resistance. Numer Heat Transf A **53**, 577–604 (2008)
38. Matt, CF, Cruz, ME.: Heat conduction in two-phase composite materials with three-dimensional microstructures and interfacial thermal resistance. Springer, Heidelberg (2011). doi:10.1007/8611_2010_10
39. Maxwell, J.: Treatise on Electricity and Magnetism. Clarendon, Oxford (1873)
40. Michel, J., Moulinec, H., Suquet, P.: Effective properties of composite materials with periodic microstructure: a computational approach. Comput Methods Appl Mech Eng **172**, 109–143 (1999)

41. Miller, C., Torquato, S.: Effective conductivity of hard sphere suspensions. J Appl Phys **68**, 5486–5493 (1990)
42. Milton, G.: The Theory of Composites. Cambridge University Press, Cambridge (2002)
43. Mori, T., Tanaka, K.: Average stress in the matrix and average elastic energy of materials with misfitting inclusions. Acta Metall **21**, 571–574 (1973)
44. Nogales, S.: Numerical Simulation of the Thermal and Thermomechanical Behavior of Metal Matrix Composites. Reihe **18**, Nr.317, VDI–Verlag, Düsseldorf, Germany (2008)
45. Nogales, S., Böhm, H.: Modeling of the thermal conductivity and thermomechanical behavior of diamond reinforced composites. Int J Eng Sci **46**, 606–619 (2008)
46. Nye, J.: Physical Properties of Crystals, Their Representation by Tensors and Matrices. Clarendon, Oxford (1957)
47. Ostoja-Starzewski, M.: Random field models of heterogeneous materials. Int J Solids Struct **35**, 2429–2455 (1998)
48. Ostoja-Starzewski, M., Schulte, J.: Bounding of effective thermal conductivities of multiscale materials by essential and natural boundary conditions. Phys Rev B **54**, 278–285 (1996)
49. Persson, L.: Computing effective thermal conductivities of composite materials by the homogenization method. PhD thesis, Luleå Tekniska Universitet, Luleå, Sweden (1986)
50. Phan-Tien, N., Pham, D.: Differential multiphase models for polydispersed spheroidal inclusions: thermal conductivity and effective viscosity. Int J Eng Sci **38**, 73–88 (2000)
51. Ponte Castañeda, P., Willis, J.: The effect of spatial distribution on the effective behavior of composite materials and cracked media. J Mech Phys Solids **43**, 1919–1951 (1995)
52. Progelhof, R., Throne, R., Ruetsch, R.: Methods for predicting the thermal conductivity of composite systems: a review. Polym Eng Sci **16**, 615–625 (1976)
53. Rintoul, M., Torquato, S.: Reconstruction of the structure of dispersions. J Colloid Interface Sci **186**, 467–476 (1997)
54. Ruch, P., Beffort, O., Kleiner, S., Weber, L., Uggowitzer, P.: Selective interfacial bonding in Al(Si)–diamond composites and its effect on thermal conductivity. Compos Sci Technol **66**, 2677–2685 (2006)
55. Segurado, J.: Micromecánica computacional de materiales compuestos reforzados con partículas. PhD thesis, Universidad Politécnica de Madrid, Spain (2004)
56. Smit, R., Brekelmans, W., Meijer, H.: Prediction of the mechanical behavior of non-linear heterogeneous systems by multi-level finite element modeling. Comput Method Appl Mech Eng **155**, 181–192 (1998)
57. Terada, K., Kikuchi, N.: Microstructural design of composites using the homogenization method and digital images. Mater Sci Res Int **2**, 65–72 (1996)
58. Torquato, S.: Effective electrical conductivity of two-phase disordered composite media. J Appl Phys **58**, 3790–3797 (1985)
59. Torquato, S.: Random Heterogeneous Media. Springer, New York, NY (2002)
60. Torquato, S., Rintoul, D.: Effect of the interface on the properties of composite media. Phys Rev Lett **75**, 4067–4070 (1995)
61. Weng, G.: The theoretical connection between Mori–Tanaka theory and the Hashin–Shtrikman–Walpole bounds. Int J Eng Sci **28**, 1111–1120 (1990)
62. Wiener, O.: Die Theorie des Mischkörpers für das Feld der stationären Strömung. Abh Math-Phys Kl Königl Sächs Ges Wiss **32**, 509–604 (1912)
63. Willis, J.: Bounds and self-consistent estimates for the overall moduli of anisotropic composites. J Mech Phys Solids **25**, 185–202 (1977)

Thermal Residual Stresses in Aluminium Matrix Composites

F. Teixeira-Dias and L.F. Menezes

Abstract It is well known that residual stresses strongly influence the behaviour of most materials and, in particular, of composite materials. This chapter presents one approach to the numerical determination of thermal residual stresses in metal matrix composites (MMC). The subject of residual stresses is introduced and the corresponding mathematical and constitutive models are described in detail. It is considered that the reinforcement material is elastic and that the metallic matrix may exhibit thermoelastic-viscoplastic behaviour. A progressive gradient based time-integration algorithm is described that leads to the implementation of the proposed constitutive models in a finite element analysis code. The corresponding variational formulation and discretisation into finite elements is also described. In order to guarantee stabilised convergence and to increase the precision of results, the authors also propose a time-step optimisation algorithm. All the formalisms are tested measuring the influence of the reinforcement volume fraction and cooling rate on the resulting residual stresses.

1 Introduction

Nowadays, metal matrix composite materials (MMC) are highly relevant materials in the scope of engineering applications mostly due to their mechanical properties and characteristics. In general terms, these materials often have very high specific stiffness, strength and low density. Most of the manufacturing processes associated

F. Teixeira-Dias (✉)
GRIDS–DAPS Division of Armour & Protection Systems, Department of Mechanical Engineering, Universidade de Aveiro, Campus Universitário de Santiago, 3810-193 Aveiro, Portugal
e-mail: ftd@ua.pt

L.F. Menezes
Department of Mechanical Engineering, Universidade de Coimbra, 3030 Coimbra, Portugal
e-mail: luis.menezes@dem.uc.pt

A. Öchsner and G.E. Murch (eds.), *Heat Transfer in Multi-Phase Materials,*
Adv Struct Mater 2, DOI 10.1007/8611_2010_1,

to MMC imply that the material must, at some stage, go through high temperatures and temperature gradients. These temperatures are most of the time close, or even above, the melting temperature of the metallic matrix material. This fact determines most of the mechanical properties of the final material. Consequently, it is of utmost importance that adequate numerical methods and models are developed that can represent the behaviour of MMC over the whole temperature range. Additionally, due to the fact that the constituent materials will have distinct coefficients of thermal expansion (CTE) and their constitutive behaviour will be different it is possible that residual stresses may arise when the MMC is subjected to high amplitude temperature changes. These residual stresses (and consequent residual strains) will affect the final properties and behaviour of the MMC [1–8].

The high specific stiffness and strength of metal matrix composite materials, and their thermal properties are good enough reasons to justify the high strategic interest in these engineering materials. Presently there is a wide variety of modelling approaches to the behaviour of MMC. Most approaches are based on the distinct properties of the constituent materials – the matrix and the reinforcement materials [9–11]. Most of these models are micromechanical models and are thus also based on the topology and geometrical distribution of the reinforcement components.

The overall behaviour of a metal matrix composite material is often dependent on the temperature and is highly sensitive to its variations. This is due to two main reasons: (1) the behaviour of the metallic matrix is temperature dependent and (2) temperature changes induce residual stress and strain fields within the MMC, due to the mismatch of coefficients of thermal expansion [12, 13]. Nonetheless, residual stresses in MMC may also have a mechanical origin: these stresses may be due to the non-homogeneous flow of matrix material around the reinforcement elements.

In general terms, metallic materials used as matrix in metal matrix composites have coefficients of thermal expansion that are often one order of magnitude higher than the CTE of the ceramic reinforcement material. It is then predictable that when cooling down the MMC from fabrication temperature thermal residual stress fields may arise [14, 15].

2 Mathematical Modelling and Algorithms

The following paragraphs introduce the continua kinematics approach used to describe the thermal and mechanical behaviour of metal matrix composites. The constitutive models used for both MMC constituent materials are also presented and detailed. The section ends with a description of the time-integration procedures.

2.1 Constitutive Modelling

When a multiphase material is submitted to changes in temperature residual stress fields may be generated. These may be generated by the differences in the coefficients of thermal expansion (CTE) of the constituent materials. The behaviour of materials under these conditions has been studied by many researchers often using experimental techniques [16–27].

2.1.1 Continua Kinematics

Within this text, an updated Lagrangian approach is proposed in order to continuously follow the evolution and movement of all material points in a particular medium. The goal is to determine the present configuration of the material, $\mathcal{C}_t$, starting from the reference configuration on a previous instant, $\mathcal{C}_0$. Let P be a material point in the continuous medium $\boldsymbol{\Omega}$ and $\mathbf{p}$ and $\mathbf{x}$ the position vectors of this material point in the configurations $\mathcal{C}_0$ and $\mathcal{C}_t$, respectively, then

$$\mathbf{p} = \bar{\mathbf{p}}(\mathbf{x}, t), \tag{1}$$

$$\mathbf{x} = \bar{\mathbf{x}}(\mathbf{p}, t), \tag{2}$$

$$\mathbf{x} = \mathbf{p} + \mathbf{u}(\mathbf{p}, t), \tag{3}$$

where $\mathbf{u}(\mathbf{p}, t)$ represents the displacement of the material point P between the configurations $\mathcal{C}_0$ and $\mathcal{C}_t$. The gradient $\mathbf{F}$ of the point transformation $\mathbf{x}$ can be defined as:

$$\mathbf{F}(\mathbf{p}, t) = \mathbf{F} \doteq \frac{\partial}{\partial \mathbf{p}}\bar{\mathbf{x}}(\mathbf{p}, t) = \mathbf{I} + \frac{\partial}{\partial \mathbf{p}}\mathbf{u}(\mathbf{p}, t), \tag{4}$$

where $\mathbf{I}$ is the second order identity tensor. The velocity field associated with this transformation is

$$\dot{\mathbf{x}}(\mathbf{p}, t) = \dot{\mathbf{x}}(\bar{\mathbf{p}}(\mathbf{x}, t), t) \doteq \frac{\partial}{\partial t}\bar{\mathbf{x}}(\mathbf{p}, t) \tag{5}$$

to which corresponds the velocity gradient $\mathbf{L}$, defined as

$$\mathbf{L}(\mathbf{x}, t) \doteq \mathrm{grad}[\mathbf{v}(\mathbf{x}, t)] = \frac{\partial}{\partial \mathbf{x}}\mathbf{v}(\mathbf{x}, t) = \frac{\partial}{\partial \mathbf{p}}\dot{\mathbf{x}}(\mathbf{p}, t)\frac{\partial}{\partial \mathbf{x}}\bar{\mathbf{p}}(\mathbf{x}, t) \tag{6}$$

where grad is the gradient operator relative to $\mathbf{x}$, keeping t constant. According to (4), the previous relation can also be written as

$$\mathbf{L} = \dot{\mathbf{F}}\mathbf{F}^{-1}. \tag{7}$$

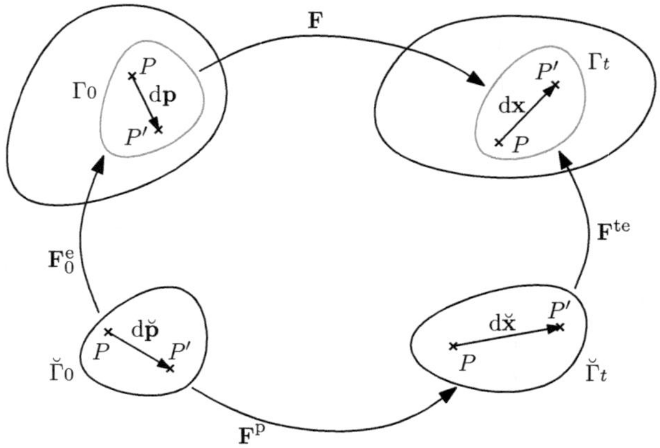

Fig. 1 Schematic representation of the decomposition of transformation gradient **F**

From this point it is important to distinguish the thermoelastic and viscoplastic contributions to the transformation gradient **F**, between instants t_0 and t. In order to do this one must analyse the infinitesimal neighbourhoods of material point P in both configurations $\mathcal{C}_0$ and $\mathcal{C}_0$, as shown in Fig. 1. The following configurations must also be defined: Γ_0 – the configuration of the infinitesimal neighbourhood of P at instant t_0; $\breve{\Gamma}_0$ – the configuration obtained after elastic relaxation of Γ_0; Γ_t – the configuration of the infinitesimal neighbourhood of P at instant t; and $\breve{\Gamma}_t$ – the configuration obtained after elastic relaxation of Γ_t.

Let $\mathbf{dp}$ be the position vector of material point P', in the infinitesimal neighbourhood of P, relative to P itself. Consequently, $\mathrm{d}\breve{\mathbf{p}}$, $\mathrm{d}\mathbf{x}$ and $\mathrm{d}\breve{\mathbf{x}}$ are the transformations of vector $\mathrm{d}\mathbf{p}$ within the configurations $\breve{\Gamma}_0$, Γ_t and $\breve{\Gamma}$, respectively. Thus, the following three transformation gradients can be defined in the neighbourhood of P: $\mathbf{F}_0^{\mathrm{e}}$ – the transformation gradient $\breve{\Gamma}_0$; $\mathbf{F}^{\mathrm{p}}$ – the transformation gradient $\breve{\Gamma}_0$; and $\mathbf{F}^{\mathrm{te}}$ – the transformation gradient $\breve{\Gamma}_t$. It is now possible to relate vectors $\mathrm{d}\mathbf{p}$ and $\mathrm{d}\mathbf{x}$ based on these definitions, that is,

$$\mathrm{d}\mathbf{x} = \mathbf{F}^{\mathrm{te}}\mathbf{F}^{\mathrm{p}}(\mathbf{F}_0^{\mathrm{e}})^{-1}\mathrm{d}\mathbf{p}. \tag{8}$$

Relating the previous equation to (4) gradient **F** can be redefined as

$$\mathbf{F} = \mathbf{F}^{\mathrm{te}}\mathbf{F}^{\mathrm{p}}(\mathbf{F}_0^{\mathrm{e}})^{-1}. \tag{9}$$

Replacing **F** in (7) it is possible to obtain

$$\mathbf{L} = \dot{\mathbf{F}}^{\mathrm{te}}(\mathbf{F}^{\mathrm{te}})^{-1} + \mathbf{F}^{\mathrm{te}}\dot{\mathbf{F}}^{\mathrm{p}}(\mathbf{F}^{\mathrm{p}})^{-1}(\mathbf{F}^{\mathrm{te}})^{-1} = \mathbf{L}^{\mathrm{te}} + \mathbf{L}^{\mathrm{p}}, \tag{10}$$

where $\mathbf{L}^{\mathrm{te}} = \dot{\mathbf{F}}^{\mathrm{te}}(\mathbf{F}^{\mathrm{te}})^{-1}$ and $\mathbf{L}^{\mathrm{p}} = \mathbf{F}^{\mathrm{te}}\dot{\mathbf{F}}^{\mathrm{p}}(\mathbf{F}^{\mathrm{p}})^{-1}(\mathbf{F}^{\mathrm{te}})^{-1}$ are the thermoelastic and viscoplastic parts of the velocity gradient tensor $\mathbf{L}$, respectively. However, $\mathbf{L}$ can still be decomposed to its strain rate ($\mathbf{D}$) and rotation rate ($\mathbf{W}$) tensors [28], as

$$\mathbf{L} = \mathbf{D} + \mathbf{W}. \tag{11}$$

These tensors correspond to the symmetric ($\mathbf{L}^{\mathrm{S}}$) and anti-symmetric ($\mathbf{L}^{\mathrm{A}}$) parts of $\mathbf{L}$, that is,

$$\mathbf{D}(\mathbf{x}, t) \doteq \mathbf{L}^{\mathrm{S}} = \frac{1}{2}\left(\mathbf{L} + \mathbf{L}^{\mathrm{T}}\right), \tag{12}$$

$$\mathbf{W}(\mathbf{x}, t) \doteq \mathbf{L}^{\mathrm{A}} = \frac{1}{2}\left(\mathbf{L} - \mathbf{L}^{\mathrm{T}}\right). \tag{13}$$

2.1.2 Material Behaviour

The goal of this text is to model the development of residual stresses in metal matrix composites (MMC). These stress fields can arise, for example, from the cooling down stage that is imposed on the MMC during the manufacturing process. Let $\dot{T}$ be the cooling rate, considered constant. It will also be assumed that the temperature field $T(t)$ is homogeneous within the material. Its evolution can then be described by

$$T(t) = T_0 + \dot{T}t. \tag{14}$$

T_0 is the initial temperature value, at time instant $t = t_0$. The material is considered to be stress free at $t = t_0$, which is a reasonable consideration due to the manufacturing temperatures of most metal matrix composites [29]. In technological processes involving the cooling down of two or more distinct materials (e.g. reinforcement and matrix) it is important to distinguish each behaviour model. This distinction will be made relating the melting (decomposition) temperature of each material (T^{m}) and the process temperature. This relation is designated by homologous temperature (T^{h}) and can be defined as

$$T^{\mathrm{h}} = \frac{T}{T^{\mathrm{m}}}. \tag{15}$$

2.1.3 Constitutive Modelling of the Reinforcement

The maximum temperature levels reached during the manufacturing processes of most MMCs are much lower than the decomposition temperature of the reinforcement materials, thus the homologous temperature is often $T^{\mathrm{h}} < 1$. Facing this it is reasonable to consider that the reinforcement material exhibits thermoelastic behaviour. As a consequence, the following hypotheses are assumed [30]: (1) elastic strains are small; (2) elastic behaviour is isotropic; and (3) the influence of plastic strain on the elastic constants can be neglected. The reinforcement behaviour can then be described by Hooke's hyperelastic law [31], i.e.

$$\dot{\sigma} = \mathbf{C}^{\mathrm{e}}_{\mathrm{R}} : \mathbf{D}^{\mathrm{e}}, \tag{16}$$

where $\dot{\sigma}$ is the time derivative of Cauchy's stress tensor and $\mathbf{C}^{\mathrm{e}}_{\mathrm{R}}$ is the elastic tensor of the reinforcement material which, assuming isotropic behaviour, can be defined as [32]

$$\mathbf{C}^{\mathrm{e}}_{\mathrm{R}} = 2\mu_{\mathrm{R}}\mathbf{1} + \lambda_{\mathrm{R}}\mathbf{I} \otimes \mathbf{I}, \tag{17}$$

where $\mathbf{1}$ is the fourth order identity tensor. μ_{R} and λ_{R} are Lamé's coefficients for the reinforcement material and $\mathbf{D}^{\mathrm{e}}$ is the elastic part of the thermoelastic strain rate tensor, given by

$$\mathbf{D}^{\mathrm{te}} = \mathbf{D}^{\mathrm{e}} + \mathbf{D}^{\mathrm{t}}. \tag{18}$$

The thermal part of the strain rate tensor is $\mathbf{D}^{\mathrm{t}} = \alpha_{\mathrm{R}}\dot{T}\mathbf{I}$ and α_{R} is the thermal expansion coefficient of the reinforcement. Combining and manipulating (16)–(18) it can be determined that

$$\dot{\sigma} = 2\mu_{\mathrm{R}}\mathbf{D}^{\mathrm{te}} + \left[\left(k_{\mathrm{R}} - \frac{2}{3}\mu_{\mathrm{R}}\right)\mathrm{trace}(\mathbf{D}^{\mathrm{te}}) - 3k_{\mathrm{R}}\alpha_{\mathrm{R}}\dot{T}\right]\mathbf{I}, \tag{19}$$

with

$$k_{\mathrm{R}} = \lambda_{\mathrm{R}} + \frac{2}{3}\mu_{\mathrm{R}}. \tag{20}$$

2.1.4 Constitutive Modelling of the Matrix

The maximum temperature reached during the manufacturing process of the MMC is often of the same order of magnitude as the melting temperature of the matrix material. For this reason the matrix homologous temperature is $T^{\mathrm{h}} \approx 1$. It is then admitted that the material behaviour is both temperature and strain rate dependent. Several constitutive models have been proposed for such situations. However, the authors will focus on a material model that uses an internal state variable that will allow the description of the static and dynamic recovery of the matrix material.

This model is developed based on the assumptions that plastic strains are highly dependent on the strain rate and that the internal state of the material determines its behaviour. The proposed model will allow the modelling of phenomena such as: (1) the effects of strain rate and temperature; (2) static and dynamic recovery and recrystallisation processes; (3) internal damage and its evolution; and (4) crystalline structure and its evolution.

When formulating the constitutive model it is necessary to identify the internal state variables. The proposed approach uses only one internal state variable, s, representing the resistance to plastic flow. This is clearly a limiting approach but it is nonetheless possible to model both hardening and sensitivity to strain rate and temperature with such a law [33]. The proposed model is based on the following decomposition of the strain rate tensor:

$$\mathbf{D} = \mathbf{D}^{\mathrm{e}} + \mathbf{D}^{\mathrm{t}} + \mathbf{D}^{\mathrm{vp}}, \tag{21}$$

where $\mathbf{D}^{\mathrm{e}}$ and $\mathbf{D}^{\mathrm{t}} = \alpha_{\mathrm{M}}\dot{T}\,\mathbf{I}$ are the elastic and thermal parts of the strain rate tensor, respectively. α_{M} is the temperature dependent thermal expansion coefficient of the matrix material. $\mathbf{D}^{\mathrm{vp}}$ is the viscoplastic strain rate tensor considered to be isochoric, that is, $\mathrm{trace}(\mathbf{D}^{\mathrm{vp}}) = 0$.

In the technological processes to be modelled it can be considered that strains are small when compared to 1 and that rotations can be neglected. Thus, the small strains approach is used and the evolution law of Cauchy's stress tensor is considered identical to the reinforcement material law, that is,

$$\dot{\sigma} = \mathbf{C}^{\mathrm{e}}_{\mathrm{M}} : \mathbf{D}^{\mathrm{e}}, \tag{22}$$

where the evolution of the material tensor $\mathbf{C}^{\mathrm{e}}_{\mathrm{M}}$ will be described by

$$\dot{\sigma} = 2\mu_{\mathrm{M}}(\mathbf{D} - \mathbf{D}^{\mathrm{vp}}) + \left[\left(k_{\mathrm{M}} - \frac{2}{3}\mu_{\mathrm{M}}\right)\mathrm{trace}(\mathbf{D}) - 3k_{\mathrm{M}}\alpha_{\mathrm{M}}\dot{T}\right]\mathbf{I}. \tag{23}$$

Lamé's coefficients for the matrix material are temperature dependent, i.e., $\mu_{\mathrm{M}} = \mu_{\mathrm{M}}(T)$ and $\lambda_{\mathrm{M}} = \lambda_{\mathrm{M}}(T)$. The constitutive relation for $\mathbf{D}^{\mathrm{vp}}$, or flow law, is described by [33]

$$\mathbf{D}^{\mathrm{vp}} = \frac{3\dot{\bar{\varepsilon}}^{\mathrm{p}}}{2\bar{\sigma}}\sigma'. \tag{24}$$

In the previous relation, σ' is Cauchy's stress deviatoric tensor and $\bar{\sigma}$ is von Mises' equivalent stress. The equivalent plastic strain rate, $\dot{\bar{\varepsilon}}^{\mathrm{p}}$, is

$$\dot{\bar{\varepsilon}}^{\mathrm{p}} = f(\bar{\sigma}, s, T). \tag{25}$$

It is necessary to define the function governing the evolution of the internal state variable s. This function is itself dependent on $\bar{\sigma}$, s and T, that is,

$$\dot{s} = g(\bar{\sigma}, s, T). \tag{26}$$

The state variable in (25) can be directly related to the equivalent stress in such a way that

$$\dot{\bar{\varepsilon}}^{\mathrm{p}} = f(\frac{\bar{\sigma}}{s}, T). \tag{27}$$

This condition agrees with previous dislocation slipping activation models [34]. It is also assumed that the evolution of the internal state variable can be defined by [35–37]

$$\dot{s} = g(\bar{\sigma}, s, T) = \dot{\bar{\varepsilon}}^{\mathrm{p}} h(\bar{\sigma}, s, T) - \dot{r}(s, T). \tag{28}$$

In this equation, $h(\bar{\sigma}, s, T)$ is associated with the hardening and dynamic restoration phenomena. Static restoration phenomena are described by $\dot{r}(s, T)$.

The independence of function $\dot{r}(s, T)$ from the equivalent stress reflects the fact that this function represents phenomena occurring within the material in the absence of applied stress. After experimental analyses it can be concluded that different materials have distinct functional dependences of $h(\bar{\sigma}, s, T)$ and $\dot{r}(s, T)$. Experimental tests done on aluminium show that a power law is the most adequate. As for Fe-2%Si the best dependence of the strain rate with the stress is exponential. Based on these observations the following combined exponential and power dependence between strain rate and stress can be stated:

$$\dot{\bar{\varepsilon}}^{\mathrm{p}} = A\exp\left(-\frac{Q}{R_g T}\right)\left[\sinh\left(\xi\frac{\sigma}{s}\right)\right]^{1/m}, \tag{29}$$

where A, Q, m and ξ are material constants: A is the pre-exponential factor, m is the strain rate sensitivity, Q is the activation energy and R_g is the perfect gas constant. A relevant consequence of (29) is the proportional relation between the internal variable s and the equivalent stress $\bar{\sigma}$, that is, $\bar{\sigma} = cs$ where

$$c = \frac{1}{\xi}\sinh^{-1}\left\{\left[\frac{\dot{\bar{\varepsilon}}^{\mathrm{p}}}{A}\exp\left(\frac{Q}{R_g T}\right)\right]^m\right\}. \tag{30}$$

With the previous relation it is now possible to indirectly determine variable s. The proportionality between s and $\bar{\sigma}$ can be used to determine the static restoration function $\dot{r}(s, T)$ and the hardening function $h(\bar{\sigma}, s, T)$. However, previous experimental studies have proved that the influence of static restoration phenomena on the strain resistance is less than 1% of the overall strain resistance. Based on these results, $\dot{r}(s, T)$ will be neglected on the following considerations. The adopted form for the hardening function is

$$h(\bar{\sigma}, s, T) = h_0 \left|1 - \frac{s}{s^*}\right|^a \operatorname{sgn}\left(1 - \frac{s}{s^*}\right) \tag{31}$$

with $a > 1$. Introducing (31) in relation (28) results in

$$\dot{s} = \dot{\bar{\varepsilon}}^{\mathrm{p}} \left[h_0 \left|1 - \frac{s}{s^*}\right|^a \operatorname{sgn}\left(1 - \frac{s}{s^*}\right)\right] \tag{32}$$

where h_0 is the hardening rate and s^* is a saturation value for the scalar variable s, associated with a determined temperature and strain rate, such that

$$s^* = \bar{s}\left[\frac{\dot{\bar{\varepsilon}}^{\mathrm{p}}}{A} \exp\left(\frac{Q}{R_g T}\right)\right]^n. \tag{33}$$

The constants h_0, a, $\bar{s}$ and n are once more material parameters.

2.2 *Time-Integration of the Constitutive Model*

Differential equations describing the evolution of state variables of most physical processes with identical complexity of the ones described in this text can only be integrated numerically [38]. Several numerical integration approaches can be used with strain rate constitutive laws. These integration methods are often referred to as: (1) totally explicit or progressive methods; (2) totally implicit or regressive methods; or (3) semi-implicit or progressive gradient methods.

The constitutive relations used herein are numerically stable. However, when using a fully explicit Euler-type time integration scheme it is often necessary to significantly reduce the time step in order to guarantee numerical stability [35]. Using a semi-implicit integration algorithm is one way to overcome this question. These algorithms approximate the fully-implicit methods using Taylor series' developments of the constitutive functions and tolerate significantly larger time-steps [16]. The most relevant disadvantage of the progressive gradient methods is the fact that their precision deteriorates when large time increments are used in high gradient stages of the simulation. As a consequence, it is necessary to associate careful time-step control algorithms with these integration procedures [28, 38].

2.2.1 Progressive Gradient Integration Method

The set of constitutive equations proposed here has, among other, the advantage of using a scalar parameter s that represents the internal state of the material and models the resistance of the material to plastic flow. The time integration of these equations can be made with a progressive gradient integration method [28, 39].

The main goal of the integration method is to determine the configuration $\mathcal{C}_{n+1}$ starting from the initial step configuration $\mathcal{C}_n$, at a time interval Δt. The known state variables at $\mathcal{C}_n$ are σ_n, s_n and T_n. The state variables at $\mathcal{C}_{n+1}$ are σ_{n+1}, s_{n+1} and T_{n+1}. The first step is to integrate the viscoplastic strain rate tensor $\mathbf{D}^{\mathrm{vp}}$ along the time increment Δt in order to determine the plastic strain increment $\Delta\varepsilon^{\mathrm{p}}$, that is,

$$\Delta\varepsilon^{\mathrm{p}} = \int_{t_n}^{t_{n+1}} \mathbf{D}^{\mathrm{vp}} \mathrm{d}t. \tag{34}$$

This integration can be made using the approximation

$$\Delta\varepsilon^{\mathrm{p}} = \bar{\mathbf{D}}^{\mathrm{vp}} \Delta t \tag{35}$$

where $\bar{\mathbf{D}}^{\mathrm{vp}}$ is the viscoplastic strain rate tensor weighted from the limit time instants of the increment, that is,

$$\bar{\mathbf{D}}^{\mathrm{vp}} = \mathbf{D}_n^{\mathrm{vp}} + \Phi\big(\mathbf{D}_{n+1}^{\mathrm{vp}} - \mathbf{D}_n^{\mathrm{vp}}\big). \tag{36}$$

Φ is a scalar parameter defined in the interval [0, 1] and represents the weighting factor for the integration. $\mathbf{D}_{n+1}^{\mathrm{vp}}$ is determined using a first order truncated Taylor approximation in such a way that

$$\mathbf{D}_{n+1}^{\mathrm{vp}} = \mathbf{D}_n^{\mathrm{vp}} + \frac{\partial \mathbf{D}_n^{\mathrm{vp}}}{\partial f}\Delta f + \frac{\partial \mathbf{D}_n^{\mathrm{vp}}}{\partial \bar{\sigma}}\Delta\bar{\sigma} + \frac{\partial \mathbf{D}_n^{\mathrm{vp}}}{\partial \sigma'}\Delta\sigma'. \tag{37}$$

Performing the partial derivatives in the previous relation leads to

$$\mathbf{D}_{n+1}^{\mathrm{vp}} = \frac{3f_n}{2\bar{\sigma}}\sigma'_n + \frac{3\sigma'_n}{2\bar{\sigma}_n}\Delta f + \frac{3}{2}f_n\sigma'_n\left(-\frac{1}{\bar{\sigma}_n^2}\right)\Delta\bar{\sigma} + \frac{3f_n}{2\bar{\sigma}_n}\Delta\sigma'. \tag{38}$$

where the equivalent viscoplastic strain increment is the increment of the function $f(\bar{\sigma}, s, T)$, that is,

$$\Delta f = \frac{\partial f_n}{\partial \bar{\sigma}}\Delta\bar{\sigma} + \frac{\partial f_n}{\partial s}\Delta s + \frac{\partial f_n}{\partial T}\Delta T \tag{39}$$

and the equivalent stress increment is

$$\Delta\bar{\sigma} = \frac{\partial \bar{\sigma}}{\partial \sigma'} : \Delta\sigma'. \tag{40}$$

Performing the derivatives with respect to σ' one obtains

$$\frac{\partial\bar{\sigma}}{\partial\sigma'} = \frac{1}{2}\left(\frac{3}{2}\sigma'_n : \sigma'_n\right)^{-1/2}(3\sigma'_n) = \frac{3}{2\bar{\sigma}_n}\sigma'_n. \tag{41}$$

Introducing the previous relation in (40) results in

$$\Delta\bar{\sigma} = \frac{3}{2\bar{\sigma}_n}\sigma'_n : \Delta\sigma'. \tag{42}$$

In order to determine the increment of the deviatoric Cauchy stress tensor, $\Delta\sigma'$, it is first necessary to calculate the deviatoric part of the stress rate tensor $\dot{\sigma}$. Introducing tensor $\dot{\sigma}$ and calculating its deviatoric part results in

$$\dot{\sigma}' = 2\mu_{\mathrm{M}}\left(\mathbf{D}' - \mathbf{D}^{\mathrm{vp}}\right). \tag{43}$$

Integrating along Δt the increment of the deviatoric Cauchy stress tensor becomes

$$\Delta\sigma' = 2\mu_{\mathrm{M}}(\Delta\varepsilon' - \Delta\varepsilon^{\mathrm{p}}), \tag{44}$$

where the total strain increment and its deviatoric part are

$$\Delta\varepsilon = \int_{t_n}^{t_n+\Delta t}\mathbf{D}\,\mathrm{d}t \quad \text{and} \quad \Delta\varepsilon' = \Delta\varepsilon - \frac{1}{3}\mathrm{trace}(\Delta\varepsilon)\mathbf{I}, \tag{45}$$

respectively. Manipulating relations (35)–(38) it is possible to derive the following relation to determine the plastic strain increment:

$$\Delta\varepsilon^{\mathrm{p}} = \left(\frac{3\sigma'_n}{2\bar{\sigma}_n}\right)\left[f_n + \Phi\left(\frac{\partial f_n}{\partial\bar{\sigma}}\Delta\bar{\sigma} + \frac{\partial f_n}{\partial s}\Delta s + \frac{\partial f_n}{\partial T}\Delta T\right)\right]\Delta t \tag{46}$$

$$+\Phi\frac{3f_n}{2\bar{\sigma}_n}\left[\Delta\sigma' - \left(\sigma'_n : \Delta\sigma'\right)\frac{3\sigma'_n}{2\bar{\sigma}_n^2}\right]\Delta t, \tag{47}$$

or, in a more compact form

$$\Delta\varepsilon^{\mathrm{p}} = \Delta\bar{\varepsilon}^{\mathrm{p}}\left(\frac{3\sigma'_n}{2\bar{\sigma}_n}\right) + \frac{3}{2h_1}\left[\Delta\sigma' - \left(\sigma'_n : \Delta\sigma'\right)\frac{3\sigma'_n}{2\bar{\sigma}_n^2}\right], \tag{48}$$

where the scalar h_1 is

$$h_1 = \frac{\bar{\sigma}_n}{\Phi f_n \Delta t} \tag{49}$$

and the equivalent plastic strain increment is

$$\Delta\bar{\varepsilon}^{\mathrm{p}} = \left[f_n + \Phi\left(\frac{\partial f_n}{\partial \bar{\sigma}} \Delta\bar{\sigma} + \frac{\partial f_n}{\partial s} \Delta s + \frac{\partial f_n}{\partial T} \Delta T \right) \right] \Delta t. \tag{50}$$

Performing the scalar product between the deviatoric stress tensor σ'_n and the plastic strain increment $\Delta\varepsilon^{\mathrm{p}}$, defined in (48), leads to

$$\sigma'_n : \Delta\varepsilon^{\mathrm{p}} = \Delta\bar{\varepsilon}^{\mathrm{p}} \left(\frac{3\sigma'_n : \sigma'_n}{2\bar{\sigma}_n} \right) + \frac{3}{2h_1} \left[\sigma'_n : \Delta\sigma' - \left(\sigma'_n : \Delta\sigma' \right) \frac{3\sigma'_n : \sigma'_n}{2\bar{\sigma}_n^2} \right]. \tag{51}$$

Combining relations (42) and (44) it is possible to derive the following expression for the equivalent stress increment:

$$\Delta\bar{\sigma} = \frac{3\mu_{\mathrm{M}}}{\bar{\sigma}_n} \sigma'_n : \Delta\varepsilon' - 3\mu_{\mathrm{M}} \Delta\bar{\varepsilon}^{\mathrm{p}}. \tag{52}$$

The scalar product $\sigma'_n : \Delta\varepsilon'$ can be simplified replacing σ'_n and $\Delta\varepsilon'$ by its definitions, that is,

$$\sigma'_n : \Delta\varepsilon' = \left[\sigma_n - \frac{1}{3} \mathrm{trace}(\sigma_n) \mathbf{I} \right] : \left[\Delta\varepsilon - \frac{1}{3} \mathrm{trace}(\Delta\varepsilon) \mathbf{I} \right] \tag{53}$$

leading to

$$\sigma'_n : \Delta\varepsilon' = \sigma_n : \Delta\varepsilon - \frac{1}{3} \mathrm{trace}(\sigma_n) \mathbf{I} : \Delta\varepsilon \quad \text{or} \quad \sigma'_n : \Delta\varepsilon' = \sigma'_n : \Delta\varepsilon. \tag{54}$$

Consequently, relation (52) can be rewritten in the form

$$\Delta\bar{\sigma} = \frac{3\mu_{\mathrm{M}}}{\bar{\sigma}_n} \sigma'_n : \Delta\varepsilon - 3\mu_{\mathrm{M}} \Delta\bar{\varepsilon}^{\mathrm{p}}. \tag{55}$$

Replacing the equivalent stress increment $\Delta\bar{\sigma}$ (55) in the formulation of the equivalent plastic strain increment (50), recalling that the increment of the internal state variable is $\Delta s = \Delta\bar{\varepsilon}^{\mathrm{p}} h$ and performing some algebraic manipulation leads to

$$\Delta\bar{\varepsilon}^{\mathrm{p}} = \frac{\Delta t}{1 + v_n} \left(f_n + \Phi \frac{3\mu_{\mathrm{M}}}{\bar{\sigma}_n} \frac{\partial f_n}{\partial \bar{\sigma}} \sigma'_n : \Delta\varepsilon + \Phi \frac{\partial f_n}{\partial T} \Delta T \right), \tag{56}$$

where the scalar variable v_n is

$$v_n = \Phi \frac{\partial f_n}{\partial \bar{\sigma}} G_n \Delta T \quad \text{with} \quad G_n = 3\mu_{\mathrm{M}} - h_n \left(\frac{\partial f_n}{\partial s} \Big/ \frac{\partial f_n}{\partial \bar{\sigma}} \right). \tag{57}$$

Inserting the deviatoric stress increment $\Delta\sigma'$, given by relation (44), in (48), that defines the plastic strain increment, results in

$$\Delta\varepsilon^{\mathrm{p}} = \Delta\bar{\varepsilon}^{\mathrm{p}} \left(\frac{3\sigma'_n}{2\bar{\sigma}_n} \right) + \frac{3}{2h_1} \left[2\mu_{\mathrm{M}} (\Delta\varepsilon' - \Delta\varepsilon^{\mathrm{p}}) - \frac{3\mu_{\mathrm{M}}}{\bar{\sigma}_n^2} (\sigma'_n : \Delta\sigma') \sigma'_n \right]. \tag{58}$$

Finally replacing the deviatoric strain increment $\Delta\varepsilon'$ and rearranging all terms leads to

$$\begin{aligned} \Delta\varepsilon^{\mathrm{p}} = & \frac{3\bar{\mu}_n}{\mu_{\mathrm{M}}} \left(1 + \frac{3\mu_{\mathrm{M}}}{h_1} \right) \frac{\sigma'_n}{2\bar{\sigma}_n} \Delta\bar{\varepsilon}^{\mathrm{p}} + \frac{3\bar{\mu}_n}{h_1} \left[\Delta\varepsilon - \frac{1}{3} \mathrm{trace}(\Delta\varepsilon) \right] \\ & - \frac{9\bar{\mu}_n}{h_1 \bar{\sigma}_n^2} (\sigma'_n : \Delta\varepsilon) \sigma'_n, \end{aligned} \tag{59}$$

where

$$\bar{\mu}_n = \frac{\mu_{\mathrm{M}}}{1 + 3\mu_{\mathrm{M}}/h_1}. \tag{60}$$

2.2.2 Elasto-Plastic Secant Modulus

In order to determine the increment of the stress tensor it is necessary to integrate the previously defined constitutive law along the whole time increment $\Delta t \equiv [t_n, t_{n+1}]$, that is,

$$\dot{\sigma} = 2\mu_{\mathrm{M}} \mathbf{D} + \left(k_{\mathrm{M}} - \frac{2}{3}\mu_{\mathrm{M}} \right) \mathrm{trace}(\mathbf{D}) \mathbf{I} - 2\mu_{\mathrm{M}} \mathbf{D}^{\mathrm{vp}} - 3k_{\mathrm{M}} \alpha_{\mathrm{M}} \dot{T} \mathbf{I}. \tag{61}$$

Linearising the previous differential equation along the time increment Δt, replacing the plastic strain increment (59) and recalling relations (34) and (45), it is possible to obtain

$$\Delta\sigma = 2\mu_{\mathrm{M}}\Delta\varepsilon - 2\bar{\mu}_n\left(\frac{3\sigma'_n}{2\bar{\sigma}_n} + \frac{9\bar{\mu}_n}{2h_1}\frac{\sigma'_n}{\bar{\sigma}_n}\right)\left[f_n + \Phi\left(\frac{3\mu_{\mathrm{M}}}{\bar{\sigma}_n}\frac{3\mu_{\mathrm{M}}}{\bar{\sigma}_n^2}\sigma'_n : \Delta\varepsilon\right)\frac{\partial f_n}{\partial\bar{\sigma}}\right.$$
$$\left. + \Phi\frac{\partial f_n}{\partial T}\Delta T\right]\frac{\Delta t}{1+v_n} - 2\bar{\mu}_n\frac{3\mu_{\mathrm{M}}}{h_1}\left[\Delta\varepsilon - \frac{1}{3}\mathrm{trace}(\Delta\varepsilon)\right] \tag{62}$$
$$+ 2\bar{\mu}_n\frac{9\mu_{\mathrm{M}}}{2h_1\bar{\sigma}_n^2}(\sigma'_n : \Delta\varepsilon)\sigma'_n + \left(k_{\mathrm{M}} - \frac{2}{3}\mu_{\mathrm{M}}\right)\mathrm{trace}(\Delta\varepsilon)\mathbf{I} - 3k_{\mathrm{M}}\alpha_{\mathrm{M}}(\Delta T)\mathbf{I}.$$

or, in a more compact way,

$$\Delta\sigma = 2\mu_{\mathrm{M}}\Delta\varepsilon + \bar{\lambda}_n\ \mathrm{trace}(\Delta\varepsilon)\mathbf{I} - K_1(\sigma'_n : \Delta\varepsilon)\sigma'_n - K_2\sigma'_n - 3k_{\mathrm{M}}\alpha_{\mathrm{M}}(\Delta T)\mathbf{I}. \tag{63}$$

In the previous relation

$$K_1 = \frac{3}{\bar{\sigma}_n^2}\left[\frac{v_n}{1+v_n}\frac{3\mu_{\mathrm{M}}^2}{G_n} - (\mu_{\mathrm{M}} - \bar{\mu}_n)\right], \tag{64}$$

$$K_2 = \frac{\Delta t}{1+v_n}\left(f_n + \Phi\frac{\partial f_n}{\partial T}\Delta T\right)\frac{3\mu_{\mathrm{M}}}{\bar{\sigma}_n} \text{ and } \bar{\lambda}_n = k_{\mathrm{M}} - \frac{2}{3}\bar{\mu}_n. \tag{65}$$

Finally, performing some algebraic rearrangement of (63) leads to

$$\Delta\sigma = \mathbf{M}^{\mathrm{sec}} : \Delta\varepsilon - K_2\sigma'_n - 3k_{\mathrm{M}}\alpha_{\mathrm{M}}(\Delta T)\mathbf{I}. \tag{66}$$

In the previous relation, $\mathbf{M}^{\mathrm{sec}}$ is the elasto-plastic secant modulus, that can be defined by its components as

$$M_{ijkl}^{\mathrm{sec}} = \bar{\lambda}_n\delta_{ij}\delta kl + \bar{\mu}_n(\delta_{ik}\delta jl + \delta_{il}\delta jk) - K_1\sigma'_{ij}\sigma'_{kl}, \tag{67}$$

where δ_{ij} is the Kronecker delta.

2.3 Variational Formulation and Discretisation

Performing numerical simulations using the finite element method (FEM) is essentially an approximation to determine the behaviour of a real system. This task can be done solving a limited set of equations that describe the real system.

2.3.1 Equilibrium Equations and Boundary Conditions

The system to be modelled is a solid deformable body that occupies a physical space designated by Ω delimited by an exterior surface Σ. As this work concerns metal matrix composites (MMC), Ω is built from zones from distinct materials, one metallic – the metal matrix, Ω_M – and the other ceramic – the reinforcement, Ω_R. Ω can thus be defined by the following relations:

$$\Omega_\mathrm{R} = \cup_i \Omega_\mathrm{R}^i \text{ and } \Omega = \Omega_\mathrm{M} \cup \Omega_\mathrm{R}. \tag{68}$$

It is supposed that at time instant t, Ω is subjected to a set of diverse external loads: volume and surface loads, temperature variations, etc. The exterior surface of Ω is divided in a set of surfaces Σ_i such that $\Sigma = \cup_i \Sigma_i$ in which velocities and/or loads are known and prescribed. Thus, neglecting the effect of volumetric loads, it is possible to formulate the equilibrium of Ω as follows:

$$\mathrm{div}\sigma = \mathbf{0}, \text{ in } \Omega. \tag{69}$$

The boundary conditions to which the deformable body Ω is subjected can also be listed as follows:

$$\mathbf{v} = \mathbf{v}^* \text{ on } \Sigma_v, \quad \mathbf{t} = \mathbf{t}^* \text{ on } \Sigma_t \quad \text{and} \quad \mathbf{v} = \mathbf{v}^* \wedge \mathbf{t} = \mathbf{t}^* \text{ on } \Sigma_{v,t} \tag{70}$$

where $\mathbf{v}$ and $\mathbf{v}^*$ are generic and prescribed velocities, respectively, and $\mathbf{t}$ and $\mathbf{t}^*$ are generic and prescribed Cauchy stress vectors, respectively. If $\mathbf{n}$ is the unit external normal vector that defines Σ, then $\mathbf{t} = \sigma\mathbf{n}$. Boundary conditions are considered to be unaltered during the whole duration of the process.

2.3.2 Variational Formulation

The generic problem defined by (69) and (70) is satisfied only on the condition that the principle of virtual work (PVW) is also satisfied whatever the virtual velocity field $\delta\mathbf{v}$, that is,

$$\int_\Omega \sigma : \delta\mathbf{D}\mathrm{d}\Omega = \int_{\Sigma_t} \mathbf{t}^* \delta\mathbf{v}\mathrm{d}\Sigma + \int_{\Sigma_{v,t}} \mathbf{t}^* \delta\mathbf{v}\mathrm{d}\Sigma. \tag{71}$$

However, once this work concerns only the development of residual stresses in MMC due to the cooling down process, the effects of external loads are not considered. Consequently, the PVW becomes

$$\int_\Omega \sigma : \delta\mathbf{D}\mathrm{d}\Omega = \mathbf{0}, \tag{72}$$

where $\delta\mathbf{D}$ is the symmetric part of the virtual velocity gradient tensor. The virtual velocity field is continuously differentiable and confines to the boundary conditions defined in Σ_v.

The principle of virtual work is written in two distinct forms: (1) total Lagrangian and (2) updated Lagrangian. On the first approach all integrations of the PVW are relative to the initial configuration of the system, $\mathcal{C}_0$. The updated Lagrangian approach is computationally more expensive because the integrations are performed relative to a reference configuration $\mathcal{C}_t$, which is the initial configuration of each time increment $[t, t+\Delta t]$. This last approach is the one adopted within this work and will be described on the following paragraphs.

The configuration of the deformable body Ω at time instant t is the reference configuration, $\mathcal{C}_0$, for the current time step $[t, t+\Delta t]$. The final configuration at the end of the current time increment, $\mathcal{C}$, is then the reference configuration for the next step. Thus, (72) becomes

$$\int_{\Omega(C_0)} \sigma_t : \delta\mathbf{D}\mathrm{d}\Omega = \mathbf{0}, \tag{73}$$

where σ_t is Cauchy's stress tensor at the start of the increment (instant t). Assuming that only small strains and rotations occur between two consecutive configurations it is possible to subtract the PVW from the initial and ending instant of the time step, resulting in

$$\int_{\Omega(C)} \sigma_{t+\Delta t} : \delta\mathbf{D}\mathrm{d}\Omega - \int_{\Omega(C_0)} \sigma_t : \delta\mathbf{D}\mathrm{d}\Omega = \mathbf{0}. \tag{74}$$

The previous relation can be simplified as

$$\int_{\Omega(C_0\to C)} \Delta\sigma : \delta\mathbf{D}\mathrm{d}\Omega = \mathbf{0}. \tag{75}$$

2.3.3 Discretisation of the Principle of Virtual Work

The implementation of the principle of virtual work (PVW) with the finite element method (FEM) starts by performing the discretisation of the virtual strain rate tensor $\delta\mathbf{D}$, in the form

$$\delta\mathbf{D} = \frac{1}{2}\left\{\frac{\partial(\delta\mathbf{v})}{\partial\mathbf{x}_0} + \left[\frac{\partial(\delta\mathbf{v})}{\partial\mathbf{x}_0}\right]^{\mathrm{T}}\right\} \quad \text{or} \quad \delta\mathbf{D} = [\mathrm{grad}(\delta\mathbf{V})]^{\mathrm{S}}. \tag{76}$$

Once the deformable body Ω is discretised in finite elements the virtual velocity field $\delta\mathbf{v}$ is discretised as follows:

$$\delta\mathbf{v}(\mathbf{x},t) = \sum_{A=1}^{NN} \mathbf{N}_A(\mathbf{x})\delta\mathbf{v}_A(t). \tag{77}$$

NN is the total number of nodes in the finite element mesh, $\mathbf{N}_A(\mathbf{x})$ are the shape functions and $\delta\mathbf{v}_A$ are the virtual velocities determined at the mesh nodes. Inserting the velocity field $\delta\mathbf{v}$ in relation (76) leads to

$$\delta\mathbf{D} = \left\{ \text{grad} \left[\sum_{A=1}^{NN} \mathbf{N}_A(\mathbf{x}) \delta\mathbf{v}_A(t) \right] \right\}^{\text{S}} = \sum_{A=1}^{NN} \mathbf{B}_A(\mathbf{x}) \delta\mathbf{v}_A(t), \tag{78}$$

where $\mathbf{B}_A(\mathbf{x})$ is the derivative of the shape function matrix.

Introducing the virtual strain rate tensor $\delta\mathbf{D}$ in the principle of virtual work, and after some algebraic manipulation, results in

$$\sum_{A=1}^{NN} \delta\mathbf{v}_A^{\text{T}} \int_{\Omega} \mathbf{B}_A^{\text{T}} \Delta\sigma \text{d}\Omega = \mathbf{0}. \tag{79}$$

Once the principle of virtual work is valid whatever the virtual velocity field $\delta\mathbf{v}_A$, (79) becomes

$$\int_{\Omega} \mathbf{B}_A^{\text{T}} \Delta\sigma \text{d}\Omega = \mathbf{0} = \mathbf{Q}_A \quad \text{with} \quad A = 1, \ldots, NN. \tag{80}$$

However, once the present work concerns sets of distinct materials, the stress increment $\Delta\sigma$ must be determined separately for $\mathbf{x} \in \Omega_{\text{R}}$ or $\mathbf{x} \in \Omega_{\text{M}}$. Consequently, the following two relations are valid distinctly for the two materials:

$$\mathbf{Q}_A^{\text{R}} = \int_{\Omega_{\text{R}}} \mathbf{B}_A^{\text{T}} \Delta\sigma_{\text{R}} \text{d}\Omega \quad \text{and} \quad \mathbf{Q}_A^{\text{M}} = \int_{\Omega_{\text{M}}} \mathbf{B}_A^{\text{T}} \Delta\sigma_{\text{M}} \text{d}\Omega. \tag{81}$$

2.3.4 Finite Elements

As an example, the deformable body Ω is discretised in isoparametric 8-node hexaedric elements with trilinear interpolation functions. This particular finite element has eight integration points for full integration and one for reduced integration [40, 41]. However, this isoparametric finite element has deficient behaviour when used to solve problems involving plastic strain [42, 43]. Using this finite element with full integration – using all the integration points – significantly increases the global stiffness of the element, leading to abnormal hydrostatic stresses and deteriorating the final solution. This particular phenomenon is associated with the fact that plastic strain is isochoric. Nonetheless, it is possible to correct this by reducing the number of integration points used in each element. This technique is designated by uniform reduced integration.

The discretisation of the velocity and virtual velocity fields $\mathbf{v}$ and $\delta\mathbf{v}$, respectively, are given by

$$\mathbf{v} = \sum_{a=1}^{NE} \mathbf{N}_a^e \mathbf{v}_a^e \quad \text{and} \quad \delta\mathbf{v} = \sum_{a=1}^{NE} \mathbf{N}_a^e \delta\mathbf{v}_a^e. \tag{82}$$

$\mathbf{v}_a^e$ and $\delta\mathbf{v}_a^e$ on the previous relations are the nodal and virtual nodal velocities of node a, respectively. From relations (82) it is possible to write the velocity gradient tensor $\mathbf{L}$ and the virtual velocity gradient tensor $\delta\mathbf{L}$ as

$$\mathbf{L} = \frac{\partial\mathbf{v}}{\partial\mathbf{x}} = \sum_{a=1}^{NE} \frac{\partial\mathbf{N}_a^e}{\partial\mathbf{x}} \mathbf{v}_a^e \quad \text{and} \quad \delta\mathbf{L} = \frac{\partial(\delta\mathbf{v})}{\partial\mathbf{x}} = \sum_{a=1}^{NE} \frac{\partial\mathbf{N}_a^e}{\partial\mathbf{x}} \delta\mathbf{v}_a^e, \tag{83}$$

respectively. When using full integration, tensors $\mathbf{L}$ and $\delta\mathbf{L}$ are determined on all Gauss integration points. When using uniform reduced integration $\mathbf{L}$ and $\delta\mathbf{L}$ are calculated only on the reduced integration point – the center of the finite element. However, this technique leads to reduced stiffness, may lead to the appearance of eigenmodes and deteriorates the final solution. One way to avoid all these problems is to implement a reduced selective integration approach [44–47]. In the process, frequently designated by $\bar{\mathbf{B}}$, reduced integration is selectively applied only to some terms of the stiffness matrix. The hydrostatic components of tensors $\mathbf{L}$ and $\delta\mathbf{L}$ are considered constant within the whole element and are thus calculated only on the reduced integration point. For this purpose, $\mathbf{L}$ is decomposed on its deviatoric and hydrostatic components as

$$\mathbf{L} = \mathbf{L}' + \mathbf{L}^{\mathrm{h}} \quad \text{where} \quad \mathbf{L}^{\mathrm{h}} = \frac{1}{3}\mathrm{trace}(\mathbf{L})\mathbf{I}. \tag{84}$$

The tensor $\mathbf{L}$ can then be replaced by a tensor designated by $\bar{\mathbf{L}}$, such that

$$\bar{\mathbf{L}} = \mathbf{L} + \bar{\mathbf{L}}^{\mathrm{h}} - \mathbf{L}^{\mathrm{h}}, \tag{85}$$

and $\bar{\mathbf{L}}^{\mathrm{h}}$ is determined only on the central integration point. Thus, the discretisation of $\bar{\mathbf{L}}$ leads to

$$\bar{\mathbf{L}} = \sum_{a=1}^{NE} \left\{ \frac{\partial\mathbf{N}_a^e}{\partial\mathbf{x}} \mathbf{v}_a^e + \frac{1}{3}\left[\frac{\partial\mathbf{N}_a^e}{\partial\mathbf{x}} \mathrm{trace}\left(\mathbf{v}_a^e\right) - \frac{\partial\mathbf{N}_a^e}{\partial\mathbf{x}} \mathrm{trace}\left(\mathbf{v}_a^e\right) \right] \right\} \tag{86}$$

where $\bar{\mathbf{N}}_a^e$ corresponds to $\mathbf{N}_a^e$ determined at the central point of the finite element.

2.3.5 Element Equations

Integrating (19) along the whole time increment $[t, t + \Delta t]$ and considering relation (45) it becomes possible to determine the stress increment in the reinforcement material $\Delta\sigma_{\mathrm{R}}$, that is,

$$\Delta\sigma_{\mathrm{R}} = 2\mu_{\mathrm{R}}\Delta\varepsilon + \left[\left(k_{\mathrm{R}} - \frac{2}{3}\mu_{\mathrm{R}}\right)\mathrm{trace}(\Delta\varepsilon) - 3k_{\mathrm{R}}\alpha_{\mathrm{R}}(\Delta T)\right]\mathbf{I}. \tag{87}$$

After some algebraic manipulation the previous relation becomes

$$\Delta\sigma_{\mathrm{R}} = \mathbf{C}^{e}_{\mathrm{R}}\Delta\varepsilon - 3k_{\mathrm{R}}\alpha_{\mathrm{R}}(\Delta T)\mathbf{I} \tag{88}$$

where $\mathbf{C}^{e}_{\mathrm{R}}$, defined in (17), is time- and temperature-independent. Inserting the stress increment $\Delta\sigma_{\mathrm{R}}$ in the formulation of the principle of virtual work leads to

$$\mathbf{Q}^{\mathrm{R}}_{a} = \int_{\Omega_{\mathrm{R}}} \mathbf{B}^{\mathrm{T}}_{a}\left[\mathbf{C}^{e}_{\mathrm{R}}\Delta\varepsilon - 3k_{\mathrm{R}}\alpha_{\mathrm{R}}(\Delta T)\mathbf{I}\right]\mathrm{d}\Omega. \tag{89}$$

The discretisation of the incremental strain field $\Delta\varepsilon$ results in

$$\Delta\varepsilon = \sum_{b=1}^{NE} \mathbf{B}_{b}\Delta\mathbf{u}_{b}. \tag{90}$$

Introducing $\Delta\varepsilon$ given by the previous relation in expression (89) one obtains

$$\mathbf{Q}^{\mathrm{R}}_{a} = \sum_{b=1}^{NE}\left(\int_{\Omega_{\mathrm{R}}} \mathbf{B}^{\mathrm{T}}_{a}\mathbf{C}^{e}_{\mathrm{R}}\mathbf{B}_{b}\mathrm{d}\Omega\right)\Delta\mathbf{u}_{b} - 3k_{\mathrm{R}}\alpha_{\mathrm{R}}(\Delta T)\int_{\Omega_{\mathrm{R}}} \mathbf{B}^{\mathrm{T}}_{a}\mathbf{I}\mathrm{d}\Omega, \tag{91}$$

or, written in a compacted form,

$$\mathbf{Q}^{\mathrm{R}}_{a} = \mathbf{0} = \sum_{b=1}^{NE} \mathbf{k}^{\mathrm{R}}_{ab}\Delta\mathbf{u}_{b} - \Delta\mathbf{f}^{\mathrm{R}}_{a}. \tag{92}$$

Expression (92) is the generic equilibrium system of equations of a finite element of reinforcement material, where

$$\mathbf{k}^{\mathrm{R}}_{ab} = \int_{\Omega_{\mathrm{R}}} \mathbf{B}^{\mathrm{T}}_{a}\mathbf{C}^{e}_{\mathrm{R}}\mathbf{B}_{b}\mathrm{d}\Omega \quad \text{and} \quad \Delta\mathbf{f}^{\mathrm{R}}_{a} = 3k_{\mathrm{R}}\alpha_{\mathrm{R}}(\Delta T)\int_{\Omega_{\mathrm{R}}} \mathbf{B}^{\mathrm{T}}_{a}\mathbf{I}\mathrm{d}\Omega. \tag{93}$$

are the stiffness matrix and the second member for the reinforcement material finite elements, respectively.

According to previous sections, the matrix material exhibits thermoelastic-viscoplastic behaviour. Thus, the increments of plastic strain $\Delta\varepsilon^{\mathrm{p}}$ and stress $\Delta\sigma_{\mathrm{M}}$ were determined in Sect. 2.2.1. Inserting this stress increment in the principle of virtual work, as given in (81), leads to

$$\mathbf{Q}_a^{\mathrm{M}} = \int_{\Omega_{\mathrm{M}}} \mathbf{B}_a^{\mathrm{T}} \left[\mathbf{M}^{\mathrm{sec}} \Delta\varepsilon - K_2 \sigma_n' - 3k_{\mathrm{M}}\alpha_{\mathrm{M}}(\Delta T)\mathbf{I} \right] \mathrm{d}\Omega. \tag{94}$$

Performing once again the discretisation of the incremental strain field $\Delta\varepsilon$ in accordance to (90) and accounting for relation (94) leads to

$$\mathbf{Q}_a^{\mathrm{M}} = \sum_{b=1}^{NE} \left(\int_{\Omega_{\mathrm{M}}} \mathbf{B}_a^{\mathrm{T}} \mathbf{M}^{\mathrm{sec}} \mathbf{B}_b \mathrm{d}\Omega \right) \Delta\mathbf{u}_b \tag{95}$$

$$- K_2 \int_{\Omega_{\mathrm{M}}} \mathbf{B}_a^{\mathrm{T}} \sigma_n' \mathrm{d}\Omega - 3k_{\mathrm{M}}\alpha_{\mathrm{M}}(\Delta T) \int_{\Omega_{\mathrm{M}}} \mathbf{B}_a^{\mathrm{T}} \mathbf{I} \mathrm{d}\Omega, \tag{96}$$

or, written in a compacted form,

$$\mathbf{Q}_a^{\mathrm{M}} = \mathbf{0} = \sum_{b=1}^{NE} \mathbf{k}_{ab}^{\mathrm{M}} \Delta\mathbf{u}_b - \Delta\mathbf{f}_a^{\mathrm{M}}. \tag{97}$$

The previous expression is the generic equilibrium system of equations of a finite element of matrix material, where

$$\mathbf{k}_{ab}^{\mathrm{M}} = \int_{\Omega_{\mathrm{M}}} \mathbf{B}_a^{\mathrm{T}} \mathbf{M}^{\mathrm{sec}} \mathbf{B}_b \mathrm{d}\Omega \tag{98}$$

and

$$\Delta\mathbf{f}_a^{\mathrm{M}} = K_2 \int_{\Omega_{\mathrm{M}}} \mathbf{B}_a^{\mathrm{T}} \sigma_n' \mathrm{d}\Omega + 3k_{\mathrm{M}}\alpha_{\mathrm{M}}(\Delta T) \int_{\Omega_{\mathrm{M}}} \mathbf{B}_a^{\mathrm{T}} \mathbf{I} \mathrm{d}\Omega \tag{99}$$

are the stiffness matrix and the second member for the matrix material finite elements, respectively.

According to the previously described formulations and considering the element equilibrium equation systems (92) and (97), the global system of equations that must be solved is

$$\mathbf{Q}^{\mathrm{R}} + \mathbf{Q}^{\mathrm{M}} = \mathbf{0}. \tag{100}$$

This system of equations can still be solved as

$$\sum_{B=1}^{NN} \mathbf{K}_{AB} \Delta \mathbf{u}_{\mathrm{B}} = \Delta \mathbf{F}_{\mathrm{A}} \tag{101}$$

where $\mathbf{K}_{AB}$ is the global stiffness matrix, $\Delta\mathbf{u}_{\mathrm{B}}$ is the increment displacement vector at the current time instant and $\Delta\mathbf{F}_{\mathrm{A}}$ is the global second member vector. A and B designate the global numbering of the nodes in the finite element mesh.

2.4 Time-Step Optimisation

The constitutive equations described in previous sections, relative to the matrix material, are highly non-linear. As a consequence, using a constant time-step size can be an inadequate approach to obtain good numerical results. Doing so, the time-step should be sufficiently small to guarantee the stability of the numerical approach during the whole process simulation. Thus, it is of utmost importance to use a variable time-step size algorithm when implementing the formulations described [35].

The progressive gradient integration method is relatively simple to implement due to the fact that it is not necessary to iterate to determine the state variables on each increment. However, when rapid changes in the plastic strain rate occur, the results obtained with the progressive gradient integration method can become inaccurate. Nonetheless, this problem can be minimised by implementing an additional and complementary algorithm to automatically control the time-step size on each increment. The proposed control algorithm is based on the variations of the plastic strain rate.

The control parameter in this automatic time-step algorithm is the maximum increment of plastic strain in all the integration points of the finite element discretisation during the current time increment. Two distinct criteria are defined in order to do so. The first criterion is based on r_C, which is the parameter that controls the variations on the equivalent plastic strain rate, that is,

$$r_C = \frac{C_{\max}}{C_{\mathrm{tol}}} \quad \text{with} \quad C_{\max} = \max_{i=1,ng} \left| \dot{\bar{\varepsilon}}^{\mathrm{p}}_{i,n+1} - \dot{\bar{\varepsilon}}^{\mathrm{p}}_{i,n} \right| \tag{102}$$

where $C_{\max}$ is the maximum equivalent plastic strain rate variation in the current time increment among all the integration Gauss points. C_{tol} is a predefined (user-defined) control parameter that can be determined as

$$C_{\mathrm{tol}} = \delta \frac{s_0}{E_{\mathrm{M}}}. \tag{103}$$

ng and $\delta \in]0, 1]$ on the previous relations are the total number of Gauss points of the domain and a scalar parameter, respectively. E_{M} and s_0 are the elastic modulus and the initial value of the state variable s, respectively.

The second criterion limits the equivalent plastic strain rate increment directly in such a way as to avoid too large plastic strain increments. To achieve this r_D is defined as

$$r_D = \frac{D_{\mathrm{max}}}{D_{\mathrm{tol}}} \quad \text{with} \quad D_{\mathrm{max}} = \max_{i=1,ng} \Delta\bar{\varepsilon}_i^{\mathrm{p}} \tag{104}$$

where D_{tol} is a predefined (user-defined) control parameter.

Based on the previously defined criteria, it is now possible to choose the dominant criterion, calculating the parameter $r_{\mathrm{max}} = max\{r_C, r_D\}$ and determining if the solution from the current increment is acceptable or not, modifying the time-step size whenever needed.

In order to perform this optimisation automatically it is possible to use an algorithm in which each increment is optimised for the following increment. The parameters that control this optimisation algorithm are the scalars $w_0 < 1$, $w_i > 1$ and $q_j < 1$ with $i = 1, \ldots, n$. If these parameters are chosen correctly it is possible to reach a compromise between total CPU simulation of the process and the precision of the obtained results.

With the proposed time step optimisation algorithm it is possible to adjust the increment size to be small in the stages where strong gradients are developed and to be large enough when the strain rate does not change significantly. Scalar values q_i define a finite set of intervals in which the correction factors w_i are applied, with $i = 1, \ldots, n$. Additionally, it is even possible to develop a continuous step optimisation process, performing a numerical fitting to the set of values (q_i, w_i). The following generic equation leads to the referred numerical adjustment and to the generation of several optimisation profiles controlling only the scalar parameter F_{c}

$$w = w_1 + (w_n - w_1)\left\{\frac{exp\left(F_{\mathrm{c}}\frac{r_{\mathrm{max}}-q_1}{q_n-q_1}\right)}{exp(F_{\mathrm{c}}) - 1}\right\}. \tag{105}$$

3 Implementation and Results

The models and approaches proposed in previous sections are now tested in the determination of residual stresses in specific metal matrix composite materials. For the sake of example, these models are tested using an Aluminium matrix composite, with SiC reinforcement.

It is natural to suppose that technological questions associated with the manufacturing of metal matrix composites (MMC) may influence their response in service within a particular structural application. Thus, it is of utmost importance and relevance to study and investigate the role of the manufacturing technological parameters on the final properties and characteristics of MMC and in particular on the development and elimination of residual thermal stress fields. The particular aspects studied and presented on this section are: (1) the effect of the reinforcement volume fraction, F_v, on the final distribution of residual stresses and other state variables; and (2) the effect of the cooling rate, $\dot{T}$, on the levels of residual stresses. Other tecnological issues such as, for example, those related to the distribution, orientation, and morphology of the reinforcement are also very important and have been studied by other authors, such as Stautter et al. [48], Sorensen et al. [49], Watt et al. [50] and Pettermann et al. [51].

As an example of application of the models proposed and described in previous sections a set of numerical simulations was performed to test their the numerical efficiency. Results are shown concerning (1) the effect of the metal matrix composite reinforcement volume fraction and (2) the effect of the cooling rate. Numerical simulations are performed on a unidirectional fibre reinforced MMC. The geometrical model of this MMC and the corresponding representative unit cell are schematically represented in Fig. 2a, b, respectively.

Boundary conditions are such that the coordinate planes in the representative unit cell (RUC) are planes of symmetry. All finite element simulations were performed considering an Al-SiC composite. The material properties – elastic modulus, Lamé coefficients and CTE – for the aluminium matrix are temperature dependent and given as [14]

$$\begin{aligned} E_{\mathrm{M}}(T) &= 73474 - 43.48T[\mathrm{MPa}], \\ \mu_{\mathrm{M}}(T) &= 27041 - 17.057T[\mathrm{MPa}], \\ \alpha_{\mathrm{M}}(T) &= 28.7 \times 10^{-6} + 2.47 \times 10^{-8}T[\mathrm{K}^{-1}], \end{aligned} \tag{106}$$

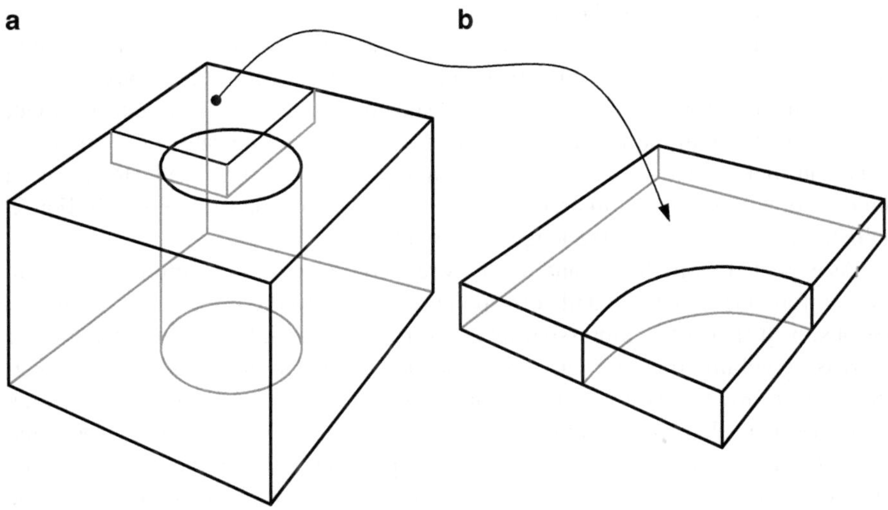

Fig. 2 Unidirectional fibre reinforced metal matrix composite (MMC): (**a**) geometrical model and (**b**) representative unit cell (RUC)

respectively, where T is the temperature in degrees Kelvin. The elastic modulus, Lamé coefficients and CTE for the reinforcement material are

$$\begin{aligned} E_{\mathrm{R}} &= 41 \times 10^4\,\mathrm{MPa}, \\ \mu_{\mathrm{R}} &= 16.532 \times 10^4\,\mathrm{MPa}, \\ \alpha_{\mathrm{R}} &= 0.43 \times 10^{-5}\,\mathrm{K}^{-1}, \end{aligned} \tag{107}$$

respectively. Material parameters used in the constitutive model are given by Tei-xeira-Dias and Menezes [14]. A constant cool rate is considered in all simulations and the resulting homogeneous temperature field is given by

$$T(t) = T_{\mathrm{i}} + \dot{T}t, \tag{108}$$

where t is the time. The initial and final temperatures are $T_{\mathrm{i}} = 933$ K and $T_{\mathrm{f}} = 293$ K, respectively.

3.1 Reinforcement Volume Fraction

The reinforcement volume fraction F_{v} is one of the manufacturing and technological parameters that influences most the mechanical and thermal characteristics of the composite material [26, 29, 52, 53]. Thus, the study of the influence of this parameter is particularly interesting as it can be defined and controlled both on the design and on the manufacturing stages in such a way as to achieve a set of desired mechanical properties for the MMC.

Numerical simulations were made on the referred representative unit cell using the constitutive models and numerical approaches described on previous sections. The σ_{xx} and σ_{yy} stress profiles along Oy are shown in Figs. 3 and 4 for reinforcement volume fractions in the range of 5–35%. A constant cooling rate $|\dot{T}| = 100$ Ks^{-1} was considered. These results concern a representative unit cell (RUC) of a unidirectional fibre metal matrix composite, such as the one represented in Fig. 2. Directions Ox and Oy are the length and width of the RUC.

As expected, both the σ_{xx} and the σ_{yy} stress components are compressive within the reinforcement material and correspond to an hydrostatic stress state. It can also be observed that the compressive levels of σ_{xx} in the reinforcement decrease for increasing volume fraction. The same tendency can be observed for σ_{yy}. However, the levels of σ_{xx} increase with the volume fraction, that is, as the distance between reinforcement elements decreases. It can be seen that the equivalent stress that arises from the results presented is almost independent of the reinforcement volume fraction, reaching its maximum values near the matrix-reinforcement interface. The maximum value of the equivalent stress is close to 70 MPa at room temperature. This value is above the yield stress of the matrix material, which is close to 40 MPa.

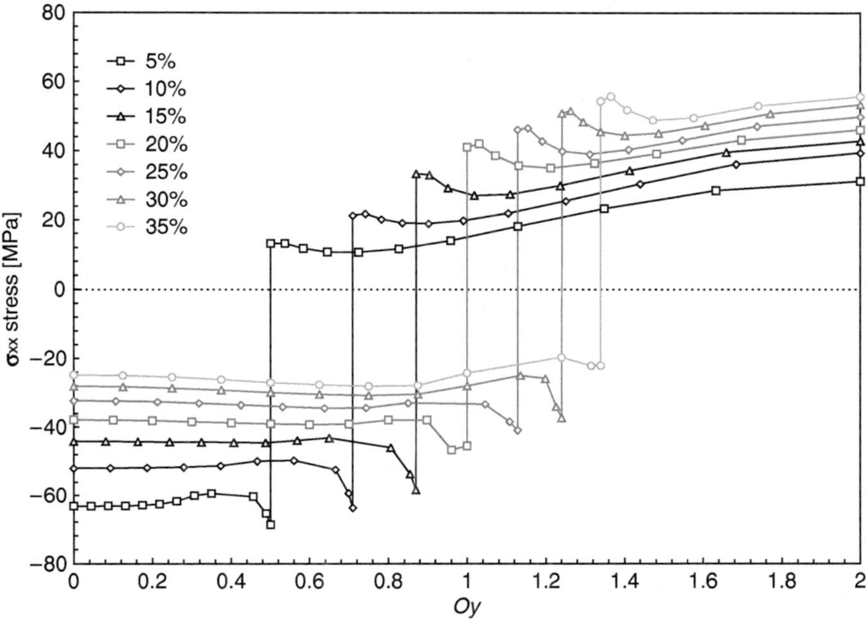

Fig. 3 σ_{xx} stress profiles along *Oy* for reinforcement volume fractions in the range of 5–35% at a constant cooling rate $|\dot{T}| = 100\ \mathrm{Ks}^{-1}$

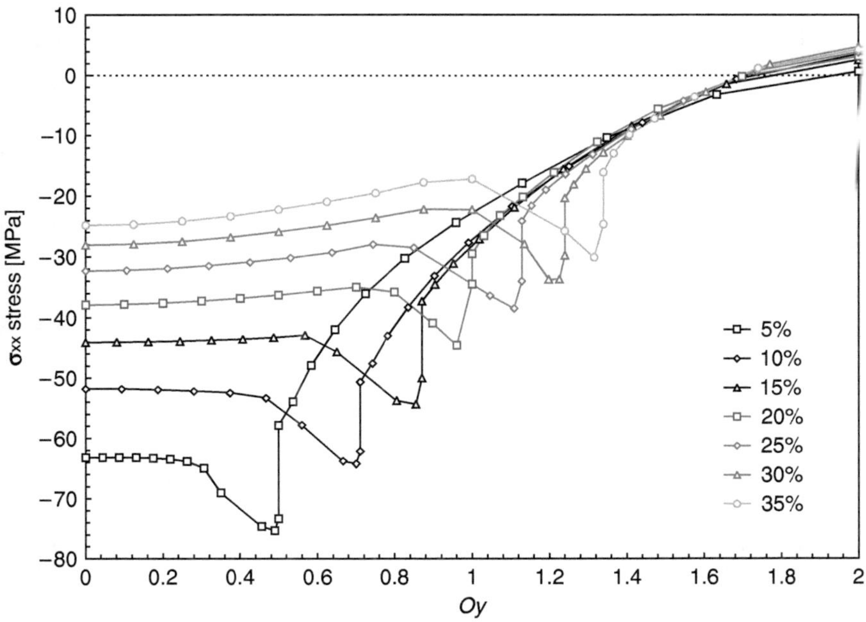

Fig. 4 σ_{yy} stress profiles along *Oy* for reinforcement volume fractions in the range of 5–35% at a constant cooling rate $|\dot{T}| = 100\ \mathrm{Ks}^{-1}$

This fact reflects the presence of hardening effects during the cooling down stages of the manufacturing process, as a consequence of the development of plastic strain.

3.2 Cooling Rate

As was stated before, most of the manufacturing processes used in the production of metal matrix composites (MMC) induce high temperature levels and gradients in the metallic matrix, often close to (or even above) its melting temperature. During the posterior cooling down stage, residual stresses may arise due to the mismatch between the coefficients of thermal expansion (CTE) of the constituent materials. Consequently, the cooling down stage is expected to have a determinant effect on the levels of residual stresses at room temperature within the MMC. Absolute cooling rates in the range of 0.1 to 500 Ks^{-1} were tested considering that the material is stress-free at fabrication temperature (T_i). The dependence of the equivalent stress σ_{eq}, internal parameter s and equivalent plastic strain $\dot{\bar{\varepsilon}}^p$ on the cooling rate are shown in Fig. 5.

In these results it can be clearly observed that there is a gradual decrease of both σ_{eq} and s with $|\dot{T}|$ and also that this effect increases for absolute cooling rates below

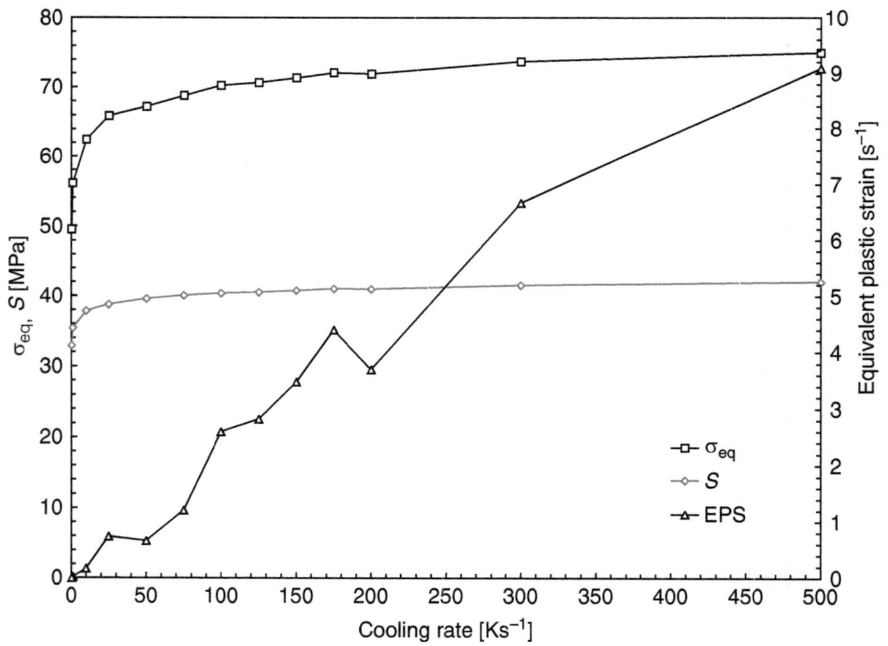

Fig. 5 Dependence of the equivalent stress σ_{eq}, state variable s and equivalent plastic strain $\dot{\bar{\varepsilon}}^p$ on the cooling rate ($\dot{T}$)

$50\ Ks^{-1}$. Nonetheless, in global terms, it can be stated that the cooling rate has a limited effect on the residual stress levels at room temperature. The effect of changing the cooling rate is mostly felt on the relatively low stresses generated at higher temperatures.

4 Final Remarks

The authors presented a full numerical approach to the determination of residual stresses in dual-phase microstructured materials, with applications to metal matrix composites (MMC). The model is applied in a finite element algorithm and tested with some numerical examples in order to prove its effectiveness and evaluate the effect of the reinforcement volume fraction and of the cooling rate on the levels of residual stresses at room temperature. Residual stress fields are determined in a representative unit cell (RUC) associated with a unidirectional fibre reinforced Al-SiC composite with volume fractions ranging from 5 to 35%.

It is shown that both normal stress components σ_{xx} and σ_{yy} are compressive in nature resulting in a hydrostatic stress state in the reinforcement. The levels of compressive stress decrease for the higher volume fractions. Nonetheless, the tendency is opposite in the metallic matrix with higher values of σ_{xx} for increasing reinforcement volume fractions. The equivalent stress reaches values above the yield limit, leading to the development of plastic strain near the matrix-reinforcement interface.

The influence of the cooling rate on the residual stresses at room temperature is not evident. This influence is only significant for absolute cooling rates under $50\ Ks^{-1}$. This can be explained by the fact that the residual stresses at room temperature are mostly generated at the lower temperature range, that is, under 600 K, where the viscoplastic behaviour of the aluminium matrix is less sensitive to the strain rate.

References

1. Parlevliet, P.P., Bersee, H.E.N., Beukers, A.: Residual stresses in thermoplastic composites – a study of the literature – Part I: Formation of residual stresses. Compos. Part A **37**(11), 1847–1857 (2006)
2. Parlevliet, P.P., Bersee, H.E.N., Beukers, A.: Residual stresses in thermoplastic composites – a study of the literature – Part II: Experimental techniques. Compos. Part A **38**(3), 651–665 (2006)
3. Liu, H.T., Sun, L.Z.: Effects of thermal residual stresses on effective elastoplastic behavior of metal matrix composites. Int. J. Solids Struct. **41**(8), 2189–2203 (2004)
4. Meijer, G., Ellyin, F., Xia, Z.: Aspects of residual thermal stress/strain in particle reinforced metal matrix composites. Compos. Part B **31**(1), 29–37 (2000)
5. Huang, Z.-M.: Cyclic response of metal matrix composite laminates subjected to thermo-mechanical fatigue loads. Int. J. Fatigue **24**, 463–475 (2002)

6. Gungor, S.: Residual stress measurements in fibre reinforced titanium alloy composites. Acta Mater. **50**(8), 2053–2073 (2002)
7. Altan, G., Topcu, M.: Thermo-elastic stress of a metal-matrix composite disc under linearly-increasing temperature loading by analytical and FEM analysis. Adv. Eng. Softw. **41**(4), 604–610 (2010)
8. Rudajevova, A., Milicka, K.: Residual and thermal strain in AX41 magnesium alloy reinforced with short Saffil fibres. Compos. Sci. Technol **68**(12), 2474–2478 (2008)
9. Sayman, O.: An elastic-plastic thermal stress analysis of aluminum metal-matrix composite beams. Compos. Struct. **53**(4), 419–425 (2001)
10. Sayman, O., Ozer, M.R.: Elastic-plastic thermal stress analysis of aluminum-matrix composite beams under a parabolically temperature distribution. Compos. Sci. Technol. **61**(14), 2129–2137 (2001)
11. Zhang, H., Ramesh, K.T., Chin, E.S.C.: A multi-axial constitutive model for metal matrix composites. J. Mech. Phys. Solids **56**(10), 2972–2983 (2008)
12. Ju, D.-Y.: Simulation of the thermo-mechanical and interfacial stress of metal matrix composites under thermal shock process. Compos. Struct. **48**, 113–118 (2000)
13. Choo, H., Bourke, M.A.M., Daymond, M.R.: A finite-element analysis of the inelastic relaxation of thermal residual stresses in continuous-fibre-reinforced composites. Compos. Sci. Technol **61**(12), 1757–1772 (2001)
14. Teixeira-Dias, F., Menezes, L.F.: Numerical determination of the influence of the cooling rate and reinforcement volume fraction on the levels of residual stresses in Al-SiC composites. Comput. Mater. Sci. **21**(1), 26–36 (2001)
15. Teixeira-Dias, F., Andrade-Campos, A., Pinho-da-Cruz, J.: On the effect of the orientation of the reinforcement on the overall behaviour of Al-SiC_p composites. Comput. Struct. **82**, 1323–1331 (2004)
16. Ledbetter, H.M., Austin, M.W.: Internal strain (stress) in an SiC–Al particle-reinforced composite: an X-ray diffraction study. Mater. Sci. Eng. **89**, 53–61 (1987)
17. Kolhe, R., Hui, C.Y., Ustundag, E., Sass, S.L.: Residual thermal stresses and calculation of the critical metal particle size for interfacial crack extension in metal-ceramic matrix composites. Acta Mater. **44**(1), 279–287 (1996)
18. Fernandez, P., Fernandez, R., Gonzalez-Doncel, G., Bruno, G.: Correlation between matrix residual stresses and composite yield strength in PM 6061-Al-15 vol.% SiC_w. Scripta Mater. **52**(8), 793–797 (2004)
19. Jiang, Z., Li, G., Lian, J., Ding, X., Sun, J.: Elastic-plastic stress transfer in short fibre-reinforced metal-matrix composites. Compos. Sci. Technol. **64**, 1661–1670 (2004)
20. Fang, Q., Sidky, P.S., Hocking, G.M.: Residual stresses in titanium matrix composites (TMC) in thermomechanical cycling using matrix etching. Mater. Sci. Eng. **288**(2), 293–297 (2000)
21. Bruno, G., Fernandez, R., Gonzalez-Doncel, G.: Relaxation of the residual stress in 6061 Al-15 vol.% SiC_w composites by isothermal annealing. Mater. Sci. Eng. **382**, 188–197 (2004)
22. Michel, J.C., Suquet, P.: An analytical and numerical study of the overall behaviour of metal-matrix composites. Model. Simulat. Mater. Sci. Eng. **2**(3A), 637–658 (1994)
23. Ramakrishnan, N.: An analytical study on strengthening of particulate reinforced metal matrix composites. Acta Mater. **44**(1), 69–77 (1996)
24. Shen, Y.-L., Finot, M., Needleman, A., Suresh, S.: Effective elastic response of two-phase composites. Acta Mater. **42**(1), 77–97 (1994)
25. Zahl, D.B.: The effect of interfacial sliding on the strength of metal matrix composites. Comput. Mater. Sci. **1**(3), 249–258 (1993)
26. Ho, S., Saigal, A.: Thermal residual stresses and mechanical behavior of cast SiC/Al composites. Mater. Sci. Eng. A **183**, 39–47 (1994)
27. Shaw, L.L., Miracle, D.B.: Effects of an interfacial region on the transverse behavior of metal-matrix composites – a finite element analysis. Acta Mater. **44**(5), 2043–2055 (1996)
28. Anand, L.: Constitutive equations for hot-working of metals. Int. J. Plast. **1**(3), 213–231 (1985)

29. Shen, Y.-L., Finot, M., Needleman, A., Suresh, S.: Effective plastic response of two-phase composites. Acta Mater. **43**(4), 1701–1722 (1995)
30. Suéry, M., Teodosiu, C., Menezes, L.F.: Thermal residual stresses in particle-reinforced viscoplastic metal matrix composites. Mater. Sci. Eng. A **167**, 97–105 (1993)
31. Hughes, T.J.R.: Numerical implementation of constitutive models: rate independent deviatoric plasticity, theoretical foundations for large scale computations of nonlinear material behaviour. Martinus Nighoff, The Netherlands (1994)
32. Zywicz, E., Parks, D.M.: Thermo-viscoplastic residual stresses in metal matrix composites. Compos. Sci. Technol. **33**(4), 295–315 (1988)
33. Brown, S.B., Kim, K.H., Anand, L.: An internal variable constitutive model for hot working of metals. Int. J. Plast. **5**(2), 95–130 (1989)
34. Kocks, U.F., Argon, A.S., Ashby, M.F.: Thermodynaics and kinetics of slip. Pergamon, Oxford (1975)
35. Lush, A.M., Weber, G., Anand, L.: An implicit time-integration procedure for a set of internal variable constitutive equations for isotropic elasto-viscoplasticity. Int. J. Plast. **5**(5), 521–549 (1989)
36. Weber, G., Anand, L.: Finite deformation constitutive equations and a time integration procedure for isotropic, hyperelastic-viscoplastic solids. Comput. Methods Appl. Mech. Eng. **79**(2), 173–202 (1990)
37. Weber, G., Lush, A.M., Zavaliangos, A., Anand, L.: An objective time-integration procedure for isotropic rate-independent and rate-dependent elastic-plastic constitutive equations. Int. J. Plast. **6**(6), 701–744 (1990)
38. Teixeira-Dias, F., Menezes, L.F.: A kinematic and incremental integration model for the micromechanical numerical analysis of dual-phase materials. Comput. Mater. Sci. **25**, 237–245 (2002)
39. Pierce, D., Shih, C.F., Needleman, A.: A tangent modulus method for rate dependent solids. Comput. Struct. **18**(5), 875–887 (1984)
40. Dhatt, G., Touzout, G.: The finite element method displayed. Wiley, New York (1984)
41. Hughes, T.J.R.: The finite element method – linear static and dynamic finite element analysis. Prentice-Hall, Englewood Cliffs (1987)
42. Nagtegaal, J.C., Parks, D.M., Rice, J.R.: On numerically accurate finite element solutions in the fully plastic range. Comput. Methods. Appl. Mech. Eng. **4**(2), 153–177 (1974)
43. Takizawa, H., Makinouchi, A., Santos, A., Mori, N.: FE-Simulation of 3D Sheet Metal Forming Processes in Automotive Industry, VDI Berichte, vol 894, pp 167–184. VDI Verlag (1991)
44. Malkus, D.S., Hughes, T.J.R.: Mixed finite element methods – reduced and selective integration techniques: a unification of concepts. Comput. Methods Appl. Mech. Eng. **15**(1), 63–81 (1978)
45. Hughes, T.J.R.: Generalization of selective integration procedures to anisotropic and nonlinear media. Int. J. Numer. Methods Eng. **15**(9), 1413–1418 (1980)
46. Shimodaira, H.: Equivalence between mixed models and displacement models sing reduced integration. Int. J. Numer. Methods Eng. **21**(1), 89–104 (1985)
47. Menezes, L.F., Teodosiu, C., Makinouchi, A.: FE-Simulation of 3D Sheet Metal Forming Processes in Automotive Industry, VDI Berichte, vol 894, pp 381–403. VDI Verlag (1991)
48. Sautter, M., Dietrich, Ch, Poech, M.H., Schmauder, S., Fischmeister, H.F.: Finite element modelling of a transverse-loaded fibre composite effects of section size and net density. Comput. Mater. Sci. **1**(3), 225–233 (1993)
49. Sorensen, N.J., Suresh, S., Tvergaard, V., Needleman, A.: Effects of reinforcement orientation on the tensile response of metal-matrix composites. Mater. Sci. Eng. A **197**(1), 1–10 (1995)
50. Watt, D.F., Xu, X.Q., Lloyd, D.J.: Effects of particle morphology and spacing on the strain fields in a plastically deforming matrix. Acta Mater. **44**(2), 789–799 (1996)

51. Pettermann, H.E., Boehm, H.J., Rammerstorfer, F.G.: Some direction-dependent properties of matrix-inclusion type composites with given reinforcement orientation distributions. Compos. Part B **28**(3), 253–265 (1997)
52. Jain, M., MacEwen, S.R., Wu, L.: Finite element modelling of residual stresses and strength differential effect in discontinuously reinforced metal matrix composites. Mater. Sci. Eng. A **183**(1–2), 111–120 (1994)
53. Manoharan, M., Gupta, M.: Effect of silicon carbide volume fraction on the work hardening behaviour of thermomechanically processed aluminium-based metal matrix composites. Compos. Part B **30**(1), 107–112 (1999)

Heat Conduction in Two-Phase Composite Materials with Three-Dimensional Microstructures and Interfacial Thermal Resistance

Carlos Frederico Matt and Manuel Ernani Cruz

Abstract The goals envisioned for the current chapter are threefold. First, it gives a general overview of heat conduction in two-phase composite materials with three dimensional microstructures and interfacial thermal resistance. Second, it describes the application of homogenization theory to the multiscale heat conduction problem in the composite medium in order to derive the boundary-value problem defined on a representative volume element of the composite microstructure (the cell problem) and an expression for the composite effective thermal conductivity. Third, it describes a finite-element-based computational scheme to calculate the effective thermal conductivity of composite materials with general 3-D microstructures and interfacial thermal resistance. Numerical results for the effective conductivity are presented and, when possible, compared with available analytical predictions. The numerical results reported here confirm that computational approaches are a helpful tool for understanding the complex macroscopic thermal behavior of composite materials.

1 Introduction

Two-phase composite materials are a special class of multicomponent or heterogeneous media defined as systems composed of two constituents (or phases) with distinct macroscopic properties. In general, one of the constituents may be dispersed as particles (as in particulate composites) or fibers (as in fibrous or

C.F. Matt (✉)
Electric Power Research Center Av. Horácio Macedo 354 CP 68007, Rio de Janeiro 21941-590 RJ, Brazil
e-mail: cfmatt@cepel.br

M.E. Cruz
Federal University of Rio de Janeiro, Politécnica/COPPE/UFRJ, Av. Horácio Macedo 2030, 68503, 21941-972, Rio de Janeiro, RJ, Brazil
e-mail: manuel@mecanica.coppe.ufrj.br

A. Öchsner and G.E. Murch (eds.), *Heat Transfer in Multi-Phase Materials*,
Adv Struct Mater 2, DOI 10.1007/8611_2010_10,

short-fiber composites) inside a continuous matrix of the other constituent; the former constituent is referred to as the dispersed phase whereas the latter one is simply referred to as the matrix. The constituents may be a solid, a liquid, a gas or a combination thereof; nevertheless, only two-phase composites with solid constituents are discussed in this chapter.

The size, shape, distribution and orientation of the dispersed phase inside the matrix define the microstructure of a composite material. The dispersed phase in particulate composites are generally modeled as spheres. The spheres may be orderly or randomly distributed within the matrix. Hence, depending on the particle distribution, the microstructure for a particulate composite is represented by an ordered or a random array of spheres. The dispersed phase in fibrous or short-fiber composites comprises geometrical shapes slender in one direction; it is usually approximated by cylinders or ellipsoids of revolution. The fibers may also be orderly or randomly distributed within the matrix as in particulate composites; nevertheless, in fibrous or short-fiber composites, fiber orientation also plays an important role in the composite microstructure. During the manufacturing process of a fibrous or short-fiber composite, the matrix and the fibers are compressed together such that the fibers tend to distribute themselves in planes normal to the applied pressure. If all the fibers have the same orientation in those planes, one says that they are longitudinally-aligned; otherwise, they are transversely-aligned.

Generally, three characteristic length scales may be identified in a composite material, namely, the size of the dispersed phase (the microscale); the size of the microstructure (the mesoscale) and the physical dimensions of the composite itself (the macroscale). The size of the microstructure is also referred to as the characteristic length of the representative volume element of the composite microstructure. Frequently, these length scales possess widely different orders of magnitude; one says that they are well-separated. There are a variety of transport phenomena in composite materials with wide engineering applicability, such as heat conduction, fluid dynamics, forced and natural convection, and radiation, to name just a few. The discussion addressed here will be restricted to heat conduction. Heat conduction in composite materials with well-separated length scales are mathematically described by partial differential equations with rapidly-varying coefficients. Even with the actual computational resources, classical analytical and/or numerical treatments to these equations are difficult. Hence, developments of analytical and/or numerical schemes that alleviate the difficulties (stiffness) associated with the well-separated length scales are desirable and have been the subject of research efforts during recent years.

The application of composite materials in industry has remarkably increased in recent years, due to their ease and low cost of fabrication, and the tailorability of their mechanical and thermal properties. Composite materials are generally classified as organic matrix composites, metal matrix composites or ceramic matrix composites, depending on the matrix material. In organic matrix composites, reinforcements with high thermal conductivity are embedded in an electrically insulating matrix, such as polymers and thermoplastics. They have become attractive materials for electronic packaging applications due to a combination of properties like high thermal conductivity, low thermal expansion coefficient, thermal stability and low dielectric constant [18]. Metal matrix composites have also been widely used in electronic packaging applications because of their high thermal conductivity and the tailorability of their coefficient of thermal expansion [9, 38]. An adjustable coefficient of thermal expansion is a desirable feature because a mismatch of that property between heat sink and substrate gives rise to thermally induced stresses, which contribute to device failure. Furthermore, the recent miniaturization of electronic components with increasingly higher energy densities requires highly conducting materials as the performance of those electronic devices depends on maintaining appropriate environment temperatures [9]. Ceramic matrix composites represent a new class of refractory composites for high-temperature environments in which glasses, glass-ceramics and ceramic matrices are reinforced with fibers [16]. Ceramic matrix composites are attractive materials because they possess improved fracture toughness, impact resistance, greatly increased tolerance to mechanical damage compared to many single brittle materials and wide versatility of thermal properties [16]. In a high-temperature environment, materials with superior thermal properties become desirable. For instance, heat exchangers operating at high temperatures require materials with high thermal conductivity and high thermal diffusivity in order to improve resistance to thermal shock; on the other hand, other equipments require materials with extremely low thermal conductivity in order to minimize heat losses.

As stated previously, owing to the usual complex microstructures, the detailed (local) study of heat conduction in composite materials is a hard task. Therefore, engineering analyses focus on the macroscopic behavior of such materials, dictated by the so-called effective properties, such as the effective thermal conductivity. The determination of the latter in terms of the microstructure, individual properties of the phases, and other relevant physical parameters has scientific and practical importance [13, 27, 37]. Because most fabrication processes of composite materials do not ensure a perfect thermal contact between the constituent phases, the effective thermal conductivity depends on the interfacial thermal resistance between the matrix and the dispersed phase. This interfacial, or contact, resistance is due to poor chemical and/or mechanical adherence, caused by different thermal expansion coefficients of the phases, and presence of roughness, waving and impurities at the interface [6, 9, 10, 14, 28]. Physically, the contact resistance tends to thermally insulate the dispersed phase, and may dramatically reduce the composite effective conductivity. Manufacturing processes of composite materials may also lead to other defects such as voids inside the matrix which also contribute to reduce their effective conductivities.

Recent reviews of analytical [10, 13, 27, 28], computational [7, 19, 22, 33], and experimental [8, 28] techniques for the study of heat conduction in composite materials are available. Analytical [15, 27] and phenomenological [5, 11] treatments of heat conduction in composites are generally restricted to single spheroidal particles embedded in an infinite matrix, to dilute dispersed-phase volume fractions, or to the validity of 'ad-hoc' assumptions; the majority of analytical and phenomenological approaches neglects the interfacial thermal resistance and the ones which account for it are usually restricted to the above-mentioned simple microstructures. Accurate results for the effective conductivity of ordered arrays of spheres (low-to-moderate volume fractions) without interfacial thermal resistance may be obtained from the works of Sangani and Acrivos [34] and McPhedran and McKenzie [26]. For ordered and random arrays of spheres with uniform interfacial thermal resistance the reader should consult, respectively, the works of Cheng and Torquato [6] and Benveniste [5]. The works of Hatta and Taya [17], Furmañski [12] and Dunn et al. [8] provide analytical expressions for the effective thermal conductivity of arrays of perfectly-aligned and misoriented short fibers without interfacial resistance. The vast majority of computational approaches [19, 21, 22, 33], despite their increased flexibility to treat more complex physics and microstructural geometries, calculate the effective thermal conductivity exclusively in terms of the individual properties of the constituents, and the volume fraction, shape, and spatial distribution of the dispersed phase. There are numerical approaches which account exactly for an interfacial thermal resistance (see, for example, the works of Duschlbauer and his co-workers [9, 10]); however, some of them are not flexible enough to also account for other complexities, such as dispersed voids in the matrix and large volume fractions, close to the maximum packing.

To sum up, purely analytical treatments of heat conduction in composite materials provide important expressions for the effective conductivity, from which the effects of dispersed-phase volume fraction, shape, and spatial distribution are easily quantified, but they are restricted in most cases to simple microstructures, and to dilute and moderate volume fractions. Phenomenological approaches are a valuable aid to investigate complex physics and microstructural geometries, nevertheless they adopt heuristic assumptions regarding the temperature distribution inside the matrix, which are not easily verified in practice. Computational methodologies, in principle, appear now to be the most appropriate means to treat heat conduction in composites, because their inherent flexibility permits gradual improvement of the physical and geometrical modeling of such materials. On the other hand, the composite microstructure must be prescribed a priori, and it remains nontrivial to reproduce computationally all the details of an actual composite microstructure.

The goals envisioned for the current chapter are threefold. First, it gives a general overview of heat conduction in two-phase composite materials with three dimensional microstructures and interfacial thermal resistance. Second, it describes the application of homogenization theory to the multiscale heat conduction problem in the composite medium in order to derive the boundary-value problem defined on a representative volume element of the composite microstructure (the cell problem) and an expression for the composite effective thermal conductivity. Third, it describes a finite-element-based computational scheme to calculate the effective thermal conductivity of composite materials with general 3-D microstructures and interfacial thermal resistance. Numerical results for the effective conductivity are presented and, when possible, compared with available analytical predictions. The numerical results reported here confirm that computational approaches are a helpful tool for understanding the complex macroscopic thermal behavior of composite materials.

2 Formulation of Heat Conduction Problem

Here a statistically homogeneous composite [37] is considered, illustrated in Fig. 1, whose continuous and dispersed phases are, respectively, a solid homogeneous matrix of thermal conductivity k^c occupying domain Ω_c, and solid homogeneous particles/fibers of tensorial thermal conductivity k^d_{ij}, $i, j = 1, 2, 3$, occupying domain Ω_d. The particles/fibers have arbitrary shapes and orientations, and are orderly or randomly distributed within the matrix. An interfacial thermal resistance function R_I is present at the interface $\partial\Omega_s$ (a disconnected set) between the matrix and the dispersed phase. The composite extends throughout a macroscale region $\Omega = \Omega_c \cup \Omega_d$ of characteristic dimension L, over which an external temperature gradient $\Delta T/L$ is imposed. The representative volume element (RVE) of the composite microstructure is the locally-periodic cell Ω_{pc}, which contains several particles and/or fibers of characteristic dimension ℓ (the microscale). The characteristic dimension of the RVE is referred to as the mesoscale, and denoted by λ. The composite length scales are assumed to be well-separated and, given statistical homogeneity, one can define the small parameter $\varepsilon \equiv \lambda/L \ll 1$ for the medium.

2.1 Strong Form

For steady state heat conduction in the medium described above, the non-dimensional strong form of the boundary value problem is given by [32]

$$-\frac{\partial}{\partial y_i}\left(\frac{\partial \theta^c}{\partial y_i}\right) = G_c \quad \text{in } \Omega_c, \tag{1}$$

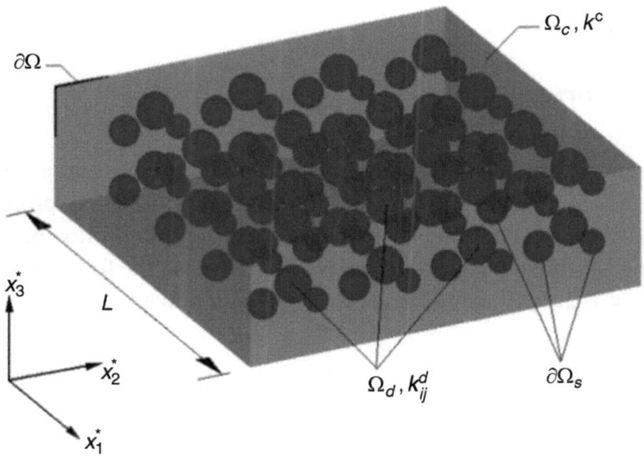

Fig. 1 Statistically homogeneous composite with 3-D microstructure

$$-\frac{\partial}{\partial y_i}\left(\kappa_{ij}\,\frac{\partial\theta^d}{\partial y_j}\right) = G_d \quad \text{in } \Omega_d, \tag{2}$$

$$-\frac{\partial\theta^c}{\partial y_i}\,n_i^c = -\kappa_{ij}\,\frac{\partial\theta^d}{\partial y_j}\,n_i^c \quad \text{on } \partial\Omega_s\,, \tag{3}$$

$$-\frac{\partial\theta^c}{\partial y_i}\,n_i^c = \text{Bi}\big(\theta^c - \theta^d\big) = \text{Bi}[\theta_\varepsilon]_{\partial\Omega_s} \quad \text{on } \partial\Omega_s\,, \tag{4}$$

where summation over repeated indices is implied; $\mathbf{y} \equiv \mathbf{x}^*/\lambda$, and $\mathbf{x}^*$, $\mathbf{x}^* \in IR^3$, is the macroscale coordinate or slow space variable, whose components are $x_j^*, j = 1,2,3$; $\theta_\varepsilon \equiv T_\varepsilon/\Delta T$, $G \equiv \dot{g}\,\lambda^2/(k^c\,\Delta T)$, and T_ε and $\dot{g}$ are, respectively, the temperature field and the volumetric rate of heat generation at the microscale; $\theta^c = \theta_\varepsilon|_{\Omega_c}$, $\theta^d = \theta_\varepsilon|_{\Omega_d}$, $G_c = G|_{\Omega_c}$, and $G_d = G|_{\Omega_d}$; $\mathbf{n}^c$ is the unit vector locally normal to $\partial\Omega_s$ and pointing to the outside of Ω_c; $\kappa_{ij} \equiv k_{ij}^d/k^c$ is the conductivity ratio, $i,j = 1,2,3$; $\text{Bi} \equiv h\,\lambda/k^c$ is the Biot number, and $h \equiv 1/R_I$ is the interfacial thermal conductance function (also referred to in the literature as the interface parameter or skin constant), which specifies locally the ratio of the heat flux to the temperature jump at the interface $\partial\Omega_s$. The current definition of the Biot number is based on λ rather than on ℓ as done by other researchers [5, 6]; moreover, it is related to the Kapitza radius, k^c/h, defined in [11]. The notation $[\phi]_{\partial\Omega_s}$ denotes the jump $\phi^c - \phi^d$ of the function ϕ at $\partial\Omega_s$. It is typical in measurements of thermal conductivity with macroscopic samples of composite materials, to impose different temperatures at $\partial\Omega$, the external boundaries of Ω, such that the field θ_ε can be regarded as subjected to Dirichlet boundary conditions. Later on, it will be demonstrated that the composite effective conductivity does not depend upon the boundary conditions applied at $\partial\Omega$, which is physically reasonable.

2.2 *Weak Form*

For subsequent treatment by the finite element method, it is more convenient to work with the equivalent weak form of the heat conduction problem, which naturally enforces the continuity of the heat flux at $\partial\Omega_s$, (3). Following the detailed derivation in [32] for isotropic composites, one arrives at the following weak form: given ζ_{ij}, $i,j = 1,2,3$, Bi and G, find $\theta_\varepsilon \in X'(\Omega)$ such that

$$\int_\Omega \zeta_{ij}\,\frac{\partial v}{\partial y_i}\,\frac{\partial\theta_\varepsilon}{\partial y_j}\,d\mathbf{y} + \int_{\partial\Omega_s} \text{Bi}\,[v]_{\partial\Omega_s}[\theta_\varepsilon]_{\partial\Omega_s}\,ds = \int_\Omega v\,G\,d\mathbf{y} \quad \forall v \in X(\Omega)\,, \tag{5}$$

where $\zeta_{ij}(\mathbf{y}) = \delta_{ij}$ if $\mathbf{y} \in \Omega_c$ and $\zeta_{ij}(\mathbf{y}) = \kappa_{ij}$ if $\mathbf{y} \in \Omega_d$, $i,j = 1,2,3$; $X'(\Omega) = \{w\,|\,w|_{\Omega_c} = w^c \in H^1(\Omega_c), w|_{\Omega_d} = w^d \in H^1(\Omega_d), [w]_{\partial\Omega_s} = \beta \in IR\}$; $X(\Omega) = \{w \in$

$X'(\Omega)\} \cap H_0^1(\Omega)$; $H^1(\Omega)$ is the space of all functions for which both the function and derivative are square-integrable over Ω [31], and the functions in $H_0^1(\Omega) \subset H^1(\Omega)$ vanish on $\partial\Omega$. It is clear from (5) that the formulation permits that the Biot number, in general, will vary spatially over the interface $\partial\Omega_s$, i.e., $\mathrm{Bi} = \mathrm{Bi}(\mathbf{y})$.

2.3 Application of Homogenization Theory

Even with the actual computational resources, traditional analytical and/or numerical methods have difficulties in solving the weak form (5). The coefficients $\zeta_{ij}(\mathbf{y})$, $\mathrm{Bi}(\mathbf{y})$ and $G(\mathbf{y})$ are rapidly-varying functions of the space coordinate $\mathbf{y}$ due to the well-separated length scales and to the large number of particles/fibers within the matrix. Analytical and/or numerical methods able to decouple such a multiscale boundary-value problem into a boundary-value problem defined on an equivalent homogeneous medium that behaves identically as the original multiscale medium are thus suitable to handle partial differential equations with rapidly-varying coefficients, as (5); homogenization theory is one of such methods. To apply homogenization theory [1, 4] to (5), asymptotic developments for the temperature field in terms of characteristic length scales of the phenomenon under investigation have revealed itself as an attractive strategy.

The first task in applying the homogenization theory to the weak form (5) is to write θ_ε and v as functions of two space variables, namely, the fast or microscopic variable $\mathbf{y}$ and the slow or macroscopic variable $\mathbf{x} \equiv \mathbf{x}^*/L = \varepsilon\mathbf{y}$. Note that for the problem under investigation, only spatial scales enter the formulation. The second task is to introduce multiple-scale asymptotic expansions, here needed up to the second-order term in ε only; in other words, the temperature field and the weight function are expanded as

$$\theta_\varepsilon = \sum_{k=0}^{2} \varepsilon^k \theta_k(\mathbf{x},\mathbf{y}) \quad \text{and} \quad v = \sum_{k=0}^{2} \varepsilon^k v_k(\mathbf{x},\mathbf{y}). \tag{6}$$

Substituting the asymptotic expansions given by (6) into the weak form (5) and then applying the chain rule for differentiation, one derives the following expression [32]

$$\begin{aligned}
&\int_\Omega \zeta_{ij}\left(\frac{\partial v_0}{\partial y_i} + \varepsilon\frac{\partial v_0}{\partial x_i} + \varepsilon\frac{\partial v_1}{\partial y_i} + \varepsilon^2\frac{\partial v_1}{\partial x_i} + \varepsilon^2\frac{\partial v_2}{\partial y_i}\right)\left(\frac{\partial \theta_0}{\partial y_j} + \varepsilon\frac{\partial \theta_0}{\partial x_j} + \varepsilon\frac{\partial \theta_1}{\partial y_j}\right.\\
&\quad \left. + \varepsilon^2\frac{\partial \theta_1}{\partial x_j} + \varepsilon^2\frac{\partial \theta_2}{\partial y_j}\right)d\mathbf{y} + \int_{\partial\Omega_s} \mathrm{Bi}\left[v_0 + \varepsilon v_1 + \varepsilon^2 v_2\right]_{\partial\Omega_s}\left[\theta_0 + \varepsilon\theta_1 + \varepsilon^2\theta_2\right]_{\partial\Omega_s} ds\\
&= \int_\Omega \left(v_0 + \varepsilon v_1 + \varepsilon^2 v_2\right) G\, d\mathbf{y} \quad \forall v_0, v_1, v_2 \in X(\Omega).
\end{aligned} \tag{7}$$

Five different homogenization processes are possible [1], corresponding to five different orders of magnitude for the Biot number: $\mathrm{Bi} = O(\varepsilon^a), a \in \{-1, 0, 1, 2, 3\}$. Here, only the physically relevant model corresponding to $a = 0$ will be discussed; such a model is associated with a finite value of the interfacial thermal resistance R_I, i.e., $0 \leq \mathrm{Bi} < \infty$. Note that $\mathrm{Bi} = 0$ represents a perfect insulating interface, while $\mathrm{Bi} \to \infty$ represents perfect interfacial thermal contact.

The third task in applying the homogenization theory is to collect systematically equal powers of ε in (7). For instance, by collecting equal powers of ε^0 in (7) one arrives at the following equation

$$\int_\Omega \zeta_{ij} \frac{\partial v_0}{\partial y_i} \frac{\partial \theta_0}{\partial y_j} d\mathbf{y} + \int_{\partial\Omega_s} \mathrm{Bi} [v_0]_{\partial\Omega_s} [\theta_0]_{\partial\Omega_s} ds = 0 \quad \forall v_0 \in X(\Omega). \tag{8}$$

Note that (8) is valid for all weight functions $v_0 \in X(\Omega)$. Hence, one may choose a weight function such that $[v_0]_{\partial\Omega_s} = 0$. By substituting $[v_0]_{\partial\Omega_s} = 0$ into (8) one immediately concludes that $\partial\theta_0/\partial y_j = 0$. Now, by substituting $\partial\theta_0/\partial y_j = 0$ into (8), one immediately has $[\theta_0]_{\partial\Omega_s} = 0$, from which one concludes that $\theta_0 = \theta_0(\mathbf{x})$, i.e., θ_0 does not depend on the microscopic space variable $\mathbf{y}$. Physically, the temperature expansion coefficient $\theta_0(\mathbf{x})$ represents the macroscopic temperature field defined on the equivalent homogeneous medium.

By collecting equal powers of ε^1 and by taking into account the previous conclusions, one arrives at the trivial identity $0 = 0$; in other words, no new information is gained when collecting equal powers of ε^1. Finally, by collecting equal powers of ε^2 in (7) and by taking into account all the aforementioned conclusions, one derives the following equation

$$\int_\Omega \zeta_{ij} \left(\frac{\partial \theta_0}{\partial x_j} \frac{\partial v_0}{\partial x_i} + \frac{\partial \theta_1}{\partial y_j} \frac{\partial v_0}{\partial x_i} + \frac{\partial \theta_0}{\partial x_j} \frac{\partial v_1}{\partial y_i} + \frac{\partial \theta_1}{\partial y_j} \frac{\partial v_1}{\partial y_i} \right) d\mathbf{y} + \int_{\partial\Omega_s} \mathrm{Bi} [v_1]_{\partial\Omega_s} [\theta_1]_{\partial\Omega_s} ds = \int_\Omega v_0 G \, d\mathbf{y} \tag{9}$$

$\forall v_0 \in X''(\Omega), \forall v_1 \in X(\Omega)$. The function space $X''(\Omega)$ is a subset of $X(\Omega)$ in which a member function has no jump across the interface $\partial\Omega_s$. Equation (9) may be broken into two equations. Firstly, by arbitrarily choosing $v_1 = 0 \in X(\Omega)$ one derives

$$\int_\Omega \zeta_{ij} \left(\frac{\partial \theta_0}{\partial x_j} + \frac{\partial \theta_1}{\partial y_j} \right) \frac{\partial v_0}{\partial x_i} d\mathbf{y} = \int_\Omega v_0 G \, d\mathbf{y} \quad \forall v_0 \in X''(\Omega). \tag{10}$$

Secondly, by arbitrarily choosing $v_0 = 0 \in X''(\Omega)$ one derives

$$\int_\Omega \zeta_{ij} \left(\frac{\partial \theta_0}{\partial x_j} + \frac{\partial \theta_1}{\partial y_j} \right) \frac{\partial v_1}{\partial y_i} d\mathbf{y} + \int_{\partial\Omega_s} \mathrm{Bi} [v_1]_{\partial\Omega_s} [\theta_1]_{\partial\Omega_s} ds = 0 \quad \forall v_1 \in X(\Omega). \tag{11}$$

As one shall demonstrate in the next paragraphs, (10) and (11) will give rise to the homogenized and cell problems, respectively. The next step is to assume a separation of variables for $\theta_1(\mathbf{x},\mathbf{y})$ in the form

$$\theta_1(\mathbf{x},\mathbf{y}) = -\chi_p(\mathbf{y})\,\frac{\partial\theta_0}{\partial x_p}(\mathbf{x}). \tag{12}$$

By substituting (12) into (11) yields

$$\int_\Omega \zeta_{ij}\left(\frac{\partial\theta_0}{\partial x_j} - \frac{\partial\chi_p}{\partial y_j}\frac{\partial\theta_0}{\partial x_p}\right)\frac{\partial v_1}{\partial y_i}\,d\mathbf{y} - \int_{\partial\Omega_s} \mathrm{Bi}\,\frac{\partial\theta_0}{\partial x_p}[v_1]_{\partial\Omega_s}\left[\chi_p\right]_{\partial\Omega_s} ds = 0 \tag{13}$$

or, alternatively,

$$\int_\Omega \zeta_{ij}\,\frac{\partial\theta_0}{\partial x_p}\left(\delta_{jp} - \frac{\partial\chi_p}{\partial y_j}\right)\frac{\partial v_1}{\partial y_i}\,d\mathbf{y} - \int_{\partial\Omega_s} \mathrm{Bi}\,\frac{\partial\theta_0}{\partial x_p}[v_1]_{\partial\Omega_s}\left[\chi_p\right]_{\partial\Omega_s} ds = 0 \tag{14}$$

$\forall v_1 \in X(\Omega)$. If the composite microstructure is further assumed to be periodic then one may apply the periodicity property [1, 32]. The periodicity property relates the integral of a quantity over the composite multiscale domain Ω with the integral of the average of such quantity over a representative volume element of the composite microstructure (the periodic cell). The periodicity property for composites with finite interfacial thermal resistance is mathematically described by [32]

$$\lim_{\varepsilon\to 0}\left(\int_\Omega f(\mathbf{x},\mathbf{y})\,d\mathbf{y} + \int_{\partial\Omega_s} g(\mathbf{x},\mathbf{y})\,ds\right) = \int_\Omega \frac{1}{|\Omega_{pc}|}\left[\int_{\Omega_{pc}} f(\mathbf{x},\mathbf{y})\,d\mathbf{y} + \int_\Gamma g(\mathbf{x},\mathbf{y})\,ds\right] d\mathbf{y} \tag{15}$$

where the functions f and g are periodic in $\mathbf{y} \in IR^3$; the function g has discontinuities across $\partial\Omega_s$; Ω_{pc} denotes the periodic cell and Γ is the portion of $\partial\Omega_s$ inside Ω_{pc}. The symbol $|\Omega_{pc}|$ denotes the volume of the periodic cell, i.e., $|\Omega_{pc}| = \int_{\Omega_{pc}} d\mathbf{y}$.

By subsequently applying the periodicity property given by (15) into (14) yields

$$\int_\Omega \left\{\frac{1}{|\Omega_{pc}|}\int_{\Omega_{pc}} \zeta_{ij}\,\frac{\partial\theta_0}{\partial x_p}\left(\delta_{jp} - \frac{\partial\chi_p}{\partial y_j}\right)\frac{\partial v_1}{\partial y_i}\,d\mathbf{y}\right\} d\mathbf{y}$$
$$= \int_\Omega \left\{\frac{1}{|\Omega_{pc}|}\int_\Gamma \mathrm{Bi}\,\frac{\partial\theta_0}{\partial x_p}[v_1]_\Gamma\left[\chi_p\right]_\Gamma ds\right\} d\mathbf{y} \tag{16}$$

$\forall v_1 \in X(\Omega)$. Note that the macroscopic temperature gradient $\partial\theta_0/\partial x_p$ depends only on the macroscopic space variable $\mathbf{x}$; hence, it may be put outside the integrals over Ω_{pc} and Γ. Equation (16) must hold for all weight functions

$v_1 \in X(\Omega)$; hence, one must have (henceforth, the subscript under the weight function is abandoned)

$$\int_{\Omega_{pc}} \zeta_{ij}\left(\delta_{jp} - \frac{\partial \chi_p}{\partial y_j}\right)\frac{\partial v}{\partial y_i}\,d\mathbf{y} = \int_{\Gamma} \mathrm{Bi}\,[v]_\Gamma\,[\chi_p]_\Gamma ds \qquad \forall v \in Y(\Omega_{pc}) \tag{17}$$

or, alternatively,

$$\int_{\Omega_{pc}} \zeta_{ij}\frac{\partial \chi_p}{\partial y_j}\frac{\partial v}{\partial y_i}\,d\mathbf{y} + \int_{\Gamma} \mathrm{Bi}\,[v]_\Gamma\,[\chi_p]_\Gamma ds = \int_{\Omega_{pc}} \zeta_{ip}\frac{\partial v}{\partial y_i}\,d\mathbf{y} \qquad \forall v \in Y(\Omega_{pc}). \tag{18}$$

The unknown function χ_p, $p \in \{1,2,3\}$, is a periodic solution corresponding to a unit temperature gradient imposed in the x_p^* direction; the function space $Y(\Omega_{pc})$ defined as

$$Y(\Omega_{pc}) = \{w\,|\,w|_{\Omega_{pc,c}} = w^c \in H^1_\#(\Omega_{pc,c}), w|_{\Omega_{pc,d}} = w^d \in H^1_\#(\Omega_{pc,d}), [w]_\Gamma = \beta \in IR\}$$

is the space of all triply-periodic functions in Ω_{pc} with period $C_j\lambda$ along the y_j direction, $C_j \in IR, j = 1,2,3$, for which both the function and derivative are square-integrable over Ω_{pc}; $\Omega_{pc,c}$ and $\Omega_{pc,d}$ are the continuous and dispersed subdomains of Ω_{pc}. Equation (18) represents mathematically the cell problem for composites with finite interfacial thermal resistance. Note that due to the interfacial thermal resistance both the temperature and weight functions have jumps across the interface Γ. Note that the coefficients $\zeta_{ij}(\mathbf{y})$ and $\mathrm{Bi}(\mathbf{y})$ are not rapidly-varying functions inside Ω_{pc}. Note also that the solution of the cell problem depends neither on the boundary conditions prescribed at $\partial\Omega$ nor on the rate of volumetric heat generation $G(\mathbf{y})$; it depends only on the geometry of the periodic cell, the particles/fibers-to-matrix conductivity ratios and the magnitude of the interfacial thermal resistance. The function χ_p in (18) is determined up to a free constant; thus, it is further required for uniqueness that χ_p has zero volumetric average, i.e.,

$$\int_{\Omega_{pc}} \chi_p\,d\mathbf{y} = 0. \tag{19}$$

Now, substituting (12) into (10) one arrives at the following expression

$$\int_{\Omega} \zeta_{ij}\frac{\partial \theta_0}{\partial x_p}\left(\delta_{jp} - \frac{\partial \chi_p}{\partial y_j}\right)\frac{\partial v_0}{\partial x_i}\,d\mathbf{y} = \int_{\Omega} v_0 G\,d\mathbf{y} \qquad \forall v_0 \in X''(\Omega). \tag{20}$$

Finally, by applying the periodicity property to (20) one derives the following expression for the weak form of the homogenized problem

$$\int_{\Omega}\left\{\frac{1}{|\Omega_{pc}|}\int_{\Omega_{pc}} \zeta_{ij}\left(\delta_{jp} - \frac{\partial \chi_p}{\partial y_j}\right)d\mathbf{y}\right\}\frac{\partial \theta_0}{\partial x_p}\frac{\partial v_0}{\partial x_i}\,d\mathbf{y} = \int_{\Omega}\left\{\frac{1}{|\Omega_{pc}|}\int_{\Omega_{pc}} v_0 G\,d\mathbf{y}\right\}d\mathbf{y} \tag{21}$$

$\forall v_0 \in X''(\Omega)$. From inspection of (21), one identifies the term inside brackets on the left-hand side as the (tensorial) effective thermal conductivity $\kappa_{e,pq}$, $p,q = 1,2,3$; hence,

$$\kappa_{e,pq} \equiv \frac{k_{e,pq}}{k^c} = \frac{1}{|\Omega_{pc}|}\int_{\Omega_{pc}} \zeta_{pi}\left(\delta_{iq} - \frac{\partial \chi_q}{\partial y_i}\right) d\mathbf{y} = \frac{1}{|\Omega_{pc}|}\left\{\int_{\Omega_{pc,c}} \delta_{pi}\left(\delta_{iq} - \frac{\partial \chi_q^c}{\partial y_i}\right) d\mathbf{y} + \int_{\Omega_{pc,d}} \kappa_{pi}\left(\delta_{iq} - \frac{\partial \chi_q^d}{\partial y_i}\right) d\mathbf{y}\right\} \tag{22}$$

where $\chi_q^c = \chi_q|_{\Omega_{pc,c}}$ and $\chi_q^d = \chi_q|_{\Omega_{pc,d}}$. Note that, once the cell problem is solved, the composite effective conductivity is computed with the aid of (22). The main goal of the next section is to discuss in detail the finite element solution of cell problem (18). The solution of the homogenized problem, given by (21), will not be addressed here. Once the composite effective conductivity has been computed, the solution of the homogenized problem for a given volumetric rate of heat generation G and boundary conditions at $\partial\Omega$ is straightforward.

3 Finite Element Solution

Numerical solution of the cell problem, a boundary-value problem with periodic boundary conditions, by the finite element method requires three steps: (1) domain and mesh generation, (2) finite-element discretization of (18), and (3) solution of the resultant linear system of algebraic equations. Each of these three steps is described in what follows.

3.1 *Domain and Mesh Generation*

Domain generation comprises the construction of the geometry of the periodic cell. The microstructures discussed in this chapter are ordered arrays of spheres, ordered arrays of perfectly-aligned prolate ellipsoids of revolution, ordered arrays of circular cylinders, random arrays of spheres, and random arrays of misoriented circular cylinders. Mesh generation comprises the subdivision of the cell domain Ω_{pc} into N_{E} nonoverlapping conforming finite elements, each with domain Ω_e, $e = 1,\ldots,N_{\mathrm{E}}$, such that $\Omega_{pc,h} \equiv \cup_{e=1}^{N_{\mathrm{E}}}\Omega_e$ is the discrete approximation to the domain Ω_{pc}; the total number of global mesh nodes is denoted by N_{GN}.

Over each finite element, the solution of the cell problem is approximated by the interpolation of its values at the nodes. An isoparametric discretization implies that both the geometry Ω_{pc} and the periodic temperature field χ_p are approximated by the same interpolation functions [31]. Here, quadratic polynomials are employed as the interpolants; hence, the finite element meshes are composed of ten-node quadratic tetrahedra, a generic example of which is illustrated in Fig. 2a. The use of quadratic interpolation functions provides an improved representation of 3-D curved surfaces (spherical, ellipsoidal, and cylindrical) and an accurate numerical solution, especially for composites with a high thermal contrast between the constituent phases [23]. The adopted domain and mesh generation procedures,

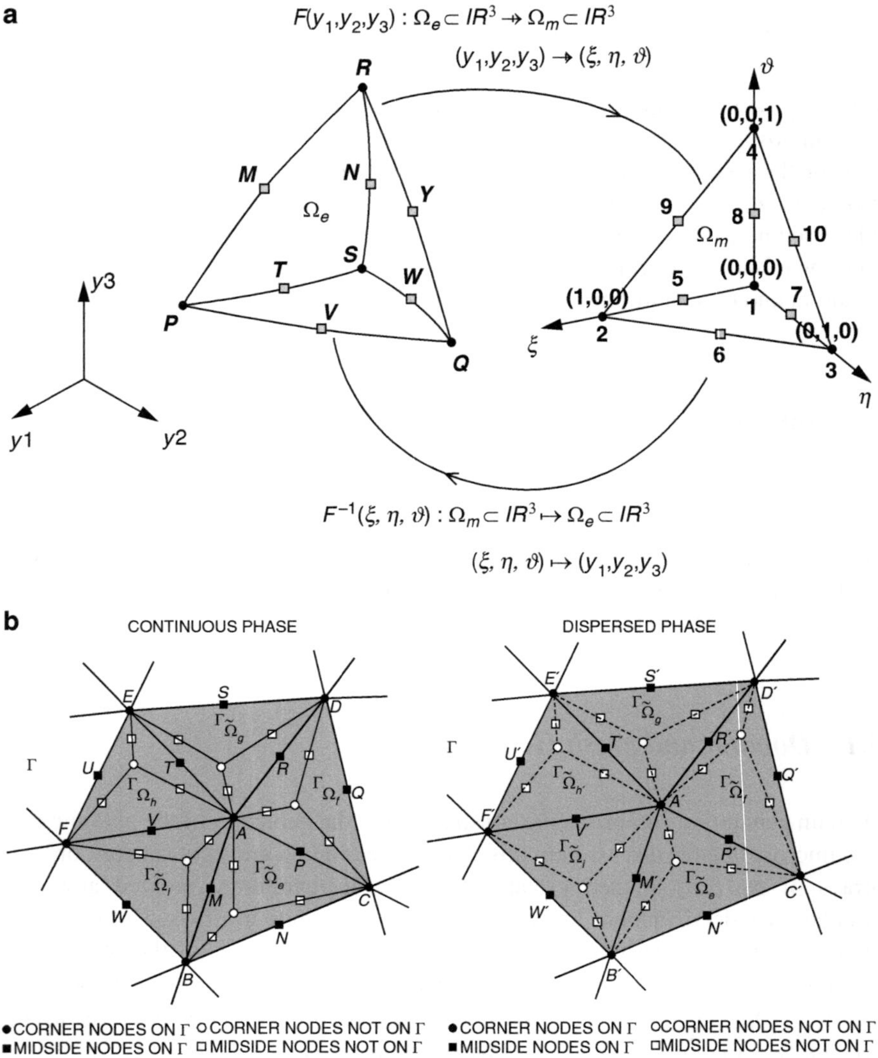

Fig. 2 (**a**) Schematic illustration of the isoparametric mapping between a generic ten-node quadratic tetrahedron (*left*) and the master quadratic tetrahedron (*right*) (**b**) Duplication of degrees of freedom corresponding to global mesh nodes at Γ. Nodes *A* and A', *B* and B', and so on are geometrically coincident

and mesh quality evaluation, are described in detail elsewhere [20–22], and exploit the resourceful third-party 2-D/3-D mesh generator NETGEN [35].

3.2 Isoparametric Second-Order Discretization

The isoparametric second-order discretization applied to solve the cell problem is a modification of the usual Galerkin method, in order to account for the discontinuities of v and χ_p across the interface Γ; the cell problem weak form is repeated here for convenience,

$$\int_{\Omega_{pc}} \zeta_{ij} \frac{\partial v}{\partial y_i} \frac{\partial \chi_p}{\partial y_j}\, d\mathbf{y} + \int_{\Gamma} \mathrm{Bi}\, [v]_\Gamma \left[\chi_p\right]_\Gamma ds = \int_{\Omega_{pc}} \zeta_{ip} \frac{\partial v}{\partial y_i}\, d\mathbf{y} \quad \forall v \in Y(\Omega_{pc}). \tag{23}$$

Two remarks should be highlighted. First, the usual continuity of the primary variable (here, the temperature) at nodes shared by adjoining tetrahedra must be enforced, only if those nodes are not at Γ; on the other hand, the balance of the secondary variable (the heat flux) at a surface shared by two adjoining tetrahedra must always be enforced. Second, in order to account for the interfacial resistance effect, each global mesh node at Γ must store two values of χ_p: one corresponding to the continuous phase, χ_p^c, and the other corresponding to the dispersed phase, χ_p^d, such that the jump $\left[\chi_p\right]_\Gamma$ is given by $\chi_p^c - \chi_p^d$. Thus, it is necessary to duplicate the degrees of freedom corresponding to global nodes (corner and midside) at Γ: as shown in Fig. 2b, a duplicate A' of the generic node A is created, such that A and A' correspond to the same geometric point in 3-D space, but store two different values (degrees of freedom) for the temperature, $\chi_A = \chi_p|_{\mathrm{node}\, A}$ and $\chi_{A'} = \chi_p|_{\mathrm{node}\, A'}$. The connectivities of the tetrahedra belonging to $\Omega_{pc,d}$ and having at least one node at Γ must then be modified accordingly, to correctly assemble the elemental equations. After the duplication procedure, the number of degrees of freedom increases by N_Γ, where N_Γ is the number of global finite element mesh nodes at Γ; the total number of degrees of freedom (i.e., values of the solution χ_p to be calculated) is N_{DOF}.

3.2.1 Numerical Treatment of the Volume Integrals

The cell problem may be written on an elemental level as

$$\int_{\Omega_e} \zeta_{ij}^e \frac{\partial v^e}{\partial y_i} \frac{\partial \chi_p^e}{\partial y_j}\, d\mathbf{y} + \int_{\Gamma_{\Omega_e}} \mathrm{Bi}^e \left[v^e\right]_{\Gamma_{\Omega_e}} \left[\chi_p^e\right]_{\Gamma_{\Omega_e}} ds = \int_{\Omega_e} \zeta_{ip}^e \frac{\partial v^e}{\partial y_i}\, d\mathbf{y}\,, \tag{24}$$

where Ω_e, $e = 1, 2, \ldots, N_{\mathrm{E}}$, is a quadratic tetrahedron of the mesh, $\zeta_{ij}^e = \zeta_{ij}|_{\Omega_e}$, $i,j = 1,2,3$ ($\zeta_{ij}^e = \delta_{ij}$ if $\Omega_e \subset \Omega_{pc,c}$ and $\zeta_{ij}^e = \kappa_{ij}$ if $\Omega_e \subset \Omega_{pc,d}$); $\mathrm{Bi}^e = \mathrm{Bi}|_{\Gamma_{\Omega_e}}$; $v^e = v|_{\Omega_e}$ and $\chi_p^e = \chi_p|_{\Omega_e}$; and Γ_{Ω_e} is the portion of the boundary of Ω_e on Γ (i.e.,

$\Gamma_{\Omega_e} = \partial\Omega_e \cap \Gamma$). In the finite element solution of (24), one seeks to approximate v^e and χ_p^e by

$$v^e = \sum_{b=1}^{10} v_b^e \psi_b^e \quad \text{and} \quad \chi_p^e = \sum_{a=1}^{10} \chi_{p,a}^e \psi_a^e \tag{25}$$

where $\psi_a^e(\mathbf{y})$, $\psi_b^e(\mathbf{y})$, $a, b = 1, 2, \ldots, 10$, are the Lagrange interpolation functions for Ω_e, and $\chi_{p,a}^e$ is the value of the temperature χ_p at the local node a of Ω_e. The usual practice with finite elements [31], as shown in the following paragraphs, is to express the Lagrange interpolation functions in terms of a reference coordinate system, related explicitly to the original system $\mathbf{y}$. Each local node a of each ten-node tetrahedron Ω_e is associated with a corresponding global node A, $A \in \{1, 2, \ldots, N_{\mathrm{GN}}\}$. Equation (24) applies for all tetrahedra of the mesh (with zero, one, two, or three corner nodes at Γ); the surface integral in (24), however, must be computed only for those tetrahedra with exactly three corner nodes at Γ.

For tetrahedra with zero, one, or two corner nodes at Γ (i.e., tetrahedra for which, respectively, Γ_{Ω_e} is an empty set, a point, or a straight line segment in 3-D), $ds = 0$; hence, (24) is simplified to

$$\int_{\bar{\Omega}_e} \zeta_{ij}^e \frac{\partial v^e}{\partial y_i} \frac{\partial \chi_p^e}{\partial y_j} \, d\mathbf{y} = \int_{\bar{\Omega}_e} \zeta_{ip}^e \frac{\partial v^e}{\partial y_i} \, d\mathbf{y} \,, \tag{26}$$

where $\bar{\Omega}_e$, $e = 1, 2, \ldots, \bar{N}_E$, is a generic tetrahedron with at most two corner nodes at Γ, and $\bar{N}_{\mathrm{E}}$ ($\bar{N}_{\mathrm{E}} < N_{\mathrm{E}}$) is the number of such tetrahedra in the mesh. Substituting the finite element approximations for v^e and χ_p^e, one arrives at the following system of linear algebraic equations for each tetrahedron $\bar{\Omega}_e$,

$$\sum_{a=1}^{10} k_{ba}^{\prime e} \chi_{p,a}^e = f_{p,b}^e \,, \tag{27}$$

$b = 1, 2, \ldots, 10$. In (27), $k_{ba}^{\prime e}$ and $f_{p,b}^e$ are, respectively, the elemental stiffness matrix and elemental forcing vector, given by

$$k_{ba}^{\prime e} = \int_{\bar{\Omega}_e} \zeta_{ij}^e \frac{\partial \psi_b^e}{\partial y_i} \frac{\partial \psi_a^e}{\partial y_j} \, d\mathbf{y} \quad \text{and} \quad f_{p,b}^e = \int_{\bar{\Omega}_e} \zeta_{ip}^e \frac{\partial \psi_b^e}{\partial y_i} \, d\mathbf{y} \,; \tag{28}$$

note that $k_{ba}^{\prime \mathrm{e}}$ is symmetric.

The integrals in (28) are evaluated with the help of an isoparametric mapping between the actual tetrahedron $\bar{\Omega}_e$, $e = 1, 2, \ldots, \bar{N}_{\mathrm{E}}$, and the standard (or master) quadratic tetrahedron $\Omega_m : \{(\xi, \eta, \vartheta) \in I\!R^3 \,|\, 0 \le \xi \le 1; 0 \le \eta \le 1 - \xi; 0 \le \vartheta \le$

$1-\xi-\eta\}$, as illustrated in Fig. 2a; (ξ,η,ϑ) are the volume coordinates [3] of the standard tetrahedron, related to the spatial coordinates (y_1,y_2,y_3) over $\bar{\Omega}_e$ by

$$y_1^e(\xi,\eta,\vartheta)=\sum_{a=1}^{10} y_{1(a)}^e\tilde{\psi}_a(\xi,\eta,\vartheta)\,, \tag{29}$$

$$y_2^e(\xi,\eta,\vartheta)=\sum_{a=1}^{10} y_{2(a)}^e\tilde{\psi}_a(\xi,\eta,\vartheta)\,, \tag{30}$$

$$y_3^e(\xi,\eta,\vartheta)=\sum_{a=1}^{10} y_{3(a)}^e\tilde{\psi}_a(\xi,\eta,\vartheta)\,, \tag{31}$$

where $y_{1(a)}^e$, $y_{2(a)}^e$, and $y_{3(a)}^e$ are, respectively, the y_1, y_2, and y_3 coordinates of local node a, $a=1,2,\ldots,10$, belonging to the tetrahedron $\bar{\Omega}_e$. The isoparametric mapping $F^{-1}(\xi,\eta,\vartheta):\Omega_m\mapsto\bar{\Omega}_e$ thus yields

$$\begin{aligned}k_{ba}^{\prime e}=&\int_0^1\int_0^{1-\xi}\int_0^{1-\xi-\eta}\zeta_{ij}^e\left(\frac{\partial\tilde{\psi}_b}{\partial\xi}\frac{\partial\xi}{\partial y_i}+\frac{\partial\tilde{\psi}_b}{\partial\eta}\frac{\partial\eta}{\partial y_i}+\frac{\partial\tilde{\psi}_b}{\partial\vartheta}\frac{\partial\vartheta}{\partial y_i}\right)\\&\times\left(\frac{\partial\tilde{\psi}_a}{\partial\xi}\frac{\partial\xi}{\partial y_j}+\frac{\partial\tilde{\psi}_a}{\partial\eta}\frac{\partial\eta}{\partial y_j}+\frac{\partial\tilde{\psi}_a}{\partial\vartheta}\frac{\partial\vartheta}{\partial y_j}\right)\det\mathbf{J}^e\,d\vartheta\,d\eta\,d\xi\,,\end{aligned} \tag{32}$$

$$f_{p,b}^e=\int_0^1\int_0^{1-\xi}\int_0^{1-\xi-\eta}\zeta_{ip}^e\left(\frac{\partial\tilde{\psi}_b}{\partial\xi}\frac{\partial\xi}{\partial y_i}+\frac{\partial\tilde{\psi}_b}{\partial\eta}\frac{\partial\eta}{\partial y_i}+\frac{\partial\tilde{\psi}_b}{\partial\vartheta}\frac{\partial\vartheta}{\partial y_i}\right)\det\mathbf{J}^e\,d\vartheta\,d\eta\,d\xi\,, \tag{33}$$

where the abbreviated symbols $\tilde{\psi}_a=\psi_a^e(y_1(\xi,\eta,\vartheta),y_2(\xi,\eta,\vartheta),y_3(\xi,\eta,\vartheta))$, $\tilde{\psi}_b=\psi_b^e(y_1(\xi,\eta,\vartheta),y_2(\xi,\eta,\vartheta),y_3(\xi,\eta,\vartheta))$, $a,b=1,2,\ldots,10$, are the standard Lagrange interpolation functions, or shape functions, of the master tetrahedron, and $\mathbf{J}^e$ stands for the elemental Jacobian matrix of the isoparametric mapping, $\mathbf{J}^e\equiv\frac{\partial(y_1,y_2,y_3)}{\partial(\xi,\eta,\vartheta)}$ [31]. The shape functions are given by the following expressions in the coordinate system (ξ,η,ϑ) [3]:

$$\begin{aligned}&\tilde{\psi}_1(\xi,\eta,\vartheta)=(1-2\xi-2\eta-2\vartheta)(1-\xi-\eta-\vartheta)\\&\tilde{\psi}_2(\xi,\eta,\vartheta)=\xi(2\xi-1)\,,\tilde{\psi}_3(\xi,\eta,\vartheta)=\eta(2\eta-1)\,,\tilde{\psi}_4(\xi,\eta,\vartheta)=\vartheta(2\vartheta-1)\\&\tilde{\psi}_5(\xi,\eta,\vartheta)=4\xi(1-\xi-\eta-\vartheta),\tilde{\psi}_6(\xi,\eta,\vartheta)=4\xi\eta\\&\tilde{\psi}_7(\xi,\eta,\vartheta)=4\eta(1-\xi-\eta-\vartheta)\,,\tilde{\psi}_8(\xi,\eta,\vartheta)=4\vartheta(1-\xi-\eta-\vartheta)\\&\tilde{\psi}_9(\xi,\eta,\vartheta)=4\xi\vartheta\,,\tilde{\psi}_{10}(\xi,\eta,\vartheta)=4\eta\vartheta.\end{aligned}$$

The components of the Jacobian matrix, J^e_{1j}, J^e_{2j}, and J^e_{3j}, $j = 1, 2, 3$, are thus given by:

$$J^e_{1j} = \sum_{a=1}^{10} y^e_{1(a)} \frac{\partial \tilde{\psi}_a}{\partial \xi_j}, \quad J^e_{2j} = \sum_{a=1}^{10} y^e_{2(a)} \frac{\partial \tilde{\psi}_a}{\partial \xi_j} \quad \text{and} \quad J^e_{3j} = \sum_{a=1}^{10} y^e_{3(a)} \frac{\partial \tilde{\psi}_a}{\partial \xi_j} \tag{34}$$

with $\xi_1 = \xi$, $\xi_2 = \eta$, and $\xi_3 = \vartheta$.

The integrals appearing in (32) and (33) are evaluated numerically using the Gauss quadrature rule with five points [3]. The derivatives of ξ, η and ϑ with respect to y_1, y_2 and y_3 are computed from the inverse of the Jacobian matrix. The elemental stiffness matrices and elemental forcing vectors are finally assembled to construct the global matrix $\mathbf{K}'$ and forcing vector $\mathbf{F}_p$. The assembling, or direct stiffness summation, procedure enforces (1) continuity of the temperature at nodes shared by adjoining tetrahedra, provided that such nodes are not located at Γ, (2) balance of the heat flux at the surfaces shared by adjoining tetrahedra, and (3) periodicity at corresponding boundary nodes lying on the external surfaces of the cell Ω_{pc}.

3.2.2 Numerical Treatment of the Surface Integral

For a generic tetrahedron $\tilde{\Omega}_e$ with exactly three corner nodes at Γ, $e = 1, 2, ..., \tilde{N}_E$, $N_{\mathrm{E}} = \bar{N}_{\mathrm{E}} + \tilde{N}_{\mathrm{E}}$, (24) can be rewritten as

$$\int_{\tilde{\Omega}_e} \zeta^e_{ij} \frac{\partial v^e}{\partial y_i} \frac{\partial \chi^e_p}{\partial y_j} \, d\mathbf{y} + \int_{\Gamma_{\tilde{\Omega}_e}} \mathrm{Bi}^e \, [v^e]_{\Gamma_{\tilde{\Omega}_e}} \left[\chi^e_p\right]_{\Gamma_{\tilde{\Omega}_e}} ds = \int_{\tilde{\Omega}_e} \zeta^e_{ip} \frac{\partial v^e}{\partial y_i} \, d\mathbf{y} \, . \tag{35}$$

The finite element discretization of the two volume integrals in (35) follows the procedure described in the previous section, and using (32) and (33), will give rise to the corresponding elemental stiffness matrices and elemental forcing vectors. However, for tetrahedron $\tilde{\Omega}_e$, one must still discretize the surface integral on the left-hand side of (35), which accounts for the discontinuities of the weight function and temperature at $\Gamma_{\tilde{\Omega}_e}$, $e = 1, 2, \ldots, \tilde{N}_{\mathrm{E}}$. Henceforth, the Biot number is assumed to be constant over the interface Γ and it will be written outside the integrals.

Differently from the treatment of the volume integrals, as a much smaller number of elements are involved (i.e., only those tetrahedra with three corner nodes at Γ), it is more convenient to discretize the surface integral in (23) globally, rather than elementally using (35). Therefore, it is more computationally efficient to sum directly the surface integral contributions to the appropriate entries of the global matrix $\mathbf{K}'$, rather than to incorporate elemental contributions into the elemental stiffness matrices, and then perform direct stiffness summation with these elemental data structures. Once all surface integral contributions are incorporated into $\mathbf{K}'$, one finally obtains the global stiffness matrix $\mathbf{K}$.

For the degrees of freedom corresponding to global nodes *at* Γ, discontinuous weight functions are used, for which the jump β at the interface is nonzero. It will be seen that, for a generic global node at Γ, the resulting surface integrals depend on the Biot number and on the areas of the tetrahedra faces lying on Γ and sharing that node. For example, for the generic corner node A shown in Fig. 2b, located on the continuous phase side of Γ and shared by five tetrahedra (denoted in Fig. 2b by $\tilde{\Omega}_e$, $\tilde{\Omega}_f$, $\tilde{\Omega}_g$, $\tilde{\Omega}_h$ and $\tilde{\Omega}_i$, $e,f,g,h,i \in \{1,2,3,\ldots,\tilde{N}_E\}$, the resulting surface integrals depend on Bi and on the areas of the faces $\Gamma_{\tilde{\Omega}_e}$, $\Gamma_{\tilde{\Omega}_f}$, $\Gamma_{\tilde{\Omega}_g}$, $\Gamma_{\tilde{\Omega}_h}$ and $\Gamma_{\tilde{\Omega}_i}$ shared, respectively, by the pairs of tetrahedra $\tilde{\Omega}_e$ and $\tilde{\Omega}_{e'}$, $\tilde{\Omega}_f$ and $\Gamma_{\tilde{\Omega}_{f'}}$, $\tilde{\Omega}_g$ and $\Gamma_{\tilde{\Omega}_{g'}}$, $\tilde{\Omega}_h$ and $\tilde{\Omega}_{h'}$, and $\tilde{\Omega}_i$ and $\tilde{\Omega}_{i'}$. In the unstructured tetrahedral meshes generated by NETGEN, the number of tetrahedra with exactly three nodes at a surface and sharing a common corner node varies from four to seven. In the following, for the sake of completeness, the scheme to calculate the contributions arising from the surface integral in (23) is described for both a generic corner node A and a generic midside node M at Γ (see Fig. 2b).

The finite dimensional subspace $Y_h(\Omega_{pc,h})$ of $Y(\Omega_{pc,h})$, in which the weight function v lies, is defined such that $Y_h(\Omega_{pc,h}) = Y(\Omega_{pc,h}) \cap P_2(\Omega_e)$, $e = 1,2,\ldots,$ N_{E}, and $P_2(\Omega_e)$ is the space of all quadratic polynomials defined on tetrahedron Ω_e. The weight function v_A for a generic corner node A is thus $v_A = \beta\phi_A \in Y_h(\Omega_{pc,h})$, where ϕ_A is the standard quadratic interpolant. For implementation simplicity and without loss of generality, it is convenient to choose $\beta = 1$, so that $[v_A]_\Gamma = 1$. Now, defining $\phi_A|_{\Gamma_{\tilde{\Omega}_e}} = v_A|_{\Gamma_{\tilde{\Omega}_e}}$ as the restriction of v_A to the surface $\Gamma_{\tilde{\Omega}_e}$ of tetrahedron $\tilde{\Omega}_e$, $e \in \{1,2,\ldots,\tilde{N}_{\mathrm{E}}\}$ (note that $\phi_{A|\Gamma_{\tilde{\Omega}_e}} = \psi^e_{a|\Gamma_{\tilde{\Omega}_e}}$, where a is the local node of tetrahedron $\tilde{\Omega}_e$ corresponding to global node A), the restriction of the temperature χ_p to $\Gamma_{\tilde{\Omega}_e}$ can be written as

$$\chi_p|_{\Gamma_{\tilde{\Omega}_e}} = \chi_A\phi_A|_{\Gamma_{\tilde{\Omega}_e}} + \chi_B\phi_B|_{\Gamma_{\tilde{\Omega}_e}} + \chi_C\phi_C|_{\Gamma_{\tilde{\Omega}_e}} + \chi_M\phi_M|_{\Gamma_{\tilde{\Omega}_e}} + \chi_N\phi_N|_{\Gamma_{\tilde{\Omega}_e}} + \chi_P\phi_P|_{\Gamma_{\tilde{\Omega}_e}}. \tag{36}$$

To compute the jumps of the weight function and temperature across the surface $\Gamma_{\tilde{\Omega}_e}$, one must also consider the tetrahedron $\tilde{\Omega}_{e'} \in \Omega_{pc,d}$, $e' \in \{1,2,\ldots,\tilde{N}_{\mathrm{E}}\}$, which shares with $\tilde{\Omega}_e$ the surface $\Gamma_{\tilde{\Omega}_e}$ or, equivalently, the surface $\Gamma_{\tilde{\Omega}_{e'}}$. From the definition of v_A, it follows that $v_A|_{\Gamma_{\tilde{\Omega}_{e'}}} = 0$. Similarly to (36), the expression for the temperature χ_p restricted to the surface $\Gamma_{\tilde{\Omega}_{e'}}$ is

$$\begin{aligned}\chi_p|_{\Gamma_{\tilde{\Omega}_{e'}}} &= \chi_{A'}\phi_{A'}|_{\Gamma_{\tilde{\Omega}_{e'}}} + \chi_{B'}\phi_{B'}|_{\Gamma_{\tilde{\Omega}_{e'}}} + \chi_{C'}\phi_{C'}|_{\Gamma_{\tilde{\Omega}_{e'}}} + \chi_{M'}\phi_{M'}|_{\Gamma_{\tilde{\Omega}_{e'}}} \\ &\quad + \chi_{N'}\phi_{N'}|_{\Gamma_{\tilde{\Omega}_{e'}}} + \chi_{P'}\phi_{P'}|_{\Gamma_{\tilde{\Omega}_{e'}}}.\end{aligned} \tag{37}$$

From (36) and (37), the jumps of v_A and χ_p across the surface $\Gamma_{\tilde{\Omega}_e}$ are respectively deduced to be

$$[v_A]_{\Gamma_{\tilde{\Omega}_e}} = v_A|_{\Gamma_{\tilde{\Omega}_e}} - v_A|_{\Gamma_{\tilde{\Omega}_{e'}}} = v_A|_{\Gamma_{\tilde{\Omega}_e}} = \phi_A|_{\Gamma_{\tilde{\Omega}_e}}, \tag{38}$$

$$\begin{aligned}
[\chi_p]_{\Gamma_{\tilde{\Omega}_e}} &= \chi_A\,\phi_A|_{\Gamma_{\tilde{\Omega}_e}} + \chi_B\,\phi_B|_{\Gamma_{\tilde{\Omega}_e}} + \chi_C\,\phi_C|_{\Gamma_{\tilde{\Omega}_e}} + \chi_M\,\phi_M|_{\Gamma_{\tilde{\Omega}_e}} + \chi_N\,\phi_N|_{\Gamma_{\tilde{\Omega}_e}} \\
&\quad + \chi_P\,\phi_P|_{\Gamma_{\tilde{\Omega}_e}} - \chi_{A'}\,\phi_{A'}|_{\Gamma_{\tilde{\Omega}_{e'}}} - \chi_{B'}\,\phi_{B'}|_{\Gamma_{\tilde{\Omega}_{e'}}} - \chi_{C'}\,\phi_{C'}|_{\Gamma_{\tilde{\Omega}_{e'}}} \\
&\quad - \chi_{M'}\,\phi_{M'}|_{\Gamma_{\tilde{\Omega}_{e'}}} - \chi_{N'}\,\phi_{N'}|_{\Gamma_{\tilde{\Omega}_{e'}}} - \chi_{P'}\,\phi_{P'}|_{\Gamma_{\tilde{\Omega}_{e'}}}.
\end{aligned} \tag{39}$$

As dictated by (23), one must now integrate the product of the jumps of v_A and χ_p across $\Gamma_{\tilde{\Omega}_e}$, yielding

$$\begin{aligned}
\int_{\Gamma_{\tilde{\Omega}_e}} \mathrm{Bi}\,[v_A]_{\Gamma_{\tilde{\Omega}_e}}[\chi_p]_{\Gamma_{\tilde{\Omega}_e}}\,ds = \mathrm{Bi}\Bigg\{ &\chi_A \int_{\Gamma_{\tilde{\Omega}_e}} \phi_A|_{\Gamma_{\tilde{\Omega}_e}}\,\phi_A|_{\Gamma_{\tilde{\Omega}_e}}\,ds + \chi_B \int_{\Gamma_{\tilde{\Omega}_e}} \phi_A|_{\Gamma_{\tilde{\Omega}_e}}\,\phi_B|_{\Gamma_{\tilde{\Omega}_e}}\,ds \\
&+ \chi_C \int_{\Gamma_{\tilde{\Omega}_e}} \phi_A|_{\Gamma_{\tilde{\Omega}_e}}\,\phi_C|_{\Gamma_{\tilde{\Omega}_e}}\,ds + \chi_M \int_{\Gamma_{\tilde{\Omega}_e}} \phi_A|_{\Gamma_{\tilde{\Omega}_e}}\,\phi_M|_{\Gamma_{\tilde{\Omega}_e}}\,ds \\
&+ \chi_N \int_{\Gamma_{\tilde{\Omega}_e}} \phi_A|_{\Gamma_{\tilde{\Omega}_e}}\,\phi_N|_{\Gamma_{\tilde{\Omega}_e}}\,ds + \chi_P \int_{\Gamma_{\tilde{\Omega}_e}} \phi_A|_{\Gamma_{\tilde{\Omega}_e}}\,\phi_P|_{\Gamma_{\tilde{\Omega}_e}}\,ds \\
&- \chi_{A'} \int_{\Gamma_{\tilde{\Omega}_e}} \phi_A|_{\Gamma_{\tilde{\Omega}_e}}\,\phi_{A'}|_{\Gamma_{\tilde{\Omega}_{e'}}}\,ds - \chi_{B'} \int_{\Gamma_{\tilde{\Omega}_e}} \phi_A|_{\Gamma_{\tilde{\Omega}_e}}\,\phi_{B'}|_{\Gamma_{\tilde{\Omega}_{e'}}}\,ds \\
&- \chi_{C'} \int_{\Gamma_{\tilde{\Omega}_e}} \phi_A|_{\Gamma_{\tilde{\Omega}_e}}\,\phi_{C'}|_{\Gamma_{\tilde{\Omega}_{e'}}}\,ds - \chi_{M'} \int_{\Gamma_{\tilde{\Omega}_e}} \phi_A|_{\Gamma_{\tilde{\Omega}_e}}\,\phi_{M'}|_{\Gamma_{\tilde{\Omega}_{e'}}}\,ds \\
&- \chi_{N'} \int_{\Gamma_{\tilde{\Omega}_e}} \phi_A|_{\Gamma_{\tilde{\Omega}_e}}\,\phi_{N'}|_{\Gamma_{\tilde{\Omega}_{e'}}}\,ds - \chi_{P'} \int_{\Gamma_{\tilde{\Omega}_e}} \phi_A|_{\Gamma_{\tilde{\Omega}_e}}\,\phi_{P'}|_{\Gamma_{\tilde{\Omega}_{e'}}}\,ds \Bigg\} \\
= \mathrm{Bi}\Bigg\{ &\sum_{L\in\{A,B,C,M,N,P\}} \chi_L \int_{\Gamma_{\tilde{\Omega}_e}} \phi_A|_{\Gamma_{\tilde{\Omega}_e}}\,\phi_L|_{\Gamma_{\tilde{\Omega}_e}}\,ds \\
&- \sum_{L'\in\{A',B',C',M',N',P'\}} \chi_{L'} \int_{\Gamma_{\tilde{\Omega}_e}} \phi_A|_{\Gamma_{\tilde{\Omega}_e}}\,\phi_{L'}|_{\Gamma_{\tilde{\Omega}_{e'}}}\,ds \Bigg\}
\end{aligned} \tag{40}$$

For the generic corner node A depicted in Fig. 2b, four more expressions similar to (40) must be written, due to the jumps of v_A and χ_p across the neighboring surfaces $\Gamma_{\tilde{\Omega}_f}$, $\Gamma_{\tilde{\Omega}_g}$, $\Gamma_{\tilde{\Omega}_h}$, and $\Gamma_{\tilde{\Omega}_i}$ shared, respectively, by the pairs of tetrahedra $\tilde{\Omega}_f$ and $\tilde{\Omega}_{f'}$, $\tilde{\Omega}_g$ and $\tilde{\Omega}_{g'}$, $\tilde{\Omega}_h$ and $\tilde{\Omega}_{h'}$, and $\tilde{\Omega}_i$ and $\tilde{\Omega}_{i'}$ (in general for a node at Γ, one must generate as many expressions as the number of tetrahedra in the mesh that share that node). For an interested reader, the remaining four expressions for node A are presented in Matt and Cruz [24]. The complete procedure just described for the generic corner node A must be repeated for all the corner nodes of the mesh lying at Γ, including the duplicates. The resulting surface integrals, such as the ones appearing in (40), must be summed to the appropriate components of the global matrix $\mathbf{K}'$ so as to form $\mathbf{K}$, as explained next.

Three auxiliary data arrays for the construction of the global stiffness matrix $\mathbf{K}$ and forcing vector $\mathbf{F}$ are used: (1) the array $\mathrm{IEN}(a,e)$, $a = 1,2,\ldots,10$, $e = 1,2,\ldots,N_{\mathrm{E}}$, which associates to each local mesh node a of a tetrahedron Ω_e the corresponding global node (this array is first provided by the mesh generator NETGEN, and later altered according to the duplication procedure illustrated in Fig. 2b); (2) the vector $\mathrm{ID}(A), A \in \{1,2,\ldots,N_{\mathrm{GN}}+N_{\Gamma}\}$, which associates to each global node the corresponding equation number $\mathrm{ID}(A)$, $\mathrm{ID}(A) \in \{1,2,\ldots,N_{\mathrm{DOF}}\}$, where N_{DOF} is the total number of equations or degrees of freedom (for generic periodic nodes G and H lying on distinct external surfaces of Ω_{pc}, $\mathrm{ID}(G) = \mathrm{ID}(H)$ is enforced); and (3) the array $\mathrm{LM}(a,e) = \mathrm{ID}(\mathrm{IEN}(a,e))$, which associates to each local mesh node a of a tetrahedron Ω_e the corresponding equation number.

From the equations associated with corner node A, one must sum to the components of the global matrix $K'_{\mathrm{ID}(A)\,\mathrm{ID}(A)}$ and $K'_{\mathrm{ID}(A)\,\mathrm{ID}(A')}$, respectively, the contributions

$$\begin{aligned}\mathrm{Bi}\Bigg\{&\int_{\Gamma_{\tilde{\Omega}_e}} \phi_A|_{\Gamma_{\tilde{\Omega}_e}} \phi_A|_{\Gamma_{\tilde{\Omega}_e}}\, ds + \int_{\Gamma_{\tilde{\Omega}_f}} \phi_A|_{\Gamma_{\tilde{\Omega}_f}} \phi_A|_{\Gamma_{\tilde{\Omega}_f}}\, ds\\ &+ \int_{\Gamma_{\tilde{\Omega}_g}} \phi_A|_{\Gamma_{\tilde{\Omega}_g}} \phi_A|_{\Gamma_{\tilde{\Omega}_g}}\, ds + \int_{\Gamma_{\tilde{\Omega}_h}} \phi_A|_{\Gamma_{\tilde{\Omega}_h}} \phi_A|_{\Gamma_{\tilde{\Omega}_h}}\, ds + \int_{\Gamma_{\tilde{\Omega}_i}} \phi_A|_{\Gamma_{\tilde{\Omega}_i}} \phi_A|_{\Gamma_{\tilde{\Omega}_i}}\, ds\Bigg\},\end{aligned} \tag{41}$$

$$\begin{aligned}-\mathrm{Bi}\Bigg\{&\int_{\Gamma_{\tilde{\Omega}_e}} \phi_A|_{\Gamma_{\tilde{\Omega}_e}} \phi_{A'}|_{\Gamma_{\tilde{\Omega}_{e'}}}\, ds + \int_{\Gamma_{\tilde{\Omega}_f}} \phi_A|_{\Gamma_{\tilde{\Omega}_f}} \phi_{A'}|_{\Gamma_{\tilde{\Omega}_{f'}}}\, ds\\ &+ \int_{\Gamma_{\tilde{\Omega}_g}} \phi_A|_{\Gamma_{\tilde{\Omega}_g}} \phi_{A'}|_{\Gamma_{\tilde{\Omega}_{g'}}}\, ds + \int_{\Gamma_{\tilde{\Omega}_h}} \phi_A|_{\Gamma_{\tilde{\Omega}_h}} \phi_{A'}|_{\Gamma_{\tilde{\Omega}_{h'}}}\, ds + \int_{\Gamma_{\tilde{\Omega}_i}} \phi_A|_{\Gamma_{\tilde{\Omega}_i}} \phi_{A'}|_{\Gamma_{\tilde{\Omega}_{i'}}}\, ds\Bigg\}.\end{aligned} \tag{42}$$

To the components $K'_{\mathrm{ID}(A)\,\mathrm{ID}(B)}$ and $K'_{\mathrm{ID}(A)\,\mathrm{ID}(B')}$, one must sum, respectively, the contributions

$$\mathrm{Bi}\left\{\int_{\Gamma_{\tilde{\Omega}_e}} \phi_A|_{\Gamma_{\tilde{\Omega}_e}} \phi_B|_{\Gamma_{\tilde{\Omega}_e}}\, ds + \int_{\Gamma_{\tilde{\Omega}_i}} \phi_A|_{\Gamma_{\tilde{\Omega}_i}} \phi_B|_{\Gamma_{\tilde{\Omega}_i}}\, ds\right\}, \tag{43}$$

$$-\,\mathrm{Bi}\left\{\int_{\Gamma_{\tilde{\Omega}_e}} \phi_A|_{\Gamma_{\tilde{\Omega}_e}} \phi_{B'}|_{\Gamma_{\tilde{\Omega}_{e'}}}\, ds + \int_{\Gamma_{\tilde{\Omega}_i}} \phi_A|_{\Gamma_{\tilde{\Omega}_i}} \phi_{B'}|_{\Gamma_{\tilde{\Omega}_{i'}}}\, ds\right\}. \tag{44}$$

The remaining 28 expressions to be summed to the components of the global matrix $\mathbf{K}'$ associated with corner node A are easily derived, similarly to (41)–(44). With reference to Fig. 2b, eight expressions are associated with the neighboring corner nodes C, D, E, F and corresponding duplicates; ten expressions are associated with the midside nodes M, P, R, T, V and corresponding duplicates, shared by two neighboring tetrahedra; and, finally, ten more expressions are associated with the midside nodes N, Q, S, U, W and corresponding duplicates.

For a generic midside node at Γ, the computations of the jumps of the weight function and temperature are analogous to the ones just described for a generic corner node. All midside nodes at Γ are always shared by only two tetrahedra with exactly three corner nodes at Γ, thus reducing significantly the number of equations. For the generic midside node M depicted in Fig. 2b, one must only compute the jumps of v_M and χ_p across the surfaces $\Gamma_{\tilde{\Omega}_e}$ and $\Gamma_{\tilde{\Omega}_i}$, shared, respectively, by the pairs of tetrahedra $\tilde{\Omega}_e$ and $\tilde{\Omega}_{e'}$, and $\tilde{\Omega}_i$ and $\tilde{\Omega}_{i'}$. The two resulting expressions for the integrals of the products of the jumps are

$$\int_{\Gamma_{\tilde{\Omega}_e}} \mathrm{Bi}\,[v_M]_{\Gamma_{\tilde{\Omega}_e}} [\chi_p]_{\Gamma_{\tilde{\Omega}_e}}\, ds = \mathrm{Bi}\left\{ \sum_{L\in\{A,B,C,M,N,P\}} \chi_L \int_{\Gamma_{\tilde{\Omega}_e}} \phi_M|_{\Gamma_{\tilde{\Omega}_e}} \phi_L|_{\Gamma_{\tilde{\Omega}_e}}\, ds \right.$$
$$\left. - \sum_{L'\in\{A',B',C',M',N',P'\}} \chi_{L'} \int_{\Gamma_{\tilde{\Omega}_e}} \phi_M|_{\Gamma_{\tilde{\Omega}_e}} \phi_{L'}|_{\Gamma_{\tilde{\Omega}_{e'}}}\, ds \right\}, \quad (45)$$

$$\int_{\Gamma_{\tilde{\Omega}_i}} \mathrm{Bi}\,[v_M]_{\Gamma_{\tilde{\Omega}_i}} [\chi_p]_{\Gamma_{\tilde{\Omega}_i}}\, ds = \mathrm{Bi}\left\{ \sum_{L\in\{A,F,B,V,W,M\}} \chi_L \int_{\Gamma_{\tilde{\Omega}_i}} \phi_M|_{\Gamma_{\tilde{\Omega}_i}} \phi_L|_{\Gamma_{\tilde{\Omega}_i}}\, ds \right.$$
$$\left. - \sum_{L'\in\{A',F',B',V',W',M'\}} \chi_{L'} \int_{\Gamma_{\tilde{\Omega}_i}} \phi_M|_{\Gamma_{\tilde{\Omega}_i}} \phi_{L'}|_{\Gamma_{\tilde{\Omega}_{i'}}}\, ds \right\}. \quad (46)$$

Hence, from (45) and (46), it is clear that one must sum to the components $K'_{\mathrm{ID}(M)\,\mathrm{ID}(M)}$ and $K'_{\mathrm{ID}(M)\,\mathrm{ID}(M')}$, respectively, the contributions

$$\mathrm{Bi}\left\{ \int_{\Gamma_{\tilde{\Omega}_e}} \phi_M|_{\Gamma_{\tilde{\Omega}_e}} \phi_M|_{\Gamma_{\tilde{\Omega}_e}}\, ds + \int_{\Gamma_{\tilde{\Omega}_i}} \phi_M|_{\Gamma_{\tilde{\Omega}_i}} \phi_M|_{\Gamma_{\tilde{\Omega}_i}}\, ds \right\}, \quad (47)$$

$$-\,\mathrm{Bi}\left\{ \int_{\Gamma_{\tilde{\Omega}_e}} \phi_M|_{\Gamma_{\tilde{\Omega}_e}} \phi_{M'}|_{\Gamma_{\tilde{\Omega}_{e'}}}\, ds + \int_{\Gamma_{\tilde{\Omega}_i}} \phi_M|_{\Gamma_{\tilde{\Omega}_i}} \phi_{M'}|_{\Gamma_{\tilde{\Omega}_{i'}}}\, ds \right\}. \quad (48)$$

To the components $K'_{\mathrm{ID}(M)\,\mathrm{ID}(P)}$ and $K'_{\mathrm{ID}(M)\,\mathrm{ID}(P')}$ one must sum, respectively, the contributions

$$\mathrm{Bi} \int_{\Gamma_{\tilde{\Omega}_e}} \phi_M|_{\Gamma_{\tilde{\Omega}_e}} \phi_P|_{\Gamma_{\tilde{\Omega}_e}}\, ds \quad \text{and} - \mathrm{Bi} \int_{\Gamma_{\tilde{\Omega}_e}} \phi_M|_{\Gamma_{\tilde{\Omega}_e}} \phi_{P'}|_{\Gamma_{\tilde{\Omega}_{e'}}}\, ds\,. \quad (49)$$

The remaining 14 expressions to be summed to the components of the global matrix $\mathbf{K}'$ associated with midside node M are easily derived, similarly to (47)–(49). With reference to Fig. 2b, six expressions are associated with the neighboring midside nodes V, N, W and corresponding duplicates; and eight expressions are associated with the neighboring corner nodes A, B, C, F and corresponding duplicates. The procedure described above for node M must be repeated for all the midside nodes of the mesh lying at Γ, including the duplicates.

To close this section, it is important to remark that surface integrals such as the integral over $\Gamma_{\tilde{\Omega}_e}$ are evaluated numerically through a 12-point Gauss quadrature rule [3], by employing an isoparametric mapping between $\Gamma_{\tilde{\Omega}_e}$ (actual domain of integration) and the standard quadratic triangle (computational domain of integration).

3.3 Iterative Solution of Discrete System

The resulting linear system of algebraic equations may be cast in the following form

$$\mathbf{K}\chi_p = \mathbf{F}\,, \tag{50}$$

where $\mathbf{K} = [K_{IJ}]$, $\mathbf{F} = [F_I]$, $I, J = 1, 2, \ldots, N_{\mathrm{DOF}}$, are the previously obtained global stiffness matrix and forcing vector, and $\chi_p = [\chi_{p,J}]$ is the vector of unknown nodal values of the temperature χ_p. The uniqueness condition, given in continuous form by $\int_{\Omega_{pc}} \chi_p \, d\mathbf{y} = 0$, is imposed discretely by requiring that χ_p has zero algebraic average. The system given by (50) is solved iteratively using the *minimum residual method*, described in detail in [30]. The minimum residual method is suitable for symmetric semi-definite systems of linear equations. After the incorporation of all contributions which result from the surface integral in (23) to the appropriate components of the global stiffness matrix, the latter remains symmetric, but not necessarily positive-definite for arbitrary choices of Bi. The iteration proceeds until the square of the ratio of the Euclidean norm of the residual to the Euclidean norm of the initial residual falls below a user-prescribed tolerance, σ^2.

Finally, after the numerical determination of the field χ_p, $p = 1, 2, 3$, the components of the effective thermal conductivity tensor, $\kappa^{\mathrm{N}}_{e,pq}$, are computed numerically through the equivalent discrete form of (22), i.e.,

$$\begin{aligned}
\kappa^{\mathrm{N}}_{e,pq} &= \frac{1}{|\Omega_{pc,h}|} \int_{\Omega_{pc,h}} \zeta_{pi}(\mathbf{y}) \left(\delta_{iq} - \frac{\partial \chi_q}{\partial y_i} \right) d\mathbf{y} \\
&= \frac{1}{\sum_{e=1}^{N_E} |\Omega_e|} \sum_{e=1}^{N_E} \int_{\Omega_e} \zeta^e_{pi} \left(\delta_{iq} - \frac{\partial \chi^e_q}{\partial y_i} \right) d\mathbf{y} \\
&= \frac{1}{\sum_{e=1}^{N_E} |\Omega_e|} \sum_{e=1}^{N_E} \int_{\Omega_e} \zeta^e_{pi} \left(\delta_{iq} - \sum_{a=1}^{10} \chi^e_{q,a} \frac{\partial \psi^e_a}{\partial y_i} \right) d\mathbf{y},
\end{aligned} \tag{51}$$

where $\zeta^e_{pi} = \delta_{pi}$ if $\Omega_e \subset \Omega_{pc,c}$, $\zeta^e_{pi} = \kappa_{pi}$ if $\Omega_e \subset \Omega_{pc,d}$, and $|\Omega_e|$ denotes the volume of tetrahedron Ω_e, $e = 1, 2, \ldots, N_{\mathrm{E}}$.

4 Numerical Results for the Effective Conductivity

In this section, numerical results for the effective thermal conductivity of ordered and random arrays are presented and, when possible, validated against previous analytical predictions. The results demonstrate the accuracy and flexibility of the computational approach previously discussed. For the ordered and random arrays investigated here, it is assumed that the dispersed phase is thermally-isotropic, i.e., $\kappa_{ij} = \kappa\,\delta_{ij}$, $i, j = 1, 2, 3$; thus, henceforth, the symbol κ denotes the phase conductivity ratio.

4.1 Particulate Composites

Particulate composites are a special class of composite materials designed for applications that do not require strong directionality. The dispersed phase is usually modeled as spheres. The spheres may be orderly or randomly distributed within the matrix; in the former case, one has the ordered arrays of spheres, among which the most widely known are the simple cubic (SC), the body centered (BCC) and the face centered (FCC); in the latter case, one has the random arrays of spheres. Analytical and semi-analytical expressions for the effective conductivity of ordered and random arrays of spheres are available in the literature; many of them may be encountered in the works of Batchelor and O'Brien [2], McPhedran and McKenzie [26], McKenzie et al. [25], Sangani and Acrivos [34], Hasselman and Johnson [15], Cheng and Torquato [6], to name just a few. There are expressions for the effective conductivity of ordered arrays of spheres with uniform interfacial thermal resistance; on the other hand, for the random arrays of spheres, the majority of expressions developed for the effective conductivity assumes perfect thermal contact at the particle/matrix interface. The numerical results for the effective conductivity of particulate composites are presented and validated for the simple cubic array of spheres and for random arrays of spheres.

4.1.1 Simple Cubic Array of Spheres

The first set of results has been computed for the simple cubic array of spheres with an uniform interfacial resistance. The periodic cell is composed of a cubic matrix of side λ, at which geometric center lies one sphere of diameter d, as illustrated in Fig. 3a. For a given concentration c, λ and d are related by $c = (\pi d^3)/(6\lambda^3)$.

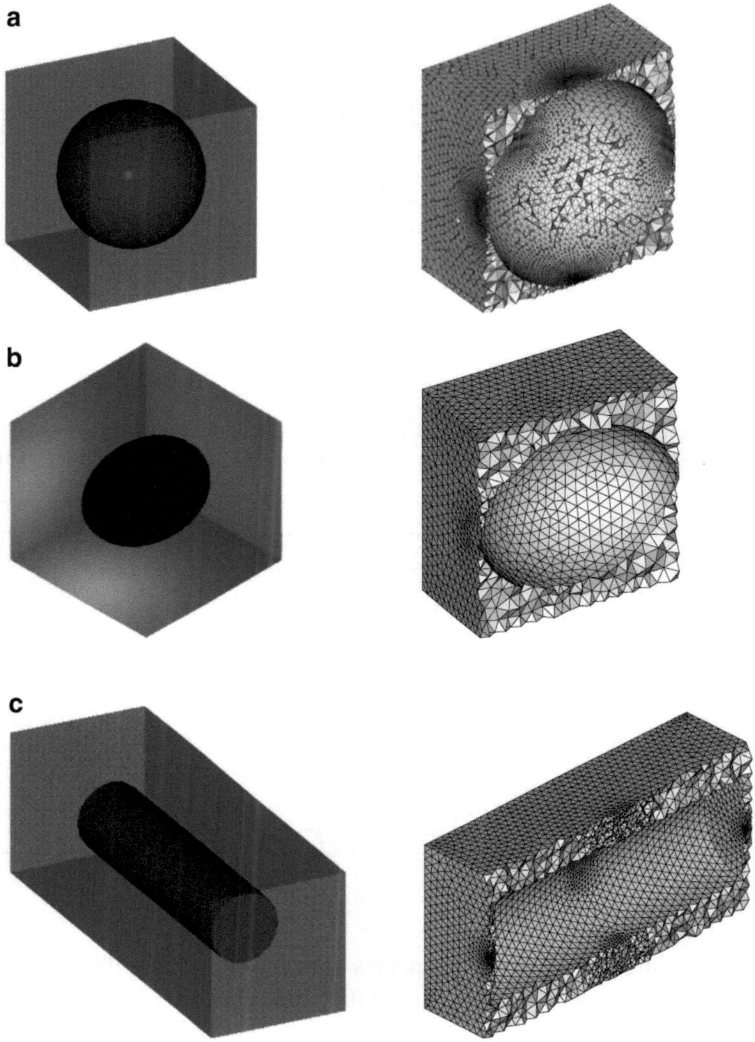

Fig. 3 Ordered arrays of (**a**) spheres: illustration of geometry (*left*), and finite element mesh for $c = 0.50$ (*right*), with 247,857 global mesh nodes and 177,502 quadratic tetrahedra; (**b**) prolate ellipsoids of revolution: geometry (*left*), and mesh for $c = 0.30$ and $\rho_f = 1.30$ (*right*), with 72,401 global nodes and 48,582 tetrahedra; (**c**) circular cylinders: geometry (*left*), and mesh for $c = 0.30$, $\rho_p = 2$ and $\rho_f = 3$ (*right*), with 237,820 global nodes and 169,361 tetrahedra

Because, by symmetry, $\kappa_{e,11} = \kappa_{e,22} = \kappa_{e,33} = \kappa_e$ and $\kappa_{e,pq} = 0$, $p \neq q$, $p, q = 1, 2, 3$, for the simple cubic array of spheres, the effective conductivity tensor is completely described by the scalar κ_e.

Table 1 shows the mesh refinement analysis performed in order to set the nominal mesh spacing, $h_0^* \equiv h_0/\lambda$, for subsequent numerical computations of the effective conductivity. The mesh-independence study reported in Table 1 has been

Table 1 Numerical convergence analysis for the finite element computation of the effective conductivity of the simple cubic array of spheres with $c = 0.30$, $\kappa = 100$ and $\mathrm{Bi} \in \{10^{-2}, 10^{0}, 10^{2}\}$

h_0^*	N_E	N_{GN}	κ_e^{N}		
			$\mathrm{Bi} = 10^{-2}$	$\mathrm{Bi} = 10^{0}$	$\mathrm{Bi} = 10^{2}$
0.20	786	1,381	0.6088	0.788	2.1450
0.15	1,344	2,405	0.6085	0.791	2.1460
0.10	3,634	6,209	0.6084	0.793	2.1485
0.08	5,938	9,891	0.6083	0.794	2.1490
0.06	26,414	38,837	0.6078	0.795	2.145
0.05	39,667	58,920	0.6078	0.795	2.145
0.04	67,002	98,233	0.6078	0.795	2.145

Table 2 Numerical, κ_e^{N}, and semi-analytical, κ_e^{CT} and κ_e^{SA}, results for the effective conductivity of the ordered array of spheres as a function of Bi, for three values of c, $c \in \{0.10, 0.30, 0.50\}$, and $\kappa = 100$

Bi	$c = 0.10$		$c = 0.30$		$c = 0.50$	
	κ_e^{N}	κ_e^{CT}	κ_e^{N}	κ_e^{CT}	κ_e^{N}	κ_e^{CT}
10^{7}	1.3227	1.3227	2.275	2.272	5.350	4.651
10^{5}	1.3227	1.3226	2.275	2.272	5.348	4.650
10^{3}	1.3190	1.3190	2.261	2.258	5.202	4.583
10^{-1}	0.8629	0.8629	0.6269	0.6269	0.412	0.412
10^{-3}	0.8572	0.8572	0.6059	0.6058	0.374	0.373
10^{-5}	0.8571	0.8571	0.6057	0.6056	0.374	0.373
10^{-7}	0.8571	0.8571	0.6057	0.6056	0.374	0.373
0	0.8571	0.8571	0.6057	0.6056	0.374	0.373
	$\kappa_e^{\mathrm{SA}}(\kappa = 0)$	$\kappa_e^{\mathrm{SA}}(\kappa = 100)$	$\kappa_e^{\mathrm{SA}}(\kappa = 0)$	$\kappa_e^{\mathrm{SA}}(\kappa = 100)$	$\kappa_e^{\mathrm{SA}}(\kappa = 0)$	$\kappa_e^{\mathrm{SA}}(\kappa = 100)$
	0.8571	1.3227	0.6057	2.274	0.376	4.892

done for the simple cubic array of spheres with $c = 0.30$, $\kappa = 100$ and for three distinct values of the Biot number, $\mathrm{Bi} \in \{10^{-2}, 10^{0}, 10^{2}\}$.

After the numerical convergence study reported in Table 1, the nominal mesh spacing chosen for all subsequent computations of the effective conductivity is 0.05. The chosen tolerance σ^2 of the minimum residual method is 10^{-8}. These values ensure the numerical results are correct to the significant digits presented, and that the incomplete-iteration error is much less than the discretization error [20–22].

In Table 2, the numerical predictions, κ_e^{N}, are shown together with the semi-analytical ones by Cheng and Torquato [6], κ_e^{CT}, and by Sangani and Acrivos [34], κ_e^{SA}, for three values of c, $c \in \{0.10, 0.30, 0.50\}$, $\kappa = 100$, and several values of Bi. Cheng and Torquato [6] derived an approximate equation for the effective conductivity of the simple cubic array of spheres as a function of c, κ, and a contact resistance parameter R, related to the Biot number by $R = (2\lambda/d)(\kappa/\mathrm{Bi})$. Sangani and Acrivos [34] also derived an approximate expression for the effective conductivity of the same array as a function of c and κ, but for composites with a perfect thermal contact between the phases (i.e., Bi $\rightarrow \infty$). It is observed from the analysis of Table 2, that the numerical predictions κ_e^{N} agree very well with κ_e^{CT} for $c = 0.10$ and $c = 0.30$, for all values of Bi. Furthermore, for these concentration values, as Bi is increased to larger and larger values, both κ_e^{N} and κ_e^{CT} approach κ_e^{SA}. Also, for $c = 0.10$ and $c = 0.30$, when $\mathrm{Bi} = 0$, both κ_e^{N} and κ_e^{CT} match κ_e^{SA} for $\kappa = 0$, i.e., for thermally-insulating spheres. This behavior is expected, because $\mathrm{Bi} = 0$ means an infinite thermal resistance at the surface of the spheres, which prohibits heat transfer between the matrix and the spheres.

On the other hand, when $c = 0.50$, the numerical predictions show that the analytical values of Cheng and Torquato [6] for $\mathrm{Bi} \in \{10^3, 10^5, 10^7\}$ significantly underestimate the effective conductivity; the maximum relative deviation encountered, $E_r \equiv |\kappa_e^{\mathrm{N}} - \kappa_e^{\mathrm{CT}}|/\kappa_e^{\mathrm{CT}}$, is approximately 15% for $c = 0.50$ and $\mathrm{Bi} = 10^7$. Also, for $\mathrm{Bi} = 10^7$, a large discrepancy is verified between κ_e^{N} and κ_e^{SA}. First, it is well known that the approximate expression of Cheng and Torquato [6] for the effective conductivity does not provide accurate results for large values of Bi in the moderate-to-high range of concentration values. Second, as previously pointed out in [21], the approximate expression of Sangani and Acrivos [34] is inaccurate for $c > 0.45$ and $\kappa > 10$. Therefore, while a fairly good agreement is verified between κ_e^{N} and κ_e^{SA} for $\kappa = 0$, the same is not observed for $\kappa = 100$. Finally, because analytical predictions are available for this array geometry, some of the data in Table 2 are plotted in Fig. 4 to further substantiate the validation of the numerical calculations.

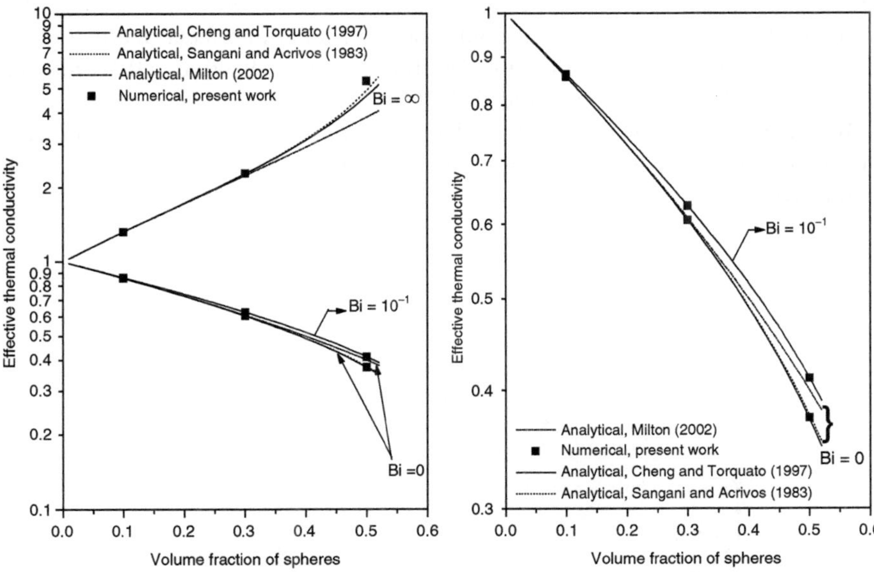

Fig. 4 Analytical and numerical values for the effective thermal conductivity as a function of the volume fraction of spheres; $\kappa = 100$ and $\mathrm{Bi} \in \{0, 10^{-1}, \infty\}$. The right plot is a magnification of the left plot for the $\mathrm{Bi} \in \{0, 10^{-1}\}$ curves

It is clearly seen from the plots that the present scheme leads to correct values for the effective thermal conductivity over the entire ranges of the volume fraction and Biot number.

4.1.2 Random Array of Spheres

The second set of results has been computed for random arrays of spheres without voids and with voids. The random array of spheres without voids comprises a cube of side λ containing in its interior three identical nonoverlapping whole spheres of diameter d', $d = O(d)$. On the other hand, the random array of spheres with voids comprises a cube of side λ containing in its interior six nonoverlapping whole spheres, three of diameter d' representing the particles, and three of diameter $d'/3$ representing the voids inside the matrix. The spatial positions of the centers of the spheres in both random arrays are sequentially and randomly chosen from a uniform distribution over the available (inner) space of the cube. The periodic cells for the random arrays of spheres without and with voids are illustrated in Fig. 5.

For both random arrays of spheres, the numerical results for the effective conductivity have been obtained for samples of ten different cell configurations each, $C \in \{1, \ldots, 10\}$. Each cell configuration corresponds to different positions of the centers of the spheres. Table 3 provides the numerical results, $\kappa_e^{\mathrm{N}}(C)$, for such arrays, together with the mean and standard deviation for each sample, respectively denoted by $\overline{\kappa_e^{\mathrm{N}}}$ and $S_{\kappa_e^{\mathrm{N}}}$. Table 3 also shows the results by Benveniste [5], κ_e^{B}, for random arrays of spheres without voids and with finite interfacial thermal resistance. The value of the concentration for all the arrays is fixed, $c = 0.15$, and the void content of the arrays with voids is 0.56%.

Based on a phenomenological model, Benveniste [5] derived the following analytical expression for the effective thermal conductivity, nondimensionalized with respect to k^c,

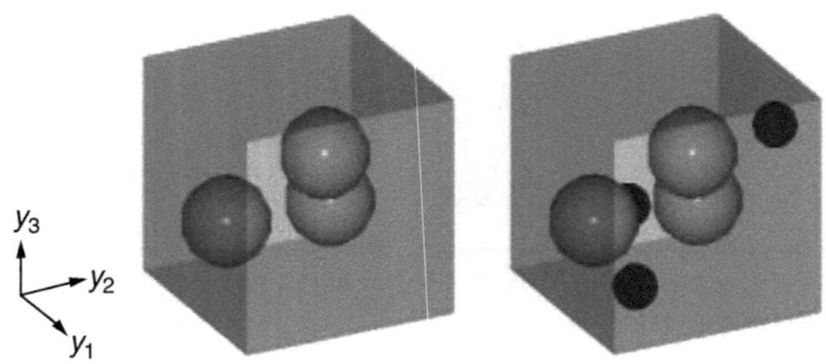

Fig. 5 Random arrays of spheres without and with voids (*smaller, darker spheres*) and associated Cartesian coordinate system

Table 3 Numerical results for the effective thermal conductivities of random arrays of spheres without voids and with 0.56% voids, for $c = 0.15$ and for various combinations of values of κ and R, $\kappa \in \{50, 10000\}$, $R \in \{30, 60, 5000, 20000\}$. Each sample is made up of ten different cell configurations, $C \in \{1, \ldots, 10\}$

Values of $\kappa_e^N(C)$ for $c = 0.15$				
$\kappa = 50,\ R_c = 49$				
	Without voids		With 0.56% voids	
C	$R = 30$	$R = 60$	$R = 30$	$R = 60$
1	1.0163	0.9253	1.0133	0.9217
2	1.0162	0.9242	1.0129	0.9213
3	1.0164	0.9269	1.0130	0.9238
4	1.0163	0.9255	1.0130	0.9225
5	1.0164	0.9265	1.0133	0.9234
6	1.0163	0.9260	1.0131	0.9227
7	1.0162	0.9240	1.0130	0.9205
8	1.0164	0.9263	1.0131	0.9223
9	1.0162	0.9228	1.0131	0.9194
10	1.0161	0.9211	1.0129	0.9180
$\overline{\kappa_e^N}$	1.0163	0.925	1.0131	0.922
$S_{\kappa_e^N}$	0.0001	0.002	0.0001	0.002
κ_e^B	1.0187	0.927	–	–

Values of $\kappa_e^N(C)$ for $c = 0.15$				
$\kappa = 10000,\ R_c = 9999$				
	Without voids		With 0.56% voids	
C	$R = 5000$	$R = 20000$	$R = 5000$	$R = 20000$
1	1.0497	0.8789	1.0469	0.8751
2	1.0492	0.8761	1.0457	0.8733
3	1.0505	0.8831	1.0470	0.8802
4	1.0498	0.8795	1.0465	0.8766
5	1.0503	0.8821	1.0472	0.8790
6	1.0500	0.8807	1.0469	0.8774
7	1.0492	0.8758	1.0460	0.8721
8	1.0502	0.8816	1.0473	0.8772
9	1.0487	0.8724	1.0457	0.8688
10	1.0480	0.8678	1.0448	0.8646
$\overline{\kappa_e^N}$	1.0496	0.878	1.0464	0.874
$S_{\kappa_e^N}$	0.0008	0.005	0.0008	0.005
κ_e^B	1.0522	0.880	–	–

$$\kappa_e^B = \frac{2(1-c) + \tilde{\beta}[1 + 2c + \frac{2-2c}{\kappa}]}{2 + c + \tilde{\beta}[1 - c + \frac{2+c}{\kappa}]}, \tag{52}$$

where $\tilde{\beta} \equiv (hd)/(2k^c)$. It should be emphasized that the parameter $\tilde{\beta}$ defined by Benveniste [5] is related to the contact resistance parameter R defined by Cheng and Torquato [6] and, consequently, to the Biot number Bi. In general terms, the results presented in Table 3 show the combined effects on the effective thermal conductivity due to varying microstructures, particle size, and the presence of voids inside the matrix; these effects are noticeable, despite the rather low values of both the particle and void volume fractions. The results in Table 3 also highlight the physical and geometrical flexibility of the previously described computational approach.

By comparison between the results reported on Table 3 and the ones for the simple cubic array of spheres with $c = 0.15$ one may verify that, for fixed values of κ and Bi (or, equivalently, R or $\tilde{\beta}$), the presence of three spheres inside the cube, as opposed to one in the simple cubic array, tends to reduce the composite effective conductivity. This tendency is expected, because, for a fixed value of c, as the number of spheres inside the cube increases the diameters of the spheres decrease, thus increasing their surface area to volume ratio, equal to $6/d$. Therefore, the contribution of the interfacial resistance towards reducing the composite effective conductivity increases. It should be emphasized that this particle size effect in composites with imperfect interface has been verified by previous researchers [11, 15]. As to the presence of voids inside the matrix, they clearly tend to reduce the effective conductivity, because they act as barriers to heat flow, not unlike the interfacial thermal resistance. Finally, it is interesting to note that, with our three-sphere random arrays of simple construction (leading to low values of $S_{\kappa_e^{\mathrm{N}}}$), the numerical results $\overline{\kappa_e^{\mathrm{N}}}$ agree rather well with the results κ_e^{B} by Benveniste [5], particularly for the higher value of the resistance parameter R for a given value of the conductivity ratio κ.

4.2 Short-Fiber Composites

Short-fiber composites are a special class of composite materials designed for applications that require strong directionality. Short-fiber composites possess a highly anisotropic thermal behavior even when both the matrix and the fibers behave as thermally isotropic materials. The fibers possess elongated shapes which are usually approximated by prolate or oblate ellipsoids of revolution, or by circular cylinders. Optical micrographies taken from fabricated short-fiber composites reveal that circular cylinders are indeed a better representation of the geometrical shape of an actual fiber [28].

Compared with particulate composites, there are fewer works in the literature that provide analytical expressions for the effective conductivity of short-fiber composites; the reader should consult the works by Nomura and Chou [29], Hatta and Taya [17], Furmañski [12], Dunn et al. [8], Duschlbauer et al. [9, 10], Mirmira and Fletcher [28] to name just a few. Although the expressions obtained for the effective conductivity account for relevant geometric parameters in short-fiber composites such as fiber distribution, orientation and aspect ratio, the majority of them neglects the interfacial thermal resistance. On the other hand, computational approaches based on finite-element schemes (see, for example, Duschlbauer et al.

[9], Matt and Cruz [24]), have the advantage of being flexible enough to account for the aforementioned geometric parameters as well as for an interfacial thermal resistance at the fibers' surface. Numerical results for the effective conductivity of ordered arrays of prolate ellipsoids of revolution (perfectly-aligned) and circular cylinders (perfectly-aligned and misoriented) are presented in the last sections of the chapter.

4.2.1 Ordered Array of Prolate Ellipsoids of Revolution

The third set of results, summarized in Table 4, has been computed for the ordered array of perfectly-aligned prolate ellipsoids of revolution with an uniform interfacial resistance. It is composed of a cubic matrix of side λ, at which geometric center lies one prolate ellipsoid, as illustrated in Fig. 3b. The semi-axes of the ellipsoid along the y_1, y_2 and y_3 directions have lengths ℓ, d and d, respectively, and the ellipsoid aspect ratio is defined by $\rho_f = \ell/d$. For a given concentration c, λ, d and ℓ are related by $c = (4\pi d^2 \ell)/(3\lambda^3)$. Because, by symmetry, $\kappa_{e,22} = \kappa_{e,33} \equiv \kappa_{e,\mathrm{T}}$, $\kappa_{e,11} \equiv \kappa_{e,\mathrm{L}}$, and $\kappa_{e,pq} = 0, p \neq q, p, q = 1, 2, 3$, the effective conductivity tensor for the ordered array of ellipsoids is completely described by the two scalars $\kappa_{e,\mathrm{L}}$ and $\kappa_{e,\mathrm{T}}$, respectively designated henceforth as the longitudinal and transverse effective conductivities.

Previous analytical expressions for the longitudinal and transverse conductivities were developed for random, rather than ordered, arrangements of perfectly-aligned ellipsoids without interfacial thermal resistance [8, 12, 17]. Thus, to check the accuracy of the present numerical predictions, the adopted procedure is to fix the concentration, and gradually decrease the ellipsoid aspect ratio from a value close to the maximum, $\rho_{f,max}$, to $\rho_f = 1$, when the ellipsoid becomes a sphere. The maximum value $\rho_{f,max}$ corresponds to the value of ρ_f for which the surface of the ellipsoid touches two opposite faces of the cube; e.g., for $c = 0.30$, $\rho_{f,max} \approx 1.32$.

Table 4 Numerical results for the longitudinal, $\kappa^{\mathrm{N}}_{e,\mathrm{L}}$, and transverse, $\kappa^{\mathrm{N}}_{e,\mathrm{T}}$, effective conductivities of the ordered array of prolate ellipsoids of revolution as functions of ρ_f, for $c = 0.30$, $\kappa = 100$, and Bi $\in \{10^7, 10^5, 10^3, 10^{-1}, 10^{-3}, 10^{-5}, 10^{-7}, 0\}$; the symbol κ^{CT}_e denotes the analytical prediction provided by Cheng and Torquato [6] for the simple cubic array of spheres

Bi	$\rho_f = 1.3$		$\rho_f = 1.2$		$\rho_f = 1.1$		$\rho_f = 1$		
	$\kappa^{\mathrm{N}}_{e,\mathrm{L}}$	$\kappa^{\mathrm{N}}_{e,\mathrm{T}}$	$\kappa^{\mathrm{N}}_{e,\mathrm{L}}$	$\kappa^{\mathrm{N}}_{e,\mathrm{T}}$	$\kappa^{\mathrm{N}}_{e,\mathrm{L}}$	$\kappa^{\mathrm{N}}_{e,\mathrm{T}}$	$\kappa^{\mathrm{N}}_{e,\mathrm{L}}$	$\kappa^{\mathrm{N}}_{e,\mathrm{T}}$	κ^{CT}_e
10^7	4.113	2.041	2.957	2.103	2.538	2.178	2.275	2.275	2.272
10^5	4.111	2.041	2.957	2.103	2.538	2.178	2.275	2.275	2.272
10^3	3.989	2.031	2.926	2.091	2.519	2.166	2.261	2.261	2.258
10^{-1}	0.6743	0.5952	0.6607	0.6067	0.6452	0.6169	0.6269	0.6269	0.6269
10^{-3}	0.6520	0.5738	0.639	0.5857	0.6239	0.596	0.6059	0.6059	0.6058
10^{-5}	0.6518	0.5736	0.6388	0.5854	0.6237	0.5958	0.6057	0.6057	0.6056
10^{-7}	0.6518	0.5736	0.6388	0.5854	0.6237	0.5958	0.6057	0.6057	0.6056
0	0.6518	0.5736	0.6388	0.5854	0.6237	0.5958	0.6057	0.6057	0.5056

In Table 4, the numerical predictions for the longitudinal, $\kappa^{\mathrm{N}}_{e,\mathrm{L}}$, and transverse, $\kappa^{\mathrm{N}}_{e,\mathrm{T}}$, conductivities are presented as functions of ρ_f, for $c = 0.30$ and $\kappa = 100$, and for several values of Bi. Also presented are the semi-analytical predictions of Cheng and Torquato [6], κ^{CT}_e. Two important behaviors are noted from the analysis of Table 4. First, as ρ_f is progressively decreased to 1, $\kappa^{\mathrm{N}}_{e,\mathrm{L}}$ approaches $\kappa^{\mathrm{N}}_{e,\mathrm{T}}$, as expected. Second, for the limiting case $\rho_f = 1$, both numerical predictions $\kappa^{\mathrm{N}}_{e,\mathrm{L}}$ and $\kappa^{\mathrm{N}}_{e,\mathrm{T}}$ agree very well with κ^{CT}_e for all values of Bi; hence, the accuracy of the numerical predictions for the ordered array of ellipsoids is ascertained.

4.2.2 Ordered Array of Circular Cylinders

The fourth set of results, summarized in Table 5, has been computed for the ordered array of circular cylinders with an uniform interfacial resistance. It is composed of a parallelepiped of sides H, λ, and λ along the y_1, y_2 and y_3 directions, respectively, and at which geometric center lies one circular cylindrical fiber of diameter d and length ℓ along the y_1-axis, as illustrated in Fig. 3c. The fiber and parallelepiped aspect ratios are defined, respectively, by $\rho_f = \ell/d$ and $\rho_p = H/\lambda$. For a given concentration c, λ, H, d and ℓ are related by $c = (\pi d^2 \ell)/(4\lambda^2 H)$. Again, the effective conductivity tensor is completely described by two scalars only: the longitudinal, $\kappa_{e,\mathrm{L}}$, and the transverse, $\kappa_{e,\mathrm{T}}$, effective conductivities. This

Table 5 Numerical results for the longitudinal, $\kappa^{\mathrm{N}}_{e,\mathrm{L}}$, and transverse, $\kappa^{\mathrm{N}}_{e,\mathrm{T}}$, effective conductivities of the ordered array of circular cylinders as functions of ρ_f, for $c = 0.30$, $\rho_p = 2$, $\kappa = 100$, and $\mathrm{Bi} \in \{10^7, 10^5, 10^3, 10^{-1}, 10^{-3}, 10^{-5}, 10^{-7}, 0\}$

Effective longitudinal conductivity $\kappa^{\mathrm{N}}_{e,\mathrm{L}}$, $c = 0.30$, $\rho_p = 2$, $\kappa = 100$								
Biot number Bi								
ρ_f	10^7	10^5	10^3	10^{-1}	10^{-3}	10^{-5}	10^{-7}	0
1.5	2.249	2.249	2.242	0.6422	0.5855	0.5849	0.5849	0.5849
2.0	2.964	2.964	2.951	0.6950	0.6308	0.6301	0.6301	0.6301
2.5	4.246	4.245	4.217	0.7343	0.6601	0.6594	0.6594	0.6594
3.0	8.165	8.164	8.046	7.727	0.6870	0.6862	0.6862	0.6862
$\rho_{f,max} \approx 3.24$	30.70	30.70	30.70	30.70	30.70	30.70	30.70	30.70
Effective longitudinal conductivity κ^{RM}_e (rule of mixtures), $c = 0.30$, $\kappa = 100$								
$\kappa^{\mathrm{RM}}_e = 30.70$								
Effective transverse conductivity $\kappa^{\mathrm{N}}_{e,\mathrm{T}}$, $c = 0.30$, $\rho_p = 2$, $\kappa = 100$								
Biot number Bi								
ρ_f	10^7	10^5	10^3	10^{-1}	10^{-3}	10^{-5}	10^{-7}	0
1.5	2.399	2.399	2.383	0.6279	0.6053	0.6051	0.6051	0.6051
2.0	2.145	2.145	2.133	0.6056	0.5834	0.5832	0.5832	0.5832
2.5	2.007	2.007	1.997	0.5881	0.5662	0.5659	0.5659	0.5659
3.0	1.892	1.892	1.884	0.5717	0.5495	0.5493	0.5493	0.5493
$\rho_{f,max} \approx 3.24$	1.836	1.836	1.827	0.5598	0.5380	0.5376	0.5376	0.5376
Effective transverse conductivity κ^{HJ}_e (square array), $c = 0.30$, $\kappa = 100$								
	1.833	1.833	1.826	0.5600	0.5387	0.5385	0.5385	0.5385

parallelepipedonal array has been studied in detail by Matt and Cruz [22] from the perspective of conductivity enhancement, when the Biot number is infinite.

An analytical expression for the effective conductivities of arrangements of finite-length circular cylinders is available for random distributions of the fibers only, and not accounting for an interfacial thermal resistance [12]. Thus, to check the accuracy of the present numerical predictions, the procedure is now to fix the concentration c and the parallelepiped aspect ratio ρ_p, while the fiber aspect ratio is gradually increased to the maximum value possible, $\rho_{f,max}$ [22]. In this limiting configuration, the two planar surfaces of the cylinder touch the parallelepiped faces normal to the y_1-axis, and the ordered array of circular cylinders becomes thermally equivalent to the square arrangement of infinite-length circular cylinders. The latter was previously studied by Hasselman and Johnson [15] and by Rocha and Cruz [32] far from the maximum-packing limit. Furthermore, for the square arrangement, the longitudinal conductivity is predicted by the rule of mixtures, $\kappa_e^{\mathrm{RM}} = 1 - c + \kappa c$, independently of the interfacial resistance.

In Table 5, the numerical predictions for the longitudinal, $\kappa_{e,\mathrm{L}}^{\mathrm{N}}$, and transverse, $\kappa_{e,\mathrm{T}}^{\mathrm{N}}$, conductivities are presented as functions of ρ_f, for $c = 0.30$, $\rho_p = 2$ and $\kappa = 100$, and for several values of Bi. Also presented are the analytical predictions derived from the rule of mixtures, κ_e^{RM} (exact), and those of Hasselman and Johnson [15], κ_e^{HJ} (approximate). Two important facts are obtained from the analysis of Table 5. First, as ρ_f approaches $\rho_{f,max}$, $\kappa_{e,\mathrm{T}}^{\mathrm{N}}$ approaches κ_e^{HJ}, as expected physically and confirmed numerically. Second, for the limiting case $\rho_f = \rho_{f,max}$, the numerical predictions $\kappa_{e,\mathrm{L}}^{\mathrm{N}}$ and $\kappa_{e,\mathrm{T}}^{\mathrm{N}}$ agree very well with κ_e^{RM} and κ_e^{HJ}, respectively, for all values of Bi. The abrupt increase in $\kappa_{e,\mathrm{L}}^{\mathrm{N}}$ when ρ_f changes from 3.0 to $\rho_{f,max} \approx 3.24$ is simple to explain: when $\rho_f = 3.0$, the global nodes belonging to the two planar surfaces of the fiber are the only ones that contribute to the interfacial thermal resistance to heat flow in the longitudinal direction; on the other hand, when $\rho_f = \rho_{f,max}$, the aforementioned global nodes now lie on the opposite faces of the parallelepiped normal to the longitudinal direction, where periodic boundary conditions are enforced, and the nodes no longer prohibit heat flow in the fiber in that direction. Therefore, $\kappa_{e,\mathrm{L}}^{\mathrm{N}}$ exactly matches κ_e^{RM} for $\rho_f = \rho_{f,max}$, even in the presence of a finite interfacial thermal resistance at the lateral surface of the fiber.

4.2.3 Random Array of Misoriented Circular Cylinders

To highlight the flexibility of the present computational scheme, numerical results are computed for the effective conductivity of a complex, random array of misoriented circular cylinders. The intention here is not to compare the numerical predictions with experimental measurements, but rather to show that the scheme can indeed be applied to realistic microstructures. The geometry and the associated finite element mesh for the random array considered here are shown in Fig. 6.

It is composed by a parallelepipedonal cell, containing in its interior eight misoriented circular cylinders and two empty spheres. The spheres are

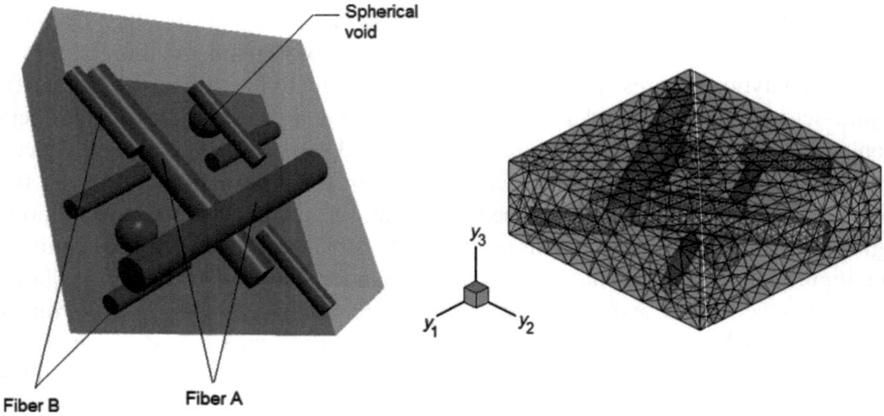

Fig. 6 Random array with misoriented fibers and voids: geometry and corresponding finite element mesh, with 50,603 global mesh nodes and 27,685 quadratic tetrahedra

Table 6 Characteristic dimensions, in μm, for the random array shown in Fig. 6

Array element	Amount	Characteristic dimension (μm)			
		Length	Height	Thickness	Diameter
Fiber A	2	800	–	–	100
Fiber B	6	300	–	–	60
Void	2	–	–	–	120
Parallelepiped	1	850	900	280	–

Table 7 Numerical results for the effective conductivities, $\kappa^{N}_{e,11}$, $\kappa^{N}_{e,22}$ and $\kappa^{N}_{e,33}$, of the random array shown in Fig. 6 as functions of the Biot number

$\kappa = 100$			
Bi	$\kappa^{N}_{e,11}$	$\kappa^{N}_{e,22}$	$\kappa^{N}_{e,33}$
10^{-5}	0.874	0.872	0.858
10^{-3}	0.874	0.873	0.858
10^{-1}	0.883	0.882	0.859
10^{0}	0.958	0.959	0.868
10^{1}	1.334	1.379	0.936
10^{3}	2.020	2.132	1.199
10^{5}	2.051	2.162	1.213

representative of voids, or microcracks, in the composite matrix, which may be induced during the manufacturing process due to a mismatch in the thermal expansion coefficients of the composite constituents [8]. There are two types of cylindrical fibers, denoted Fiber A and Fiber B, with aspect ratios equal to 8 and 5, respectively. Typical characteristic dimensions for the parallelepiped, the fibers and the spherical voids are shown in Table 6.

The random array shown in Fig. 6 may be thought of as a model microstructure for hybrid short-fiber composites, those with reinforcements of different types [8]. As pointed out in [8], hybrid short-fiber composites have become attractive materials for electronic packaging applications, because, due to their higher thermal conductivities, they may withstand very high heat fluxes while maintaining a low temperature operating environment. The higher thermal conductivities of such materials originate from the fact, that higher volume fractions can be obtained by combining different reinforcements of different characteristic sizes [36].

In Table 7, the numerical predictions for the effective thermal conductivities, $\kappa^{N}_{e,11}$, $\kappa^{N}_{e,22}$ and $\kappa^{N}_{e,33}$, are shown as functions of the Biot number, for $\kappa = 100$. The fiber volume fraction is $c = 0.082$, and the void content is 0.8%. The values of c, ρ_f, κ, and void content used in the computations are representative of typical data reported in the literature. The data in Table 7 indicate that the random array has a transversely isotropic behavior in the y_1–y_2 plane, and that the effect of the Biot number is more pronounced on the in-plane conductivities, $\kappa^{N}_{e,11}$ and $\kappa^{N}_{e,22}$, than on the out-of-plane conductivity, $\kappa^{N}_{e,33}$.

To sum up, the accuracy and flexibility of the computational scheme discussed in detail in this chapter has been demonstrated by effecting truly three-dimensional calculations of the effective thermal conductivity of composites with ordered and random (periodic) microstructures, accounting for the presence of a thermal resistance at the particle/fiber-matrix interface. The numerical predictions have been validated against analytical results for ordered and random arrays of spheres, and for the limiting cases when the ordered arrays of perfectly-aligned prolate ellipsoids of revolution and circular cylinders are thermally equivalent, respectively, to the simple cubic array of spheres and the square array of unidirectional fibers. The results show that the magnitude of the interfacial thermal resistance significantly affects the effective conductivity of composite materials. The computational approach discussed here is flexible, applicable to complex, realistic three-dimensional microstructural models, and it is also able to account for an interfacial thermal resistance (uniform or not) and for the presence of voids inside the matrix.

Up to date, reliable comparisons between analytical, phenomenological and numerical predictions with experimental measurements remain a non-trivial effort, due to the difficulty in ascertaining the composite microstructure, and to the lack of information regarding the magnitudes of various physical effects, such as the interfacial thermal resistance and void content in real composites. Hence, in order to improve the accuracy of numerical predictions, research efforts must be concerted in two important directions. First, more realistic geometric models for the composite microstructure must be developed; these geometric models must be flexible enough to include, for example, fibers with random orientations and distinct aspect ratios. Second, experimental techniques or prediction methods must be developed and/or improved, in order to obtain good estimates for the interfacial thermal resistance. Computational approaches shall be a helpful tool in the future for a better understanding of the thermal behavior of composite materials.

Acknowledgement M.E. Cruz would like to thank the Brazilian Council for Development of Science and Technology (CNPq) for Grants PQ-306592/2006-1 and APQ-471801/2004-6. The authors also thank Dr. Joachim Schöberl, from Johannes Kepler Universität Linz, Austria, for the free academic license of NETGEN 4.4.

References

1. Auriault, J.L., Ene, H.I.: Macroscopic modelling of heat transfer in composites with interfacial thermal barrier. Int. J. Heat Mass Transf. **37**, 2885–2892 (1994)
2. Batchelor, G.K., O'Brien, R.W.: Thermal or electrical conduction through a granular material. Proc. R. Soc. Lond. A **355**, 313–333 (1977)
3. Bathe, K.J.: Finite Element Procedures in Engineering Analysis, 1st edn. Prentice-Hall, Inc., Englewood Cliffs, New Jersey (1982). Chaps. 1, 3, 7
4. Bensoussan, A., Lions, J.L., Papanicolaou, G.C.: Asymptotic Analysis for Periodic Structures, 1st edn. North-Holland Publishing Co., Amsterdam (1978). Chaps. 1–2
5. Benveniste, Y.: Effective thermal conductivity of composites with a thermal contact resistance between the constituents: Nondilute case. J. Appl. Phys. **61**, 2840–2843 (1987)
6. Cheng, H., Torquato, S.: Effective conductivity of periodic arrays of spheres with interfacial resistance. Proc. R. Soc. Lond. A **453**, 145–161 (1997)
7. Cruz, M.E.: Computational approaches for heat conduction in composite materials. In: Esteve, Y.V., Carlomagno, G.M., Brebbia, C.A. (eds.) Computational Methods and Experimental Measurements X, pp. 657–668. WIT Press, Southampton, UK (2001)
8. Dunn, M.L., Taya, M., Hatta, H., Takei, T., Nakajima, Y.: Thermal conductivity of hybrid short fiber composites. J. Compos. Mater. **27**, 1493–1519 (1993)
9. Duschlbauer, D., Böhm, H.J., Pettermann, H.E.: Numerical simulation of thermal conductivity of MMCs: effect of thermal interface resistance. Mater. Sci. Technol. **19**, 1107–1114 (2003)
10. Duschlbauer, D., Pettermann, H.E., Böhm, H.J.: Heat conduction of a spheroidal inhomogeneity with imperfectly bonded interface. J. Appl. Phys. **94**(3), 1539–1549 (2003)
11. Every, A.G., Tzou, Y., Hasselman, D.P.H., Raj, R.: The effect of particle size on the thermal conductivity of ZnS/diamond composites. Acta Metallica Mater. **40**, 123–129 (1992)
12. Furmañski, P.: Influence of different parameters on the effective thermal conductivity of short-fiber composites. J. Compos. Mater. **4**, 349–362 (1991)
13. Furmañski, P.: Heat conduction in composites: homogenization and macroscopic behavior. Appl. Mech. Rev. **50**, 327–356 (1997)
14. Garnier, B., Dupuis, T., Gilles, J., Bardon, J.P., Danes, F.: Thermal contact resistance between matrix and particle in composite materials measured by a thermal microscopic method using a semi-intrinsic thermocouple. In: Proceedings of the 12th International Heat Transfer Conference, Grenoble, France, pp. 9–14 (2002)
15. Hasselman, D.P.H., Johnson, L.F.: Effective thermal conductivity of composites with interfacial thermal barrier resistance. J. Compos. Mater. **21**, 508–515 (1987)
16. Hasselman, D.P.H., Johnson, L.F., Syed, R., Taylor, M.P., Chyung, K.: Heat conduction characteristics of a carbon-fibre-reinforced lithia-alumino-silicate glass-ceramic. J. Mater. Sci. **22**, 701–709 (1987)
17. Hatta, H., Taya, M.: Equivalent inclusion method for steady state heat conduction in composites. Int. J. Eng. Sci. **24**, 1159–1172 (1986)
18. Jiajun, W., Su, Y.X.: Effects of interfacial thermal barrier resistance and particle shape and size on the thermal conductivity of AlN/PI composites. Compos. Sci. Technol. **64**, 1623–1628 (2004)
19. Kumar, S., Murthy, J.Y.: A numerical technique for computing effective thermal conductivity of fluid-particle mixtures. Numer. Heat Transf. B Fund. **47**, 555–572 (2005)

20. Matt, C.F., Cruz, M.E.: Calculation of the effective conductivity of ordered short-fiber composites. In: Proceedings of the 35th AIAA Thermophysics Conference, Anaheim, California, Paper AIAA 2001–2968 (2001)
21. Matt, C.F., Cruz, M.E.: Application of a multiscale finite-element approach to calculate the effective conductivity of particulate media. Comput. Appl. Math. **21**(2), 429–460 (2002)
22. Matt, C.F., Cruz, M.E.: Enhancement of the thermal conductivity of composites reinforced with anisotropic short fibers. J. Enhanced Heat Transf. **13**, 1–22 (2006)
23. Matt, C.F., Cruz, M.E.: Numerical prediction of the effective thermal conductivity of composite materials. In: Proceedings of the 9th AIAA/ASME Joint Thermophysics and Heat Transfer Conference, San Francisco, California, Paper AIAA 2006–54155 (2006)
24. Matt, C.F., Cruz, M.E.: Effective thermal conductivity of composite materials with 3-d microstructures and interfacial thermal resistance. Numer. Heat Transf. Appl. A **53**, 577–604 (2008)
25. McKenzie, D.R., McPhedran, R.C., Derrick, G.H.: The conductivity of lattices of spheres II. The body centred and face centred cubic lattices. Proc. R. Soc. Lond. A **362**, 211–232 (1978)
26. McPhedran, R.C., McKenzie, D.R.: The conductivity of lattices of spheres I. The simple cubic lattice. Proc. R. Soc. Lond. A **359**, 45–63 (1978)
27. Milton, G.W.: The Theory of Composites, 1st edn. Cambridge University Press, Cambridge (2002). Chaps. 2, 7
28. Mirmira, S.R., Fletcher, L.S.: Comparative study of thermal conductivity of graphite fiber organic matrix composites. In: Proceedings of the 5th ASME/JSME Joint Thermal Engineering Conference, San Diego, California, Paper AJTE99-6439 (1999)
29. Nomura, S., Chou, T.W.: Bounds of effective thermal conductivity of short-fiber composites. J. Compos. Mater. **14**, 120–129 (1980)
30. Paige, C.C., Saunders, M.A.: Solution of sparse indefinite systems of linear equations. SIAM J. Numer. Anal. **12**, 617–629 (1975)
31. Reddy, J.N., Gartling, D.K.: The Finite Element Method in Heat Transfer and Fluid Dynamics, 2nd edn. CRC Press LLC, Boca Raton (2001). Chaps. 1–3
32. Rocha, R.P.A., Cruz, M.E.: Computation of the effective conductivity of unidirectional fibrous composites with an interfacial thermal resistance. Numer. Heat Transf. A Appl. **39**, 179–203 (2001)
33. Rolfes, R., Hammerschmidt, U.: Transverse thermal conductivity of CFRP laminates: a numerical and experimental validation of approximation formulae. Compos. Sci. Technol. **54**, 45–54 (1995)
34. Sangani, A.S., Acrivos, A.: The effective conductivity of a periodic array of spheres. Proc. R. Soc. Lond. A **386**, 263–275 (1983)
35. Schöberl, J.: NETGEN – 4.4, User's Manual. Numerical and Symbolic Scientific Computing, Institute of Mathematics, Johannes Kepler Universität Linz, Austria (2001)
36. Takei, T., Hatta, H., Taya, M.: Thermal expansion behavior of particulate filled composites II: multi-reinforcing phases (hybrid composites). Mater. Sci. Eng. A **131**, 145–152 (1991)
37. Torquato, S.: Random Heterogeneous Materials, Microstructure and Macroscopic Properties, 1st edn. Springer, New York (2002). Chaps. 1, 2, 5–7
38. Yoshida, K., Morigami, H.: Thermal properties of diamond/copper composite material. Microelectron. Reliab. **44**, 303–308 (2004)

Part II
Porous and Cellular Materials (one or more solid phases and free volume)

Heat Transfer in Graphitic Foams

Anthony G. Straatman

Abstract Porous graphitic foam possesses unique characteristics that make it an interesting material for consideration in heat transfer applications. The present chapter summarizes work that has been done over the past several years to characterize graphitic foams that are permeable to fluid flow for eventual use in applications requiring convective heat transfer enhancement. The work covers: engineering models describing the structure of the foam, models of effective thermal conductivity, experimental work that quantifies enhancements in various flow orientations, and computational work that has been undertaken to explore the hydraulic and thermal behaviour of graphitic foam at both the pore-level and the macroscopic level.

1 Introduction

Porous graphitic foam (hereinafter referred to as GF) is a spherical void phase porous material that has unique thermo-physical and geometric characteristics [1] making it suitable for widespread use in heat transfer applications in microelectronics and power generation. The unique characteristics include:

1. A high effective thermal conductivity of between 40 and 180 W/m K [2, 3]. The high effective conductivity is a result of the extremely high conductivity of the graphitized carbon material (k = 800–1,900 W/m K). In comparison, similar porosity aluminum foams have effective conductivities of approximately 2–26 W/m K, which result from solid-phase conductivities of 140–237 W/m K (for aluminum alloys).

A.G. Straatman
Department of Mechanical and Materials Engineering, The University of Western Ontario, N6A 5B9, London, Canada
e-mail: astraatman@eng.uwo.ca

A. Öchsner and G.E. Murch (eds.), *Heat Transfer in Multi-Phase Materials,* Adv Struct Mater 2, DOI 10.1007/8611_2010_2,

2. An open, inter-connected void structure that permits fluid infiltration such that substantial increases in surface area are available (5,000–50,000 m^2/m^3) for enhanced convective heat exchange.
3. A low density (from 0.2 to 0.6 g/cm^3, depending upon porosity), which makes the material suitable for compact and lightweight applications. In comparison, aluminum foam has a density of 0.3–0.8 g/cm^3, depending upon porosity.
4. An increase in exposed or plan surface area and a rough open structure that promotes increased mixing at the external fluid interface.

The combination of high solid-phase conductivity and moderate porosity enables GF to entrain heat from adjacent surfaces and conduct it deep into the solid structure of the foam to be swept away by fluid that infiltrates the structure. The internal surface area serves to reduce the convective resistance thereby fostering excellent energy exchange at the pore level. It is this combination of high material conductivity, moderate porosity and high internal surface area that makes GF attractive as a heat transfer material for both single and multiphase applications.

This chapter describes recent efforts to characterize graphitic foam as a material for use in convective heat transfer applications. The chapter is organized such that geometric characterization is considered first, followed by experimental work and computational work. The final sections provide comparisons of graphitic foam to other highly-porous metal foams under forced convection conditions and provide concluding remarks for the chapter.

2 Geometric Characterization of Graphitic Foam

There are essentially two types of porous solids: those produced by packing or sintering solid particles or cylinders together and those produced by casting or foaming a material during solidification. In both cases, the final product consists of interspersed regions of solid and fluid (or void). Depending on the final structure, the void regions may be isolated from one another (trapped phase) or continuous, as in the case of an interconnected pore structure. Graphitic foam fabricated using the ORNL patented process [1] is a case of near-spherical interconnected pores, as shown in Fig. 1. The figure illustrates the open, interconnected structure of the foam and the near-homogeneous distribution of void size.

2.1 The Unit Cube Geometric Idealization

A geometric model of the foam structure is required to permit calculation of geometric parameters, thereby enabling the exploration of flow and heat transfer, and optimization of the foam structure for different applications. A geometric model is required regardless of whether it is pore-level or volume-averaged flow

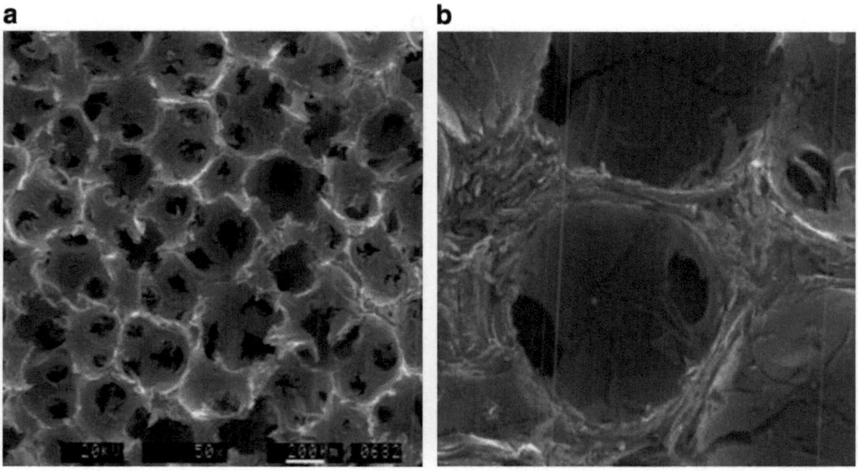

Fig. 1 (**a**) Scanning electron micrograph of the carbon foam surface [1]; (**b**) scanning electron micrograph of the carbon foam surface of a single pore

and heat transfer that is sought, or whether it is a numerical or experimental tool being used to study the foam. In any case, local calculation of the interface area and estimates of the pore window size and ligament dimensions is necessary. Based on the geometry characteristics of the GF under consideration [1–3], Yu et al. [4] used the following assumptions to formulate a geometric model:

1. The entire foam is assumed to have a single and uniform void (pore) diameter.
2. The pores are considered to be spherical and centred inside unit-cubes.
3. The pores are regularly arranged in space, and each pore connects with six adjacent pores on the six surfaces of the unit cube.

Figure 2a shows a three-dimensional CAD image of the unit-cube geometry proposed by Yu et al. [4] for a spherical void phase material. Figure 2b shows a pore block of unit-cubes with exposed pore surfaces that are cut at the center plane of the cube at the sides, front and top of the pore block. Figure 3 compares the internal geometry of the idealized geometry model with similar images of the GF obtained from ORNL.

The images illustrate that the idealized geometry model successfully captures the main features of the internal structure of the GF. Figure 4 shows the detailed dimensions of the unit-cube model proposed by Yu et al. [4]. Here, D is the pore (void phase) diameter; $H = f(D,\varepsilon)$ is the height of a unit-cube defined by the given pore diameter and the porosity; $h = (D - H)/2$ is the spherical cap height of the pore; $d = \sqrt{D^2 - H^2}$ is the interconnected pore window diameter; and $c = (H-d)/2$ is the width of the corner strut at the centre plane of the unit cube (also called the ligament width).

In this geometric model, the size of a unit-cube is not universally constant, rather it is determined by setting the desired porosity and pore diameter. By the definition

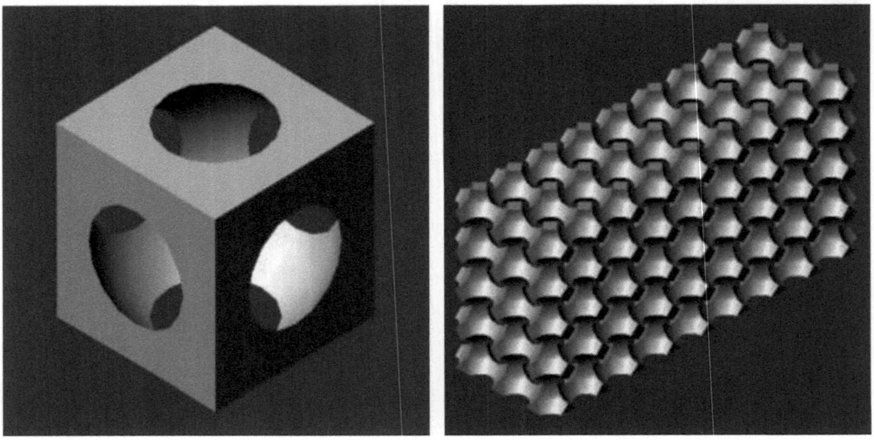

Fig. 2 CAD illustrations showing the unit-cube model developed in Yu et al. [4]: (**a**) a single unit-cube with spherical void; (**b**) a pore block containing a number of interconnected pores

of porosity (volume of void space/total volume), an expression relating the cube height H, the porosity ε and the pore diameter D was derived [4]:

$$H^3 - \frac{3\pi D^2}{(4\varepsilon + \pi)} H + \frac{4\pi D^3}{3(4\varepsilon + \pi)} = 0 \tag{1}$$

The dimension of the unit-cube H is thus obtained by solving (1) for given values of ε and D. Once the size of the unit-cube is established, the geometry is fixed and the remaining geometric parameters can be evaluated. Since we are mainly interested in using the foam for convective enhancement, the GF must have an interconnected pore structure, and thus we are only interested in the range of pore window sizes: $0 < d < H$ $(0.52 < \varepsilon < 0.96)$. When $d \leq 0$, the unit-spheres are isolated from one another and for $d \geq H$, $c \leq 0$ meaning that the ligaments connecting the solid phase are broken.

For a general application where fluid passes both across and through the porous material, information on the geometry is required for the internal structure and for the external interface between the porous material and the fluid. The internal structure is described in [4] in terms of an area to volume ratio. The external geometry is described in terms of the exposed surface area factor and the absolute roughness of the exposed surface. The external geometry parameters are only required for cases where fluid flows across the exposed surface, and the reader is referred to the original paper [4] for their description.

The interior surface area to volume ratio β quantifies the internal surface area available for convective heat transfer. For a solid unit-cube, $\varepsilon = 0$, which represents the lower limit. For porous graphitic foam made by the ORNL process [1], this value can be as high as 5,000–50,000 m^2/m^3, depending upon the porosity and void

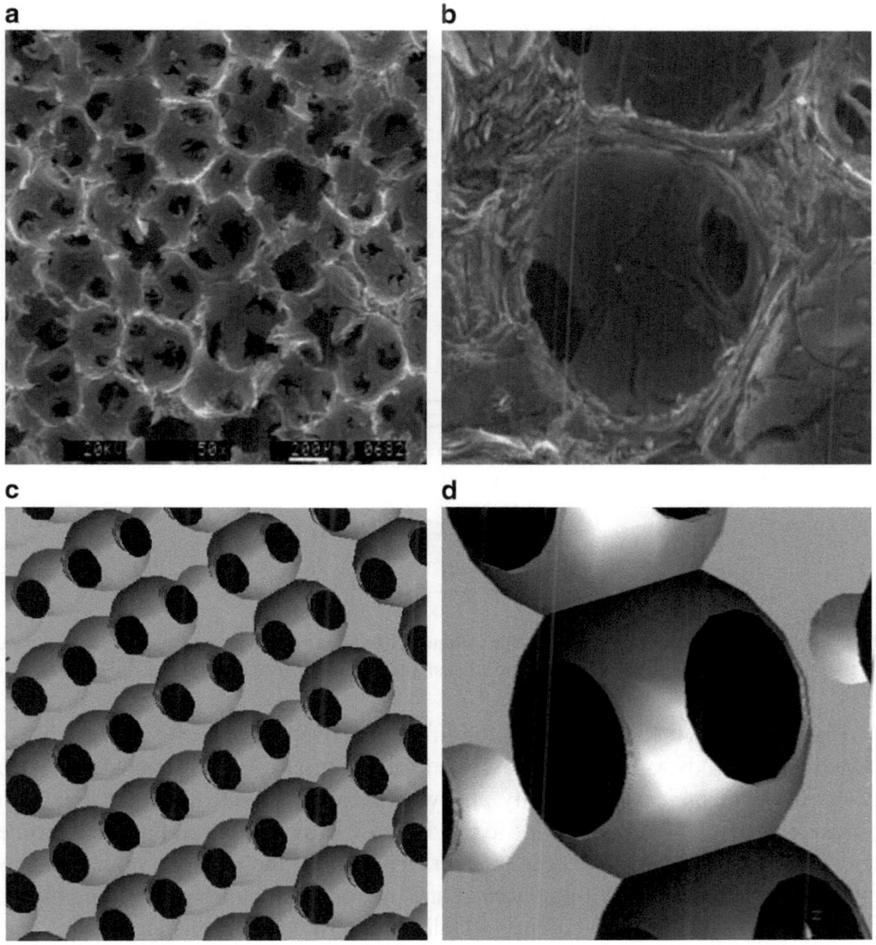

Fig. 3 A comparison of the idealized geometry proposed by Yu et al. [4] (**a** and **b**) with the structure of GF (**c** and **d**) obtained from ORNL [1]

diameter. The ratio β for the proposed unit-cube geometry is expressed as: $\beta = S_{\text{int}-wall}/H^3$, where $S_{int\text{-}wall}$ is the void interior surface area in a single unit-cube obtained from:

$$S_{\text{int}-wall} = \pi D^2 - 6(\pi D h) = \pi D(3H - 2D) \tag{2}$$

giving the final expression:

$$\beta = \frac{\pi D}{H^3}(3H - 2D) \tag{3}$$

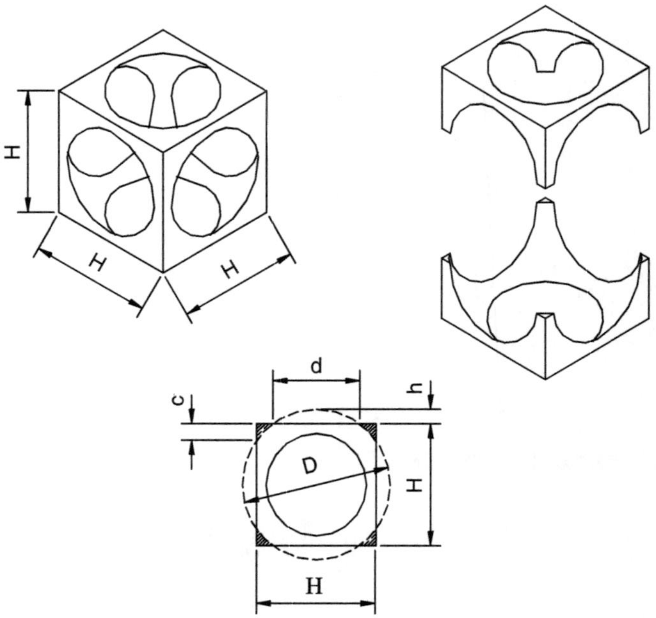

Fig. 4 Detailed dimensions of the Unit Cube Geometry Model at a cross-section cut at the centre plane of the unit cube. Used with permission from Yu et al. [4] and from the ASME Journal of Heat Transfer

Figure 5 shows the variation of β for different D over the range $0.52 < \varepsilon < 0.96$, which corresponds to the limits described following (1). Here it is evident that β decreases with increasing porosity and with increasing void diameter. For high porosities, β decreases sharply due to the decrease in available solid material inside the foam. For a convective heat transfer application, the optimal value for β is evaluated by comparison of the thermal and hydrodynamic resistances. The higher the value of β the higher the area available for internal heat exchange but this also results in higher net viscous losses resulting in a higher fluid pressure drop.

2.2 An Effective Conductivity Model

A complete thermal model consists of two parts: (a) an effective thermal conductivity model, which is required to quantify the rate at which energy can be entrained into the foam structure, and (b) a convective exchange model to quantify the rate at which energy can be removed by the infiltrated fluid. A model for the effective thermal conductivity of GF can be devised mathematically based on the unit cube geometric model, while the model for convective exchange will be given in a subsequent section related to experiments.

The effective or stagnant thermal conductivity of a porous material is a function of the thermal conductivities of the solid and the fluid, the porosity and the structure

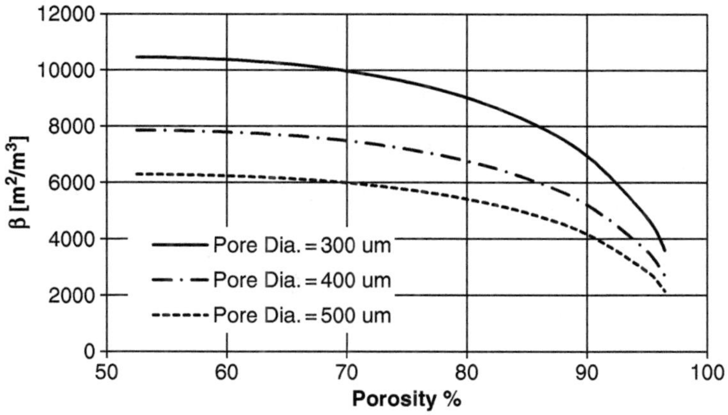

Fig. 5 Internal surface area to volume ratio, β, plotted as a function of porosity for three different spherical void (pore) diameters

of the foam, and is another important parameter characterizing the porous material. The equivalency technique has been applied by Yu et al. [4] to the solid phase volume since the solid phase is the controlling factor for the effective thermal conductivity of the GF material. The effective thermal conductivity derived by [4] is based on the following assumptions:

1. The air inside the void structure of the foam is stagnant. Thus, no convective exchange between the air and the solid takes place in the pore channel.
2. Radiation heat transfer in the inter-connected pore channel is negligible.
3. Local thermal equilibrium exists between the solid and fluid phases at the pore level.

An equivalency process was then used to simplify the pore-level geometry while preserving the volume ratio of solid/fluid (porosity), and the resulting square bar size of a simple unit cell can be determined by solving the following cubic equation [4]:

$$a^3 - \frac{3}{4}a^2 - \frac{1-\varepsilon}{16}H^3 = 0 \tag{4}$$

Figure 6 shows the details of the equivalency and electrical analogy processes used by Yu et al. [4] to obtain the effective thermal conductivity of the graphitic foam. The equivalent square bar structure is first divided into parallel and series parts as shown in Fig. 6b. The parallel and series parts are then converted into the simple forms shown in Fig. 6c by applying the equivalency method to the volume. The top view of Fig. 7d shows the equivalent heat transfer circuit that represents the parallel part, and the bottom view shows the equivalent heat transfer circuit that represents the series segment. The simplified parallel part is presented as a pure

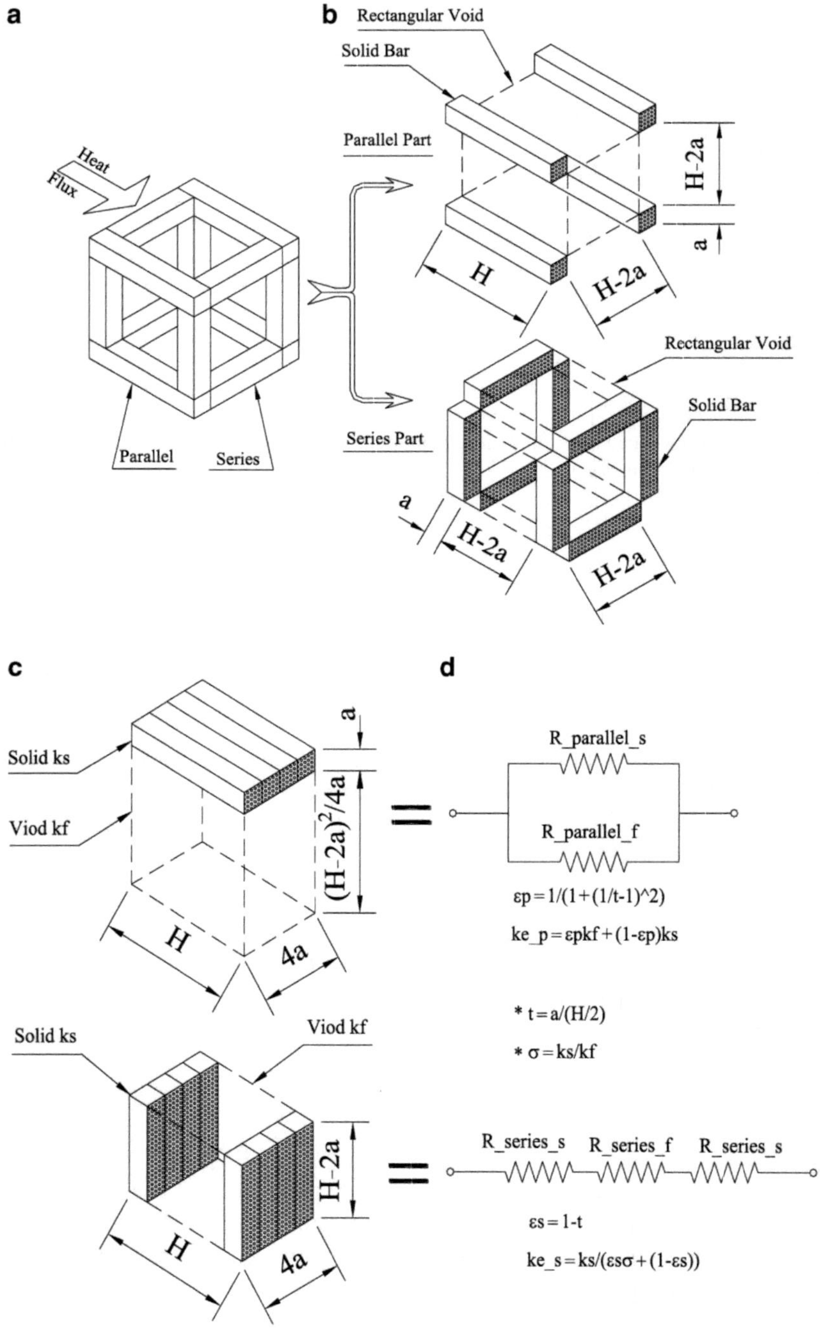

Fig. 6 (**a**) Square bar equivalent for thermal-electric analogy; (**b**) Equivalent parallel and series parts; (**c**) Simplified parallel and series parts; (**d**) Equivalent heat transfer circuits. Used with permission from Yu et al. [4] and from the ASME Journal of Heat Transfer

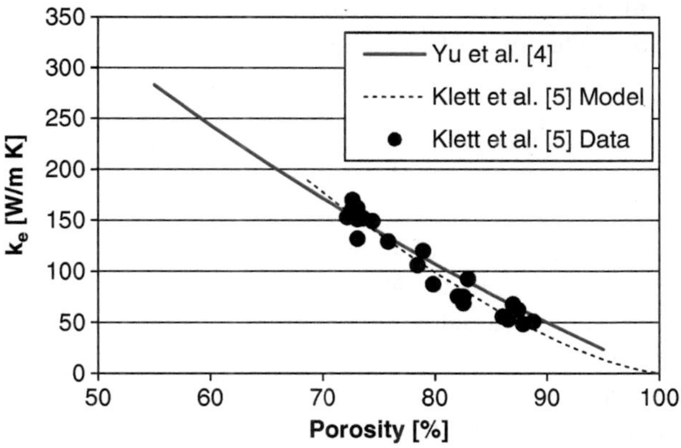

Fig. 7 Plot showing the effective thermal conductivity of GF as a function of porosity for $k_s = 1{,}300$ W/m K

parallel circuit consisting of the solid and void parts as shown in the top view of Fig. 6d, and its effective thermal conductivity k_{ep} is calculated as:

$$k_{ep} = \frac{\left(\frac{1}{t} - 1\right)^2 + \sigma}{\left(\frac{1}{t} - 1\right)^2 + 1} k_f \tag{5}$$

where k_f is the thermal conductivity of the fluid and k_s for the solid, $t = 2a/H$ is the normalized thickness of the square bar, and σ is the ratio of the thermal conductivity of the solid phase to the fluid phase: $\sigma = k_s/k_f$. The simplified series part is represented as a pure series circuit consisting of the solid and void parts as shown in the bottom view of Fig. 6d, and its effective thermal conductivity k_{es} is calculated as:

$$k_{es} = \frac{k_s}{(1 - t)\sigma + t} \tag{6}$$

The effective thermal conductivity of a porous material is given by:

$$k_e = \varepsilon_e k_{ep} + (1 - \varepsilon_e) k_{es} \tag{7}$$

where ε_e is the volume ratio of the parallel part to the sum of the series part and the parallel part, and is determined by:

$$\varepsilon_e = 1 - 2t + 2t^2 \tag{8}$$

When (5), (6) and (8) are substituted into (7), the final expression for the effective thermal conductivity of the spherical void porous material is obtained:

$$k_e = \left(1 - 2t - 2t^2\right)\left(\frac{\left(\frac{1}{t} - 1\right)^2 + \sigma}{\left(\frac{1}{t} - 1\right)^2 + 1}\right)k_f + \frac{2t(1 - t)}{(1 - t)\sigma + t}k_s \tag{9}$$

Figure 7 shows the variation of k_e with ε as predicted by (9) (solid line). The effective conductivity is seen to decrease with increasing porosity due to the reduction of solid phase material with increasing ε. Included in Fig. 7 is measured data reported by Klett et al. [5]; the data was converted from density to porosity form assuming a ligament density of 2.23 g/cm^3 [5]. Figure 7 also compares predictions of the model relation developed by Klett et al. [5]. It is clear from the figure that both models are in good agreement with the measured data. Important to note is that in the Yu et al. [4] model there are no parameters to adjust; the predicted result is obtained using the geometry of the foam and the solid and fluid phase thermal conductivities only. As such, the proposed model can be applied in engineering heat transfer models and in computational fluid dynamics codes without special tuning.

It is important to note that the heat transfer is not dictated by the effective conductivity alone. That is, for low porosity, the effective conductivity is high which means that heat is readily transferred into the porous material, but it is difficult for fluid to penetrate the foam resulting in lower convection and an imbalance in the conductive-convective resistances. For high porosity, the effective conductivity is low meaning that conduction into the foam is low, but it is easy for fluid to penetrate the foam so the convective resistance is lower. As such, the optimal porosity must be obtained by considering the rate at which heat is transferred into the foam and the rate at which heat can be removed by internal convection. Also note that this optimum is not universal, rather it is very application dependent. An analogy can be drawn here to fluid machinery, where optimal performance occurs at the design point and performance drops off for higher and lower flows.

3 Experimental Studies

Several studies exist in the literature describing the characterization of foams that could be considered for convective heat transfer enhancement. Antohe et al. [6], Paek et al. [7] and Boomsma and Poulikakos [8] all report on the hydraulic losses of as-cast and compressed aluminum foams and provide information quantifying the permeability and form drag coefficients for foams of different porosity. In this manner, pressure losses can be characterized using the classical Darcy-Forchheimer law. Calmidi and Mahajan [9] studied forced convection of air in high porosity aluminum foams using experiments combined with computational fluid dynamics. In their experiments, large (196 mm × 63 mm × 45 mm-thick) aluminum foam blocks of different porosity were mounted into a channel and heated from one side while air was forced through the foam. Since the effective conductivity of the aluminum foam was relatively low (maximum 7.4 W/m K), the extended-surface efficiency of the blocks was very low and convective heat transfer was only found to occur in a thin layer of foam directly adjacent to the heated substrate. Boomsma et al. [10] studied the effect of foam-compression on the flow and heat transfer characteristics of aluminum foams. Compression of the foam essentially increases the internal area to volume ratio of the foam and was done as an attempt to reduce the internal convective resistance and thereby reduce the volume of foam required to remove a given amount of energy. In this case, 40 mm $\times$ 40 mm $\times$ 2 mm-thick blocks of foam were heated from one side while water was forced through the foam structure. Compression of the foam was shown to increase the heat transfer effectiveness by up to a factor of two, however with an associated increase in the pressure drop.

Relatively little information is available in the open literature on the characterization of graphitic foams. Gallego and Klett [2] provided some of the first data on the pressure drop and heat transfer for graphitic foam. Their study provided estimates of the influence of configuration on the heat transfer and pressure drop, but little detail was provided to assess the influence of porosity, pore diameter and Flow. Yu et al. [4] proposed a sphere-centered unit-cube geometry model to characterize the internal structure of a spherical void-phase porous material, as described in the previous section. The model developed by Yu et al. [4] has been utilized in several subsequent studies carried out to quantify the convective heat transfer and pressure drop obtained in different flow configurations. The following sections present recent experimental data obtained for graphitic foam in parallel, impinging and forced flow arrangements.

3.1 Experiments in Parallel Flow

Experiments characterizing the heat transfer enhancement achieved by passing air across a surface of graphitic foam were carried out by Straatman et al. [11]. The study sought to establish the convective enhancement achieved by bonding layers of graphitic foam to a heated aluminum substrate. All results of the study are expressed as a ratio of the heat transfer for the foam surface with respect to the heat transfer measured for the bare (impermeable) substrate, i.e.,

$$E = \frac{\overline{Nu_{foam}}}{\overline{Nu}} = \frac{\overline{Nu_{L,foam}}}{\overline{Nu_L}} \tag{10}$$

Figure 8 shows results of enhancement as a function of foam thickness for foams 219 Top and 217 Top (see Table 1 for foam properties). The observations of heat transfer with respect to foam thickness suggest two things: that the depth of penetration of air into the foam is relatively small for parallel flow conditions, and that on average, there is no decisive advantage for using more than 3 mm of

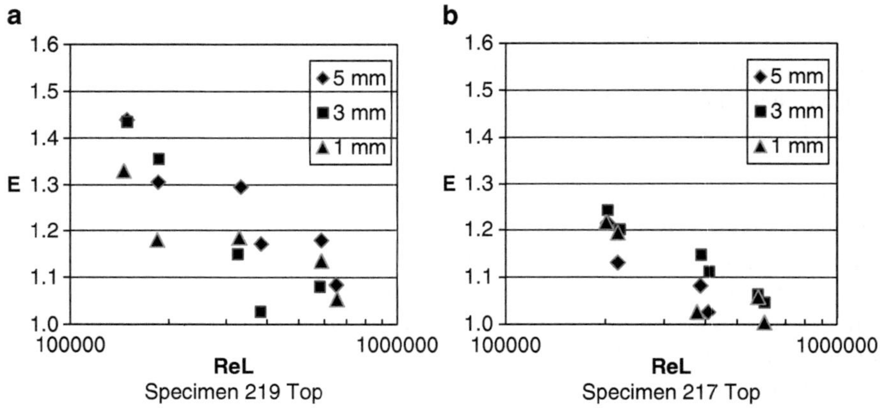

Fig. 8 Results showing the effect of foam thickness and Reynolds numbers for graphitic foams of: (**a**) 219-Top (89% porosity), and (**b**) 217-Top (78% porosity)

Table 1 Summary of properties for the carbon foam specimens tested in parallel, impinging and forced convection experiments

Specimen	Porosity (%)	Average void dia. (μm)	Highest frequency void dia.	k_{eff} (W/m K)
219 Top	89	633	800	30
217 Top	78	341	350	68
221-1	85	410	400	98
219-1	90	560	450–500	50
219-2	88	400	350–450	65
219-3	86	350	350	72
219-3-3	88	400	350–450	61
218-3	88	400	400	61
POCO™	82	500	510	120

foam. This means that the depth of penetration of air into the foam is as little as 3–5 pore-diameters (assuming that the first 1 mm of foam is filled with bonding material). Though the penetration depth of air is certainly dependent upon the pore diameter and porosity of the foam, it is difficult to resolve this influence due the non-uniformity of pore diameter and the difficulty of machining the foam thickness to within small fractions of a millimeter. Thus 3 mm serves as a first approximation for the desired thickness of GF for parallel flow arrangements. It is important to note that the depth of penetration of air into the foam is expected to be a strong function of the incidence of the foam surface with respect to the air flow, i.e. in the limit of an impinging airflow, the air would penetrate the foam surface much more deeply leading to much higher enhancements in convective heat transfer.

Concerning the dependence on Re_L, the enhancements shown in Fig. 8 are seen to be higher for low air speeds (about 1.28 on average) and lower for high air speeds (about 1.10 on average), with an approximately monotonic variation. This trend was explained in [11] in terms of the near-surface activity and the relative air flows in and across the GF. In a parallel flow, the air is not *forced* or *driven* into the foam, but rather the roughness of the foam surface produces disturbances of the sub-layer resulting in the production of near-surface eddies. The turbulent eddies actively penetrate the foam setting up weak pressure gradients near the foam surface, thereby causing air to pass through the interconnected pore structure before returning to the free stream. The air that passes through the foam is exposed to the large internal surface area thereby enhancing the net convective heat transfer of the specimen. At low air speeds, the momentum of the near-surface eddies is low, but the relative amount of air passing through the foam is significant with respect to the air flow across the exposed surface of the foam. At high air speeds, the near-surface eddies are very energetic, but the amount of air passing through the foam is small with respect to the external flow. As such, the enhancement of convective heat transfer is higher at low air speeds and lower at high air speeds. It is likely that at very high air speeds, the enhancement in heat transfer performance would be due only to the increased roughness and exposed surface area. It is also clear from Fig. 8 that the GF properties have an effect on the convective enhancement achieved. In this case, the more porous foam with the larger void diameter (219) provided increases that were nearly 20% higher than the lower porosity and smaller void diameter 221 foam. In this flow arrangement where the infiltration is passive, it is not surprising that a higher advantage is obtained from the more permeable specimen.

3.2 Experiments in Impinging Flow

As mentioned in the previous section, the effect of incidence angle on the air flow across a foam surface is of interest. A study of direct impingement has been carried out by Sultan et al. [12] to explore the heat transfer enhancement achieved by bonding layers of GF to a heated aluminum substrate. While only direct (normal) impingment was studied, enhancements were explored for both embedded and protruding layers of graphitic foam. Figure 9 compares the heat transfer enhancement obtained with the upper surface of the foam sample flush to the surface of the mounting apparatus (flat), and for the whole foam layer protruding above the plywood sheet (extended), both for unconfined impinging flow directed normal to the foam surface. The figure shows that the enhancement achieved by the extended foam layers is higher for all foam thicknesses, suggesting that the air infiltration into the foam is higher when the edges of the foam are open and a path exists for infiltrated air to pass freely out of the foam. The figure also shows that for the extended foam layers the convective enhancement is higher for thicker layers of foam indicating that the infiltration of air into the structure of the GF is substantial. Enhancements of 30–40% over that of the bare aluminum substrate were measured for extended GF layers of 2–6 mm, respectively. This is 4–12% higher than similar enhancements achieved by setting the foam surface to be flush with the impingement

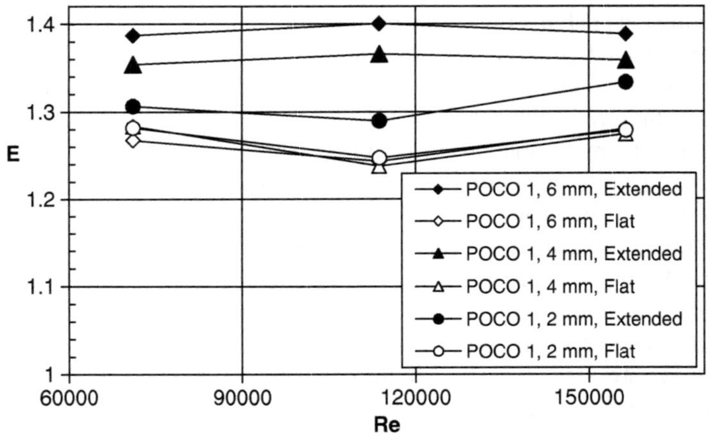

Fig. 9 Plot showing the impact of foam protrusion from the impingement surface on the convective heat transfer enhancement. See Table 1 for graphitic foam properties

surface. Figure 9 also shows that the convective enhancement for the flat cases are essentially independent of foam thickness, suggesting that very little air infiltration occurs when there is no escape route for the infiltrated air. In this case, the heat transfer enhancement is simply due to the larger surface area available at the exposed foam surface.

Figure 10 shows the influence of foam thickness and impingement air speed on convective enhancement for two of the graphitic foam specimens shown in Table 1. The plots indicate that enhancements of 30–70% are obtained over the range of air speed considered by varying the thickness of the different foam specimens. In terms of impingement air speed, the enhancement is higher at the highest velocity than the lowest, but for a significant number of the cases shown in Fig. 10, the lowest enhancement was observed at the intermediate air speed. Sultan et al. [12] explained this effect on the basis that the net convective enhancement is due to complementary influences resulting from the increased plan surface area of the foam over the bare substrate, and the exposure to internal surface area for infiltrated air.

While the plan surface area is constant for a given foam, the exposure to internal area depends upon the flow condition. Flow through the foam is driven by the pressure gradient set up by the (near) stagnation pressure at the surface and the pressure at the edge of the specimen. Since the influence of the enhanced plan area with *Re* is effectively linear (over the range of *Re* considered in [12]), the influence of flow through the foam must be larger at the low air speed. That is, if the trend for the higher air speeds is considered to be increasing linearly due to a continuous increase in the stagnation pressure, then the enhancement at the lowest air speed is higher (for all but 1 condition) than would be predicted by extrapolating backwards from the higher air speeds. This simply implies that the infiltration has a stronger influence on the net

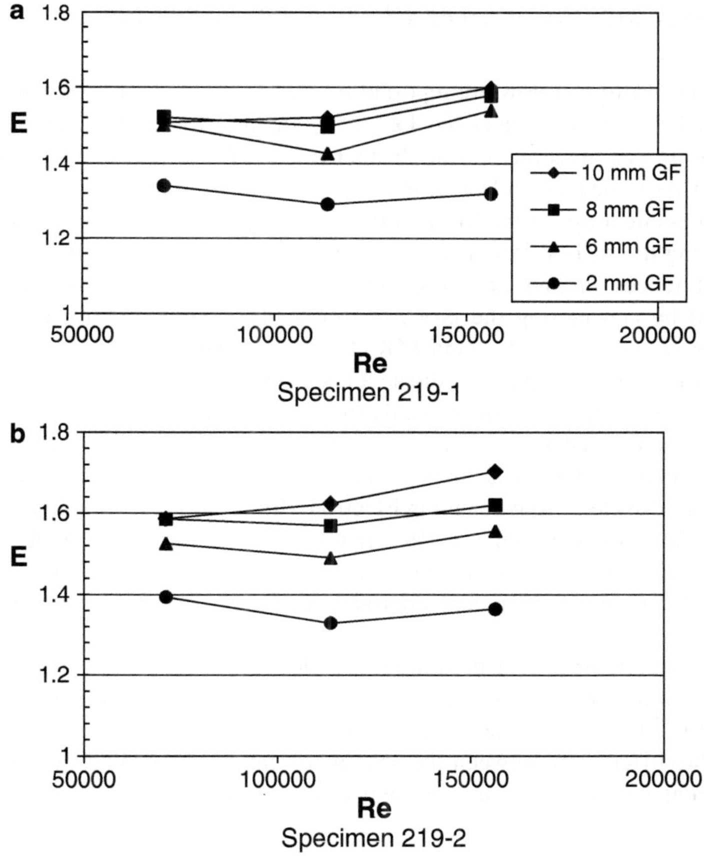

Fig. 10 Plots showing the influence of foam thickness on convective enhancement. See Table 1 for graphitic foam properties

convective enhancement at lower flow speeds. A similar phenomena was noted in [11] for the parallel flow condition, where it was argued that the relative amounts of parallel flow verses infiltrated flow affected the trend in the convective enhancement. The difference in the impingement case is that increases in the air speed eventually cause the enhancement to grow due to the higher and higher infiltration of air through the foam. In terms of the foam thickness, the trend for all specimens is that the highest convective enhancement occurs for the thickest foam layers, suggesting that air infiltrates the GF deep beneath the exposed surface. It is also observed that the largest increases occur between 2 and 6–8 mm; in most cases the difference between 8 and 10 mm of GF is relatively small, suggesting that the infiltration limit may be reached for the air speeds considered. Thicker specimens were not tested due to the limitation of the foam specimens available.

3.3 Experiments in Forced Flow: The Porous Plug

Experiments on forced flow are reported by Straatman et al. [13], in which water was passed through a channel blocked by a graphitic foam plug. Results for the pressure drop as a function of clear channel velocity U are shown in Fig. 11 for GF specimens described in Table 1. It is clear from Fig. 11 that the permeability and thus the pressure drop is very different for the four specimens tested, with the 219-3 foam having the highest fluid pressure drop and POCO™ having the lowest. The large differences in pressure drop between the foam specimens can only be reconciled by considering the properties in Table 1. Consider, for example the difference between the 219-3 and POCO™ foams. The porosity of the 219-3 is higher, which might otherwise indicate a lower pressure drop, but the void diameter of 219-3 is smaller and consequently the internal area is larger thereby increasing the internal resistance to fluid flow. However, the 20% increase in internal surface area does not alone account for the more-than twofold increase in pressure drop. It was also clear from Scanning Electron Microscope (SEM) images (given in [13]) that the cell windows in the 219-3 foam are, on average, considerably smaller than those in the POCO™ foam, and thus there is a much higher hydraulic loss in the 219-3 foam due to the sudden contraction/expansion of fluid through the pore windows.

The Nusselt number measured in [13] is plotted in Fig. 12 as a function of the Reynolds number, Re_{De}. The figure indicates that for most conditions the 219-3-3 and POCO™ foams have the highest internal heat transfer followed by 219-3 and 218-3. The differences between the specimens can be reconciled by considering the different geometric parameters and effective conductivities summarized in Table 1.

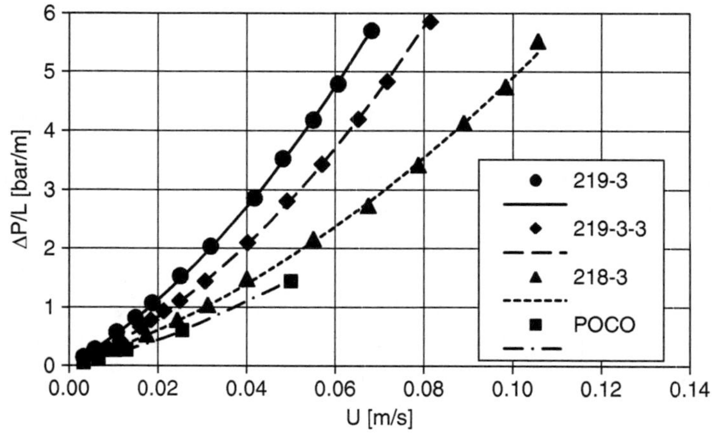

Fig. 11 Plot showing the pressure drop as a function of the filter velocity for the GF specimens considered. Adapted from Straatman et al. [13]

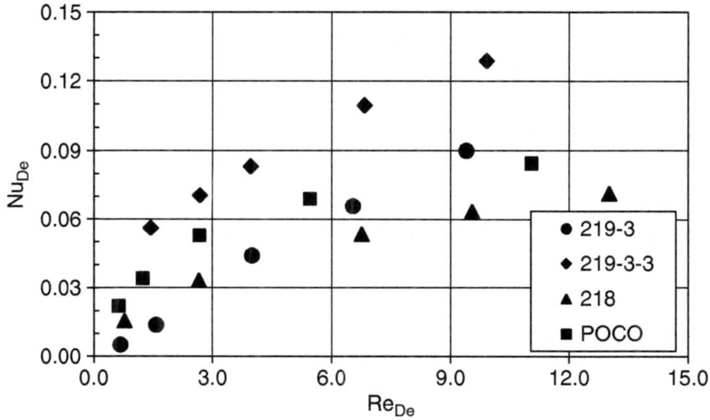

Fig. 12 Plot showing the Nusselt number as a function of Reynolds number based on D_e and A_{eff} for the four GF specimens tested experimentally. Adapted from Straatman et al. [13]

4 Computational Studies of Graphitic Foam

Computations of the flow and heat transfer in porous media are of great interest due to the advantage of being able to conduct parametric studies in a fast and efficient manner in comparison to similar activity done experimentally. In this section, two levels of computational study are presented: simulations of the pore-level fluid flow and energy transfer in GF, and simulations conducted using a volume-averaging approach. While the volume-averaging approach has seen much more widespread use in the past, computations of this nature require closure models to approximate the unresolved activity at the pore level, and such models can only be dervied by experimentation and better-resolved solutions from the pore-level.

4.1 Computations at the Pore Level

Pore level computations are done by simply solving the Navier-Stokes equations combined with the conservations of mass and energy on a domain that represents the pore structure of the porous media in question. The conservation equations for mass and momentum are given, respectively as:

$$\rho_f(\nabla \cdot u) = 0 \tag{11}$$

$$\rho_f \left[\frac{\partial u}{\partial t} + \nabla \cdot (uu) \right] = -\nabla P + \mu_f \nabla^2 u + \rho_f f \tag{12}$$

Under the additional assumption that the effects of viscous dissipation and heat generation may be neglected, the heat transfer portion of the problem is governed by the following form of the energy equation:

$$\rho_f c_{p,f}\left[\frac{\partial T}{\partial t} + \nabla \cdot (uT)\right] = k_f \nabla^2 T \tag{13}$$

These equations were solved in the idealized spherical foam geometric model furnished by Yu et al. [4] by Karimian and Straatman [14, 15]. Computations were carried out for unidirectional flow [14] and for arbitrary flow direction [14] for a small section of pores considered to be inside a generic section of an idealized GF foam block, which is ideally composed of hundreds of identical cells in each direction. It is reasonable to assume that the flow in a generic section inside the foam is periodic in nature and thus, only a representative portion of the domain required modeling. In the unidirectional case, the pore-windows were assumed to be aligned with the inlet flow leading to a flow field that loosely mimicked that of flow through a duct with geometric periodicity. To simulate the most generic multi-directional, periodic flow [15], a domain consisting of a cluster of idealized cells was (mathematically) constructed such that a spatially periodic flow could be simulated for an incoming velocity of 45° in all principle directions with respect to the pore-window planes. It was felt that 45° represents the average orientation of the windows with respect to the incoming flow. The resulting domain was comprised of two consecutive, interconnected cells in each direction x, y and z, with three pore windows as periodic sections in each direction. To complete the formulation, the periodic boundary conditions of Karimian and Straatman [16] based on the double-periodic cell geometry formulation of El Soukkary and Straatman [17] were extended to three directions. Figure 13 shows an outline of the computational domain for an idealized graphitic foam cell with porosity of $\varepsilon = 0.80$. The flow and temperature fields shown in Figs. 14 and 15 show the periodicity of the flow and temperature fields predicted using the formulation described. Symmetric flow and temperature fields can be clearly seen in Figs. 14 and 15 where streamlines are illustrated, coloured by the temperature variation.

More quantitatively, the results of the simulations are formulated in terms of the Reynolds number, Re_H, the normalized pressure drop, $_{\mathrm{H}}$, and pore-level Nusselt number, Nu_H, which are defined as:

$$Re_H = \frac{\rho V_{int} H}{\mu}, \Pi_H = \frac{\Delta P}{H} \times \frac{H^2}{\varepsilon \mu V_{int}}, Nu_H = \frac{hH}{k}$$

where V_{int} is the intrinsic velocity magnitude, H is the side length of the unit cube, ΔP is the pressure drop across a single pore, ε is the porosity of the foam, h is the average convective heat transfer coefficient, and k is the fluid conductivity.

Two major regimes were defined for steady laminar flow in porous media: Darcy or creeping flow, which is dominated by viscous forces, and inertial flow, in which

Fig. 13 Computational domain and grid for simulations done at the pore-level (Karimian and Straatman [15])

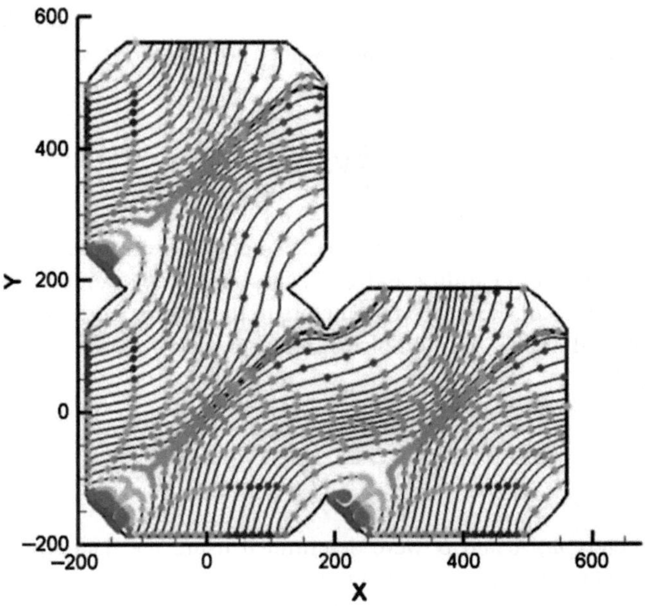

Fig. 14 Two-dimensional streamlines at a cross-section in the main-flow direction, the lines color variation indicates the local temperature

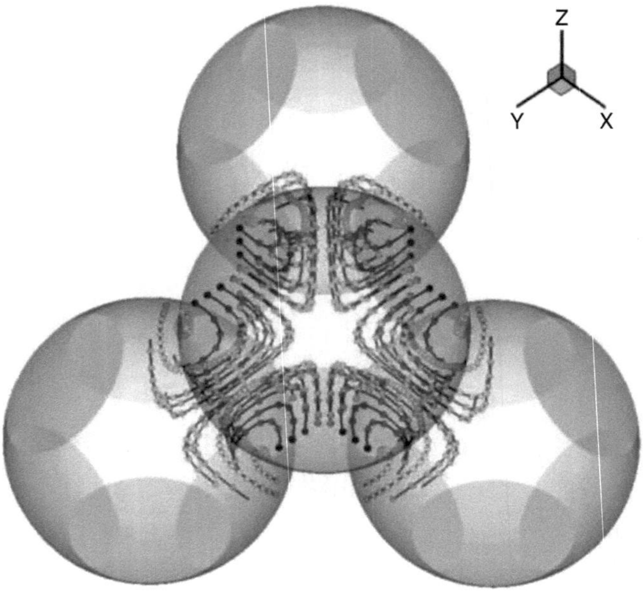

Fig. 15 Three-dimensional rakes of streamlines starting from inlets of the first cell until the flow exits the domain

the effect of inertia forces start to dominate. The Darcy-Forchheimer law governs the relation between pressure drop and velocity in terms of viscous and inertia forces:

$$-\frac{dP}{ds} = \frac{\mu}{\kappa} V_{ext} + \frac{Cf}{\sqrt{\kappa}} \rho V_{ext}^2$$

Here, V_{ext} is the extrinsic (clear channel) velocity, s is the axis parallel to flow direction, κ is the permeability and C_f is the inertia coefficient. To define the form of dependency of these two parameters to the internal structure of graphitic foams, an Ergun-like [8], semi-empirical relation was proposed by Karimian and Straatman [15]:

$$-\frac{dP}{dx}\left(\frac{\varepsilon^3}{1-\varepsilon}\right) = \frac{A(1-\varepsilon)^m}{d^2} \mu V_{ext} + \frac{B}{d} \rho V_{ext}^2$$

where parameters m, A and B were determined using the data from a series of numerical simulations across a wide range of Reynolds numbers in the stationary laminar regime. Also d, a proper length scale, was determined as a function of equivalent particle diameter (Yu et al. [4]).

To model heat transfer in spherical void phase foams, a general form of Nusselt number proposed by Kaviany [8] is used:

$$Nu = D + EPr^nRe^m$$

where D and E are functions of internal geometry. Note that the constant D is the asymptote of the model in the stagnant condition where conduction is the only means of heat transfer. D, E, n and m are once again determined by fitting the data from the numerical simulations for a range of Reynolds numbers in the stationary laminar regime.

Figures 16 and 17 show that the form of the models given in the above expressions is correct. In both cases, the model shows an excellent agreement with the experimental data over the range shown. However, there was still a considerable difference between the simulations and the experimental results for the porous plug (see Sect. 3.3), two coefficients were introduced by Karimian and Straatman [15] to reconcile the difference and to enable the use of the model expressions in the characterization of experimental data. The model was "calibrated" using an accessibility factor and a blockage ratio. Introduction of these coefficients (see [15] for details) resulted in a very good fit between the calibrated model and the experimental data for several different graphitic foams, as seen in Fig. 18.

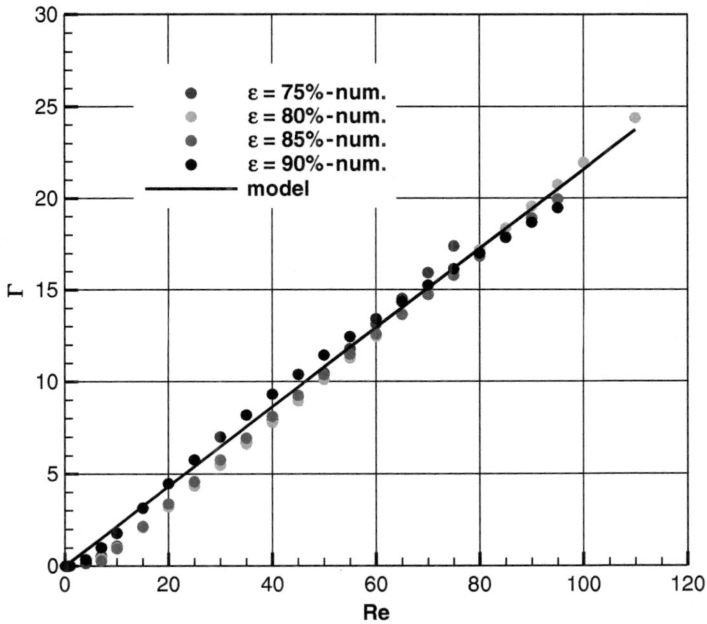

Fig. 16 Comparison of the normalized pressure drop as a function of Reynolds number. *Symbols* represent experimental results while the *lines* correspond to the model of Karimian and Straatman [15]

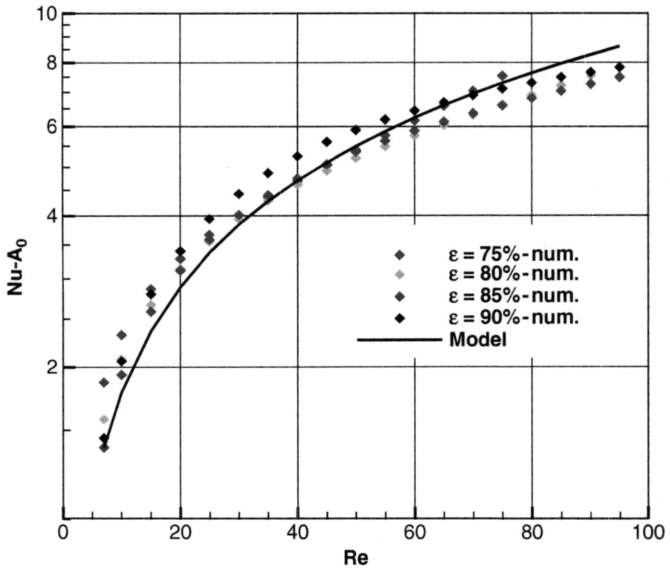

Fig. 17 Comparison of the heat transfer model with available experimental data. *Symbols* represent experimental results while the *lines* correspond to the model of Karimian and Straatman [15]

4.2 Computations at the Volume-Averaged Level

Comparisons of the thermal and hydraulic performance of porous metals is most efficiently done in a computational framework by solving the volume-averaged form of the governing transport equations. While much more information on the details of the flow and energy transfer is provided from the pore-level calculations described in Sect. 4.1, the volume-averaged framework enables investigations of much larger foam domains. Energy transfer in porous materials is typically studied in the volume-averaged framework by invoking the assumption of local thermal equilibrium between the solid and fluid phases, however for porous metals and graphitic foams, this assumption is not valid because of the large difference between the solid and fluid phase thermal conductivities. To this end, Calmidi and Mahajan [9] used a thermal non-equilibrium model to carry out calculations of forced convection in uncompressed aluminum foams. In their study, models for interstitial convective exchange and thermal dispersion were proposed and then used to simulate the heat transfer for several different specimens of aluminum foam. Betchen et al. [18] developed a complete three-dimensional conjugate heat transfer code for studying fluid/porous/solid domains, which also incorporates a thermal non-equilibrium model inside the porous domain. The models for interstitial exchange and thermal dispersion proposed by Calmidi and Mahajan [9] were implemented in the code and simulations of uncompressed aluminum foam confirmed the validity of the code and the accuracy of the models.

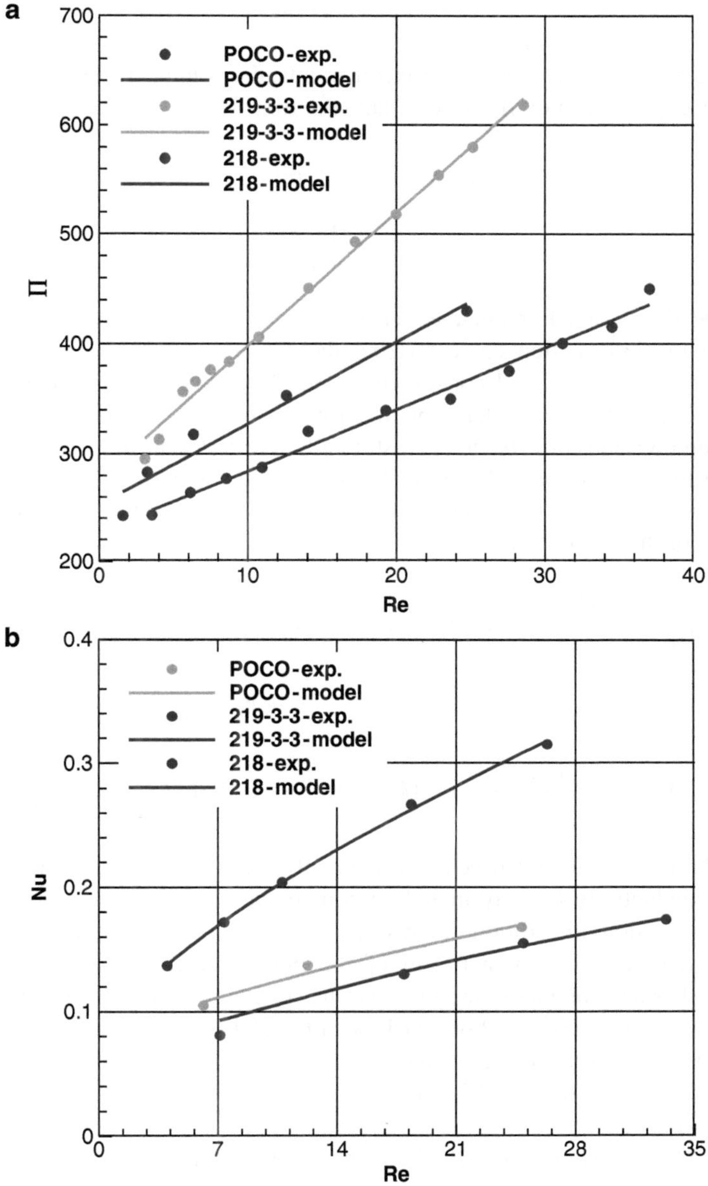

Fig. 18 Comparison of calibrated pressure (**a**) and heat transfer (**b**) models with experimental data

The framework for the volume-averaged formulation generally considers the laminar, incompressible flow of a single-phase fluid with constant thermophysical properties. The volume-averaged continuity equation may be expressed in extrinsic form as [18]:

$$\rho_f(\nabla \cdot \langle u \rangle) = 0 \tag{14}$$

Subject to appropriate length scale constraints, the volume averaged momentum equations expressed in extrinsic form take the form [18]:

$$\begin{aligned} \rho_f \frac{\partial \langle u \rangle}{\partial t} + \frac{\rho_f}{\varepsilon} \nabla \cdot (\langle u \rangle \langle u \rangle) = & - \varepsilon \nabla \langle P \rangle^f + \mu_B \nabla^2 \langle u \rangle - \frac{\varepsilon \mu_f}{K} \langle u \rangle \\ & - \frac{\varepsilon \rho_f c_E}{\sqrt{K}} |\langle u \rangle| \langle u \rangle + \varepsilon \rho_f f \end{aligned} \tag{15}$$

where the Darcy and Forchheimer terms are used to close the set of equations, replacing the information characterizing the viscous and form drag interaction between the fluid and solid constituents, which is lost in volume averaging the velocity field. The second term on the right hand side of (15) represents the macroscopic viscous effects, and is historically referred to as the Brinkman term.

Under the assumption of local thermal non-equilibrium, $\langle T_f \rangle^f = \langle T_s \rangle^s = \langle T \rangle$ is not assumed at a given point. This gives rise to extrinsic volume averaged energy equations of the form [15]:

$$\begin{aligned} \rho_f c_{p,f} \left[\varepsilon \frac{\partial \langle T_f \rangle^f}{\partial t} + \nabla \cdot \left(\langle u \rangle \langle T_f \rangle^f \right) \right] = & \nabla \cdot \left(k_{fe} \nabla \langle T_f \rangle^f \right) \\ & + h_{sf} A_{sf} \left(\langle T_s \rangle^s - \langle T_f \rangle^f \right) \end{aligned} \tag{16}$$

$$(1 - \varepsilon) \rho_s c_s \frac{\partial \langle T_s \rangle^s}{\partial t} = \nabla \cdot (k_{se} \nabla \langle T_s \rangle^s) - h_{sf} A_{sf} \left(\langle T_s \rangle^s - \langle T_f \rangle^f \right) \tag{17}$$

for the fluid and solid constituents, respectively. In (16), the fluid phase effective thermal conductivity k_{fe} may include a component accounting for the effects of thermal dispersion, in addition to the stagnant portion of the conductivity which is typically determined from a model of the pore geometry. Note also that the effective thermal conductivities in a porous medium are often anisotropic, and thus k_{fe} and k_{se} may in general be matrices, although in such a case we shall assume that all off-diagonal elements of these matrices are zero. The second term on the right hand side of (16) and (17) represent the heat transfer between the fluid and solid constituents.

The heat exchange between the fluid and solid constituents is modeled using an interstitial convective exchange model. It is worthy of note that interstitial exchange is not equivalent to bulk convective heat transfer unless thermal dispersion is negligible. Thermal dispersion arises in the fluid energy equation as a result of volume-averaging and accounts essentially for transport effects due to fluctuations in the temperature field about the volume-averaged temperature. Thermal dispersion is modeled as an enhancement to the molecular conductivity of the fluid and can thereby contribute significantly to the bulk heat transfer within a porous media. Thermal dispersion is typically modeled using the expression developed by

Calmidi and Mahajan [9]. The interstitial convective exchange coefficient h_i is obtained locally from an expression of the form:

$$Nu_L = \frac{h_i L}{k_f} = C\mathrm{Re}^m \mathrm{Pr}^n, \tag{18}$$

which can be calibrated using experimental results to give the correct overall heat transfer for any type of porous media. In (18), L is simply the length scale used for the definition of the Nusselt number. For simulations of aluminum foam, the expression for internal surface area and the parameters C, m and n were obtained from Calmidi and Mahajan [9]; the length scale used was the ligament diameter. For simulations of graphitic foam, the internal surface area was obtained using $A = \beta V$ (Yu et al. [4]), where V is the volume of the cell, and the parameters C, m and n were obtained by calibration with the results for POCO™ foam (see Table 1). In this case the length scale used was the equivalent particle diameter, D_e. Darcy and Forchheimer coefficients for aluminum foam were obtained from Boomsma et al. [10] and for GF from Straatman et al. [13, Table 2].

The calibrated model was used to explore differences between POCO™ foam and the aluminum foams considered by Calmidi and Mahajan [9] and by Boomsma et al. [10]. Figures 19 and 20 show the pressure drop and heat transfer as a function of extrinsic velocity for POCO™ foam and for an uncompressed Aluminum T-6201 foam with a porosity of 90% and effective conductivity 7.19 W/m K, as reported in Calmidi and Mahajan. The plots are shown in terms of the extrinsic velocity instead of Re_{De} to make the differences at a given flow condition more clear; differences between the internal structures of GF and aluminum foam yield very different ranges of values, which make direct comparisons difficult.

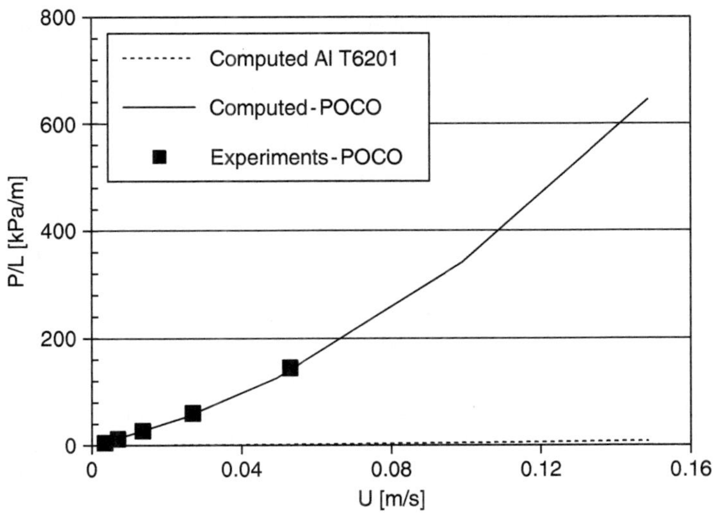

Fig. 19 Plot of the pressure drop as a function of the extrinsic velocity for POCO™ foam and for Al T6201 foam. Adapted from [13]

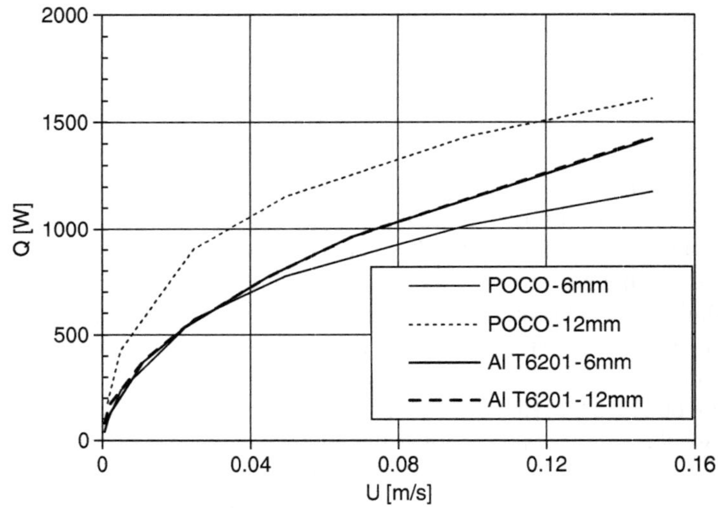

Fig. 20 Plot of the heat transfer as a function of extrinsic velocity for POCO™ foam and for Al T6201 foam. Adapted from [13]

The results shown in Figs. 19 and 20, are for the same heating condition (set using a specified base and inlet temperature) over a range of bulk velocities. On the basis of Fig. 19, it is clear that the aluminum foam has a considerably lower pressure drop than GF over the entire range tested [13]. This is largely due to differences between the porosities and internal structures of the aluminum foam and GF: the aluminum foam has a high porosity and is comprised of thin fused strands and large open cell windows, but relatively little internal surface area. The GF has a moderate porosity and is comprised of spherical voids and small(er) cell windows, but with a lower porosity and much larger internal surface area.

Figure 20 shows the heat transfer for 6 mm-thick and 12 mm-thick blocks of POCO™ and Al T-6201 foams computed using the same bulk velocity and heating conditions. At a 6 mm thickness, the Al T-6201 foam has a higher heat transfer across the full range of conditions considered. However, when the thickness of the Al T-6201 foam is doubled, no increases in heat transfer are realized at any flow rate. In contrast, doubling the thickness of the POCO™ foam leads to increases in heat transfer from 85% at low velocity to 50% at high velocity. These observations can be explained by considering the balance between heat conduction into the foam and the convective exchange between the foam and the fluid. The fact that doubling the thickness results in no increases in heat transfer in the Al foams confirms that heat is not entrained deeply into the foam and that the maximum possible heat transfer for a given base surface area is achieved using a very thin layer of aluminum foam (as described in Sect. 3 with respect to the Calmidi and Mahajan [9] experiments). For POCO™ foam, the pore-level convective heat transfer coefficient is considerably smaller than that for aluminum foam, but the lower porosity and higher conductivity serve to conduct heat much more deeply into the foam, thereby making use of much more interior surface area. At 6 mm thick, POCO™ foam removes less heat than Al T-6201 foam, but at 12 mm, the heat transfer has increased above that of aluminum foam, without yet reaching its maximum possible heat dissipation.

Comparisons were also made in Straatman et al. [13] between the compressed aluminum specimens considered by Boomsma et al. [10] by conducting simulations of $40 \times 40 \times 2$ mm-thick blocks of POCO™ foam and processing the predicted pressure drops and Nusselt numbers as done in [10]. Comparisons were done in this manner because no expressions for interstitial exchange exist to facilitate computational modeling of compressed aluminum foams. Figure 21 compares the pressure drop of POCO™ with that of three different compressed aluminum foams: 92-02, 92-03 and 92-06, where the -0X indicates the compression ratio based on the volume. Figure 21 makes it clear that compression decreases the permeability of the foam, but the pressure drop is still significantly lower than that of POCO™ over the same range of bulk velocities, again owing to the large hydraulic loss associated with the contraction/expansion of fluid through the cell windows present in GF. Figure 22 shows the Nusselt number based upon the heated area and the temperature difference between the heated base and the fluid inlet. The heat transfer for the compressed aluminum samples is seen to approach that of POCO™ foam for high compression ratios. It is expected that under compression, the local convective heat transfer coefficient decreases, but this is off-set by the increase in available internal surface area, thereby giving a lower net thermal resistance. On the basis of Figs. 21 and 22, it appears that the highly compressed aluminum offers an equivalent heat transfer for a significantly lower pressure drop, but it is important to note that this will

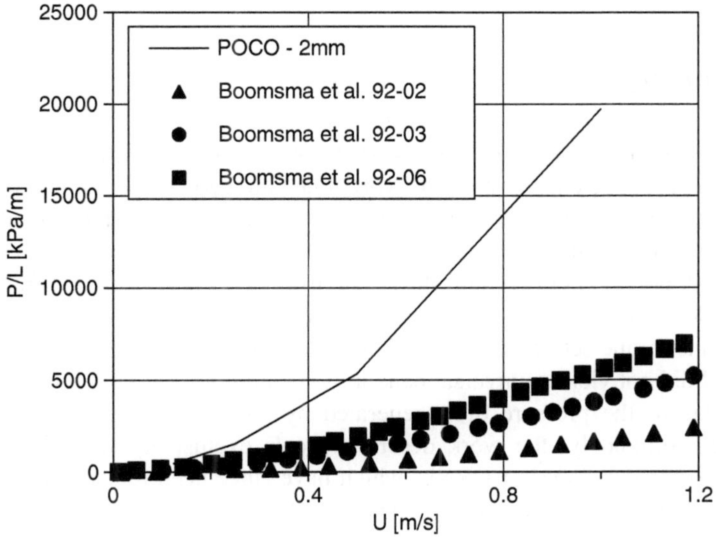

Fig. 21 Plot of the pressure drop as a function of extrinsic velocity for POCO™ foam compared to three different compressed aluminum foams as reported in Boomsma et al. [10]

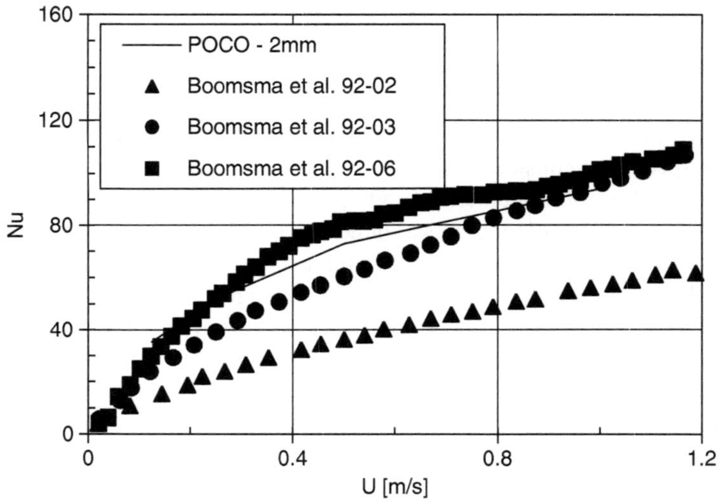

Fig. 22 Plot of the Nusselt number (as defined in Boomsma et al. [10]) as a function of extrinsic velocity for POCO^{TM} foam compared to three different compressed aluminum foams as reported in Boomsma et al. [10]

only be the case for very thin layers of foam. As shown in the comparison with uncompressed aluminum foams, if the thickness of the samples was doubled or tripled, the POCO^{TM} foam would yield significantly higher heat transfer than any derivative of compressed aluminum foam due to its high extended surface efficiency.

5 Summary

A brief survey of the most recent work on spherical void phase graphitic foams has been presented. The chapter presents work related to geometric characterization of the GF structure, experiments exploring convective heat transfer enhancements obtained in parallel, impinging and forced flow arrangements, and recent computational work done at both the pore-level and using a volume-averaged framework. It is evident from the body of work presented that while the graphitic foam possesses a unique set of material and geometric properties, the foams tested and described did not yield a universal benefit over commonly used aluminium foams, mostly due to the pressure drop generated by passing a fluid through the GF structure. Fortunately, the work described deals exclusively with some of the original foams developed at ORNL, which have since been replaced by graphitic foams that retain the spherical void phase while having a much more permeable structure that is more amenable to fluid flow. Such foams are being tested and characterized using the tools and models presented herein and will undoubtedly see far-reaching application in industry.

References

1. Klett, W.J.: Process for making carbon foam. US Patent 6,033,506, 2000
2. Klett, W.J., Hardy, R., Romine, E., Walls, C., Burchell, T.: High-thermal conductivity, mesophase-pitch-derived carbon foam: effect of precursor on structure and properties. Carbon **38**, 953–973 (2000)
3. Gallego, C.N., Klett, W.J.: Carbon foams for thermal management. Carbon **41**, 1461–1466 (2003)
4. Yu, Q., Thompson, B.E., Straatman, A.G.: A unit-cube based model for heat transfer and pressure drop in porous carbon foam. ASME J. Heat Transfer **128**, 352–360 (2006)
5. Klett, J.W., McMillan, A.D., Gallego, N.C., Walls, C.: The role of structure on the thermal properties of graphitic foams. J. Mater. Sci. **39**, 3659–3676 (2004)
6. Antohe, B.V., Lage, J.L., Price, D.C., Weber, R.M.: Experimental determination of permeability and inertia coefficients of mechanically compressed aluminum porous matrices. ASME J. Fluids Eng. **119**, 404–412 (1997)
7. Paek, W.J., Kang, H.B., Kim, Y.S., Hyum, M.J.: Effective thermal conductivity and permeability of aluminum foam materials. Int. J. Thermophys. **21**(2), 453–464 (2000)
8. Boomsma, K., Poulikakos, D.: The effects of compression and pore size variations on the liquid flow characteristics in metal foams. ASME J. Fluids Eng. **124**, 263–272 (2002)
9. Calmidi, V.V., Mahajan, R.L.: Forced convection in high porosity metal foams. ASME J. Heat Transfer **122**, 557–565 (2000)
10. Boomsma, K., Poulikakos, D., Zwick, F.: Metal foams as compact high performance heat exchangers. Mech. Mater. **35**, 1161–1176 (2003)
11. Straatman, A.G., Gallego, N.C., Thompson, B.E., Hangan, H.: Thermal characterization of porous carbon foam – convection in parallel flow. Int. J. Heat Mass Transfer **49**, 1991–1998 (2006)
12. Sultan, K., DeGroot, C., Straatman, A.G., Gallego, N.C., Hangan, H.: Thermal characterization of porous graphitic foam – convection in impinging flow. Int. J. Heat Mass Transfer (2009). doi:10.1002/fld.2050
13. Straatman, A.G., Gallego, N.C., Yu, Q., Betchen, L.J., Thompson, B.E.: Forced convection heat transfer and hydraulic losses in porous carbon foam. ASME J. Heat Transfer **129**(9), 1237–1245 (2007)
14. Karimian, S.A.M., Straatman, A.G.: CFD Study of the hydraulic and thermal behaviour of spherical void phase porous materials. Int. J. Heat Fluid Flow **29**(1), 292–305 (2007)
15. Karimian, S.A.M., Straatman, A.G.: Numerical modeling of multi–directional flow and heat transfer in graphitic foams. ASME J. Heat Transfer **131**(5). http://dx.doi.org/10.1115/1.3084122 (2009)
16. Karimian, S.A.M., Straatman, A.G.: A thermal periodic boundary condition for heating and cooling processes. Int. J. Heat Fluid Flow **28**, 329–339 (2007)
17. El-Soukkary, T.M., Straatman, A.G.: The prediction of spatially periodic flows using a finite-volume model. Int. J. Numer. Methods Fluids **41**, 303–317 (2003)
18. Betchen, L.J., Straatman, A.G., Thompson, B.E.: A non-equilibrium finite-volume model for conjugate fluid/porous/solid domains. Numer. Heat Transfer Part A **49**, 543–565 (2006)

Heat Transfer in Polyolefin Foams

Marcelo Antunes, José Ignacio Velasco, Eusebio Solórzano, and Miguel Ángel Rodríguez-Pérez

Abstract This chapter is dedicated to the study of heat transfer in polyolefin-based foams, particularly thermal conductivity, as a function of their structure and chemical composition. A small review of the main experimental techniques used to measure the thermal conductivity of these materials is also given, focusing on the transient plane source method (TPS), as well as different theoretical models commonly used for estimating its value. Alongside cellular structure (cell size, anisotropy, etc) and composition considerations, particular importance is given to the analysis of the presence of micrometric and nanometric-sized fillers in the resulting cellular composite thermal properties. This is a novel research field of particular interest, thought to extend the application range of these lightweight materials by tailoring their conductivity.

1 Introduction

It is well known that heat transfer is one the most important fields of research for cellular polymers due to the wide number of applications and uses of these materials as thermal insulators. Heat transfer in these materials strongly depends on relative density, cellular characteristics such as cell size, cell density, cell anisotropy, etc, and presence of additional phases and/or fillers (concentration, orientation and dispersion of these additional phases) [1–3].

M. Antunes (✉) and J.I. Velasco
Centre Català del Plàstic, Departament de Ciència dels Materials i Enginyeria Metal·lúrgica, Universitat Politècnica de Catalunya, C/Colom 114, 08222 Terrassa, Barcelona, Spain
e-mail: marcelo.antunes@upc.edu; jose.ignacio.velasco@upc.edu

E. Solórzano and M.A. Rodríguez-Pérez
Cellular Materials Laboratory (CellMat), Condensed Matter Physics Department, University of Valladolid, 47011 Valladolid, Spain
e-mail: esolo@fmc.uva.es; marrod@fmc.uva.es

A. Öchsner and G.E. Murch (eds.), *Heat Transfer in Multi-Phase Materials*,
Adv Struct Mater 2, DOI 10.1007/8611_2010_44,

This chapter is focused on the study of heat transfer in polyolefin-based foams, although most of the concepts and trends presented are applicable to most of the polymeric cellular materials available in the market. It displays the main tendencies of heat transfer, focusing on thermal conductivity, as a function of the structure and chemical composition for different types of polyethylene and polypropylene foams with densities ranging from 20 to 600 kg/m^3. It also shows some strategies to modify the thermal conductivity in terms of structure, compounding and production techniques. A small review of the main experimental techniques to measure the thermal conductivity of these materials is also given and different theoretical models commonly used for determining the thermal conductivity of polymer foams have been applied.

Part of the chapter is dedicated to the analysis of the presence of third phases (micrometric and nanometric-sized fillers) in the resulting cellular composite thermal properties. This is a novel research field of particular interest, thought to extend the application range of lightweight materials by tailoring their conductivity, and actually scarce information about the thermal behaviour of thermoplastic foams with conductive fillers has been published [3–5].

1.1 The Concept of Cellular Solid

A cellular solid is a two-phase material in which a gas has been dispersed in a solid continuous matrix. If the matrix is polymeric in nature, the material is known as cellular polymer or polymer foam.

Among the most important parameters that modify the physical and transport properties of these materials are the nature and morphology of the base material, type of gas entrapped inside the cells, density, and the cellular architecture and topology, such as cell connectivity (closed, open or partially interconnected cells), cell size (ϕ) and distribution of cell sizes, cell wall thickness (δ) and respective distribution, fraction of solid in the cell struts (f_s) and cell geometry and shape [6]. Some of these basic parameters used to characterize the cellular structure are related by the following expression:

$$\phi(1-f_s)\frac{\rho}{\rho_s}=C\delta, \tag{1}$$

where ρ/ρ_s is the so-called relative density of the cellular material (ρ: density of the foam and ρ_s: density of the respective unfoamed solid matrix) and C is a constant that depends on the cell's shape and geometry. For instance, this constant has a value of 3.46 for pentagonal dodecahedron [1] and 3.35 for tetrakaidecahedral cells [7].

The concept of foam as a two-phase material is important to understand their behaviour as that resulting from the combination of the properties of both phases and their relative content. Due to this reason, the relative density, and analogously the expansion ratio, ER ($ER = \rho_s/\rho$), is a crucial parameter when studying these materials, being directly related to the relative volume fraction of both phases: $V_{gas} = (\rho_s - \rho)/(\rho_s - \rho_{gas}) \approx 1 - \rho/\rho_s$ and $V_{sol} = (\rho - \rho_{gas})/(\rho_s - \rho_{gas}) \approx \rho/\rho_s$. It is important to mention that these approximations are only valid for relative densities over 5%.

1.2 Polyolefin Foams: Production Techniques

In order to fully understand how the thermal properties vary with the structure, it is also necessary to explain the production methods of these materials. A small introduction on the most common foaming processes and typical resulting cellular structures is presented here. Polyolefins such as polypropylene (PP) or polyethylene (PE) are commercially foamed using one of the basic foaming processes described in this section [8]. All results and microstructures shown here for the PP foams come from lab-produced materials, whereas PE foams results were mainly obtained from commercially available materials. As we will see, cellular structure variations induced by the different processing techniques or by process parameter modifications may slightly modify the thermal conduction behaviour of the whole system.

- *Foaming by direct extrusion*, the foam is directly obtained by a sudden decompression at the exit of an extrusion die, normally using a physical blowing agent (PBA) such as CO_2 or n-butane [9, 10] as seen in Fig. 1a. The physically foamed extruded foams tend to show a rather anisotropic cellular structure with cells elongated in the flow direction due to the stresses applied during the extrusion process.
- *Foaming by injection moulding*, the polymer expansion is adjusted by controlling chemical blowing agent (CBA) thermal decomposition or PBA expansion inside a closed injection mould. A variation of the conventional injection moulding is the microcellular injection moulding or *Mucell®* technique. In this method, supercritical N_2 or CO_2 is introduced in the plasticizing injection unit and mixed with the melted polymer before injecting in the mould [11] (see Fig. 1b). Small cell sizes are typically obtained, although this foaming technique is limited to rather high density materials (>300 kg/m^3).
- *Foaming by compression-moulding*, the material is foamed by simultaneously applying heat and pressure in order to decompose the CBA, nucleate the cells and subsequently expand the material by sudden decompression, as depicted in Fig. 1c. This process commonly uses exothermic CBAs such as azodicarbonamide (ADC) [12]. Generally speaking, foams produced using this technique present small cell sizes and certain cell-size gradients, with smaller cells close to the mould's surface.
- *Gas dissolution foaming*, the material is foamed inside an autoclave by a high-pressure gas dissolution process. This process, commercially developed by

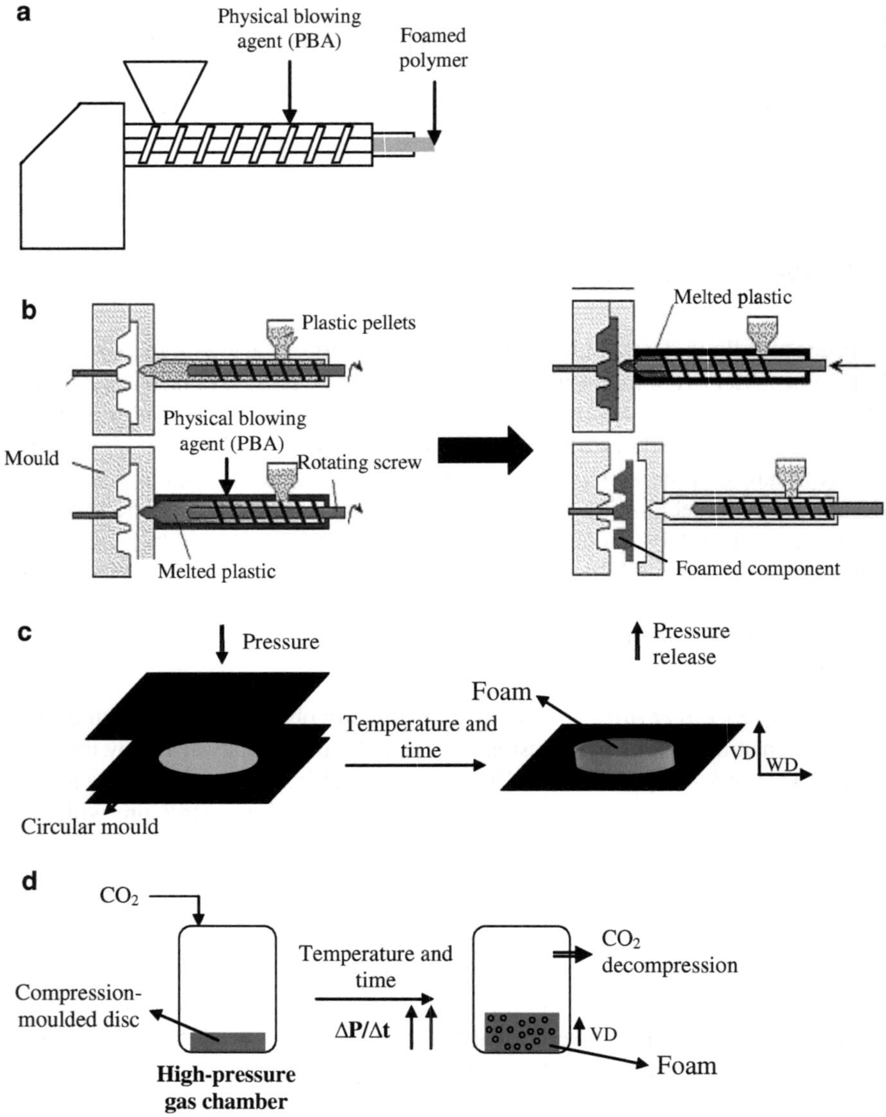

Fig. 1 Schematics showing (**a**) direct extrusion, (**b**) injection moulding, (**c**) compression-moulding and (**d**) CO_2 dissolution pressure-quench foaming processes

Zotefoams [13], uses N_2 or CO_2 as physical blowing agent dissolving the gas in the polymer in a semi-solid state and afterwards allowing the material to expand by heating at a temperature above the softening temperature of the polymer–gas mixture. Although not available commercially, a second strategy, known as the pressure-quench method (see Fig. 1d), considers a one-step gas dissolution process. In this method, the material is nucleated and foamed by carefully

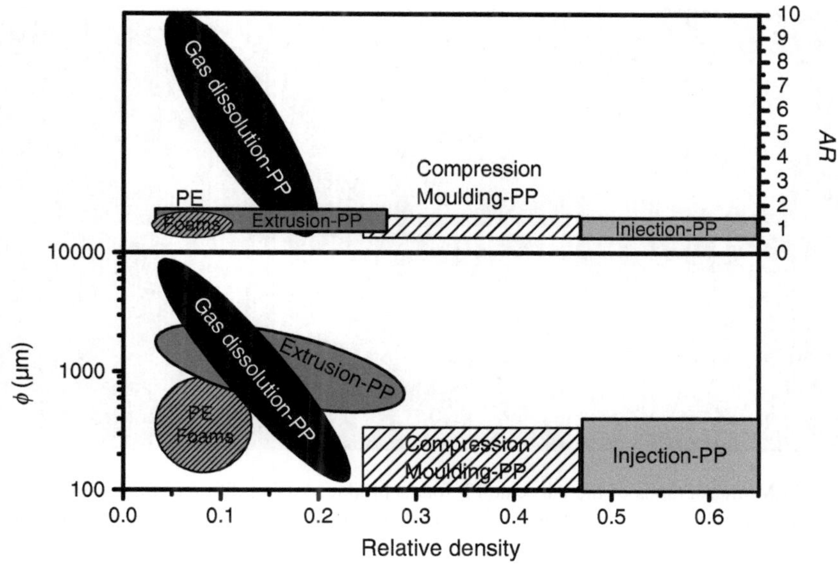

Fig. 2 Cell size, ϕ, and anisotropy ratio, *AR* ($AR = \phi_{VD}/\phi_{WD}$, VD: vertical direction of foaming; WD: width direction) versus relative density for the analyzed foams

controlling the sudden pressure drop and pressure drop rate applied during gas decompression [14]. This was the strategy used for preparing the PP lab-produced foams analyzed in this chapter.

The main cellular characteristics (average cell size, ϕ, and anisotropy ratio, *AR*) of the polypropylene and polyethylene based foams analyzed in this chapter are summarized in Fig. 2.

Regarding the characteristic cellular anisotropies exhibited in Fig. 2, it is important to mention that lab scale-produced PP foams were particularly conditioned to this kind of anisotropies (with special mention to the ones prepared by the pressure quench method). Similar processes in other laboratories or industrial-scaled ones based on similar techniques may not exhibit such anisotropic cellular structures.

Scanning electron micrographs showing typical cellular structures for the different processing techniques are shown in Fig. 3 for the lab-produced PP-based foams and in Fig. 4 for the PE-based commercial ones.

1.3 Composite Polyolefin Foams: Production Techniques

Polymeric cellular materials can also incorporate fillers, i.e., secondary solid phases, commonly inorganic, with the intention of extending their applicability window. We will focus our attention on the effects of the incorporation of these fillers in the thermal conduction behaviour of the foams.

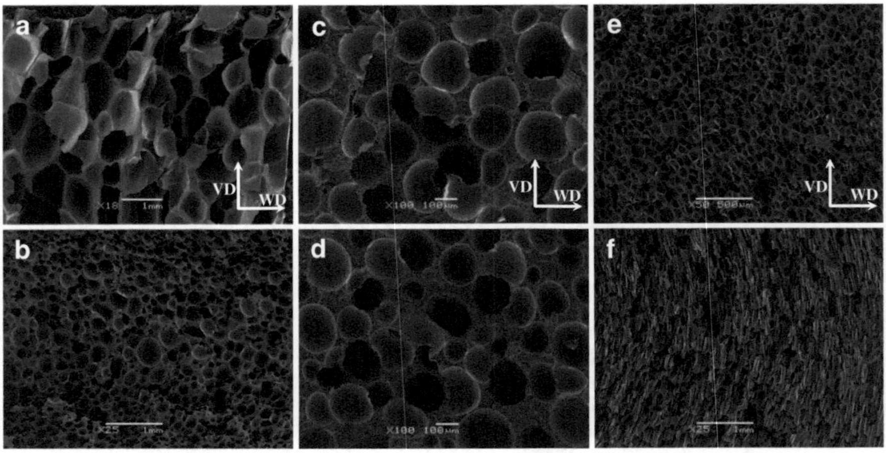

Fig. 3 SEM micrographs of the lab scale PP foams produced using different foaming processes: (**a**) direct extrusion, (**b**) injection-moulding, (**c**, **d**) compression-moulding, and (**e**, **f**) pressure-quench CO_2 batch foaming

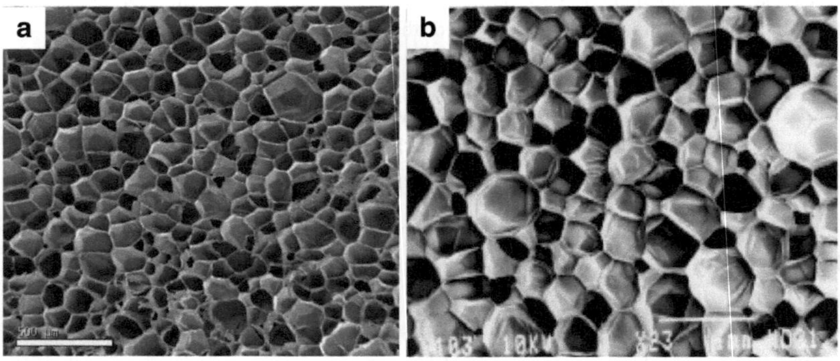

Fig. 4 SEM micrographs showing typical cellular structures of commercial PE foams prepared by (**a**) compression-moulding and (**b**) N_2 dissolution

Polyolefin composites were initially prepared by melt-compounding the different fillers in a twin-screw extruder with a polypropylene-based matrix and a CBA (azodicarbonamide). The resulting composites were subsequently foamed by compression-moulding chemical foaming. Particularly, the influence of two different highly conductive fillers was experimentally evaluated:

1. Incorporation of high amounts (50 and 70 wt.%) of a micrometric-sized filler, magnesium hydroxide ($Mg(OH)_2$) (Fig. 5), commonly used as flame retardant [15, 16]. This kind of filler typically exhibits a particle size in the range of a few micrometers (<10 μm). The maximum theoretical thermal conductivity is assumed to be approximately 130 W m^{-1} K^{-1} [17], although this value depends highly on the crystalline orientation.

Fig. 5 SEM micrograph showing the typical hexagonal shape of $Mg(OH)_2$ particles

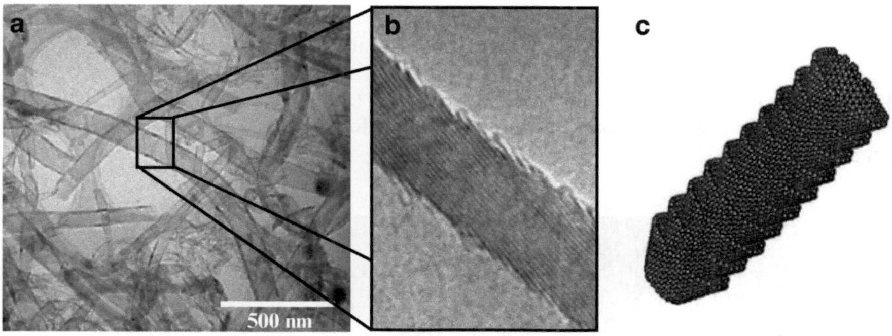

Fig. 6 (**a**, **b**) Transmission electron micrographs and (**c**) schematic showing the stacked-cup structure of the sub-micron vapour grown carbon nanofibres

2. Incorporation of different amounts of carbon nanofibres (from 5 to 20 wt.%), a theoretically highly thermally conductive filler (>2,000 W m^{-1} K^{-1}) [18]. This fibrous-like nanometric filler was added with the objective of obtaining PP-based cellular materials with improved thermal conductivities [19, 20]. The carbon nanofibres used here were sub-micron vapour grown carbon fibres (s-VGCF) with a stacked-cup structure produced using a floating catalyst technique with a diameter of 20–80 nm, a fibre length higher than 30 μm and a graphitization degree of 70%. These nanofibres were kindly supplied by *Grupo Antolín* (Burgos, Spain). Figure 6 presents two different magnification transmission electron micrographs, as well as a schematic displaying the stacked-cup structure of the carbon nanofibres.

A graph indicating the main cellular structure characteristics, cell size and anisotropy, of these materials is shown in Fig. 7.

It can be observed that the materials filled with $Mg(OH)_2$ showed an expansion ratio between 2 and 3 and anisotropy ratios up to 3. The PP-CNF composite foams were intentionally produced with a fixed density ($ER \approx 3$), although foams with lower densities could have been produced.

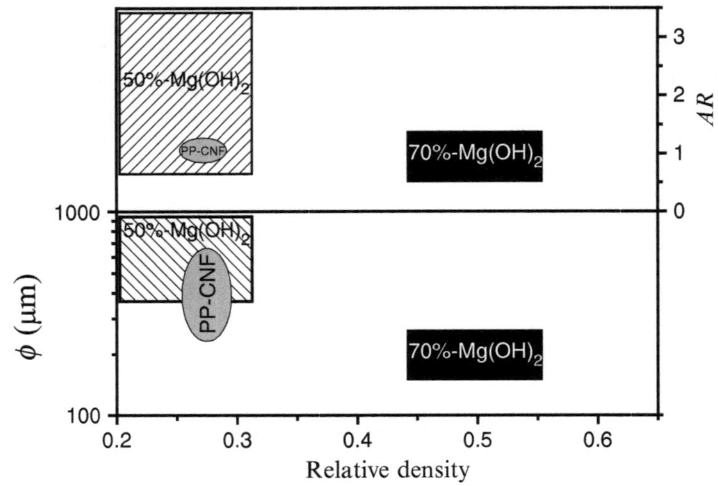

Fig. 7 Characteristic cell size, ϕ, and anisotropy ratio, *AR* ($AR = \phi_{VD}/\phi_{WD}$, VD: vertical direction of foaming; WD: width direction) versus relative density for the composite foams analyzed in this chapter

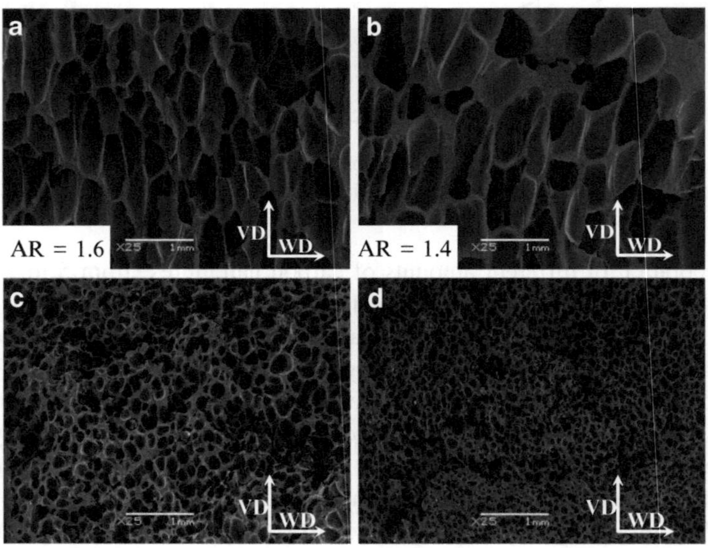

Fig. 8 SEM pictures of the 50 wt.% $Mg(OH)_2$-PP foams: (**a**) 0.20 and (**b**) 0.23 relative density and 70 wt.% $Mg(OH)_2$-PP foams: (**c**) 0.47 and (**d**) 0.55 of relative density

Figure 8 shows some examples of the cellular structure of $Mg(OH)_2$ foams. Contrary to the foams filled with 70 wt.% $Mg(OH)_2$ that show isometric-like cell structures with small cell sizes (≈180 μm), the foams with a 50 wt.% $Mg(OH)_2$

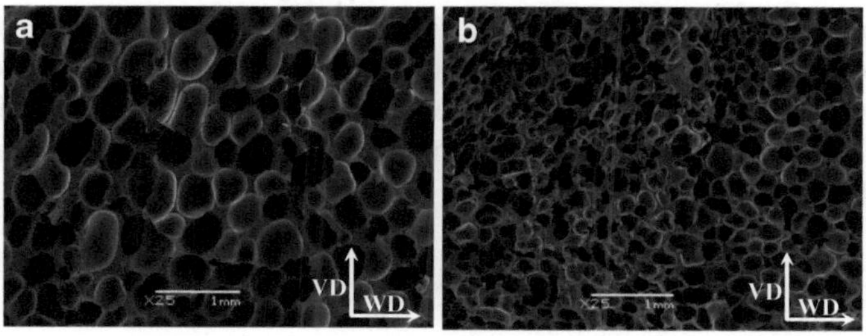

Fig. 9 Characteristic SEM pictures of (**a**) 5 and (**b**) 20 wt.% carbon nanofibre-reinforced polypropylene foams

present higher cell sizes (from 700 to almost 1,000 μm) and increasingly higher cell anisotropies for lower relative densities.

Some examples of the typical cellular structures of PP-CNF foams are shown in Fig. 9. The foamed nanocomposites were prepared with closed-cell structures and typical expansion ratios of 3. Isometric-like cellular structures ($AR \approx 1$) with increasingly smaller cell sizes with gradually increasing the concentration of carbon nanofibres were obtained. For instance, the 5 wt.% CNF foam displayed an average cell size slightly above 500 μm, its value decreasing to 400 and around 250 μm respectively for the 10 and 20 wt.% CNF foams.

2 Experimental Methods to Determine Thermal Conductivity

The use of polymer foams is widespread in thousands of industrial applications and there is a continuous interest in regulating their thermal properties, in most cases with the objective of reducing the thermal conductivity. A wide variety of different experimental techniques to measure this property have been developed for different experimental conditions and materials [21, 22]. The process of measuring this property is complicated by the fact that in several practical situations most of the heat transfer mechanisms (conduction, convection and radiation) have to be considered. Thus, for each material it is necessary to identify the ideal measuring procedure considering factors such as the expected conductivity, shape of the material, available sample size, density, etc.

Generally speaking, in order to measure the thermal conductivity or a related property by a steady state or a transient method, the experimental arrangement must simulate a solution of the basic heat conduction equation:

$$\frac{1}{k}\frac{\partial T}{\partial t} = \nabla^2 T + \frac{A(x,y,z,t)}{\lambda}, \tag{2}$$

where k and λ are respectively the thermal diffusivity and conductivity, T is the temperature, t is the time, and A (x,y,z,t) is the heat generated per volume and time.

For longitudinal unidirectional heat flow, no radial losses and disregarding the existence of a heat source into the solid, the analysis of the steady-state term leads to a linear dependence between temperature and dimension in the heat flow direction. The heat per area and time through a sample can be determined using Fourier's law:

$$Q_a = G\lambda \frac{\Delta T}{d}, \quad (3)$$

where Q_a is the heat flow generated by the application of a temperature difference between the two sides of the material (ΔT), separated over a distance d (the material thickness), and G is a constant, determined by calibration for each given apparatus. Figure 10 shows a schematic diagram of the typical plate-like steady state equipment for determining the thermal conductivity.

The steady state techniques are the most commonly used methods, and there are several standard methods (ISO and ASTM) based on this procedure [23, 24]. Nonetheless, in some cases not all the heat generated in the upper plate is conducted to the lower one, thus being necessary to account for heat losses. Moreover, the heat flow is not always normal to the heat surfaces and there is a small gap between both heater surfaces and the surfaces of the sample. This gap contributes to the reduction of the effective transferred heat (interfacial heat transfer resistance). On the other hand, although this equipment is relatively fast in operation, there is still a need to reduce such times, particularly for quality control applications.

Alternatively, transient methods based on the analysis of the transient term solution of (2) which relates change in temperature with time, are used. Transient hot wire, transient hot strip, transient plane source and laser flash methods are probably the most important techniques based on measuring the sample's thermal behaviour under a transient heat flow regime. The laser flash method differs from the others since it is a non contact method and determines the thermal diffusivity of the sample instead of the thermal conductivity. Transient Plane Source (TPS) can be considered as the evolution of both transient hot wire and transient hot strip (by combining some aspects of the transient hot probe method, not mentioned before).

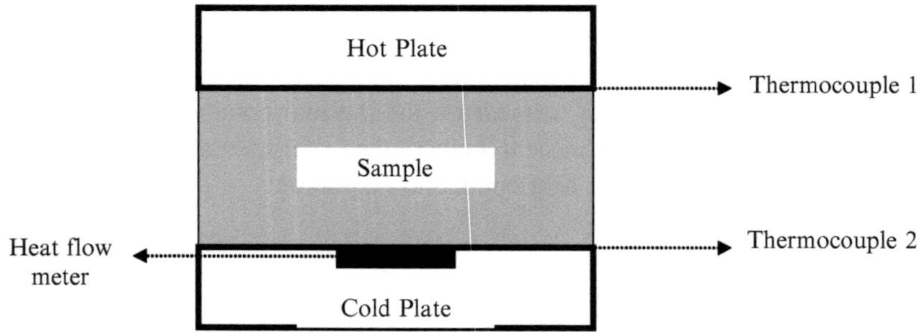

Fig. 10 Schematic diagram of typical standard equipment for determining the thermal conductivity based on Fourier's law

Therefore we can consider the TPS method as the most representative of these transient contact methods.

In the TPS method a round and plane heat source is used (see Fig. 11, left). It acts as a transient plane source working simultaneously as a temperature sensor. This element consists of an electrical conducting pattern of thin nickel foil in the form of a double spiral inserted between two insulating plastic layers. The total thickness of this sensor is only a few tens of micrometers. The TPS element is placed between two samples with both sensor faces in contact with the two samples surfaces as depicted in Fig. 11, left. Two samples of similar characteristics are required for this purpose. During testing a constant electric power is supplied to the sensor and the temperature increase is recorded. To relate the change in temperature with time, the equation for the heat conduction assuming the conditions reported by Log et al. and Gustavsson et al. is applied [25, 26]. Although less known, this method also enables the possibility of detecting possible thermal anisotropies [27].

Finally, the laser flash system is based in a laser beam pulse (typically <1 ms) focused on one of the faces of a relatively thin sample while the temperature increase in the opposite parallel face is recorded by a contactless method (IR pyrometer). From this temperature increase it is possible to determine the thermal diffusivity of the sample. The schematic description of the measurement procedure is shown in Fig. 11, right [28].

Transitory methods present several advantages compared to steady state ones. For example, it is possible to simultaneously obtain values of the thermal conductivity, thermal diffusivity and specific heat. The range of measurement comprises five orders of magnitude (0.01–400 W m^{-1} K^{-1}). These methods are also faster, can be used to determine the influence of material inhomogeneities and/or anisotropic characteristics and offer the ability to measure in small samples compared to the thick samples conventionally needed for the steady state methods.

On the other hand, it is important to remark that transient methods are not fully standardized. Nevertheless, in the last years some standard procedures have been approved for determining the thermal properties by the laser flash method [29, 30] and several efforts are being developed to standardize the TPS method [31]. It is

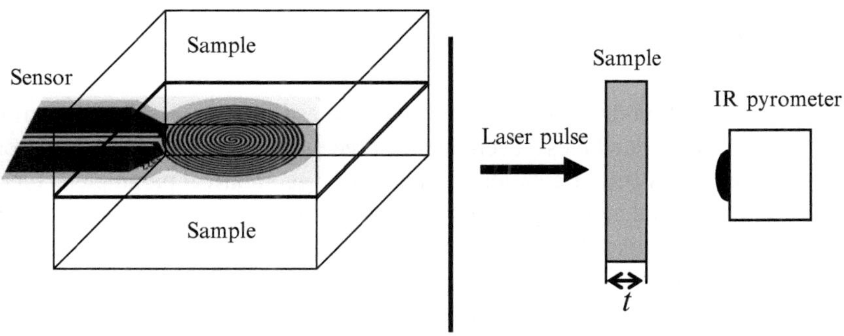

Fig. 11 Schematic diagrams for transient plane source (*left*) and laser flash (*right*) methods

also interesting to comment that the thermal properties of non-conventional new materials have been studied in the last years by using transient methods [32–40]. In addition, several works have shown that these transient techniques give comparable results to the steady state ones [41, 42].

3 Mechanisms and Models of Heat Transfer in Polymer Foams

Generally speaking, the heat transfer in any cellular material is the result of a contribution of three different mechanisms, conduction, convection and radiation, and therefore the overall thermal conductivity can be depicted as the result of four additive terms:

$$\lambda_{\text{foam}} = \lambda_{\text{cnd}}^{\text{s}} + \lambda_{\text{cnd}}^{\text{g}} + \lambda_{\text{rd}} + \lambda_{\text{cnv}}, \tag{4}$$

where $\lambda_{\text{cnd}}^{\text{s}}$ and $\lambda_{\text{cnd}}^{\text{g}}$are respectively the thermal conductivities due to conduction through the solid and gas phases and λ_{rd} and λ_{cnv} the radiation and convection terms.

3.1 *Convection*

Convection due to gas movement inside the cells may be disregarded for cellular structures with cell sizes of less than 4–5 mm [1]. Considering that almost all polymer-based foams, independently of the final relative density, present cell sizes that are clearly below these values, heat transfer due to the movement of the gas molecules entrapped inside the cells (convection) can be considered minimal when compared to conduction and radiation.

3.2 *Conduction*

Several theoretical models have been proposed to estimate the conduction term of cellular polymers. Nonetheless, even the most recent ones, which take into account for instance arbitrary cell orientations or anisotropy geometrical parameters, tend to consider rather simple geometrical-shaped arrays representing the cellular structure (cubic or polyhedral-like) [1, 41, 43–45]. On the other hand, foams exhibit in many cases cell imperfections, gradients and inhomogeneities, far away from regular geometries, especially in the case of low density foams. Nevertheless, the following formula has been proven to give reasonable compliance for the prediction of the thermal conduction term [1]:

$$\lambda_{\text{foam}} = \lambda_{\text{gas}} V_{\text{gas}} + \lambda_{\text{sol}} \frac{V_{\text{sol}}}{3} \left[f_s \sqrt{\frac{1}{AR}} + 2(1 - f_s) \left(\frac{1}{AR} \right)^{\frac{1}{4}} \right], \tag{5}$$

where f_s is the fraction of solid in the cell struts and *AR* is the geometrical anisotropy ratio, i.e., the quotient between the highest and smallest cell size in the direction of the heat flux (Fig. 12).

In the particular case of isotropic-like cellular structures, (5) is reduced to the following equation:

$$\lambda = \lambda_{\text{gas}} V_{\text{gas}} + \left(\frac{2}{3} - \frac{f_s}{3} \right) \lambda_{\text{sol}} V_{\text{sol}} \tag{6}$$

According to some other authors the thermal conductivity of a cellular solid can be modelled as:

$$\lambda = \lambda_{\text{gas}} V_{\text{gas}} + \xi \lambda_{\text{sol}} V_{\text{sol}}, \tag{7}$$

where ξ is the tortuosity, a parameter directly related to the foam's inherent irregularity. The concept of tortuosity goes beyond the conventional geometrical tortuosity [46] and implicitly considers the effect of cellular structure (cell size, cell density and cell wall thickness, f_s). This last equation is particularly interesting in the case of materials with an unknown f_s parameter. On the other hand, in the particular case of anisotropic structures it is possible to consider the effect of anisotropic tortuosity in the different material directions, so under anisotropic conditions we could also talk about equivalency between (5) and (7).

The influence of this mechanism can be modified by incorporating second phase constituents with different thermal conductivities, thus varying the expected conductivity, especially in the case of high density foams (higher contribution of the solid phase). Such is the case of the fillers considered in some of the next sections of this chapter. Particularly, the model used for these composite foams will be based on the experimental thermal conductivity of the solid composite (i.e., replacing λ and *V* variables shown in previous equations by those of the corresponding composites).

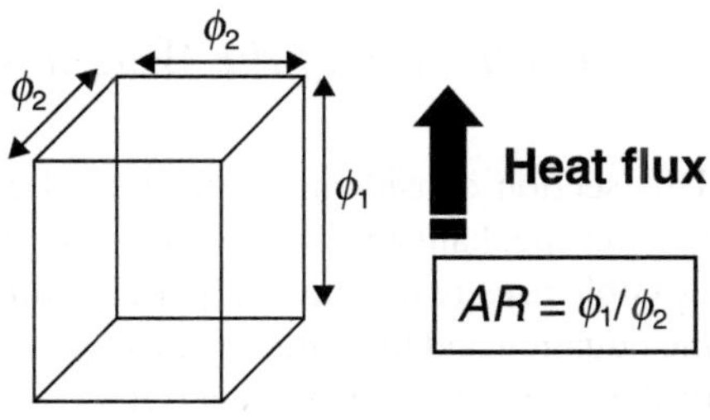

Fig. 12 Schematic showing cell anisotropy for a cubic cell geometry and the definition of *AR*

3.3 Radiation

To estimate this mechanism, the model proposed by Williams and Aldao will be adopted in this chapter. The reason is that this model has a high compliance with experimental results as has been shown in previous investigations [47–49]. One of its main advantages lies on the use of measured values for the cellular structure characteristics and polymer matrix properties instead of non-intuitive adjustable parameters. The model is based on a radiation term predicted as follows:

$$\lambda_r = \frac{4\sigma T^3 L}{1 + \left(\frac{L}{\phi}\right)\left(\frac{1}{T_N} - 1\right)}, \tag{8}$$

where σ is the Stefan–Boltzmann constant, T is the temperature, L is the material thickness, ϕ is the cell size and T_N is the fraction of radiant energy sent forward by a solid membrane of thickness L_s. This energy fraction is given by:

$$T_N = \frac{(1 - r)}{(1 - rt)}\left\{\frac{(1 - r)\cdot t}{(1 + rt)} + \frac{(1 - t)}{2}\right\}, \tag{9}$$

where r is the fraction of incident energy reflected by each gas–solid interface. This quantity is related to the refractive index of the solid matrix (ω):

$$r = \left\{\frac{\omega - 1}{\omega + 1}\right\}^2 \tag{10}$$

The coefficient t is the fraction of energy transmitted through the solid membrane of thickness L_s (cell wall thickness), which is given by:

$$t = \exp(-aL_s), \tag{11}$$

where a is the absorption coefficient of the solid matrix.

3.4 Evaluation of the Weight of the Different Terms

This section considers the theoretical evaluation of the contribution of each heat transfer mechanism for PE foams as a function of relative density. The particular example uses real values, experimentally obtained for these materials. The following equation and the values presented in Table 1 were considered for the predictions of the overall thermal conductivity.

Table 1 Properties of the foam considered for predicting the overall thermal conductivity

Property	Value
λ_{gas}	0.0263 (W m^{-1} K^{-1})
λ_{sol}	0.30 (LDPE) (W m^{-1} K^{-1})
Cell type	Closed cell
Cell size, ϕ	150–500 μm
Cell wall, L_s (μm)	$(1 - f_s) \cdot \phi \cdot 3.5347\rho_r$
f_s	0.2–0.4
Total material thickness	10 mm
Temperature	300 K
ω	1.51
a	661 cm^{-1}

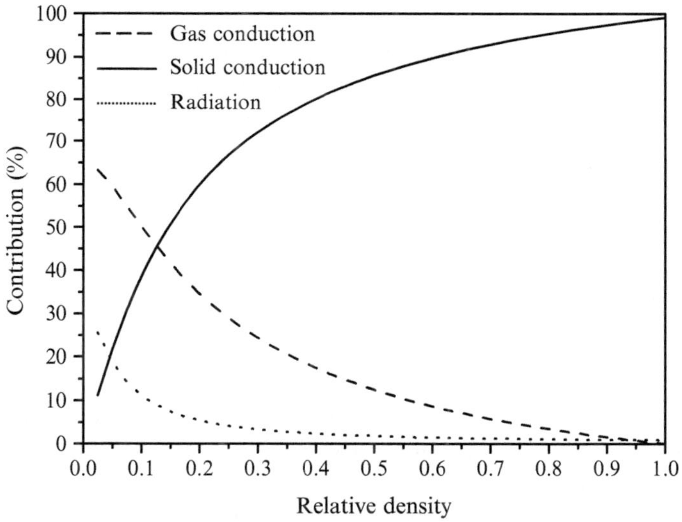

Fig. 13 Contribution of each heat transfer mechanism in the thermal conductivity for LDPE based foams

$$\lambda = \lambda_{gas} V_{gas} + \left(\frac{2}{3} - \frac{f_s}{3}\right) \lambda_{sol} V_{sol} + \frac{4\sigma T^3}{1 + \left(\frac{L}{\phi}\right)\left(\frac{1}{T_N} - 1\right)} \quad (12)$$

A relation between cell size and cell wall thickness with density such as the one shown in (1) was taken into account [1, 48, 50]. All the calculations were considered at room temperature by selecting those values for the properties of the solid materials and air at this temperature. An increase of temperature would change the predicted contributions for each heat transfer mechanism. This evaluation could be done using the same equations proposed here by introducing the variation of the properties of both gas and solid phases with temperature.

The predicted results are shown in Fig. 13. All three contributions play a significant role for relative densities below 0.2, and the contribution of the radiation mechanism becomes significant below this density. For higher relative densities only conduction (both gas and solid contributions) should be considered because the expected contribution of radiation is below 5%.

Taking into account these results it is reasonable to consider different models for relative densities below and above 0.2. For low densities, all the main mechanisms (except convection) have to be considered and for higher densities only conduction plays a significant role.

4 Methodologies Employed for Tailoring the Thermal Conductivity

The need of new cellular plastics for novel very specific applications forced the development of advanced materials with tailored properties adjusted for each particular application. This section summarizes most of the possibilities in terms of thermal conductivity modification. We will explain them, from the simplest to the most advanced methods.

4.1 Modification of the Gas or Polymer Matrix

One of the easiest ways to control the final thermal properties comes through the modification of the raw materials employed, i.e., solid and/or gaseous phase. Figure 14 shows both the thermal conductivity of several polymers and different gases used as blowing agent.

In this sense, the final density of the foam will determine the most effective strategy in order to modify the thermal conductivity. Considering the results plotted

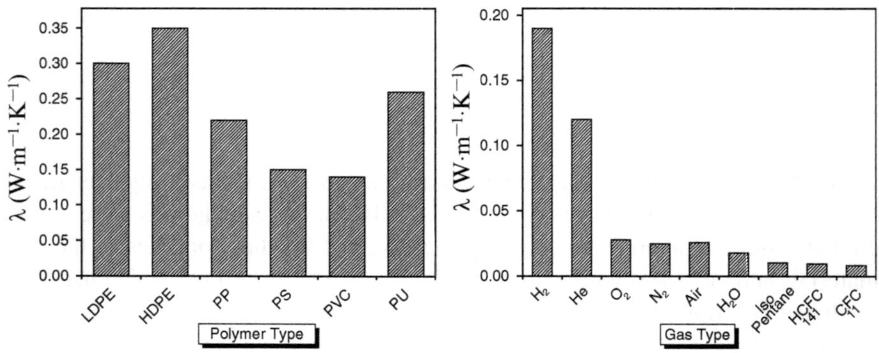

Fig. 14 Thermal conductivity of different polymeric materials (*left*) and some of the most used blowing agents (*right*)

in Fig. 13, it seems evident that it is strongly recommended to modify the gas thermal conductivity for relative densities below 0.2. Even in the 0.2–0.4 range it could also be effective to use a different gas as blowing agent in order to regulate these properties, whereas conduction through the solid phase begins to be predominant for higher relative densities.

The gas modification is a well known topic in the production of ultralow density XPS and rigid PU foams, since a significant part of the research dedicated to these materials has been focused on reducing conduction through the gas phase. The initial solution consisted in using high molecular weight gases with low thermal conductivities, traditionally CFCs and HCFCs (Fig. 14, right). Nonetheless and due to environmental restrictions, their use has been limited and the development of new environmentally friendly gases with low Global Warming Potential (GWP) is on the rise. On the other hand, two of the most used gases to produce polymer foams are N_2 and CO_2, although their permeability in polymers cause a more rapid gas escape from the cells when compared to CFC and HCFCs.

As mentioned, the other possibility lies in the modification of the solid polymer matrix. Figure 15 evaluates the influence of the polymer on the solid phase conduction through the second term of (6) and considering a particular model with an f_s variation similar to the one shown in Table 1 adapted for higher densities. It is important to take into account that for relative densities over 0.2 this term is the one with the highest influence on the global conductivity. It can be observed that for relative densities above 0.6 the differences between polymers become important, meaning that in this range it is really important to select a proper polymer in order to regulate the thermal conductivity.

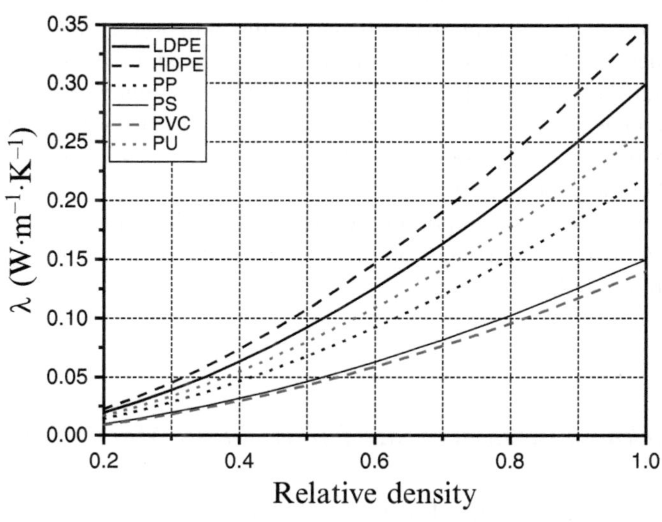

Fig. 15 Effect of the polymer on the conduction term for high density polymer foams (relative density > 0.2)

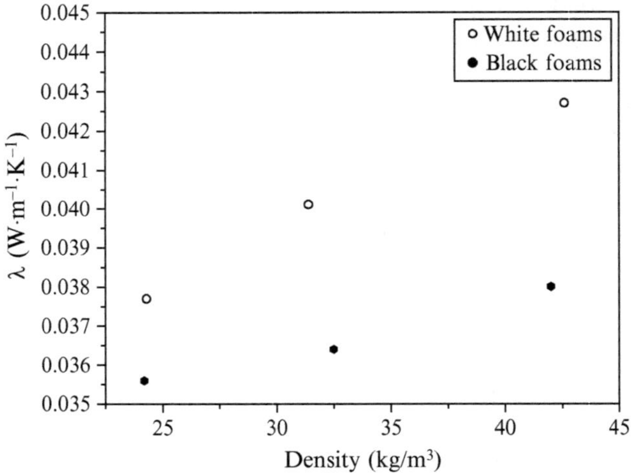

Fig. 16 Experimental data for low density polyethylene (*white* and *black* foams) with similar cellular structures and densities [48]

The limitation of the use of low thermal conductivity blowing agents and the small influence of solid matrix properties for low and ultralow density foams has made it necessary to search for alternative solutions handling the remaining active heat transfer mechanism: radiation. To this end, the thermal conductivity of low density foams is commonly modified by additives that alter the optical properties of the solid matrix (i.e., its refractive index and/or the absorption coefficient). A strong effect of the absorption coefficient on the radiation term is to be expected. Among others, carbon black is one of the most used additives. Figure 16 shows experimental data [48] for low density polyethylene foams with similar cellular structures and densities. The main difference between the two series is the presence of 2% carbon black content in the so-called black foams, significantly reducing the effective thermal conductivity.

Further examples of the importance of the solid matrix formulation can be found in Sect. 4.4, where the effects of the incorporation of a secondary phase are discussed in more detail.

4.2 Thickness Influence: Size Effects in the Macro-Scale

Contrary to the expected trend, it is possible to alter the effective thermal conductivity by modifying the dimensions of the material used, particularly its thickness, especially in the case of low density foams. Examples of this behaviour can be found in Fig. 17.

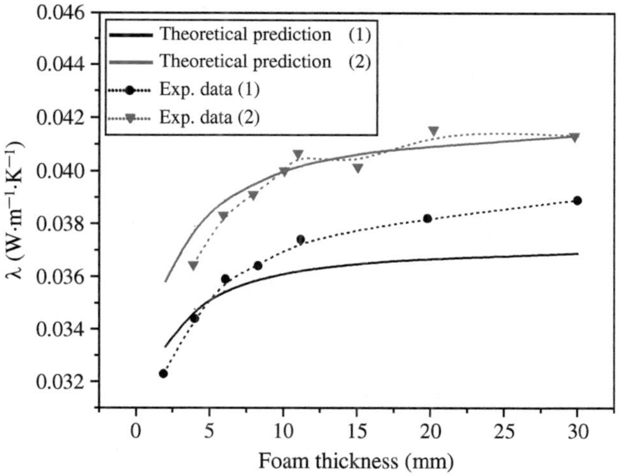

Fig. 17 Effect of foam thickness on the effective thermal conductivity for several low density polyethylene foams. (1) – relative density = 0.017 and $\phi = 313$ μm; (2) – relative density = 0.036 and $\phi = 424$ μm

Both experimental and theoretical data (calculated using (12)) follow a similar trend. The conductivity value rapidly increases with the thickness up to 10 mm and then stabilizes, reaching a nearly constant value. This interesting phenomenon demonstrates again the strong influence of radiation in the thermal conductivity, as radiation is the only mechanism depending on sample size (see (8)).

4.3 *Effect of Processing*

4.3.1 Influence of the Processing Technique

From the different foaming preparation methods described in the introduction, it is possible to identify suitable techniques for each relative density range (Fig. 2) and, as is well known, the relative density is the main parameter influencing the overall thermal conductivity. For the sake of illustration, we will analyze the thermal conductivity results of the polypropylene-based cellular materials produced by the four different lab techniques explained in Sect. 1.2.

Figure 18 displays the thermal conductivity as a function of relative density for foams produced using different methods. There is clear change in the trend for relative densities below 0.2, corresponding to the density threshold below which conduction through the gas phase and radiation become significant. As can be observed, different processing techniques give very similar thermal conductivities in the high density range (relative density > 0.2).

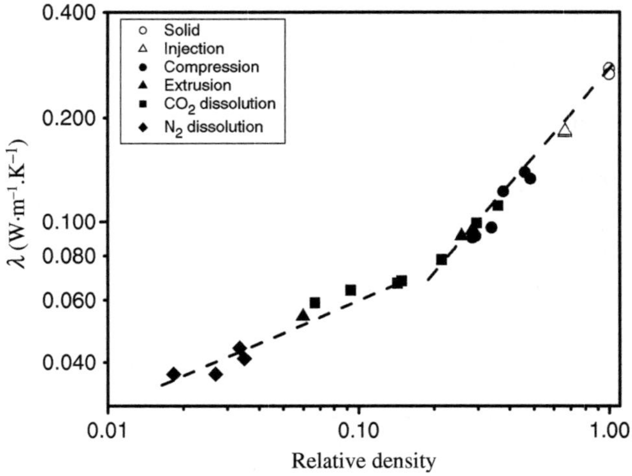

Fig. 18 Thermal conductivity versus relative density for the unfilled polypropylene foams produced using different foaming processes

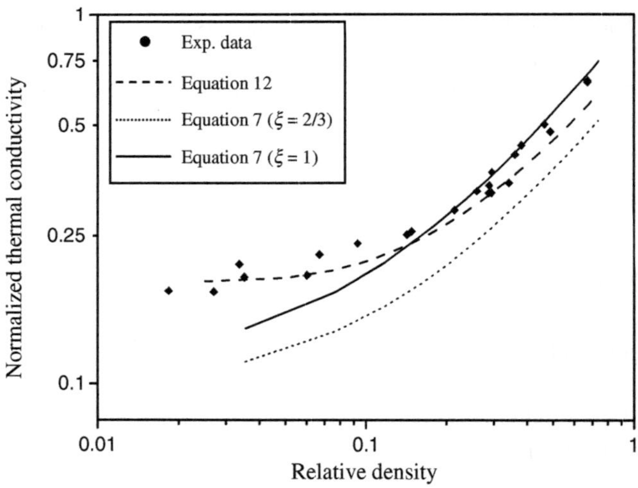

Fig. 19 Experimental normalized data and theoretical model fitting according to (7) and (12)

Model fitting using (7) and (12) are plotted together with the normalized thermal conductivity experimental values in Fig. 19. In the case of (7) (tortuosity, ξ, dependent) two different constant values for tortuosity, $\xi = 1$ and $\xi = 2/3$, are proposed. Please note that in the case of $\xi = 1$ (7) is reduced to the simple parallel model or mixtures rule [2]. In contrast to (7), (12) also considers the radiation contribution.

A value of $\xi = 2/3$ clearly underestimates the conductivity data, while a value of 1 fits with good accuracy the conductivity values for the foams in the relative density range above 0.2, thus enabling the use of the mixtures rule to predict the conductivity of high relative density foams.

Nevertheless, for relative densities below 0.2, (7) clearly underestimates the conductivity data even for a tortuosity value of 1. This is due to the fact that this equation does not consider the radiation term. On the other hand, (12) exhibits a reasonable compliance with experimental data for the whole density range although the prediction seems to be better in the case of low density foams. This fact is based on theoretical model considerations, developed for low density cellular plastics [1]. Although not presented here, other models such as Russell's or Maxwell's would have offered a good compliance with the experimental data [39].

4.3.2 Influence of the Process Parameters: Structure Modification

It is known that the thermal conductivity of low density polymer foams (relative densities below 0.2) is influenced by their cellular structure [1, 2, 51–58]. The effect of cell size, fraction of mass in the cell struts and anisotropy is explained in this section.

Figure 20 shows the theoretical thermal conductivity evolution with relative density for foams based on closed-cell low density polyethylene with the following characteristics: $f_s = 0.2$, $\phi = 880$ µm, and total thickness, $L = 10$ mm. The calculations were performed based on (12) at room temperature. Contrary to the expected

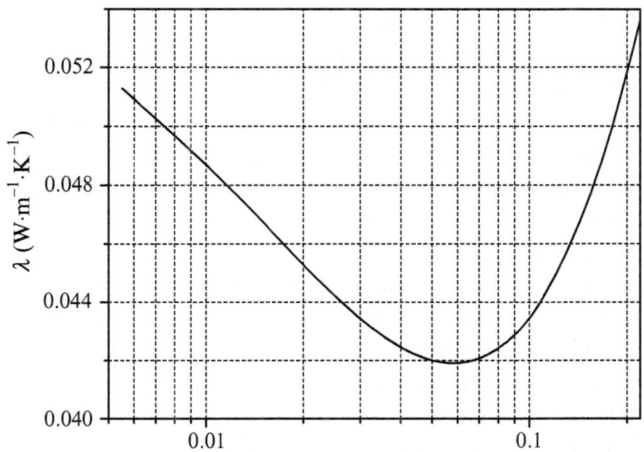

Fig. 20 Theoretical thermal conductivity versus relative density for closed-cell low density polyethylene foams with constant cell size and fraction of mass in the cell struts

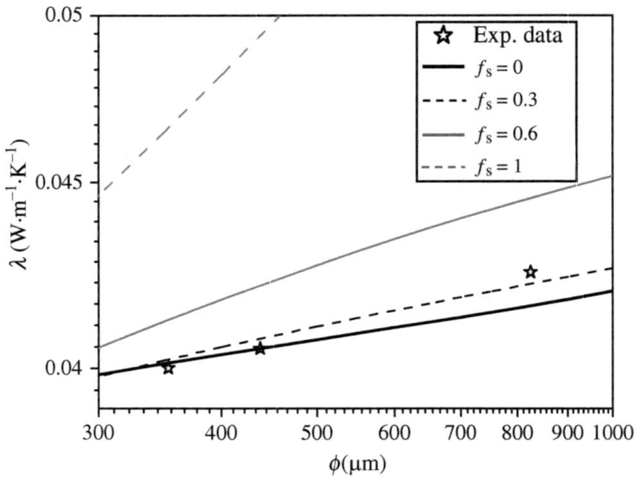

Fig. 21 Theoretical and experimental thermal conductivities versus cell size for low density polyethylene foams with the same chemical composition and relative density

trend, below a relative density of 0.05 the overall conductivity increases. This behaviour can be explained due to the strong influence of radiation at very low densities. This unexpected behaviour has to be taken into account in the case of real thermal insulation applications, selecting the optimum conditions corresponding to the minimum in the curve.

On the other hand, Fig. 21 shows the theoretical predictions of the influence of the fraction of material in the cell struts (f_s) and cell size compared to experimental data for low density polyethylene foams with a relative density of 0.046. An increase in cell size and f_s increases the total conductivity, since both factors contribute to an enhancement of the radiation term. Reducing the cell size keeping density constant introduces a higher number of cell walls in the heat flow path and walls are the main cause of scattering and absorption of the radiative heat flow. The plotted predictions are in close agreement with the experimental data [38] (relative density $= 0.046$ and $f_s \approx 0.3$), validating the above statements.

The influence of f_s is more complicated to explain, since it plays a role in solid conduction (6) and radiation (8), this last due to the dependence of L_s with f_s (Table 1). Reducing f_s, i.e., increasing cell wall thickness and building a more homogeneous distribution of the solid phase, reduces the radiative contribution and increases the conduction contribution. Since in the low density range radiation has a higher weight than conduction, the overall effect of f_s manifests reducing conductivity when this value decreases.

Figure 22 (bottom) shows the experimental thermal conductivity data obtained using the anisotropy mode of the TPS method versus relative density for the PP foams produced by the CO_2 dissolution pressure quench foaming process. While foams with high relative densities are thermally isotropic, i.e., $\lambda_{axial} \approx \lambda_{radial}$, the ones with low relative densities (<0.2) are clearly anisotropic (see Fig. 22, top).

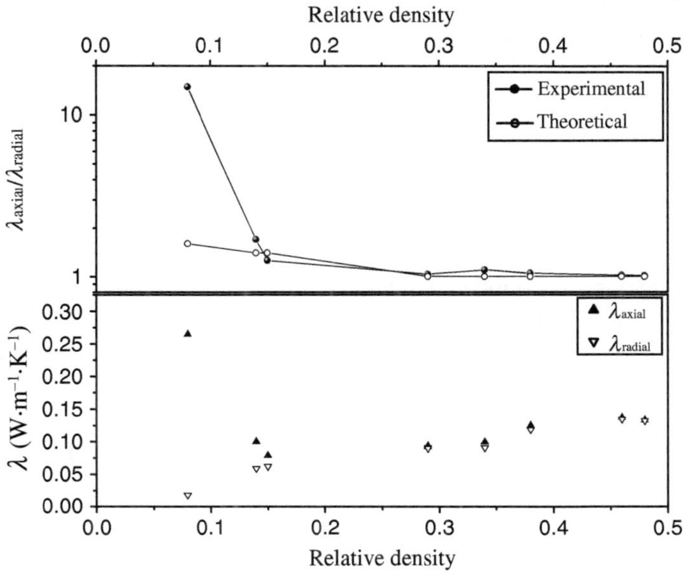

Fig. 22 Radial and axial experimental thermal conductivities (*bottom*) and ratio between the axial and radial thermal conductivities (*top*) versus relative density for the PP foams produced by the CO_2 dissolution pressure quench foaming process

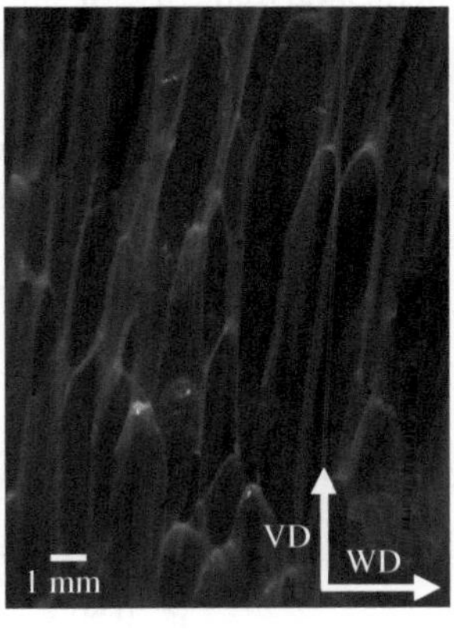

Fig. 23 Honeycomb-like cellular structure of a low density foam produced by the CO_2 dissolution pressure quench foaming process ($AR \approx 9$)

The lower the density, the higher the difference between radial and axial experimental thermal conductivities. Figure 22 (top) shows the experimental and theoretical data, determined based on (5), for the ratio between the axial and radial thermal conductivities. In this case, it can be appreciated that the mismatch among experimental and theoretical values is higher for the foam with the lower density, corresponding to the one with the higher cellular anisotropy ratio ($AR = 9$, see Fig. 23). This is due to the model used to predict these conductivities, which only accounts for conduction. As has previously been shown, for relative densities below 0.2, radiation plays a significant role in the heat transfer, and thus should be taken into account.

4.4 Incorporation of Secondary Phases: Micrometric and Nanometric-Sized Fillers

As stated previously, another strategy to modify the thermal conductivity of polymeric cellular materials and extend their applicability window consists in the incorporation of secondary solid phases, conventionally inorganic in nature.

4.4.1 Incorporation of High Amounts of a Micrometric-Sized Filler ($Mg(OH)_2$)

Mineral reinforcements are frequently incorporated into solid polymers. Depending on the type and amount, these would allow adjusting properties such as mechanical behaviour, flame retardancy or thermal and electrical conductivities. This is the case of magnesium hydroxide ($Mg(OH)_2$), a well know PP flame retardant mineral filler.

This filler is typically added to solid polymers and only a few research works dedicated to the use of similar metal hydroxides in foams with the main objective of improving their flame retardancy can be found [59]. In order to understand its influence on the thermal transport behaviour of foamed materials it is necessary to initially evaluate the evolution of thermal conductivity with filler content in the solid polymer. Experimental results are shown in Fig. 24. As can be seen, the

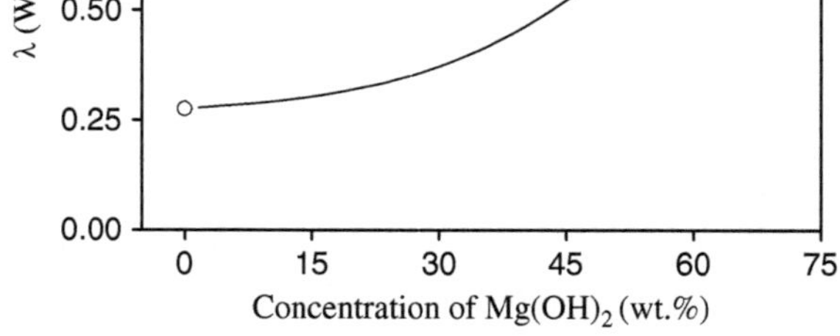

Fig. 24 Evolution of the experimental thermal conductivity of solid PP-$Mg(OH)_2$ composites as a function of $Mg(OH)_2$ concentration

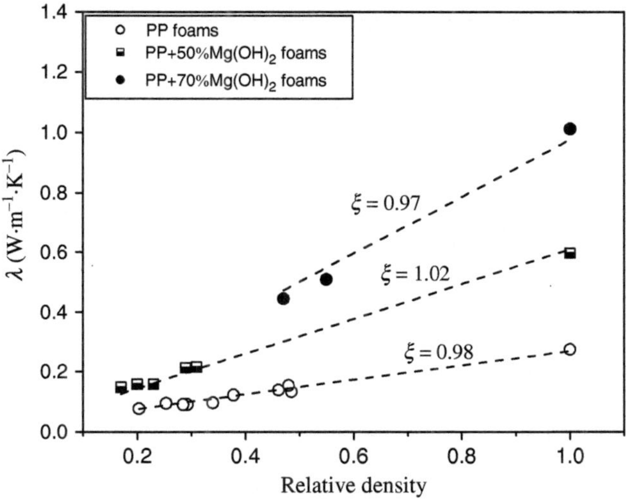

Fig. 25 Comparative experimental thermal conductivities for different composite foams filled with $Mg(OH)_2$ and fitting to (7)

evolution is not linear; this kind of non-linear behaviour is expected according to percolation theory [60]. The values measured for each solid material will be later on used to model the thermal conductivity of the foamed composites.

Figure 25 shows the experimental data for PP foams and PP-$Mg(OH)_2$ as a function of relative density. Using 50 and 70 wt.% of this filler it is possible to increase to a significant extent the conductivity of these materials, especially when adding the highest amount of filler. For instance, a three time increase is detected for foams with relative densities of 0.5 when a 70 wt.% content is used. Equation (7) has been used to predict the values, in this particular case substituting V_s and λ_s by the respective values corresponding to the solid composites (Fig. 24). As shown in Fig. 25, the optimum fitting value for tortuosity is again approximately 1, in accordance with the results obtained for the unfilled foams (Fig. 19).

4.4.2 Incorporation of Carbon Nanofibres

Nanometric-sized fillers are coming more and more important as polymer fillers, despite having started to be commercially used in polymers just a few years ago. On the other hand, nanocomposite foams are relatively new compared to solid nanocomposites.

The influence of incorporating fibrous-like conductive nanofillers (carbon nanofibres, CNF) on the transport properties of solid PP is again evaluated prior to the results for foamed composites (Fig. 26). As the amount of filler incorporated is rather low compared to magnesium hydroxide (see previous section), the influence

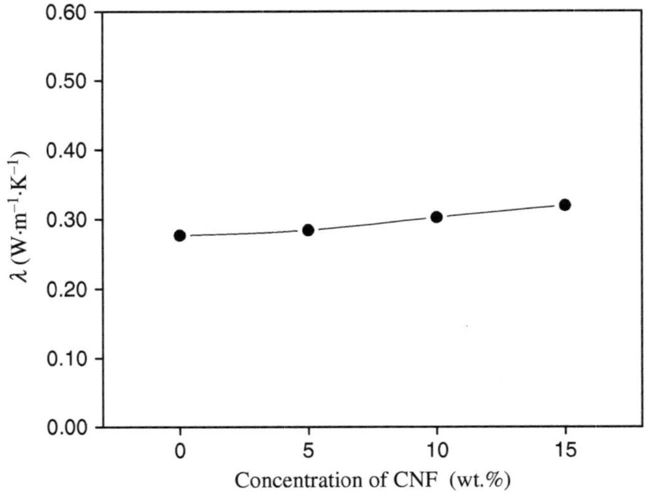

Fig. 26 Evolution of the experimental thermal conductivity of solid PP-CNF composites as a function of CNF concentration

on the overall thermal conductivity of the composite is quite poor, which was not expected considering the extremely high theoretical thermal conductivity assumed for this kind of filler (>2,000 W m^{-1} K^{-1}).

In this case, the trend observed for the thermal conductivity as a function of CNF concentration is almost linear. A linear fitting of the data (mixtures rule) lead to a very small value for the filler thermal conductivity (around 20 W m^{-1} K^{-1}). These discrepancies could be due to both a failure of the model and a probable real value of the conductivity of these CNFs one or two orders of magnitude lower than that expected theoretically. Actually, although theoretical predictions have shown that the axial thermal conductivity of carbon nanotubes is between 2,000 and 6,000 W m^{-1} K^{-1} [18, 61–63], depending on the type of structure and wall disposition, there are considerable discrepancies in the literature, with some authors reporting values as low as 20 W m^{-1} K^{-1} [64]. In our case, the thermal conductivity of the carbon nanofibres is thought to be less than that expected for CNTs due to their inner less perfection and surface, commonly displaying defects due to graphitic edge terminations [65]. Therefore, a conservative value of 20 W m^{-1} K^{-1}, seems to be proper for our analysis.

In Fig. 27 we have plotted the comparative behaviour of both unfilled and CNF composite foams as a function of relative density. The addition of this type of highly conductive nanofiller in percentages below 15 wt.% does not significantly modify the thermal properties of the original polymeric foams.

It is not simple to support this kind of behaviour by only focusing on conventional models (even assuming rather low thermal conductivity values); therefore, additional effects have to be considered. The large difference between single nanotube measurements and the ones done in the as-produced composites suggests

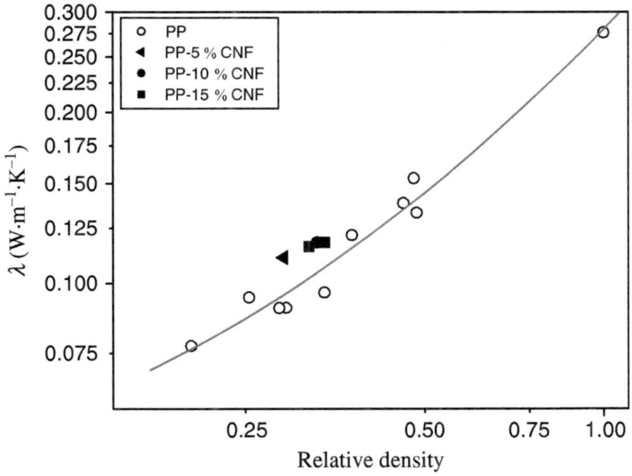

Fig. 27 Thermal conductivity versus relative density for the unfilled and CNF-reinforced PP foams

that highly resistive thermal junctions between the nanotubes due to entanglements and cluster formations dominate the thermal conduction in the composites, thus reducing the expected thermal conductivity. Therefore, low values of thermal conductivity of the nanofibres, possible mechanical damage of this filler during melt-compounding, bad dispersion and interfacial bonding between filler and matrix could be the main reasons behind the unexpected low thermal conductivity for these composites. On the other hand, the electrical conductivity of these materials increases with the carbon nanofibres content [19], suggesting that mechanisms controlling these two properties are very different.

4.4.3 Comparative Analysis

The previous analysis has proven that the addition of highly conductive fillers does not significantly improve the thermal conductivity. Fillers used are representative of both micro and nanoscale and in both cases the observed enhancement is not high compared to the thermal conductivity obtained for the unfilled reference materials. Apart from the reasons exposed above for nanofibres (part of them also applicable to magnesium hydroxide), we would like to mention the thermal resistance problem associated with the physical contact between filler and polymer. As a first approach, considering that the used fillers present a different thermal conductivity it was to be expected solid composites with different thermal conductivities for the same filler content or different trends for both types of materials. Nevertheless, as is shown in Fig. 28, this is not the case, and the values of the thermal conductivity for all solid composites follow the same trend. This seems to confirm that due to thermal resistance the low thermally conductive matrix controls the overall thermal

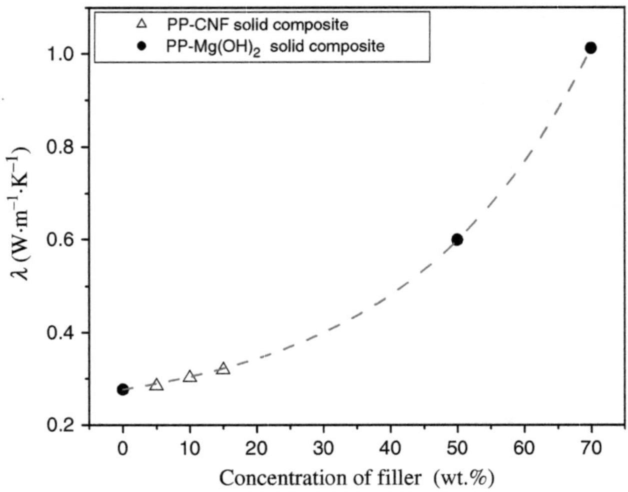

Fig. 28 Solid PP-CNF and PP-$Mg(OH)_2$ composites thermal conductivity versus filler concentration

conductivity, and that the influence of fillers only starts to be relevant when high amounts are added.

Composite foams exhibited a similar behaviour, although the comparison is not so simple due to density variations between materials.

5 Conclusions

This chapter shows that the possibility of varying density, cellular structure, raw materials employed (polymer matrix and gas), and even the incorporation of fillers, creates an enormous combination of possibilities that enables to regulate the thermal conductivity of polymer foams. The main aim was to offer a first approach to understand the mechanisms controlling this property and the methods that can be used to tailor it. With that in mind, the theoretical analysis carried out was based on simple models to predict the different thermal conductivity trends, despite much more advanced models can be found in the literature. The last part of this chapter has been intentionally focused on new emerging types of composite and nanocomposite foams, which are nowadays in the cutting edge of the research in this field.

Acknowledgements Financial assistance from the Local Government (Junta of Castile and Leon, Excellence Group GR39), Spanish Ministry of Science and Innovation and FEDER program (projects MAT 2007-62956, MAT 2009 14001-C02-01 and postdoctoral grant 2008-0946) is gratefully acknowledged.

References

1. Glicksman, L.R.: Heat transfer in foams. In: Hilyard, N.C., Cunningham, A. (eds.) Low Density Cellular Plastics: Physical Basis of Behaviour, 1st edn. Chapman and Hall, UK (1994)
2. Leach, A.G.: The thermal conductivity of foams. I. Models for heat conduction. J Phys D Appl Phys **26**, 733–739 (1993)
3. Shen, J., Han, X., Lee, L.J.: Nanoscaled reinforcement of polystyrene foams using carbon nanofibers. J Cell Plast **42**, 105–126 (2006)
4. Yang, Y., Gupta, M.C., Dudley, K.L., Lawrence, R.W.: Conductive carbon nanofiber–polymer foam structures. Adv Mater **17**, 1999–2003 (2005)
5. Yang, Y., Gupta, M.C., Dudley, K.L., Lawrence, R.W.: Novel carbon nanotube–polystyrene foam composites for electromagnetic interference shielding. Nano Lett **5**, 2131–2134 (2005)
6. Gibson, L.J., Ashby, M.F.: Cellular Solids, 2nd edn. Cambridge University Press, Cambridge (1997)
7. Mills, N.J., Gilcrist, A.: Creep and recovery of polyolefin foams – deformation mechanisms. J Cell Plast **33**, 264–292 (1997)
8. Rodríguez-Pérez, M.A.: Crosslinked polyolefin foams: production, structure, properties, and applications. Adv Polym Sci **184**, 97–126 (2005)
9. Mills, N.: Polymer Foams Handbook, Engineering and Biomechanics Applications and Design Guide, 1st edn, pp. 46–47. Elsevier, Oxford (2007)
10. Klempner, D., Sendijarevic, V.: Polymeric Foams and Foam Technology, 2nd edn, pp. 275–288. Hanser, Munich (2004)
11. Martini, J.E., Suh, N.P., Waldman, F.A.: US Patent 4,473,665, 1984
12. Puri, R.R., Collington, K.T.: The production of cellular crosslinked polyolefins. 2. The injection-molding and press molding techniques. Cell Polym **7**, 219–231 (1988)
13. UK Zotefoams: High performance polymers 2 (1999)
14. Goel, S.K., Beckman, E.J.: Generation of microcellular polymeric foams using supercritical carbon dioxide 1. Effect of pressure and temperature on nucleation. Polym Eng Sci **34**, 1137–1147 (1994)
15. Antunes, M., Haurie, L., Velasco, J.I.: Characterization of highly filled magnesium hydroxide-polypropylene composite foams. J Cell Plast. doi:10.1177/0021955X10370186 (2010)
16. Antunes, M., Realinho, V., Martínez, A.B., Solórzano, E., Rodríguez-Pérez, M.A., Velasco, J.I.: Heat transfer of mineral-filled polypropylene foams. Def Diff Forum **297–301**, 990–995 (2010)
17. Perry, D.L., Phillips, S.L.: Handbook of Inorganic Compounds. CRC, Boca Raton (1995)
18. Kim, P., Shi, L., Majumdar, A., McEuen, P.L.: Thermal transport measurements of individual multiwalled nanotubes. Phys Rev Lett **87**, 215502 (2001)
19. Antunes, M., Velasco, J.I., Realinho, V., Arencón, D.: Characterization of carbon nanofibre-reinforced polypropylene foams. J Nanosci Nanotechnol **10**, 1241–1250 (2010)
20. Antunes, M., Realinho, V., Solórzano, E., Rodríguez-Pérez, M.A., de Saja, J.A., Velasco, J.I.: Thermal conductivity of carbon nanofibre–polypropylene composite foams. Def Diff Forum **297–301**, 996–1001 (2010)
21. Tye, R.P.: Proceedings of the "Cellular Polymers, an International Conference". Rapra Technology Ltd, London (1991)
22. Tye, R.P., Coumou, K.G.: High Temp High Press **13**, 695–704 (1981)
23. ASTM E1225 – 04 Standard test method for thermal conductivity of solids by means of the guarded–comparative–longitudinal heat flow technique
24. ISO 8302:1991; Thermal insulation – determination of steady-state thermal resistance and related properties – guarded hot plate apparatus
25. ISO 13787:2003 Thermal insulation products for building equipment and industrial installations
26. Log, T., Gustafsson, S.E.: Transient plane source (TPS) technique for measuring thermal transport properties of building materials. Fire Mater **19**, 43–49 (1995)

27. Miller, M.G., Keith, J.M., King, J.A., Edwards, B.J., Klinkenberg, N.: Measuring thermal conductivities of anisotropic synthetic graphite–liquid crystal polymer composites. Polym Compos **27**, 388–394 (2006)
28. Baba, T., Ono, A.: Improvement of the laser flash method to reduce uncertainty in thermal diffusivity measurements. Meas Sci Technol **12**, 2046–2057 (2001)
29. ASTM E1461-07 Standard test method for thermal diffusivity by the flash method
30. E1461-92 Standard test method for thermal diffusivity of solids by the flash method
31. Tye, R.P., Kubicàr, L., Lockmuller, N.: The development of a standard for contact transient methods of measurement of thermophysical properties. Int. J. Thermophy **26**, 1917–1938 (2005)
32. Bouguerra, A., Ait-Mokhtar, A., Amiri, O., Diop, M.B.: Measurement of thermal conductivity, thermal diffusivity and heat capacity of highly porous building materials using transient plane source technique. Int Commun Heat Mass Transf **28**, 1065–1078 (2001)
33. Saxena, N.S.G., Pradeep, P., Mathew, G., Thomas, S., Gustafsson, M., Gustafsson, S.E.: Thermal conductivity of styrene butadiene rubber compounds with natural rubber prophylactics waste as filler. Eur Polym J **35**, 1687–1693 (1999)
34. Grujicic, M., Zhao, C.L., Biggers, S.B., Morgan, D.R.: Experimental investigation and modeling of effective thermal conductivity and its temperature dependence in a carbon-based foam. J Mater Sci **41**, 2309–2317 (2006)
35. Nishi, T., Shibata, H., Waseda, Y., Ohta, H.: Thermal conductivities of molten iron, cobalt, and nickel by laser flash method. Metall Mater Trans A **34**, 2801–2807 (2003)
36. Almanza, O., Rodríguez-Pérez, M.A., de Saja, J.A.: The measurement of the thermal diffusivity and heat capacity of polyethylene foams using the transient plane source technique. Polym Int **53**, 2038–2044 (2004)
37. Solórzano, E., Rodríguez-Pérez, M.A., Reglero, J.A., de Saja, J.A.: Density gradients in alumium foams: characterisation by computed tomography and measurements of the effective thermal conductivity. J Mater Sci **42**, 2557–2564 (2007)
38. Solórzano, E., Rodríguez-Pérez, M.A., de Saja, J.A.: Thermal conductivity of cellular metals measured by the transient plane source method. Adv Eng Mater **10**, 596–602 (2008)
39. Solórzano, E., Reglero, J.A., Rodríguez-Pérez, M.A., Lehmhus, D., Wichmann, M., de Saja, J.A.: An experimental study on the thermal conductivity of aluminium Foams by using the transient plane source method. Int J Heat Mass Transf **51**, 6259–6267 (2008)
40. Solórzano, E., Hirschmann, M., Rodríguez-Pérez, M.A., Körner, C., de Saja, J.A.: Thermal conductivity of AZ91 magnesium integral foams measured by the transient plane source method. Mater Lett **62**, 3960–3962 (2008)
41. Gustavsson, M., Karawacki, E., Gustafsson, S.E.: Thermal conductivity, thermal diffusivity and specific heat of thin samples from transient measurement with hot disk sensors. Rev Sci Instrum **65**, 3856–3859 (1994)
42. Almanza, O., Rodríguez-Pérez, M.A., de Saja, J.A.: Applicability of the transient plane source method to measure the thermal conductivity of low density polyethylene foams. J Polym Sci B Polym Phys **42**, 1226–1234 (2004)
43. Coquard, R., Baillis, D.: Numerical investigation of conductive heat transfer in high-porosity foams. Acta Mater **57**, 5466–5479 (2009)
44. Schuetz, M.A., Glicksman, L.R.: Heat Transfer in Foam Insulation. Massachusetts Institute of Technology, Cambridge, MA (1982)
45. Ahern, A., Verbist, G., Weaire, D., Phelan, R., Fleurent, H.: The conductivity of foams: a generalisation of the electrical to the thermal case. Colloids Surf A Phys Eng Asp **263**, 275–279 (2005)
46. Kaviany, M.: Principles of Heat Transfer in Porous Media (Mechanical Engineering Series), 2nd edn. Springer, Berlin (1999)
47. Williams, R.J.J., Aldao, C.M.: Thermal conductivity of plastic foams. Polym Eng Sci **6**, 293–298 (1983)

48. Almanza, O., Rodríguez-Pérez, M.A., de Saja, J.A.: Prediction of the radiation term in the thermal conductivity of crosslinked closed cell polyolefin foams. J Polym Sci B Polym Phys **38**, 993–1004 (2000)
49. Rodríguez-Pérez, M.A., González-Peña, J.I., Witten, N., de Saja, J.A.: The effect of cell size on the physical properties of crosslinked closed cell polyethylene foams produced by a high pressure nitrogen solution process. Cell Polym **21**, 165–194 (2002)
50. Kuhn, J., Ebert, H.P., Arduini-Schuster, M.C., Büttner, D., Fricke, J.: Thermal transport in polystyrene and polyurethane foam insulations. Int J Heat Mass Transf **35**, 1795–1801 (1992)
51. Russel, R.H.: Principles of heat flow in porous insulators. J Am Ceram Soc **18**, 1–5 (1935)
52. Bedeaux, D., Kapral, R.: The effective reaction rate and diffusion coefficients for a two-phase medium. J Chem Phys **79**, 1783–1788 (1983)
53. Boetes, R., Hoogendoorn, C.J.: Heat transfer in polyurethane foams for cold insulation. Proc. Int. Symp. Heat Mass Transf **24**, 14–31 (1987)
54. Loeb, A.L.: Thermal conductivity: VIII, A theory of thermal conductivity of porous materials. J Am Ceram Soc **37**, 96–99 (1954)
55. Francl, J., Kingery, W.D.: Thermal conductivity: IX. Experimental investigation of effect of porosity on thermal conductivity. J Am Ceram Soc **37**, 99–107 (1954)
56. Batty, W.J., Probert, S.D., O'Callaghan, P.W.: Apparent thermal conductivities of high-porosity cellulary insulants. Appl Energy **18**, 117–135 (1984)
57. Sims, G.L.A., Khunniteekool, C.: Cell size measurement of polymeric foams. Cell Polym **13**, 137–146 (1994)
58. Almanza, O., Rodríguez-Pérez, M.A., de Saja, J.A.: The thermal conductivity of polyethylene foams manufactured by a nitrogen solution process. Cell Polym **18**, 385–401 (1999)
59. Román-Lorza, S., Rodríguez-Pérez, M.A., de Saja, J.A., Zurro, J.: Cellular structure of EVA/ATH halogen-free flame-retardant foams. J Cell Plast **46**, 259–279 (2010)
60. Zhang, G., Xia, Y., Wang, H., Tao, Y., Tao, G., Tu, S., Wu, H.: A percolation model of thermal conductivity for filled polymer composites. J Compos Mater **44**, 963–970 (2010)
61. Berber, S., Kwon, Y.K., Tomanek, D.: Unusually high thermal conductivity of carbon nanotubes. Phys Rev Lett **84**, 4613–4616 (2000)
62. Che, J., Cagin, T., Goddard, W.A.: Thermal conductivity of carbon nanotubes. Nanotechnology **11**, 65–69 (2000)
63. Osman, M., Srivastava, D.: Temperature dependence of the thermal conductivity of single-wall carbon nanotubes. Nanotechnology **12**, 21 (2001)
64. Yi, W., Lu, L., Dian-lin, Z., Pan, Z.W., Xie, S.S.: Linear specific heat of carbon nanotubes. Phys Rev B **59**, R9015–R9018 (1999)
65. Shaffer, M.S.P., Sandler, J.K.W.: Carbon nanotube/nanofibre polymer composites. In: Advani, S.G. (ed.) Processing and Properties of Nanocomposites. World Scientific, Singapore (2006)

56. Eucken, W.: [illegible] Warmeleitfahigkeit [illegible]

Heat Transfer in Polymer Composites Filled with Inorganic Hollow Micro-Spheres

J.Z. Liang

Abstract The advances in studies on the heat transfer in polymer composites filled with inorganic hollow microsphere were reviewed, and the heat transfer process and mechanisms in the polymer/inorganic hollow microsphere composites was described in this chapter. On the basis of the law of minimal thermal resistance and the equal law of the specific equivalent thermal conductivity, a theoretical model of heat transfer in polymer/inorganic hollow micro-sphere composites was established and the corresponding equation of effective thermal conductivity was derived. The simulation of the heat transfer in the polymer/inorganic hollow microsphere composites was made by means of a finite element method. The measurement instrument and methodology of the effective thermal conductivity of the polymer/inorganic hollow microsphere composites were introduced, and the effective thermal conductivity (k_{eff}) of polypropylene composites filled with hollow glass bead (HGB) was measured. It was found that the effective thermal conductivity decreased linearly with an addition of the volume fraction (ϕ_f) of the beads. The simulated keff of the PP/HGB composites decreases also as a linear function with an increase of ϕ_f, and are roughly close to the experimental measured data, the k_{eff} with a 3D model are higher than those with a 2D model, and the difference between them increases with an increase of ϕ_f. Moreover, good agreement was showed between the measurements from the PP/HGB composites within the ϕ_f range from 0 to 20 % and the theoretical predictions by means of this equation.

J.Z. Liang
Research Division of Green Function Materials and Equipment, College of Industry Equipment and Control Engineering, South China University of Technology, Guangzhou 510640, P.R. China
e-mail: liangjz@scut.edu.cn

A. Öchsner and G.E. Murch (eds.), *Heat Transfer in Multi-Phase Materials,* Adv Struct Mater 2, DOI 10.1007/8611_2010_3,

1 Introduction

1.1 Outline

Heat transfer in polymer composites is a complicated process, especially for the inorganic hollow micro-sphere filled polymer composite systems. In this chapter, the heat transfer process and mechanisms in polymer composites filled with inorganic hollow micro-spheres are analyzed and studied, measurement methods of the thermal conductivity of the composite systems are described, and experimental results are discussed. Moreover, the relevant theoretical models of the heat transfer as well as mathematical models of the effective thermal conductivity for these composites are established.

This chapter includes four sections: section one is an introduction; section two is concerned with heat transfer theory in polymer/inorganic hollow microsphere composites; section three is concerned with the measurement of the effective thermal conductivity of polymer/inorganic hollow microsphere composites and section four is concerned with the simulation of heat transfer in polymer/inorganic hollow microsphere composites.

1.2 Background

Foamed or porous plastics are one of the polymers usually used as thermal insulation and sound insulation materials. However, their application as engineering structural materials is considerably limited due to their poor mechanical properties such as tensile and flexural strength, impact fracture toughness as well as stiffness. The research focus, therefore, is how to fabricate a kind of porous plastic that is light, has a good mechanical strength and toughness, as well as good thermal and sound insulation properties. Rigid hollow micro-spheres (e.g. hollow glass beads, hollow ceramic beads, rigid hollow plastic beads, and so on) contain inertia gas and have some advantages such as a low thermal conductivity coefficient, sound insulation and low weight. In addition, owing to their smooth spherical surface, these micro-particles do not generate important stress concentration in the interface between the fillers and the matrix. As a result, a new type of filler material has been developed recently and has been used in industries, especially in the polymer industry, shipbuilding industry and construction industry. If they are used to fill and modify resins, then polymer/hollow micro-sphere composites having thermal insulation, sound insulation, low weight and good mechanical properties may be prepared [1]. Now, these kind of composites are applied in building materials, space-flight and aviation industry and so on.

1.3 Research Progress

The effective thermal conductivity is an important characterization of heat transfer properties of materials. However, there have been few studies on the measurement and characterization of the effective thermal conductivity for polymer/hollow microsphere composites. In 2006, Liang and Li [2] investigated the influence of the filler size and content on the thermal conductivity of hollow glass bead filled polypropylene composites. The results showed that the thermal conductivity decreased nonlinearly with increasing volume fraction of the hollow glass beads, while it increased with reducing the particle diameter under the same test conditions. More recently, Yung and a colleague [3] measured the thermal conductivity of hollow glass microsphere-filled epoxy-matrix composites and also found that the thermal conductivity of the composites decreased nonlinearly with increasing volume fraction of the hollow glass microspheres.

As stated above, the heat transfer process and mechanisms in porous materials and polymer composites are very complicated, especially for polymer/hollow microsphere composites. It is quite important, therefore, to present some equations or expressions for predicting or estimating the effective thermal conductivity during heat transfer process in polymer composites. For porous materials, several researchers [4, 5] have recently derived respectively effective thermal conductivity equations based on the Maxwell expression, or have established a more accurate formula for calculating the effective thermal conductivity of porous materials [6]. Relatively, the models proposed respectively by Nielsen [7] and Cheng-Vochon [8] may better estimate the effective thermal conductivity of filled composite materials, while the Agari-Nagai equation can predict the effective thermal conductivity of composites with high-loading [9]. Liang [10] analyzed the thermal conductivity of a porous material with closed spherical and cylindrical holes. Suvorov et al. [11] studied the thermal conductivity of hollow emery filled composites. Recently, Hill and Supancic [12] proposed an indirect method to determine this interfacial boundary resistance by preparing large-scale "macromodel" simulations of the polymer-ceramic interface. They also investigated the effects of similar size and shape of platelet-shaped particles on the thermal conductivity of polymer/ceramic composite materials [13]. Yu et al. [14] measured the thermal conductivity of polystyrene-aluminum nitride composites and found that the thermal conductivity of the composites was higher for a polystyrene particle size of 2 mm than that for a particle size of 0.15 turn. The thermal conductivity of the composite was five times that of pure polystyrene at about 20% volume fraction containing aluminum nitride (AIN) for the composite containing 2 mm polystyrene particle size. Recently, Liang and Li [15] studied the heat transfer in polymer composites filled with inorganic hollow micro-spheres and proposed a theoretical model. On the basis of this model, they presented an equation of the thermal conductivity for polymer/hollow microsphere composites. More recently, Liang and Liu [16] researched heat transfer in phenolic composites filled respectively with aluminium powder and graphite powder and established a new theoretical model of heat transfer in the particulate-filled polymer composite systems based on the law of minimal thermal resistance and the equal law of the specific equivalent thermal conductivity.

1.4 Study Sense

In general, solid particulate-filled polymer composites are two or multiple-element composite systems, while a polymer filled with rigid hollow micro-spheres belongs to the kind of multiple-phase composite systems, because these particles contain inertia gas. Consequently, the heat transfer process and mechanisms in the polymer/rigid hollow micro-sphere composites are more complicated than those in polymer/solid particle composites. In the polymer industry, inorganic hollow micro-spheres are used as filler extensively more and more due to their low cost, good physical-chemical and mechanical performances, as well as good processing properties. As a result, it is quite meaningful for the development and for the engineering application of the heat transfer theory in polymeric materials that one understands the heat transfer process and mechanisms in the polymer/rigid hollow micro-sphere composites as well as providing for a quantitative description of the thermal properties of these materials.

2 Heat Transfer Theory in Polymer/Inorganic Hollow Micro-Spheres

2.1 Heat Transfer Mechanisms

In general, the thermal conductivity in thermal insulation materials is the synthetic effect of heat conduction, convection and radiation. According to the second law of thermodynamics, heat always transfers spontaneously from a high temperature body to a low temperature one. Namely, heat transfer will conduct where there is a difference in temperature. Generally, insulation materials only reduce the strength of heat exchange, and have a property of blocking heat transfer. As stated above, polymer/hollow micro-spheres composite is a kind of ternary composite, it includes three phases, namely resin, gas and sphere shell. During heat transfer in polymer/hollow micro-spheres composites, when the heat quantity is close to a hollow micro-sphere, only a small part of the heat quantity will conduct by it, while a greater part of the heat quantity will move around it due to its low conductivity, as shown in Fig. 1. Because of the low thermal conductive coefficient of the hollow micro-spheres and longer heat transfer route and complication in the filled systems, the thermal conductivity of these composites will be reduced.

It can be seen from Fig. 1 that heat transport in inorganic hollow micro-sphere filled polymer composites has three kinds of ways: (1) thermal conduction by solid; (2) heat radiation on the surface between neighboring hollow particles; (3) the natural thermal convection of gas in the hollow particles. After finishing the experiments, Skochdopole [17] pointed out that the natural thermal convection of the gas in a micro-bubble would not occur when the bubble diameter was less than 4 mm. Because the diameter of the hollow micro-spheres as fillers is usually less than 0.1 mm, the natural thermal convection of the gas in it may be neglected. Furthermore,

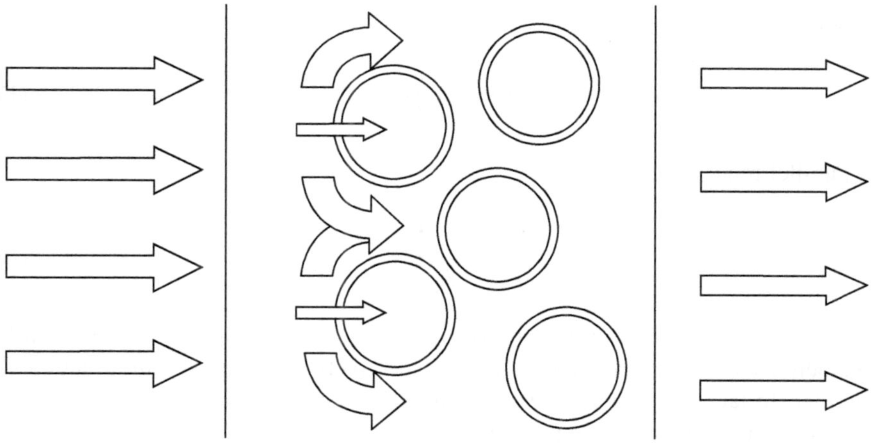

Fig. 1 Diagram of heat transfer in polymer/hollow microsphere composites

a polymer composite works usually under lower temperature conditions where the proportion of the thermal radiation in the total heat transfer is very small, hence thermal radiation may also be neglected.

Generally, the heat transfer process in inorganic hollow micro-spheres filled polymer composites is more complicated than that in other solid particulate-filled polymer composite systems because the former is a type of material with three phases, namely resin, gas and sphere shell.

2.2 Heat Transfer Analysis

As discussed above, the heat transport in inorganic hollow micro-sphere filled polymer composites has three kinds of ways. Namely solid thermal conduction, heat radiation on the surface between neighboring hollow particles, as well as the natural thermal convection in the hollow particles. According to the experiments by Skochdopole [17], the natural thermal convection of the gas in a micro-bubble would not occur when the bubble diameter was less than 4 mm, thus the natural thermal convection of the gas in it may be neglected. Furthermore, polymer composite works usually under lower temperature conditions, the proportion of the thermal radiation in the total heat is very small, the thermal radiation, therefore, may be not considered.

2.3 Basic Equations of Thermal Conduction in Composites

Now the basic equations of thermal conduction usually used in composites are mainly two kinds of types: series model and parallel model. For a binary composite system, two simple expressions can be written respectively,

Series model:

$$k_e = \left(1 - \phi_f\right)k_c + \phi_f k_d \tag{1}$$

Parallel model:

$$\frac{1}{k_e} = \frac{1 - \phi_f}{k_c} + \frac{\phi_f}{k_d} \tag{2}$$

where k_e is the thermal conductivity of composite, k_c and k_d are the thermal conductivity of the matrix and the filler respectively, ϕ_f is the volume fraction of the filler. In fact, one can use the series model or parallel model alone, or use both the two models, according to practical circumstance.

2.4 Heat Transfer Element Analysis

According to the above discussion, the theoretical analysis of the heat transfer in this section is based on the following suppositions: (1) the structure change of the composites is periodicity, and the ratio of this periodicity to the whole composite materials very small; (2) the temperature distribution along the direction of heat flow is linear. Because a periodical element is very small to the total body, the influence of this hypothesis on the total heat transfer is insignificant. Figure 2 shows a physical model of the heat transfer process of a hollow micro-sphere filled polymer composite. Now we select an element from the composite for analysis. The element is a straight cube with side length of H, and there is a hollow micro-sphere with internal radius of r and external radius of R in the element. The element is divided into polymer phase, the micro-sphere shell phase and the gas phase. The heat quantity Q transfers from the bottom to the top.

2.5 Mathematical Model

The element analysis of heat transfer in a polymer/hollow micro-sphere composite is shown in Fig. 3. The temperature in the materials is lower than the softening point temperature of the resins. The element is divided into two parts: part I and part II. Part I represents the pure polymer, and the height is h_1, namely $h_1 = H - 2R$. Part II is the compound with the polymer and a hollow micro-sphere, and the height is h_2 ($h_2 = 2R$). According to the law of minimal resistance and the equal law of the specific equivalent thermal conductivity [18], if constituting the heat resistances of these two parts, one may get the heat resistance of the whole element.

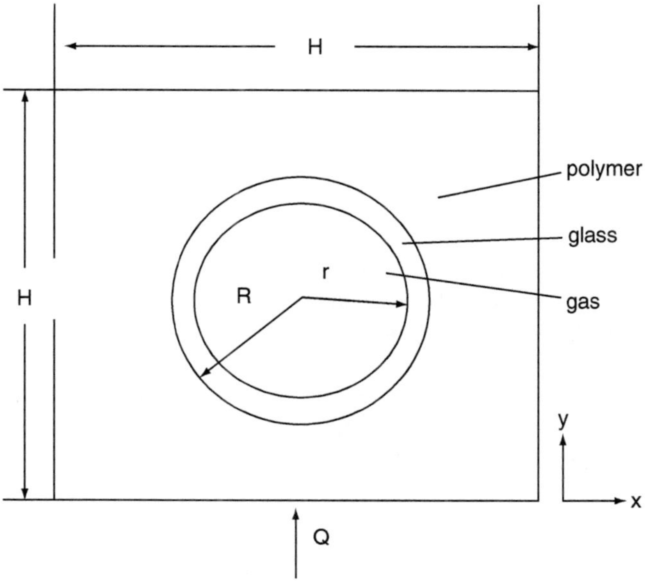

Fig. 2 Physical model of heat transfer

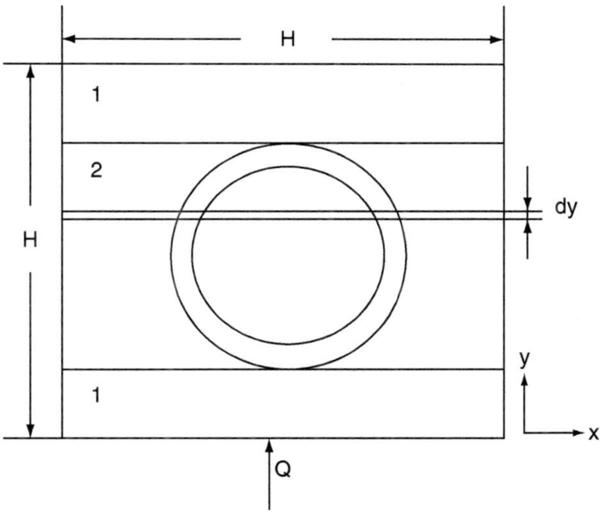

Fig. 3 Element analysis of heat transfer

Consequently, the equivalent thermal conductivity of the whole element may be determined.

Part I:

$$k_1 = k_p \tag{3}$$

Part II: Taking a thin piece with thickness of dy, according to Fourier's theorem, k_2 is given by:

$$\begin{aligned} k_2 &= (Q_p + Q_g + Q_a)/(\frac{dT}{dy} \cdot S) \\ &= k_p \frac{S_p}{S} + k_g \frac{S_g}{S} + k_a \frac{S_a}{S} \end{aligned} \tag{4}$$

where, T is the temperature, S is the area of whole across-section. k_p, k_g and k_a are the thermal conductivity of polymer matrix phase, micro-sphere shell phase and gas phase, respectively. S_p, S_g and S_a are respectively the cross section area of polymer, micro-sphere shell and internal micro-sphere. Q_p, Q_g and Q_a are the heat quantity through the polymer matrix, micro-sphere shell and gas, respectively.

Because of the linear distribution of temperature, the average thermal conductivity of each section may be first obtained as:

Part I:

$$\bar{k_1} = \int_{h_1} k_1 dy/h_1 = k_p \tag{5}$$

Part II:

$$\begin{aligned} \bar{k}_2 &= \frac{1}{h_2} \int_{h_2} \left(k_p \frac{S_p}{S} + k_g \frac{S_g}{S} + k_a \frac{S_a}{S} \right) dy \\ &= \frac{1}{h_2 S} (k_p V_p + k_g V_g + k_a V_a) \end{aligned} \tag{6}$$

where V_p, V_g and V_a are the volume of the polymer matrix, micro-sphere shell and gas, respectively.

According to the series theorem of heat resistance, the effective thermal conductivity of the composite, k_{eff} is given by:

$$\begin{aligned} k_{eff} &= \frac{H}{RS} = \frac{H}{(R_1 + R_2)S} \\ &= \frac{H}{\left(\frac{h_1}{k_p S} + \frac{h_2{}^2}{k_p V_p + k_g V_g + k_a V_a} \right) S} \end{aligned} \tag{7}$$

where R is the total heat resistance of the element, R_1 and R_2 are respectively the heat resistances of Part I and Part II. For the hollow micro-spheres, we have that:

$$V_g \rho_g + (V_s - V_g) \rho_a = V_s \rho_s \tag{8}$$

where ρ_g, ρ_a and ρ_s are the effective density of sphere shell, gas and micro-sphere respectively, V_s is the total volume of the hollow micro-spheres. Thus, one can get the expressions among the volume and density of each composition as well as the volume fraction (ϕ_f) of the hollow micro-spheres. Finally, substituting these expressions into Equation (8), we have that:

$$k_{eff} = \left[\frac{1}{k_p}\left(1 - \frac{6\phi_f}{\pi}\right)^{\frac{1}{3}} + 2\left(k_p\left(\frac{4\pi}{3\phi_f}\right)^{\frac{1}{3}} + \pi\left(\frac{2\phi_f}{9\pi}\right)^{\frac{1}{3}}\left(k_g\left(\frac{\rho_s - \rho_a}{\rho_g - \rho_a} + k_a\left(\frac{\rho_g - \rho_s}{\rho_g - \rho_a}\right) - k_p\right)\right)\right)^{-1}\right]^{-1} \tag{9}$$

2.6 *Theoretical Estimation*

The composites investigated in this section were a polypropylene (PP) composite system filled with two kinds of hollow glass beads (TK35 and TK70). These hollow glass beads (HGB) were supplied by Eco. & Chimie Co. Ltd. (Guangzhou, China), and the mean diameters and effective density of TK35 were 35 μm and 680 kg/m^3, the mean diameters and effective density of TK70 were 70 μm and 210 kg/m^3, respectively. The density of the sphere shell was 2,210 kg/m^3, and the thermal conductivity was 0.179 W/m K. The gas in the beads was an inert gas, and the density and thermal conductivity were 0.0899 kg/m^3 and 0.0228 W/m K. The thermal conductivity and density of the PP resin were 0.2 W/m K and 915 kg/m^3 respectively.

Substituting respectively the stated above data of the two filled systems into Equation (9), one may estimate the effective thermal conductivity k_{eff} of the composites corresponding to different volume fraction of hollow glass beads, and the results are shown in Fig. 4. With an addition of the HGB volume fraction, the theoretical estimations of k_{eff} of PP/HGB composites decreases linearly, and the values of k_{eff} for the PP/TK35 system are slightly higher than those for the PP/TK70 system under the same conditions.

3 Measurement of Effective Thermal Conductivity

3.1 *Raw Materials*

An injection grade of polypropylene (PP) with trade mark of CJS-700, supplied by Guangzhou petrochemical Co. Ltd in China, was used as the matrix resin, the

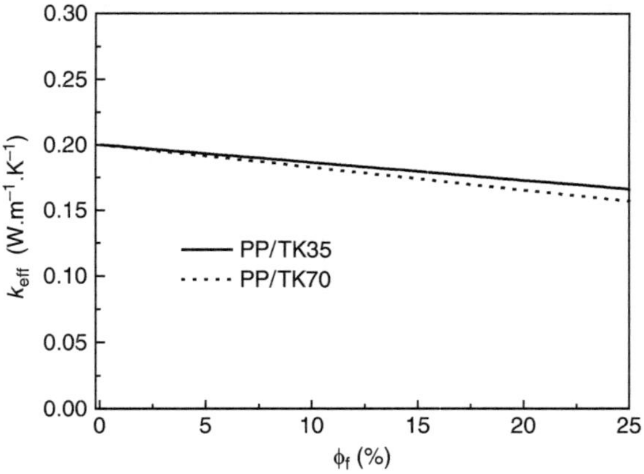

Fig. 4 Theoretical estimations of k_{eff} for PP/HGB composites

density and melt flow index (230°C, 2.16 kg) of the resin were 910 kg/m^3 and 12 g/10 min, respectively.

Two kinds of hollow glass beads (HGB) supplied by Molüs Co. Ltd in Germany, TK35 and TK70, with different size were used as the fillers in this section. The mean diameters of the fillers were 35 and 70 μm, and the density was 680 and 210 kg/m^3, respectively. The surface of the particles was pretreated with silane coupling agent. The particle size distribution of the fillers was measured by means of a laser size instrument (Model LS-C(I) supplied by Omik Co. Ltd in Zhuhai, China.

3.2 Sample Preparation

After simple mixing, the PP resin and the HGB with different proportions were compounded in a twin-screw extruder. The blending was conducted in a temperature range of 160–230°C and a screw speed of 25 r/min, and then the extrudate was granulated to produce the composites. The volume fractions of the HGB were 0, 5, 10, 15 and 20%, respectively. The specimens for thermal conductivity measurement were molded by using an injection molding machine in the temperature range of 160–240°C after drying the composites. The geometry of these specimens includes a diameter of 50 mm and a thickness of 6 mm.

3.3 Apparatus and Methodology

The thermal conductivity of the composites was measured by means of a protecting heat plate method in this test, and the main apparatus was a protecting heat flow type of thermal conductivity instrument (model NF-7) supplied by South China

University of Technology (see [2]). The environmental temperature for the test was 27°C. The specimens were plates with a length of 50 mm, a width of 50 mm and a thickness of 6 mm. Four measuring points were set up equally on a plate, and the average was reported for each specimen.

3.4 Results and Discussion

3.4.1 Experimental Results

Figure 5 shows the dependence of the effective thermal conductivity (k_{eff}) of PP/TK35 and PP/TK70 composite systems on the volume fraction (ϕ_f). It can be seen that the k_{eff} of the composites decreases with an increase of ϕ_f. When ϕ_f is less than 15%, the values of k_{eff} of PP/35 filled system are greater than those of PP/70 filled system. This indicates that the heat insulation properties for the composite systems with filled bigger diameter of hollow micro-spheres are good at lower inclusion concentrations. This comes about because when the thickness to diameter ratio of the hollow micro-spheres is fixed, the bigger the particle size, there is more gas in it (density reduction) under constant range of particle diameter, resulting in a reduction of the effective thermal conductivity. When ϕ_f is more than 15%, the values of k_{eff} of the PP/35 filled system are lower than those of the PP/70 filled system. It might be that the number of the TK35 with small particle diameter increases obviously at higher filler concentration, leading to improvement of the heat insulation properties of the material.

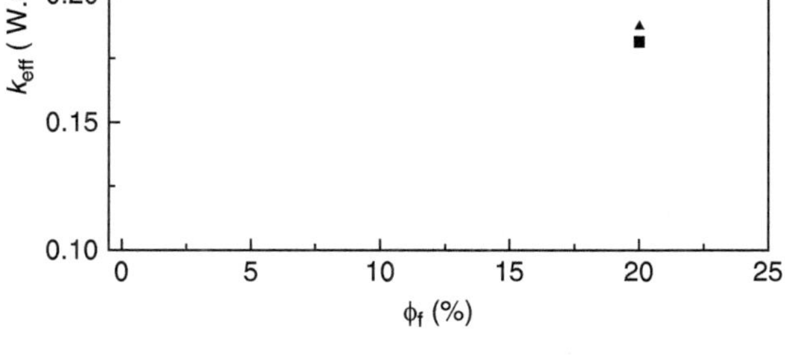

Fig. 5 Dependence of effective thermal conductivity on HGB volume fraction of composites

3.4.2 Theoretical Prediction

The composites investigated in this section were also a polypropylene (PP) filled with two kinds of hollow glass beads (TK35 and TK70). These hollow glass beads (HGB) were supplied by Eco. & Chimie Co. Ltd. (Guangzhou, China). The mean diameter and effective density were 35 μm and 680 kg/m^3 for TK35, and 70 μm and 210 kg/m^3 for TK70 respectively. The density of the sphere shell was 2,210 kg/m^3, and the thermal conductivity was 0.179 (W/mK). The gas in the beads was an inert gas, and the density and thermal conductivity were 0.0899 kg/m^3 and 0.0228 W/m K, respectively. The thermal conductivity and density of the PP resin were 0.2 W/m K and 915 kg/m^3, respectively.

Substituting this data for the two filled systems into Equation (9), one may estimate the effective thermal conductivity k_{eff} of the composites corresponding to different volume fraction of hollow glass beads, and the results are shown in Figs. 6 and 7. With an increase of the HGB volume fraction, the theoretical estimations of k_{eff} of PP/HGB composites decrease linearly, and the values of k_{eff} for PP/TK35 system are slightly higher than those for the PP/TK70 system under the same conditions.

3.5 *Comparison Between Predictions and Measurements*

Plotting respectively the measured data of the effective thermal conductivity from the experimental of these two filled PP composite systems (as shown in Fig. 5) in Figs. 6 and 7, one may verify preliminarily Equation (9). It can be seen that the theoretical estimations of the effective thermal conductivity are good and consistent

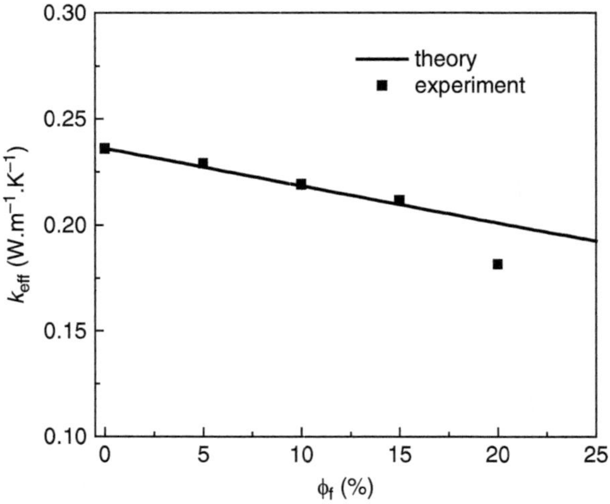

Fig. 6 Comparison between predictions and measured data of the effective thermal conductivity (PP/TK35 system)

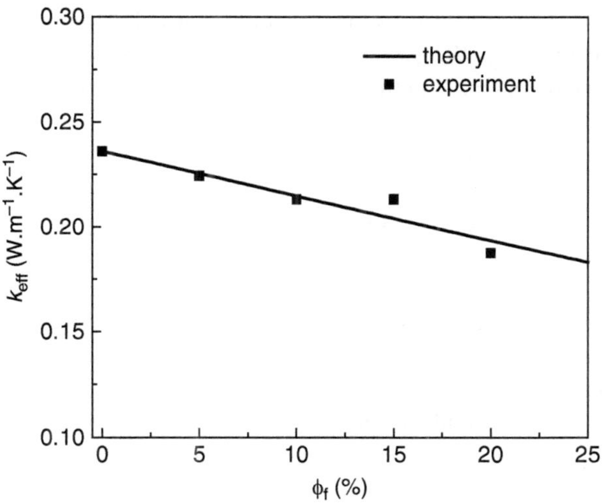

Fig. 7 Comparison between predictions and measured data of the effective thermal conductivity (PP/TK70 system)

with the measured data, and k_{eff} decreases linearly with an increase of ϕ_f, except for an individual data point. In the previous work, the authors [19] simulated the two dimensional heat transfer process in these filled systems stated above by using the finite element software ANSYS, and the results showed that the trend of the simulations are similar to the theoretical predictions. This indicates that the mathematical model (9) may describe better the relationship between the effective thermal conductivity of inorganic hollow micro-spheres filled polymer composites and material parameters when the concentration of the particles is low and the dispersion of these inclusions in the resin matrix is uniform.

4 Simulation of Heat Transfer in Polymer/Inorganic Hollow Microspheres

4.1 Basic Equation of Temperature Field

For a general 3D temperature field, if the quantity of heat required for raising the temperature by a micro-element body and the quantity of heat generated by the heat source in the micro-element body is at equilibrium, then the variables of temperature filed [*T*(*x*, *y*, *z*, *t*)] in a right angle coordinate system satisfy the following heat equilibrium equation:

$$\frac{\partial^2 T}{\partial x^2}+\frac{\partial^2 T}{\partial y^2}+\frac{\partial^2 T}{\partial z^2}+\frac{\dot{q}}{\lambda}=\frac{1}{a}\frac{\partial T}{\partial \tau} \quad (10)$$

where q is the heat flow density, W/m^2; α is the thermal diffusivity, W/m^2·K; λ is the thermal conductivity, W/m·K; τ is the heat conduction time, h. Three kinds of heat conduction boundary conditions and initial conditions can be assigned as follows [20]:

The first kind of boundary condition:

$$\begin{cases} T|_\Gamma = T_w \\ T|_\Gamma = F(x,y,t) \end{cases} \tag{11}$$

where is the body boundary, the direction is reverse clock hands, T is the known wall temperature, K, F (x, y, t) is the known temperature function (variation with time and position).

The second kind of boundary condition:

$$\begin{cases} -k\frac{\partial T}{\partial n}|_\Gamma = q_2 \\ -k\frac{\partial T}{\partial n}|_\Gamma = g(x,y,t) \end{cases} \tag{12}$$

where $g(x, y, t)$ is the heat flow density function.

The third kind of boundary condition:

$$-k\frac{\partial T}{\partial n}|_\Gamma = a(T - T_f)|_\Gamma \tag{13}$$

where α and T_f may be constants, and may also be the function varying with time and position. If α and T_f are not constants, then the average values are usually separately taken as constants in the numerical calculation.

Initial condition

$$\begin{cases} T|_{t=0} = T_0 \\ T|_{t=0} = \phi(x,y) \end{cases} \tag{14}$$

where T_0 is the beginning temperature, K; φ (x, y) is the temperature function.

4.2 *Finite Element Analysis of Heat Transfer*

4.2.1 Physical Model

The heat transfer process in a hollow micro-sphere filled polymer composite may also be investigated with a numerical simulation method based on the fundamental equations discussed above. Supposing that these hollow micro-spheres are distributed uniformly in the polymer matrix, and considering their geometry symmetry, then the 3D temperature field may be simplified as a 2D heat transfer problem. Figure 8

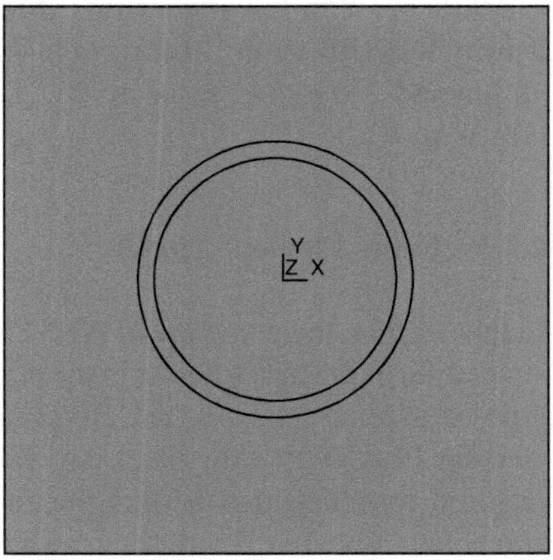

Fig. 8 An element model of heat transfer (PP/TK35)

shows the geometry of the physical real thermal model for a unit cell in the composite. It can be seen that the polymer/hollow microsphere composite is a ternary or three-phase composite system, namely, resin phase, micro-sphere wall phase and gas phase. A quantity of heat is transferred into the polymer from outside to produce thermal conduction. When the thermal current encounters the hollow sphere, a smaller part of quantity of heat carries out thermal conduction, while a greater part of it goes around the sphere due to the low conductivity of the sphere. In this case, the route of heat transfer in the composite becomes longer and complicated, leading to a reduction of the heat transfer properties of the composite. Therefore, the heat transfer process in a polymer/ hollow micro-sphere composite carries out the following three major ways: (1) thermal conduction through solid and gas; (2) thermal radiation between hollow micro-sphere surfaces; (3) natural convection of gas in hollow micro-sphere. According to the experimental studies, Skochdopole [16] pointed out that the natural convection of the gas in hollow sphere did not happened when the diameter of the hollow sphere was less than 4 mm. Because the diameter of hollow micro-sphere is from several micrometer to hundreds micrometer, the natural convection of gas may not be considered. Furthermore, the thermal radiation is not considered owing to its quite small proportion in total quantity of heat transfer under the high temperature condition.

4.2.2 Physical Property Parameter

Polypropylene composites filled respectively with two type of hollow micro-sphere (TK 35 and TK 70) were investigated in this section. These hollow micro-spheres were supplied by MOLÜS company in Germany, it was a silicate, the density was 2,210 kg/m^3 and the thermal conductivity was 0.179 (W/m·K). There was an inertia gas in the micro-sphere, the density and thermal conductivity were 0.0899 kg/m^3

and 0.0228 (W/m·K). The average diameter of TK35 was 35 μm, and the effective density was 680 kg/m^3; the average diameter and effective density of TK70 were 70 μm and 210 kg/m^3, respectively. The thermal conductivity of the PP resin was 0.2 (W/m·K).

4.2.3 Finite Element Model

In this section, the commercial ANSYS software was used to reform the numerical simulation of the heat transfer in the hollow glass bead filled PP composites. For this type of thermal analysis, the element type for the thermal analysis was the 2D element PLANE35 with six nodes. For a steady heat transfer problem, it is only required to define the thermal conductivity for the PP/TK35 and the PP/TK70 composite systems. Apart from assigning above stated physical parameters, the internal diameter and external diameter of the hollow glass bead was modeled respectively. The unit edge length was calculated according to the volume fraction ($\phi_f = 20\%$) of the bead, and then the length was assigned into the software to generate the geometry model. The quantity of heat was transferred from the bottom to the top. Bringing the first boundary conditions to the bottom, and temperature is 30°C; bringing the second boundary conditions to the top, and environment temperature is 25°C. The conductive heat transfer coefficient was 25 W/(m^2·K), the right and left boundaries were thermal insulation boundary conditions.

4.2.4 Element Mesh Division

The meshes were divided automatically by means of intelligence. The parameters were three, and the element type was a triangle. The results of the mesh division are shown in Fig. 9.

A 3D heat transfer model was established by using the similar method to better represent the heat transfer process in this kind of composite. In this chapter, a 3D model of the heat transfer in the unit cell of the PP/TK35 composite system was made.

4.3 Results and Discussion

4.3.1 Temperature Field

After solving the system of equations by a solver, this heat transfer simulation was done by means of a solver and then entered into a post-processor, the temperature cloud charts were drawn, as shown in Fig. 10. The temperature cloud chart illustrates imaginarily the temperature field in a unit cell. It can be seen in Fig. 10 that the temperature variation in the hollow glass bead is faster than outside of the bead.

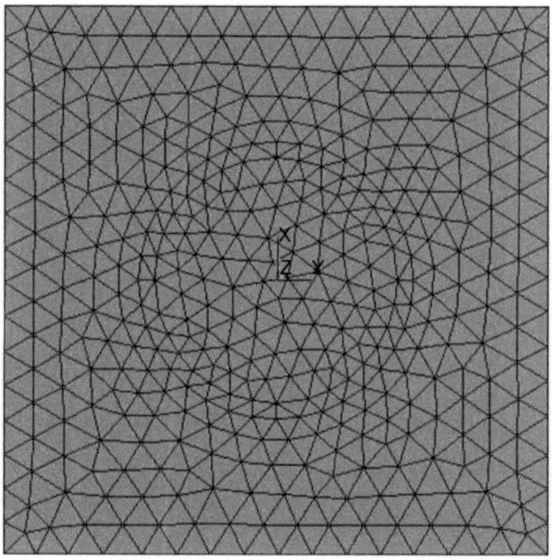

Fig. 9 Element mesh division (PP/TK35)

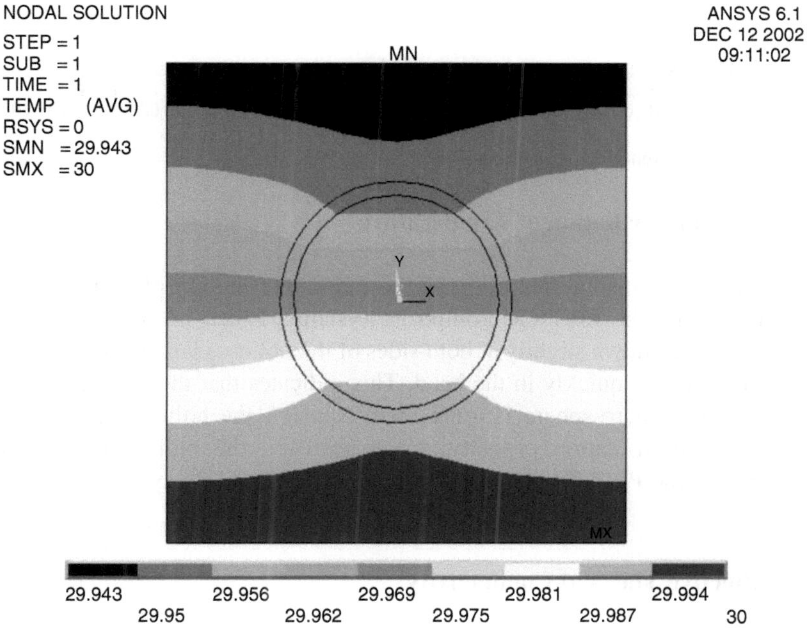

Fig. 10 Temperature cloud chart (PP/TK35)

To reflect the direction and size of the thermal current in a unit cell, a thermal current vector chart in a unit cell was drawn (see Fig. 11). The direction of the arrows represents the thermal current direction, the length of arrow represents the strength of the thermal current. It can be seen in Fig. 11 that the quantity of heat is divided as two parts when it encounters a bead: a smaller part of it flows into the sphere, the other flows along the sphere wall.

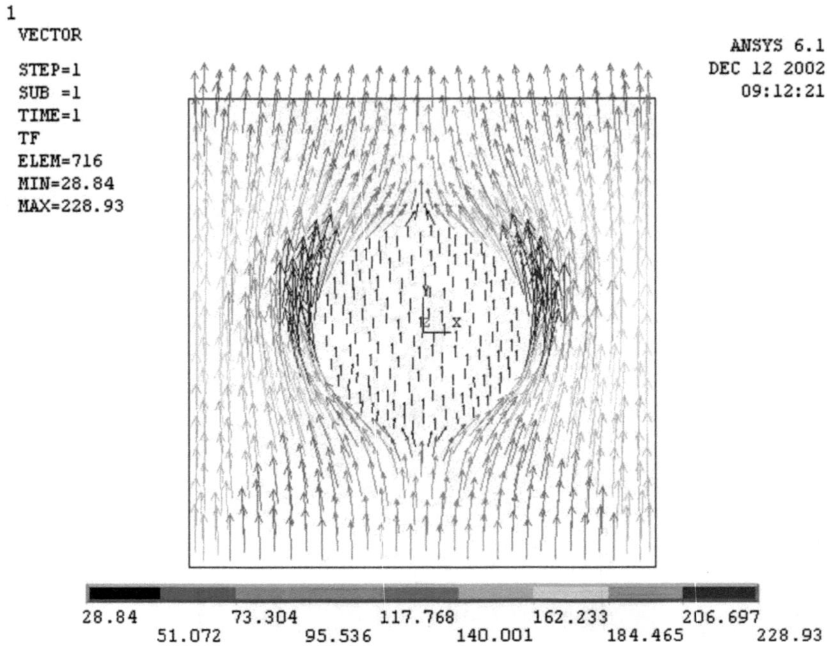

Fig. 11 Thermal current vector charts in an element (PP/TK35)

4.3.2 Radial Distribution of Temperature

Figures 12 and 13 display respectively the radial distribution of temperature for the PP/TK35 and the PP/TK70 composite systems. It can be observed that the temperature drops down slightly at both sides of the hollow glass bead, whereas the temperature reduces quickly in the bead. This indicates that the heat resistance at the inner of the micro-sphere is quite high. Namely, the hollow glass bead has a good thermal insulation property. Comparatively, the radial distribution of temperature of the PP/TK70 system is similar to the PP/TK35 system.

4.3.3 Effective Thermal Conductivity

As stated above, the effective thermal conductivity is an important parameter for characterizing the thermal properties of materials. The total quantity of heat may be obtained by applying the computer order "List Results > Reaction Solu", and then the difference in temperature is determined from the temperature cloud chart. Finally, the effective thermal conductivity (k_{eff}) for each unit cell of the composite materials could be calculated according to Fourier's law. When the volume fraction of the hollow glass beads in the composites is 5, 10, 15 and 20% respectively, the relevant geometry models are set up, determining physical parameters and boundary conditions, dividing meshes and solving, and then

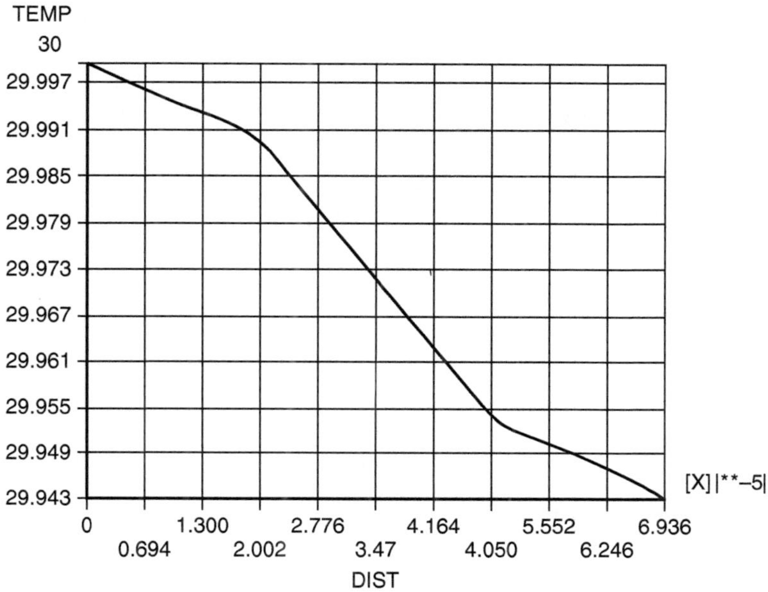

Fig. 12 Radial distribution of temperature (PP/TK35)

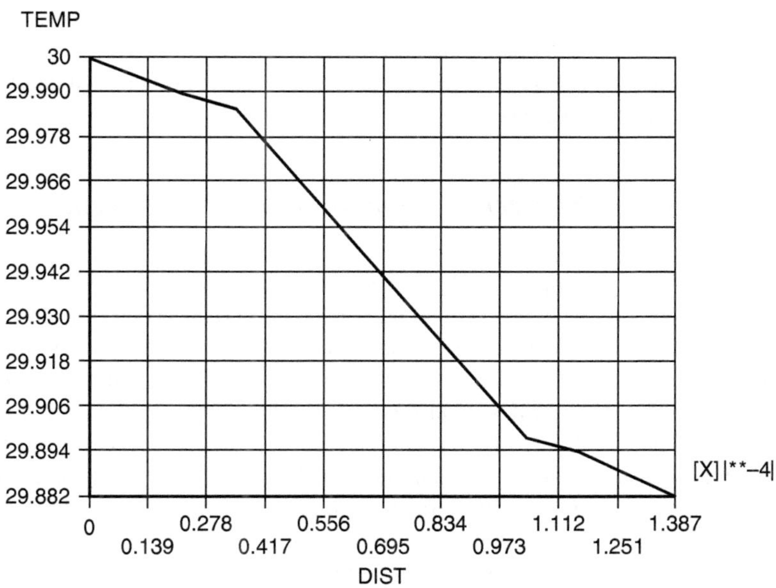

Fig. 13 Radius distribution of temperature (PP/TK70)

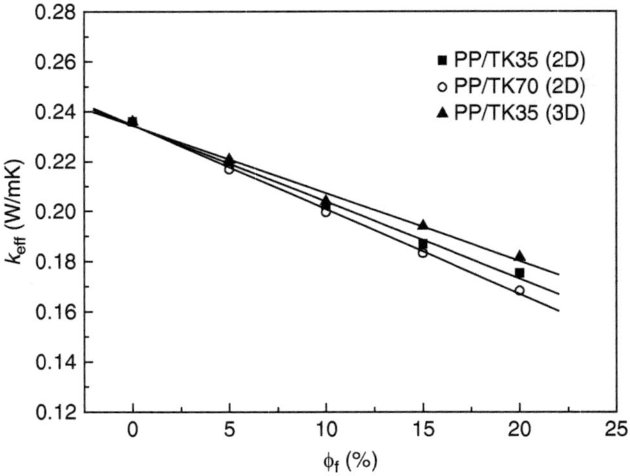

Fig. 14 Relationship between k_{eff} and ϕ_f

calculating corresponding k_{eff}. The results are shown in Fig. 14. Figure 14 illustrates the relationship between k_{eff} and ϕ_f. It can be seen that k_{eff} decreases linearly with an increase of ϕ_f. For the same value of ϕ_f, the value of k_{eff} for the PP/TK35 system is higher than that of the PP/TK70 system. This indicates that the thermal insulation and heat preservation properties of the small size hollow micro-sphere filled system is relatively good.

It can also be seen from Fig. 14 that the values of k_{eff} of the 3D model simulation are higher than those of 2D model simulation for the PP/TK35 composite system, and the difference between them increases with an increase of ϕ_f. This might be that the 3D model is relatively close to the practical heat transfer process in the polymer/hollow micro-sphere composites.

4.4 Comparison Between Simulations and Measurements

In the previous work, Liang and Li [2] measured the thermal conductivity of hollow glass-bead filled polypropylene composites by means of a protecting heat plate method. Figure 15 shows the comparison between the finite element simulations and the measured data of the thermal conductivity of PP/TK35 composite system. It can be seen that the FEM simulations are roughly close to the experimental measured data of k_{eff}. Relatively, the simulations with 3D model are closer to the experimental measured data of k_{eff} than those with the 2D model. Figure 16 shows the comparison between the finite element simulations and the measured data of the thermal conductivity of PP/TK70 composite system. Similarly, the FEM simulations are also roughly close to the experimental measured data of k_{eff}, and the simulations with 3D model are relatively closer to the experimental measured data of k_{eff} than those with the 2D model. This illustrates that the finite element

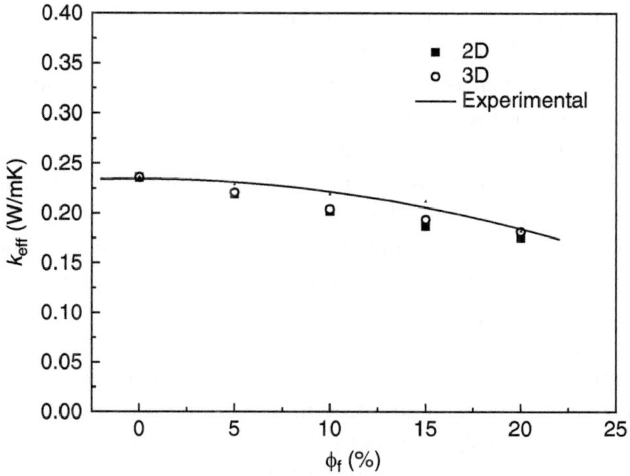

Fig. 15 Comparison between simulations and measured k_{eff} of PP/TK35 composite

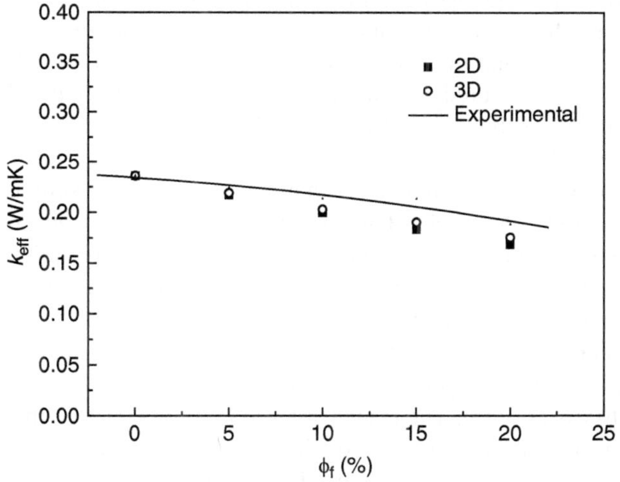

Fig. 16 Comparison between simulations and measured k_{eff} of PP/TK70 composite

simulation of the heat transfer in hollow glass bead filled PP composites is possible by using the ANSYS software in this case.

5 Conclusions

A polymer/hollow micro-sphere composite is a three-phase system, and the heat transfer process is carried out in general by the following three ways: (1) thermal conduction through solid and gas; (2) thermal radiation between hollow micro-sphere surfaces; (3) natural convection of gas in hollow micro-sphere. However, the heat transfer is made mainly by the first way. The quantity of heat is divided into two parts when the thermal current encounters a bead: the smaller part of quantity of heat flows into the sphere, the other flows along the sphere wall. The temperature variation inside the sphere is faster than that outside of the sphere.

A theoretical model of heat transfer in polymer/inorganic hollow micro-sphere composites was established and the corresponding equation of effective thermal conductivity was derived based on the law of minimal thermal resistance, the equal law of the specific equivalent thermal conductivity, and previous work.

The heat insulation property of polymeric materials may be improved by filling with inorganic hollow micro-spheres. Under the experimental conditions, the effective thermal conductivity of hollow glass bead filled polypropylene composites decreased linearly with an addition of the volume fraction of the beads.

The simulated k_{eff} of PP/Tk35 and PP/TK70 composites decreases as a linear function with an increase of the volume fraction of hollow glass beads, and it reduces somewhat with an increase of the bead diameter. In addition, the simulated values of k_{eff} with a 3D model are higher than those with a 2D model for PP/TK35 system, and the difference between them increases with an increase of ϕ_f.

Furthermore, the FEM simulations are roughly close to the experimental measured data of k_{eff}, and the simulations with the 3D model are relatively closer to the experimental measured data of k_{eff} than those with a 2D model. This indicates that the simulation of the heat transfer in PP/HGB composites made by the finite element software ANSYS is possible.

It was found that the theoretical estimations were roughly similar to the finite element simulations of the effective thermal conductivity of hollow glass bead filled polypropylene composites at lower concentration of the particles ($\phi_f \leq 20\%$). With increase of ϕ_f, k_{eff} decreased linearly, and k_{eff} decreased slightly with increase of the HGB diameter when ϕ_f was constant.

It is beneficial to improve the heat insulation property of the composite systems if the content of the hollow glass beads is not high and the distribution of the beads in the PP matrix is uniform. The results showed that the effective thermal conductivity of PP/HGB composites decreased with an increase of the HGB diameter when the concentration of the hollow glass beads was not too high.

References

1. Liang, J.Z.: Tensile and impact properties of hollow glass bead-filled PVC composites. Macrom. Mater. Eng. **287**, 588–591 (2002)
2. Liang, J.Z., Li, F.H.: Measurement of thermal conductivity of hollow glass bead filled polypropylene composites. Polym. Test. **25**, 527–531 (2006)
3. Yung, K.C., Zhu, B.L., Yue, T.M., Xie, C.S.: Preparation and properties of hollow glass microsphere-filled epoxy-matrix composites. Compos. Sci. Technol. **69**, 260–264 (2009)
4. Russell, H.W.: Principles of heat flow in porous insulation. J. Am. Ceram. Soc. **18**, 1–5 (1935)

5. Kingery, W.D.: Conductivity of multicomponent systems. J. Am. Ceram. Soc. **42**, 617–622 (1959)
6. Loeb, A.L.: A theory of thermal conductivity of porous materials. J. Am. Ceram. Soc. **37**, 96–100 (1954)
7. Nielsen, L.E.: Thermal conductivity of particulate-filled polymers. J. Appl. Polym. Sci. **29**, 3819–3825 (1973)
8. Cheng, S., Vachon, R.: A technique for predicting the thermal conductivity of suspensions: emulsions and porous materials. Int. J. Heat Mass Trans. **13**, 537–542 (1970)
9. Agari, Y., Ueda, A., Nagai, S.: Thermal conductivity of a polymer composite. J. Appl. Polym. Sci. **49**, 1625–1630 (1993)
10. Liang, X.G., Qu, W.: Effective thermal conductivity of gas-solid composite materials and the temperature difference effect at high temperature. Int. J. Heat Mass Trans **42**, 1885–1890 (1999)
11. Suvorov, S.A., Fishchev, V.N., Kapustina, S.N., Kopylova, S.V., Nekhlopochina, M.G.: Thermal conductivity of composites with a filler of hollow spherical corundum granules. Refractories **29**, 689–693 (1989)
12. Hill, R.F., Supancic, P.H.: Determination of the thermal resistance of the polymer-ceramic interface of alumina-filled polymer composites. J. Am. Ceram. Soc. **87**, 1831–1835 (2004)
13. Hill, R.F., Supancic, P.H.: Thermal conductivity of platelet-filled polymer composites. J. Am. Ceram. Soc. **85**, 851–857 (2002)
14. Yu, S.Z., Hing, P., Hu, X.: Thermal conductivity of polystyrene-aluminum nitride composite. Compos. Part A Appl. Sci. Manuf. **33**, 289–292 (2002)
15. Liang, J.Z., Li, F.H.: Heat transfer in polymer composites filled with inorganic hollow micro-spheres: a theoretical model. Polym. Test. **26**, 1025–1030 (2007)
16. Liang, J.Z., Liu, G.S.: A new heat transfer model of inorganic particulate-filled polymer composites. J. Mater. Sci. **44**, 4715–4720 (2009)
17. Skochdopole, R.E.: The thermal conductivity of foam plastics. Eng. Prog. **57**, 55–59 (1961)
18. Chen, Z.S., Qian, J., Ye, Y.H.: Theoretical derivation of effective thermal conductivity of composites. J. China Univ. Sci. Technol. **22**, 416–423 (1992)
19. Liang, J.Z., Li, F.H.: Simulation of heat transfer in hollow-glass-bead-filled polypropylene composites by finite element method. Polym. Test. **26**, 419–424 (2007)
20. Kon, X.Q.: Application of FEM in Heat Transfer, 3rd edn, p. 30. Science, Beijing (1998)

Radiative Transfer in Two-Phase Dispersed Materials

Jaona Randrianalisoa, Rémi Coquard, and Dominique Baillis

Abstract This chapter presents the treatment of radiative transfer in two-phase dispersed media in the framework of radiative transfer theory. With this aim, two modeling approaches, under the geometric optic hypothesis, are described and then compared. The first one is the traditional treatment of dispersed media as continuous and homogeneous systems, referred to here as the Homogeneous Phase Approach (HPA). The radiation propagation is characterized by effective radiative properties and modeled by the conventional Radiative Transfer Equation (RTE). The second approach is based on a separate treatment of radiative transfer in the continuous and dispersed phases, referred as the Multi-Phase Approach (MPA). In this approach, each constituting phase has its own effective radiative properties and temperatures. For each approach, the methods for predicting the radiative properties are reviewed. The radiative transfers through typical two-phase dispersed media, such as glass containing bubbles, packed bed of opaque spheres, and packed-bed of semitransparent spheres, are analyzed. The results of transmittances and reflectances from these predictive approaches are compared with available experimental data or Monte Carlo (MC) simulation.

Through this contribution, it is shown that the HPA is satisfactory for analyzing radiative transfer in two-phase dispersed media provided that the effective radiative properties are correctly predicted. For practical purpose, it is recommended to use first the well-known independent scattering theory when dispersed contents (or scatterers) are largely spaced or when their volume fraction is small. An example of these media is the glass containing bubbles studied herein. Then, the correlated

J. Randrianalisoa(✉) and D. Baillis
Université de Lyon, CNRS, INSA-Lyon, CETHIL, UMR5008, 69621 Villeurbanne, France
Université Lyon 1, 69622 Villeurbanne, France
e-mail: jaona.randrianalisoa@insa-lyon.fr, dominique.baillis@insa-lyon.fr

R. Coquard
Société « Etude Conseils Calcul en Mécanique des Structures » (ECMS), 66, boulevard Niels Bohr, 69603 Villeurbanne Cedex, France
e-mail: remi.coquard@ec2-ms.fr

A. Öchsner and G.E. Murch (eds.), *Heat Transfer in Multi-Phase Materials*,
Adv Struct Mater 2, DOI 10.1007/8611_2010_4,

scattering theory proposed by Brewster or Singh and Kaviany is the simplest model when the continuous phase is non-absorbing and the dispersed phase is constituted of opaque spheres. Finally, the ray-tracing (RT) based method can be used for arbitrary dispersed materials fulfilling the geometric optic regime.

Concerning the MPA, it is generally a suitable approach, as exemplified with glass containing bubbles and packed bed of opaque particles. It is however inaccurate for a few cases for which the scattering pattern presents strong peaks (known as rainbow peaks) due to the correlation between the rays incident on a scatterer and those transmitted through it after undergoing several internal reflections. This problem may occur only when (1) the continuum is less refracting than the scatterers; (2) the scatterers are weakly absorbing; and (3) the scatterer boundaries are specular and regular in shape. It is, for example, the case with a packed-bed of semitransparent specularly reflecting spheres.

Nomenclature

a	Particle radius, m
A	Surface area, m^{-2}
c	Interparticle distance, m
C	Particle cross section, m^{-2}
C_1, C_2	Constants in (3)
d	Particle diameter, m
d_{tr}	Transportation length of radiation inside a scatterer, m
$dist$	Sum of extinction distances traveled by N_{ray} radiation bundles, m
$dist_{bd,i}$	Mean distance between successive interactions of a ray bundle with the boundary of the substance i
ds_i	Direction vector of the ray bundle at the ith interaction point on the continuum-particle interface
$E(R,\theta)$, $E'(R,\theta)$	Proportions of ray bundles emitting from the center of a spherical medium of radius R and leaving it into a direction of angle θ
f	Fraction of ray bundles undergoing scattering or absorption
f_v	Dispersed phase volume fraction
I	Radiation intensity, W m^{-2} sr^{-1}
I_b	Blackbody radiation intensity, W m^{-2} sr^{-1}
K_1, K_2	Constants in (52)
l	Characteristic length, m
L	Sample thickness, m
l_i	Mean-free-path corresponding to the event of type i or traveled path of a ray from its emission to the point i, m
$m(a)$	Number of particle per unit volume having size in the range a and $a + da$, m^{-4}
n	Refraction index or real part of the complex refractive index
N	Number

Nt	Total number of particle per volume unit in the sample, m^{-3}
q	Radiation flux, W m^{-2}
Q	Particle efficiency factor
R	Radius of a spherical dispersed medium, in m, or hemispherical reflectance
s, s_i	Position vector, position vector at the *i*th interaction point, respectively, m
s_1'	Position vector of the ray at the absorption point in the continuum
S_R	Scaling factor of non-point scattering
t	Unit normal to surface area
T	Temperature (in K) or hemispherical transmittance
V	Volume, m^3
$W(\Theta)$	Number of ray bundles scattered in the angular interval Θ and $\Theta + d\Theta$
x	Particle size parameter
z	Abscise of the radiation intensity along the sample thickness direction, m

Greek Symbols

α	Effective absorption coefficient, m^{-1}
β	Effective extinction coefficient, m^{-1}
χ	Angle between the ray direction incoming on a particle and the outward normal to the particle surface, rad
χ'	Angle between the ray direction transmitted into a particle and the inward normal to the particle surface, rad
δ	Kroneker delta function
Δ	Direction vector
ε	Porosity or error function in (13)
ϕ	Scattering phase function of a single particle
Φ	Effective scattering phase function of the dispersed medium
γ	Scaling factor of the non-point scattering in (6)
φ	Azimuth angle, rad
κ	Absorption index or imaginary part of the complex refractive index
λ, λ_{max}	Radiation wavelength in vacuum, dominant radiation wavelength, respectively, m
μ, μ'	Cosine of angles θ and θ', respectively
$<\mu>$	Scattering anisotropy factor
η	Ratio between the real refractive indexes of the dispersed and continuous phases
Π, Π'	Cumulative probabilities of extinction or absorption
θ	Polar angle, transmittance or reflectance angle, rad
Θ	Scattering angle defined in (16), rad

ρ, ρ'	Hemispherical and directional reflectivities, respectively
σ	Effective scattering coefficient, m^{-1}
ω	Scattering albedo
Ω	Solid angle, sr

Subscripts

a	Refers to absorption
dir	Refers to the number of discrete directions in the angular interval $[0,\pi]$
e	Refers to extinction
equ	Refers to the real refractive index of the equivalent homogeneous medium
h	Refers to hemispherical transmittance or reflectance
i	Refers to parameters characterizing the radiative transfer in the ith phase
inc	Refers to incident radiation penetrating the sample
ind	Refers to radiative properties under the independent scattering hypothesis
ji	Refers to parameters characterizing the radiative transfer from the jth phase to the ith phase
N	Refers to the normal transmittance
pos	Refers to the number of discrete positions.
ray	Refers to the number of rays to be tracked in the Ray-tracing algorithm
s	Refers to scattering
tr	Refers to the radiative properties in the transport approximation defined in (29)
λ	Refers to spectral or wavelength dependent quantities
0	Refers to the continuous substance surrounding the scatterers
1	Refers to the dispersed phase
32	Refers to the mean scatterer radius defined in (24)

1 Introduction

In this contribution, we consider the case of heterogeneous two-phase materials constituted by one rigid, impermeable solid phase and one fluid (liquid or gaseous) phase. The characteristic dimension of each phase, denoted herein by d, is assumed much larger than the radiation wavelength, denoted herein by λ. The radiative heat transfer in such media is of major importance in many engineering processes and systems. Some examples of these systems are porous burners [1, 2], solid oxide fuel cells [3], thermochemical reactor [4], thermal insulation [5, 6]. While most text-books such as [7, 8] are focused on the heat and mass transfer problem in two-phase systems, few are devoted to the treatment of radiative transfer in two-phase media

[9–11]. The current chapter aims to recall the existing approaches to solve radiative transfer problems in two-phase dispersed media and to discuss their limits of applicability. Particularly, it reports the recent advances concerning the treatment of such problems. Finally, it shows the application of the discussed approaches to some types of two-phase dispersed materials.

The exact solution of the radiative transfer in two-phase media should be determined from first principles consisting in the resolution of the Maxwell equations for the electromagnetic field in the entire domain. However, such an approach is only suitable for small systems, because of the limits imposed by modern computers. The alternative method frequently encountered is to treat the radiative problem in the framework of radiative transfer theory. It consists in the resolution of the Radiative Transfer Equation (RTE). The RTE has been derived from the equations of multiple scattering of waves [12–17] although it was originally established from the energy balance of corpuscles in an elementary volume [18–22]. Recall that the RTE is suitable only if the system dimension, denoted herein by L, is much larger than the radiation wavelength λ so that the wave interferences occur in a random way. Such a criterion is fulfilled because we deal with materials characterized by $d \gg \lambda$ and $L \gg d$. There has been long experience on the modeling of radiative transfer in dispersed media, i.e. a continuous medium containing a collection of particles (or scatterers). It constitutes the starting point of modeling of radiative transfer in two-phase materials. In fact, two-phase materials can be viewed as dispersed media. For packed (or fluidized) beds, the particles are the dispersed phase while the fluid substance is the continuous phase. For low porosity closed cell foams, bubbles or cavities are the dispersed phase while the substance surrounding pores is the continuous phase. Therefore, the theories established for dispersed media can be suitable. They are discussed in the following paragraphs.

The main parameters of the RTE are the propagation constants called "radiative properties". The applicability conditions of the RTE to dispersed media are now established as can be found in the series of papers and textbooks by Mischenko et al. [17]. One of these criteria is the Far-field approximation (FFA), which imposes that (1) the scatterers must be located far from each other; and (2) the distance between the scatterers and the observation point must be much greater than the radiation wavelength [23].

1. For dispersed media with low packing fraction of scatterers, the FFA is generally fulfilled and the RTE is suitable. Moreover, the interferences between scattered waves occur in a random way, and the dispersed phase can be viewed as point scatterers. In other words, the radiation interaction with scatterers is a point scattering. Therefore, the exact solution of the radiative transfer problem is expected because the radiative properties can be also determined, for example from the interaction of a plane electromagnetic wave with an isolated scatterer known as "Independent scattering theory" [24–26].
2. For dispersed media with a high packing fraction, leading to closely spaced scatterers, the radiative transfer problem becomes more complex. Generally speaking, the RTE cannot be used due particularly to the failure of the FFA.

Other approaches such as the full wave Monte Carlo [27] and the T-matrix [28] or different techniques based on the multiple scattering of waves (such as Foldy's, [12] Twersky's, [29] and Quasicrystalline approximations [30]) can be adopted. These methods have been shown to be straightforward for scatterers with size smaller than or comparable to the radiation wavelength. However, they are not convenient when the scatterers are very large compared to the wavelength due to excessive computation time and memory consumption. Since an exact solution of the radiative transfer problem in such systems cannot be obtained, at least in the near future, coarse approximations are necessary. The most common assumption is to consider the dispersed medium as continuous and homogeneous and to use the standard RTE with "effective radiative properties" [31–36] which differs from the radiative properties of dilute media [24–26]. This approximate method is referred herein to as the "Homogeneous Phase Approach" (HPA). In thermal engineering problems, the thermal equilibrium between the continuous and dispersed phases is often assumed through the HPA. To deal with multi-temperature phases, additional terms should be introduced in the RTE which render it more complex [37, 38].

For scatterers with size much larger than the radiation wavelength, each elementary volume contains only portions of scatterers but not collections of them. An elementary volume can be therefore treated as a two-phase medium in which an energy balance can be performed in each phase. In this point of view, another modeling method detailed in this contribution is the "Multi-Phase Approach" (MPA), which consists in assigning to each phase their own transport equation (but coupled to each other) and their own effective radiative properties [39–41]. Through the MPA, each phase can thus have its own temperature. Note that such an approach is rather new in radiative transfer treatment; however, it is a common practice in multi-phase heat and mass transfer problems [7, 8].

For practical reasons, the approximate HPA and MPA appears the most useful solution method of radiative transfer in two-phase media. In these approaches, the knowledge of effective radiative properties is crucial. For convenience, they are referred to just as *radiative properties* in the following. As discussed above, the radiative properties of media constituted of sufficiently spaced scatterers can be correctly obtained through the Independent scattering theory while those of densely packed scatterers are quite complex due to "dependent scattering" phenomena. They include:

- "Interference effects," i.e. the radiation incoming on each scatterer (or particle) undergoes an interference with the radiation scattered by neighboring particles
- "Multiple scattering effects," i.e. in an elementary volume, the radiation scattered by a particle is incident on another particle to be scattered again
- "Non-point scattering," which is a purely geometric effect due especially to the proximity of particles compared to their size
- And "radiation transportation," which is also a geometric effect due to the transportation of radiation beams across substantial distances (through the particles) compared to the interparticle distance

To overcome such theoretical difficulties, the most straightforward models of radiative properties, in the large scatterer limit, are based on the Geometric Optic Approximations (GOA) by neglecting the interference of waves and the diffraction patterns as detailed later. The radiation is then treated as a "*straight-ray*."

In this framework, several prediction models of radiative properties are now available depending on the radiative transfer approaches.

- In the HPA, correlated scattering theories consisting in scaling the independent scattering results by scaling factors are suggested. Different scaling factors are given, for instance, for spheres [35, 36, 42] and cylinders [43] randomly dispersed in a transparent continuous phase. Currently, predictions based on ray-tracing techniques are developed. They have the advantage to be suitable for media with complex shapes of scatterer and in the presence of transparent [44] or semitransparent continuous phase [39, 41, 45, 46]. Other techniques of characterization of radiative properties are based on inverse (or identification) method requiring measurements of radiative quantities such as transmittance, reflectance, and/or emittance [47, 48].
- In the MPA, the ray-tracing method [39] and the correlated scattering theory [40] have been used.

This chapter is divided into two main parts. The first one details the theoretical foundation of the homogeneous phase and multi-phase approaches of the radiative transfer. The prediction of radiative properties involved in both approaches is described. The second part concerns the application of both radiative transfer approaches to two examples of two-phase materials such as glass matrix containing bubbles, and packed beds.

2 Theoretical Basis

2.1 General Hypothesis

Let us denote by, c the interparticle (i.e. surface-to-surface) distance, and n_1 and n_0 the refraction indexes of scatterers and continuous substance, respectively. The current HPA and MPA use the GOA of electromagnetic waves; thus, it neglects the wave effects such as interferences and diffraction. Therefore, the radiation is treated as a superposition of pencils of rays propagating according to a straight line. In order for this assumption to be valid, we only focus our attention on the following cases:

- The size of scatterers d is much greater than the wavelength (known as the limit of large particles). The practical criteria is $x = \pi d/\lambda \gg 1$ where x is called size parameter. The phase change of radiation passing through the scatterer is large, i.e. $x|n_1/n_0 - 1| \gg 1$ (known as the limit of hard particles). Therefore, the propagation of the radiation within the scatterer can be modeled using the geometric optic laws of reflection and refraction.

- The interparticle distance c is much greater than the wavelength, i.e. $c \gg \lambda$, and the scatterers are randomly dispersed in space, therefore there are no interference effects. Such a criterion is fulfilled when dealing with media of dispersed large particles encountered in engineering problems.
- Only the half of the total scattering corresponding to the scattering due to reflection and refraction is considered. The other half, i.e. scattering due to diffraction, is neglected. In fact, the diffraction by large scatterers is expected to be close to the incoming direction. Thus, the diffraction can be treated as transmitted radiation.

In addition to the above assumptions, we also considered that:

- The thermal conductivity of scatterers are not too high so that the radiation absorbed at one face of a scatterer is not emitted from the other face
- The scatterer size d is much smaller than the linear size of the system L. Then the radiative properties can be averaged over a representative elementary volume with a linear dimension l so that $d \ll l \ll L$
- The radiative properties are independent of the azimuth angle. Thus, radiative transfer in presence of azimuthal symmetry is assumed

2.2 *Homogeneous Phase Approach*

2.2.1 Principle

Most studies dealing with radiation propagation in two-phase materials are based on the so-called Homogeneous Phase Approach (HPA). This approach implies that the radiative behaviour of a dispersed material can be matched faithfully by an equivalent homogeneous semitransparent medium. Therefore, the radiation propagation is described using a unique homogenized spectral radiation intensity $I_\lambda(s, \Delta)$ at the abscise s along the direction Δ, defined as the energy flux in the direction Δ per units of projected area dA cosθ, solid angle $d\Omega$, and wavelength $d\lambda$ [20–22].

$$I_\lambda(s, \Delta) = \frac{d^3 q_\lambda}{dA.\cos\theta.d\Omega.d\lambda} \quad (1)$$

with θ the angle between intensity direction Δ and the normal to the surface area t (Fig. 1).

The spectral radiant intensity field is governed by the Radiative Transfer Equation (RTE) which takes into account the emission, absorption and scattering of the radiation by the two-phase material. The derivation of this equation can be found in

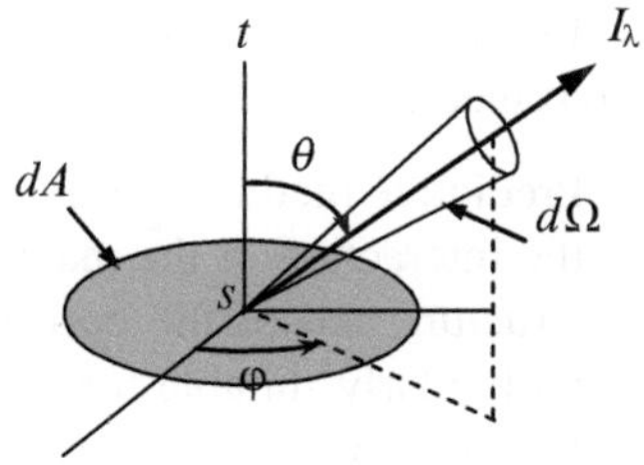

Fig. 1 Definition of the radiation intensity

standard textbooks [18, 20–22]. For a 1D radiation heat transfer in the presence of azimuthal symmetry, this equation is:

$$\mu\frac{\partial I_\lambda(z,\mu)}{\partial z}+\beta_\lambda I_\lambda(z,\mu)=\alpha_\lambda I_{b,\lambda}(T)+\frac{\sigma_\lambda}{2}\int_0^\pi \Phi_\lambda(\mu'\rightarrow\mu)\times I_\lambda(z,\mu')d\mu' \quad (2)$$

where z is the abscise of Intensity I along the z-axis of Cartesian reference, μ is the cosine of angle between I and z-axis. $I_{b,\lambda}(T)$ is the spectral radiation intensity emitted by a black body at temperature T and can be expressed as:

$$I_{b,\lambda}(T)=\frac{2n_{equ}^2C_1}{\lambda^5(e^{C_2/\lambda T}-1)}(\text{in W/m}^2) \quad (3)$$

with n_{equ} the refraction index of the equivalent homogeneous material, with $C_1 = 0.59544 \times 10^{+8}$ W μm^4 m^{-2} and $C_2 = 1.4388 \times 10^{+4}$ μm K. In (3), λ is the vacuum wavelength.

The RTE brings into play four different properties defined for each wavelength and which entirely describe the radiative behaviour of the material:

- The monochromatic extinction coefficient β_λ (in m^{-1}) describing the ability of radiation to interact with a material. This coefficient corresponds to the inverse of the mean-free-path of photons in the material before extinction $\beta_\lambda = 1/l_{e,\lambda}$.
- The monochromatic absorption and scattering coefficients α_λ (in m^{-1}) and σ_λ(in m^{-1}) describing the ability of the material to absorb and scatter radiation, respectively. Absorption and scattering are the two types of extinction, thus we have: $\beta_\lambda = \alpha_\lambda + \sigma_\lambda$.
- The scattering phase function $\Phi_\lambda(\mu' \rightarrow \mu)$ describing the angular repartition of the energy scattered by the medium from the directions with direction cosine μ' to the direction with direction cosine μ. This scattering phase function simplifies to $\Phi_\lambda(\Theta)$. Θ being the angle between the incident and scattering directions, when the assumption of azimuthal symmetry is valid.

However, it must not be forgotten that these properties are those of the equivalent homogeneous semi-transparent material whose radiative behavior must closely match that of the two-phase material.

In order to determine these characteristics, two different methodologies can be envisaged as evoked in the introduction section:

- Predictive methods based on theoretical models allowing us to quantify directly the interaction of the material with radiation from the knowledge of its internal structure and of the optical properties of the constituents. These methods are particularly suitable for the thermal optimisation of the materials.
- Experimental identification methods based on spectrometric measurements of the transmittance, reflectance or emittance by the studied material.

2.2.2 Prediction of the Equivalent Properties

Independent Scattering

The method of prediction of radiative properties which is the most widely employed for two-phase materials considers that it can be regarded as a dispersion of particles in an homogeneous purely absorbing or transparent matrix. The particles shapes and sizes must be chosen so as to reproduce faithfully the structural morphology of the two-phase material. If the distance between these particles is sufficient, they can be considered as point scatterers, which scatter radiation independently. Therefore, the key-point is to know how the incident radiation interacts with each of them. Their interaction with radiation is described by the extinction, absorption and scattering cross-sections (denoted by C_e, C_a and C_s, respectively) and the individual scattering phase function (denoted by ϕ_λ). The radiative properties of the dispersed material are then computed by simply adding the contributions of all the particles comprised in a unit-volume of material with the absorption coefficient $\alpha_{0,\lambda}$ of the homogeneous matrix [49]:

$$\begin{aligned}
\alpha_\lambda &= \alpha_{0,\lambda} + \sum_{i=1}^{Nt} C^i_{a,\lambda} \\
\sigma_\lambda &= \sum_{i=1}^{Nt} C^i_{s,\lambda} \\
\beta_\lambda &= \alpha_\lambda + \sigma_\lambda = \alpha_{0,\lambda} + \sum_{i=1}^{Nt} C^i_{e,\lambda} \\
\Phi_\lambda(\Theta) &= \frac{\sum_{i=1}^{Nt} C^i_{s,\lambda}\phi_\lambda(\Theta)}{\sigma_\lambda}
\end{aligned} \tag{4}$$

In (4), Nt denotes the total number of particles per unit volume. The individual radiative characteristics of the particle ($C_{a,\lambda}$, $C_{s,\lambda}$ and $C_{e,\lambda} = C_{a,\lambda} + C_{s,\lambda}$) and the individual scattering phase function (ϕ_λ) are obtained by analyzing their interaction with a plane incident wave. When considering the general hypothesis of Sect. 2.2.1, this interaction can be treated using the Geometric Optic (GO) laws. The particles are generally considered as randomly oriented. This method has been applied by a large number of researchers to a wide variety of two phase materials.

Fluidized or packed beds are generally represented as arrangements of spheres with constant size in a transparent fluid phase. The individual cross sections and scattering phase functions of these spheres can be calculated analytically when the GOA is valid. The independent scattering assumption is generally assumed to predict the radiative properties of high porosity media like fluidized beds of magnetite (Fe_3O_4) particles [50], beds of metallic particles [51] or water spray curtains [52]. In this latter, the fluid phase was considered as absorbing.

Finally, independent scattering hypothesis is also widely used to model radiative properties of glass foams which have been studied by numerous researchers, from a thermal point of view, since they cover the molten glass surface in glass melting furnaces. Viskanta and Fedorov [53] modeled these foams as arrangements of spherical bubbles of transparent gas in a liquid glass matrix with void fractions close to 0.7. The absorption coefficient is computed by summing the contributions of glass and bubble absorption; this results in an absorption proportional to the solid volume fraction while the scattering properties only take into account the contribution of the gas bubbles (no scattering by the glass matrix). The computation takes into account the size distribution of bubbles. The radiative characteristics of gas bubbles were derived by analyzing the radiative heat transfer through a semitransparent (absorbing and scattering) foam blanket for bubbles large compared to the wavelength of radiation ($x \gg 1$) in the limiting case of anomalous diffraction.

Beyond the Independent Scattering Limit

In numerous dispersed media with relatively low porosities, the previous assumption of independent scattering between each particle cannot be considered valid. Actually, a major challenge for two-phase materials is to determine whether the independent scattering hypothesis is valid or if dependent scattering effects have to be accounted for. The limits of applicability of the theory of independent scattering have been experimentally investigated by several authors, including Hottel et al. [54], Ishimaru and Kuga [55], and Brewster and Tien [56]. Yamada et al. [32], Brewster and Tien [56], and more recently Ivezic and Menguc [57] proposed a single criterion based on the value of the dimensionless number c/λ.Their results indicated that no far-field effect occurs as long as $c/\lambda > 0.3$ or 0.5. Other criterions have been proposed. For example, Singh and Kaviany [58] examined dependent scattering in packed beds of large-size particles ($x \gg 1$, i.e. GO domain) by carrying out Monte Carlo simulations. The Monte Carlo simulations for different porosities are compared with available experimental results and with the results of independent scattering hypothesis. The independent theory is shown to fail for porosities as high as 0.935 and even when the c/λ criterion is satisfied. For this kind of materials, the near-field effect (multiple scattering) was assumed as responsible for the dependent scattering. The failure is more drastic for transmission through beds of opaque spheres than for transparent and semi-transparent spheres. Coquard et al. conducted a similar study on beds of opaque spherical particles [44] and fibrous materials made of large cylinders [43] and stated that multiple scattering and shadowing effect only affect the particles' cross sections and become a significant (error $>10\%$) as soon as the porosity becomes lower than 94% and 90% respectively.

To evaluate the radiative properties when dependent scattering occur, a common approach is to use analytical correlations correcting the characteristics stemming from the independent scattering hypothesis. For example, Kaviany and Singh [36, 58] proposed a correlation for the dependent characteristics of beds of opaque spheres by simply scaling the optical thickness obtained from independent scattering (i.e. $\beta = \beta_{ind} S_r$) while leaving the albedo and the phase function unchanged. The scaling factor S_r depends only on the porosity and is almost independent of the emissivity:

$$S_r = 1 + 1.84 \times (1-\varepsilon) - 3.15 \times (1-\varepsilon)^2 + 7.2 \times (1-\varepsilon)^3 \text{ for } \varepsilon > 0.3 \quad (5)$$

Kamiuto [35] also proposed an heuristic correlated-scattering theory for packed of relatively large spheres that scales the extinction coefficient and the albedo while the absorption coefficient and scattering phase function are left unchanged:

$$\begin{gathered} \beta = \gamma \times \beta_{ind} \text{ and } \omega = 1 - \frac{1-\omega_{ind}}{\gamma} \\ \gamma = 1 + 3/2 \times (1-\varepsilon) - 3/4 \times (1-\varepsilon)^2 \text{ for } \varepsilon < 0.921 \end{gathered} \quad (6)$$

More recently, Brewster [42] shows through a mean-beam-length concept that the main dependent scattering mechanism of radiation in packed beds of large opaque sphere is the non-vanishing volume scattering (called also non-point scattering as opposed to point scattering in the Independent scattering theory) rather than multiple scattering stated by previous researchers. A simple scaling factor correcting the extinction coefficient from Independent scattering is obtained for this kind of particles:

$$S_r = 1/\varepsilon \quad (7)$$

Ray-Tracing (RT) Techniques

Recently, due to the rapid improvement of the performances of standard computers, there is a growing interest in ray-tracing methods. They consist in tracking the histories of a great number of radiation bundles (or ray packets) propagating through the two-phase medium under investigation. During the travel of bundles, absorption, reflections or refractions can occur and their histories are stored. They are then used to deduce the radiative properties using statistic and probability laws [39, 43–46, 59, 60]. The interaction of rays with the solid or fluid phases is dictated by the GOA, i.e. they are only applicable to materials for which the size of scatterers is much greater than the wavelength. They can be used in conjunction with X-ray tomography analysis. In this case, they permit theoretically to take into account the real morphology of the materials analyzed and thus, to avoid the simplifications generally made by the other methods. Zeghondy et al. [39, 59] were the first to apply a radiative properties computation directly to the representation of the porous structure obtained from X-ray tomography. They applied the Radiative Distribution Function Identification (RDFI) originally proposed by Tancrez and Taine [45] to a tomographic representation of open-cell mullite foams. They validated their approach by comparing measured experimental reflectance to that predicted for a homogenized semitransparent material whose anisotropic radiative properties have been obtained by the RDFI. Just about the same time, Petrasch et al. [60] applied another Monte Carlo ray tracing procedure to reticulate porous ceramics to calculate the extinction coefficients and scattering phase functions based on the newly developed probabilistic distribution functions of the extinction path-length and of the directional cosine of incident radiation. The authors take into account purely diffuse or perfectly specular surfaces.

In the RDFI method [39, 45], the extinction and absorption coefficients are retrieved by minimizing the following system of equations:

$$\sum_{0}^{N_{pos}} \left[\Pi_e\left(l_i\right) - \Pi'_e(\beta, l_i)\right]^2 \Big/ \sum_{0}^{N_{pos}} \Pi_e\left(l_i\right)^2 \tag{8}$$

$$\sum_{0}^{N_{pos}} \left[\Pi_a\left(l_i\right) - \Pi'_a(\alpha, \beta, l_i)\right]^2 \Big/ \sum_{0}^{N_{pos}} \Pi_a\left(l_i\right)^2 \tag{9}$$

where Π_e and Π_a are respectively Cumulative probabilities of extinction and absorption of ray bundles after traveling a distance l_i from their emission locations. They are constructed using ray histories from the RT on the two-phase medium. $\Pi'_{ext} = 1 - \exp(-\beta l_i)$ and $\Pi'_{abs} = \alpha[1 - \exp(-\beta l_i)]/\beta$ are respectively the theoretical cumulative probability of extinction and absorption at the distance l_i in which the unknown parameters are α and β. N_{pos} refers to the number of positions at which the cumulative probabilities are evaluated.

In the Coquard and Baillis' method, the extinction coefficient β is retrieved from the probability of rays to travel a large distance R (i.e. $R \gg d$) without being extinct [43, 44]:

$$\exp(-\beta R) = P(R) \tag{10}$$

P(R) is determined through the RT technique on the two-phase medium and corresponds, in practice, to the fraction of rays, which can travel a distance R without either being absorbed or scattered. The scattering phase function is calculated from the knowledge of the number of scattering events occurring in each scattering angle. In fact, if $W(\Theta)$ denotes the fraction of rays scattered in the angle interval Θ and Θ + dΘ measured from the incoming of rays, the scattering phase function can be given by:

$$\Phi(\Theta) = \frac{W(\Theta)}{\frac{1}{4\pi}\int_0^{4\pi} W(\Theta)d\Omega} \tag{11}$$

The integration in (11) is performed over 4π steradians (i.e. over all directions) with $d\Omega = \sin\Theta d\Theta d\varphi$ as the unit solid angle. φ refers to the azimuthal angle defined in the interval 0 to 2π. This scattering phase function is also normalized as follows:

$$\frac{1}{4\pi}\int_0^{4\pi} \Phi(\Theta)d\Omega = 1 \tag{12}$$

The scattering albedo ω is determined by minimizing the following error function:

$$\varepsilon\,(\omega) = \sum_{i=1}^{N_{dir}} \left[E\,(R,\theta_i) - E'\,(R,\beta,\omega,\Phi,\theta_i)\right]^2 \Big/ \sum_{i=1}^{N_{dir}} \left[E\,(R,\theta_i)\right]^2 \tag{13}$$

where E is the proportion of rays leaving the sphere (of the two-phase medium) of radius R in a direction of angle θ_i measured from the incoming of rays. E' refers to the same quantity as E but obtained from the RT on an equivalent continuous homogeneous spherical medium of parameters β [from (10)], Φ [from (11)], and the unknown ω. N_{dir} is the number of directions considered in the angular interval [0,π].

Recently, Randrianalisoa and Baillis suggest another algorithm so that the absorption and scattering coefficients are retrieved from analysis of absorption and scattering mean-free-paths while the scattering phase function is obtained from (11) [41, 46]. If f_{a} and f_{s} denote respectively the fraction of rays undergoing absorption and scattering events after tracking N_{ray} rays on the dispersed medium so that $f_{\mathrm{a}} + f_{\mathrm{s}} = 1$, and *dist* designates the sum of absorption and scattering paths travelled by N_{ray} bundles, the absorption and scattering coefficients can be given by:

$$\alpha = f_a \frac{N_{ray}}{dist} \text{ and } \sigma = f_s \frac{N_{ray}}{dist} \tag{14}$$

Or, in terms of extinction coefficient and scattering albedo,

$$\beta = (f_s + f_a)\frac{N_{ray}}{dist} \text{ and } \omega = \frac{f_s}{f_s + f_a} \tag{15}$$

2.2.3 Identification of the Equivalent Properties from Spectrometric Measurements

Contrary to the predictive models, these methods do not permit the optimisation of the materials, but they have the advantage of providing properties that are closely related to the real material without any assumptions concerning its structure or the optical properties of the constituents. Moreover, most of the two-phase materials can be characterized using a similar identification protocol. They are based on spectrometric measurements on samples of the material studied. The principle is to minimize the differences between these spectral values and the results obtained from numerical resolution of the RTE in a homogeneous semi-transparent material with varying radiative properties. The minimum number of measurements required is equal to the number of parameters used to model the radiative behaviour of the material studied. An extensive review of the different methodologies commonly used for the identification of the radiative properties has been conducted by Baillis and Sacadura [61]. They analyzed strategies based on collimated or diffuse incident beam shape and directional or hemispherical detection systems and paid some attention to the development of RTE solution models with increasing numbers of parameters to identify. The authors showed that the success of the identification depends on the chosen physical model, on the experimental data set (number of measurement points and angular direction choice) and on the detector noise level.

For a closed cell glass foam with 4% of void fraction, Randrianalisoa et al. [48] characterized radiative properties such as the extinction coefficient, scattering albedo, and three parameters of an approximate phase function using measurements of bidirectional transmittances and reflectances in the spectral region 1.7–4 μm. For the same material, Dombrovsky et al. [62] determined the transport extinction coefficient and scattering albedo from combination of Mie theory and hemispherical reflectance measurement. Comparison of identification results with predicting models shows that the Independent scattering approximation is suitable for analyzing radiative transfer in this low porosity closed cell porous medium.

Finally, the radiative properties of fluidized and packed beds have been investigated experimentally by Jones et al. [47] who measured the spectral and directional emission of non-isothermal packed bed of monodispersed, opaque, large spherical particles (1 mm diameter, porosity of 0.37) using a radiometric technique. In a more recent work, Baillis and Sacadura [63] measured the directional spectral emissivity of an isothermal medium made of a dispersion of large oxidised bronze (opaque) spherical particles. Comparison with predictive models tends to show that correlated scattering theory, particularly based on the Singh and Kaviany's correlation (5), is suitable and necessary.

2.3 Multi Phase Approach

Another way for treating the radiative transfer problems in dispersed materials is the so-called Multi-Phase Approach (MPA). However, although, it is widely used for the resolution of heat conduction or flow transfer in multi-phase materials, it is rather new in the field of radiative transfer. In the literature, Gusarov [40] and

Randrianalisoa and Baillis [41] have applied the MPA to investigate radiative transfer in dispersed media of opaque and semitransparent particles in transparent or absorbing fluid substance.

2.3.1 The Multiphase Radiative Transfer Equation

Let us consider a thin slab of a two phase medium constituted of a continuous phase and a dispersed phase (modeled by spherical particles for simplicity) illustrated by Fig. 2a. As before, c is the nearest distance between the particles and d is the typical size of particles. The characteristic sizes c and d are assumed much greater than the dominant wavelength λ so that the geometric optic treatment of the radiation is applicable. The main idea behind the Multiphase approach is (1) to assume that each phase (continuous or dispersed) is a homogeneous and continuous medium; and (2) the radiative transfer in each phase is modeled by a local transport equation, coupled to each other. The solutions of these transport equations can enable to retrieve the radiation intensity and temperature fields in each phase. The Fig. 2b schematizes the treatment of the radiation transfer in the two-phase medium according to the MPA. It can be noted that with this model, the thermal equilibrium between the continuous and dispersed phases is not imposed.

The different mechanisms of radiation extinctions in the continuous and dispersed substance are schematized by Fig. 3a, b, respectively. When the radiation intensity I_i propagates in the substance i (i = continuous or dispersed), it may be

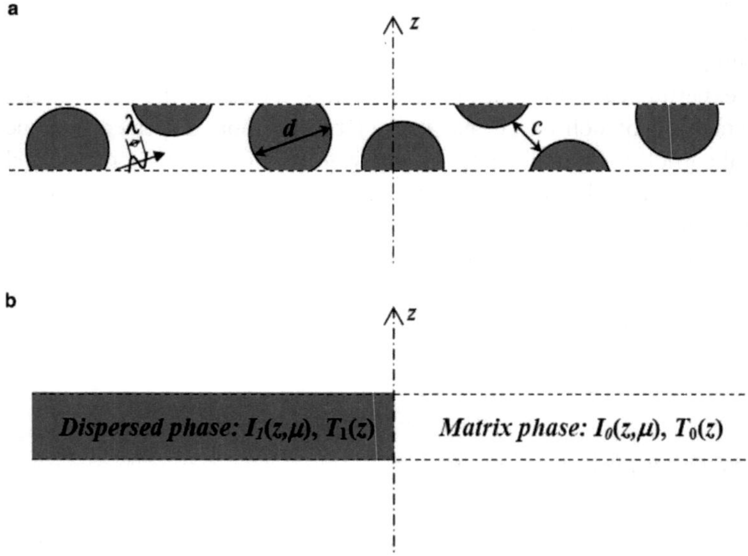

Fig. 2 (**a**) Schematization of a thin slab of a heterogeneous medium. (**b**) Modeling of a thin slab of heterogeneous medium according to the MPA

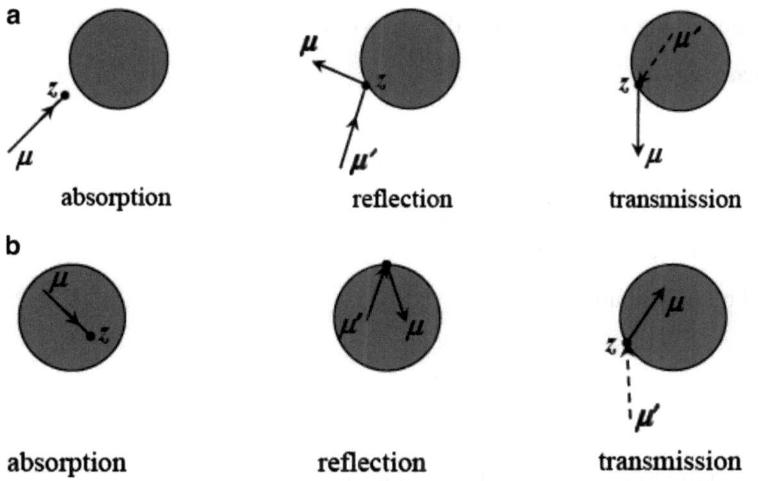

Fig. 3 (**a**) Radiation extinction in the matrix substance at the abscise z. Absorption (*left*); scattering by reflection (*center*); and scattering by transmission (*right*). (**b**) Radiation extinction in the particle substance at the abscise z. Absorption (*left*); scattering by reflection (*center*); and scattering by transmission (*right*)

absorbed (see left figure) or scattered (see center and right figures) by an interface between the phases. During scattering, if the interface between the phases is totally reflecting, the intensity I_i is entirely reflected back to the phase i (see center figure), else an amount of it is transmitted to the neighboring phase (see right figure). Assuming that the scatterers are randomly dispersed in the continuous substance, the extinction of the radiation in the substance i (continuous or dispersed) can be characterized by an absorption coefficient α_i, a scattering coefficient σ_i, and phase function Φ_i. Moreover, the energy reinforcement from the neighboring phase can be also characterized by a scattering coefficient σ_{ji} and a phase function Φ_{ji}. Such radiation transfer mechanisms are similar to that in the conventional radiative theory except for the possibility of energy exchange between phases. Therefore, the transport equation in each phase can be derived in a similar manner as for the usual RTE by taking into account the additional energy exchange between phases.

To do this, let us denotes by $I_i(z,\mu)$ the radiation intensity propagating in the phase i at a linear position z and in the direction of cosines μ with respect to the z axis (see Fig. 2a). $I_i(z + \mathrm{d}z,\mu)$ refers to the intensity of this radiation after traveling an elementary volume of linear thickness dz. During the travel of dz:

- An energy amount of $-\alpha_i I_i \mathrm{d}z$ is absorbed
- An energy amount $\alpha_i I_{\mathrm{b},i} \mathrm{d}z$ is spontaneously emitted where $I_{\mathrm{b},i}$ is the blackbody intensity defined similarly as in (3) but now T_i and n_i (instead of T and n_{equ}) are respectively the temperature and real refractive index of the phase i
- An energy amount of $-\sigma_i I_i \mathrm{d}z$ is scattered

Moreover, the intensity in the direction of cosine μ is increased by

- An amount of

$$\frac{\sigma_i}{4\pi}\int\limits_{4\pi} I_i\,(z,\mu')\,\Phi_i(\Theta)\;d\Omega'$$

because of in-scattering from all directions from the phase i,

- And by an amount of

$$\frac{\sigma_{ji}}{4\pi}\int\limits_{4\pi} I_j\,(z,\mu')\,\Phi_{ji}(\Theta)\;d\mu'$$

because of in-scattering from all directions from the neighboring phase j. As before, Θ is the angle between the incident and the scattered radiation directions. For plane parallel slab geometries as shown in Fig. 2a, the angle Θ is connected to the incidence direction (characterized by a cosine μ' and an azimuthal angle φ') and the scattering direction (characterized by a cosine of direction μ and an azimuthal angle φ) by

$$\Theta = \mu'\mu + \sqrt{1-\mu'}\sqrt{1-\mu}\cos(\varphi-\varphi') \tag{16}$$

The balance of radiation energy in volume element of linear thickness dz enables to establish the transport equation in each phase called here as the Multiphase Radiative Transfer Equation (MRTE). In the case of 1D geometry and with azimuthal symmetry, the MRTE in the steady state regime

- In the continuous substance, indexed hereafter by 0, is

$$\mu\,\frac{\partial I_0(z,\mu)}{\partial z} = -(\alpha_0+\sigma_0)I_0(z,\mu) + \alpha_0 I_{0,b}(z,T_0) + \frac{\sigma_0}{2}\int\limits_{-1}^{1} I_0\,(z,\mu')\,\Phi_0(\Theta)\;d\mu' + I_{10}(z,\mu); \tag{17}$$

- In the dispersed substance, indexed hereafter by 1, is

$$\mu\,\frac{\partial I_1(z,\mu)}{\partial z} = -(\alpha_1+\sigma_1)I_1(z,\mu) + \alpha_1 I_{1,b}(z,T_1) + \frac{\sigma_1}{2}\int\limits_{-1}^{1} I_1\,(z,\mu')\,\Phi_1(\Theta)\;d\mu' + I_{01}(z,\mu). \tag{18}$$

with

$$I_{ji}(z,\mu) = \frac{\sigma_{ji}}{2} \int_{-1}^{1} I_j\ (z,\mu')\,\Phi_{ji}(\Theta)\ d\mu' \text{ for } i=0, j=1 \text{ and } i=1, j=0 \tag{19}$$

Note that these equations have a similar form as the usual RTE except for the presence of the exchange terms I_{10} and I_{01}.

2.3.2 Determination of Radiative Properties

The determination of radiative properties for the MPA is quite new. In the literature, the radiative properties of packed bed of spheres for the MPA have been obtained from theoretical analysis of radiation intensity propagation through a thin layer of the two-phase medium [40]. For more complex geometries, the RT approach has been suggested. For example, Tancrez and Taine [45] and later Zhegondy et al. [39] used the RDFI method to calculate the radiative properties of cellular ceramics from X-ray tomography image analysis.

When the microstructure of the two-phase medium and their optical properties are known, the radiative properties involved in (17)–(19) can be determined by analyzing the propagations of radiation in the continuous and dispersed phases, respectively. When the radiation of intensity I_i travels a distance dz in the substance i bounded by the substance j, it is attenuated,

- Due to absorption according to the Beer's law: $I_i(z + \mathrm{d}z,\mu) = I_i(z,\mu)\exp(-4\pi l\kappa_i/\lambda)$ in which $l = \mathrm{d}z/\mu$ is the traveled path and κ_i is the absorption index (i.e. the imaginary part of the complex refractive index) of the substance i. It can be acknowledged that the absorption coefficient is just that of the constitutive substance because such phase is continuous and homogeneous. Hence,

$$\alpha_i = \frac{4\pi\kappa_i}{\lambda} \text{ for } i = 0 \text{ and } 1 \tag{20}$$

- Due to scattering at interfaces between the current and neighboring phases. If $dist_{\mathrm{bd},i}$ designates the mean distance between the boundaries of the substance i and ρ_i is the interface hemispherical reflectivity, the mean-free-path (mfp) in-between of scattering, $l_{\mathrm{s},i}$, is the ratio $dist_{\mathrm{bd},i}$ to ρ_i so that when the boundaries are totally transparent (i.e. $\rho_i = 0$), $l_{\mathrm{s},i} = \infty$ and when they are totally reflecting (i.e. $\rho_i = 1$), $l_{\mathrm{s},i} = dist_{\mathrm{bd},i}$. Now, remembering that the scattering coefficient is the inverse of the scattering mfp, the scattering coefficient for the radiation propagation in the substance i due to boundary reflections can be given by

$$\sigma_i = \frac{1}{dist_{bd,i}}\rho_i \text{ for } i = 0 \text{ and } 1 \tag{21}$$

The scattering coefficient for a radiation in the substance i transmitted to the substance j due to refraction can be derived in a similar manner as the scattering coefficients due to reflections in the substance i except that the hemispherical interface transmittivity $1-\rho_i$ is involved instead of the reflectivity ρ_i:

$$\sigma_{ij} = \frac{1}{dist_{bd,i}}(1 - \rho_i) \text{ for } i = 0, j = 1 \text{ and } i = 1, j = 0 \tag{22}$$

To evaluate the scattering phase functions, we can proceed by analyzing the scattering (reflection and refraction) of an intensity distributed over a thin pencil of rays of solid angle dΩ by a large scatterer, and then applying the conventional definition of the scattering phase function [20–22]:

$$\Phi(\Theta) = \frac{I(\Theta)}{\frac{1}{4\pi}\int_0^{4\pi} I(\Theta)d\Omega} \tag{23}$$

where $I(\Theta)$ refers to the intensity scattered in the angle Θ measured from the incoming intensity. Such analysis enables to obtain closed forms of scattering phase functions for regular scatterers such as spheres as given later; however, it becomes complex when the scatterer shape is irregular. Another practical approach enabling to determine the scattering phase function of arbitrary large scatterers is the ray-tracing method through the definition (11). For example, the scattering phase function due to reflections, i.e. Φ_i, is obtained from the fraction of rays initially in the substance i and reflected back to it (into the angular interval Θ and $\Theta + d\Theta$ measured from the incoming directions) by the interface between the substances i and j. The scattering phase function due to transmissions, i.e. Φ_{ij}, is determined from the fraction of rays originating from the substance i and transmitted to the substance j into the angular interval Θ and $\Theta + d\Theta$, again measured from the incoming directions.

3 Applications to Dispersed Materials

3.1 Low Porosity Closed Cell Foams (Glass Foams)

This medium is taken here to illustrate the behaviour of two phase materials where the dispersed phase is constituted of air gas while the continuous phase is constituted of solid matrix. In this case, the particle refraction index is generally less than the host medium refraction index (i.e. $n_1 < n_0$).

Glasses containing gas bubbles are encountered in practice in the glass furnaces where gas bubbles are created in the glass melt due to chemical reactions between the melt and gas combustion [53, 64]. The radiative transfer is the main energy transfer mode in this high temperature process. The understanding of the radiative transfer in such material is then of primary importance because it enables (1) to quantify the energy to be delivered by the combustion room thus reducing the energy consumption, and (2) to control the melt temperature thus improving the quality of products. The study of the radiative transfer in such medium has attracted much attention [48, 53, 62, 64, 65]. In this contribution, the radiative transfer through a sample of fused quartz containing bubbles is reported.

3.1.1 Microstructure and Optical Properties

The porous fused quartz sample was cut from a large piece of quartz collected during the shutdown of an industrial furnace in which the fused quartz is electrically heated in an inert atmosphere of helium and hydrogen. The sample surfaces were polished so that they are plane, parallel and the roughness of the quartz surface are sufficiently small compared to the dominant radiation wavelengths of interest ($\lambda_{max} \approx 2$ μm at 1,500 K). The sample void fraction is of $4 \pm 0.5\%$ obtained from density measurements. The sample thickness is 9.9 mm and the cross section is 5×5 cm so that the width-to-thickness ratio is large enough to ensure one-dimensional radiative transfer. Figure 4 shows a photograph of a typical sample. As can be noted, the bubbles are mainly spherical in shape and they have a size distribution. A scanning electron micrograph analysis shows that the roughness of bubble surfaces is also very small compared to the dominant wavelength. Figure 5 depicts the corresponding normalized size distribution determined from an image

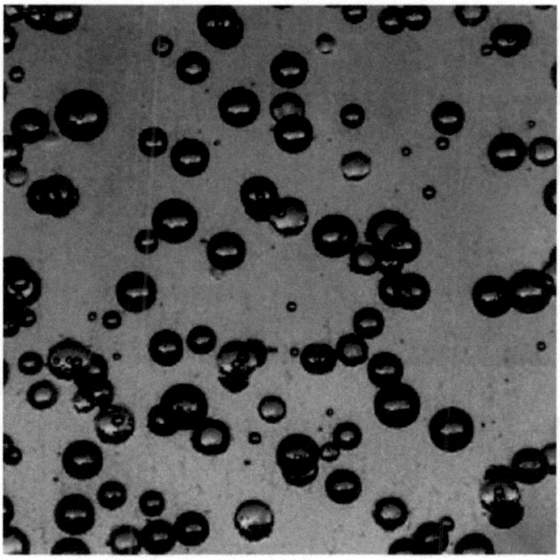

Fig. 4 Photograph of fused quartz containing bubbles

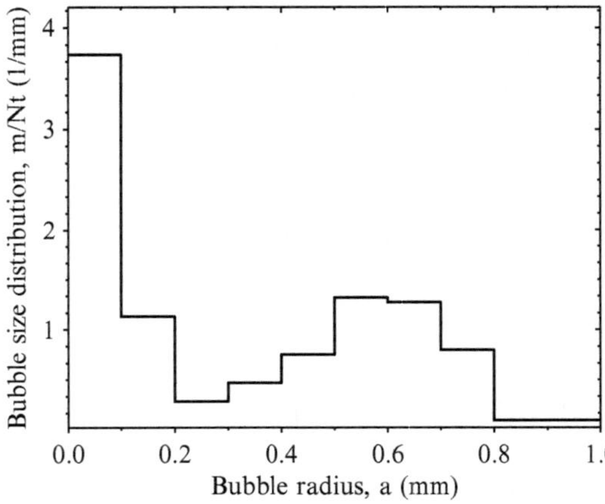

Fig. 5 Bubble normalized size distribution, *m*/*N*t

analysis of the total number of bubbles. The corresponding bubble average radius defined in (24) below is about 0.64 mm.

$$a_{32} = \frac{\int_0^\infty a^3 m(a)da}{\int_0^\infty a^2 m(a)da} \tag{24}$$

where $m(a)$ (in m^{-4}) corresponds to the number of bubbles per unit volume having size between a and $a + da$ so that the total bubble number per unit volume is $Nt = \int_0^\infty m(a)da$ (in m^{-3}).

Optical properties of fused quartz (n_0 and κ_0), required in the radiative properties calculations, have been extensively studied for different spectral regions as discussed in [48, 62]. Concerning the refraction index n_0, the three-term Sellmeier equation proposed by Malitson [66] [(25) below] is the most commonly accepted expression in the literature for the spectral range from 0.21 to 6.7 μm at ambient temperature.

$$n_0 = \sqrt{1 + \frac{0.696\,\lambda^2}{\lambda^2 - 0.068^2} + \frac{0.407\lambda^2}{\lambda^2 - 0.114^2} + \frac{0.897\lambda^2}{\lambda^2 - 9.891^2}} \tag{25}$$

The spectral value of κ_0 was recovered from the normal spectral transmittance data (performed on fused quartz sample without bubble), denoted by T_N, according to:

$$\kappa_0 = -\frac{1}{4\pi L/\lambda} \ln \left\{ \frac{\sqrt{(1-\rho')^4 + 4{T_N}^2\rho'^2} - (1-\rho')^2}{2T_N\rho'^2} \right\} \tag{26}$$

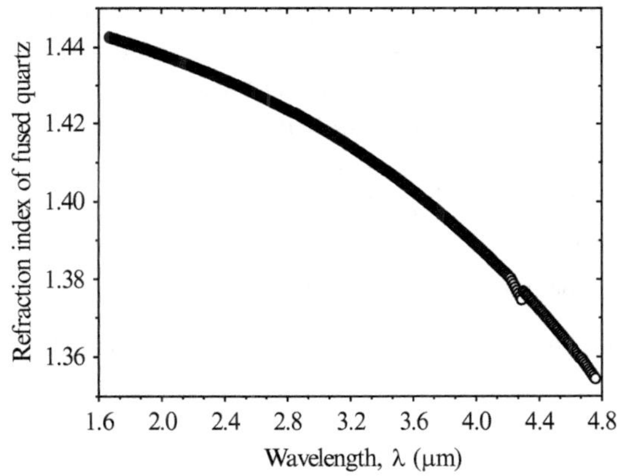

Fig. 6 Real refractive index of fused quartz n_0 calculated from (25)

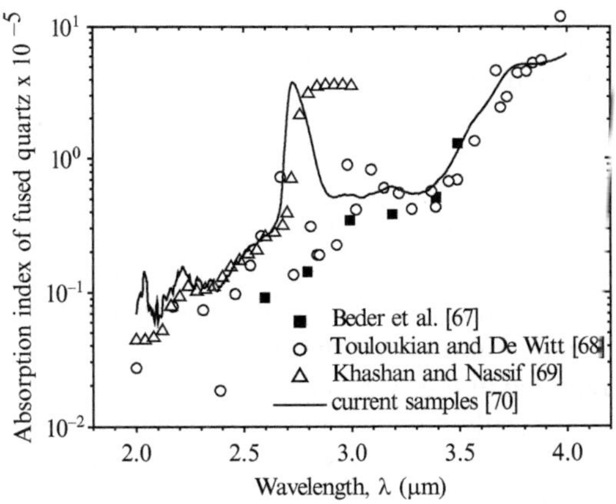

Fig. 7 Absorption index of fused quartz κ_0

where L is the sample thickness and ρ' is the Fresnel (directional) reflectivity at the air–glass interface for normally incident radiation [20–22]:

$$\rho' = \frac{(1 - n_0)^2}{(1 + n_0)^2} \tag{27}$$

Figures 6 and 7 depict the optical properties of fused quartz retrieved from relations (25) and (26). The data of κ_0 from literature [67–69] are also plotted.

3.1.2 Computation of Radiative Properties for HPA

In the literature, the effective radiative properties of glass foams were determined from the Independent scattering theory in which bubble efficiency factors and scattering phase function are either from Mie [70] or Diffraction anomalous theory

[53]. Identification methods based on measurements of bidirectional [48] or hemispherical transmittances and/or reflectances [71] were also used.

In order to investigate the competition between the MPA and HPA, radiative properties predictions using the GOA are suggested. Since the bubble volume fraction is small, we adopt Independent scattering theory.

Remembering that for an optically large scatterer ($x \gg 1$), the single particle properties C_s, C_a, and ϕ become independent of size [24, 26] and using the definitions of the bubble average radius a_{32} and bubble volume fraction $f_v = (4/3)\pi \int_0^\infty a^3 m(a)da$, it can be shown that the Independent scattering model (4) reduces to [49, 62, 72]:

$$\begin{aligned} \alpha_\lambda &= \alpha_{0,\lambda} + 0.75\, f_v Q_a / a_{32} \\ \sigma_\lambda &= 0.75 f_v Q_s / a_{32} \\ \Phi_\lambda(\Theta) &= \phi_\lambda(\Theta) \end{aligned} \tag{28}$$

with $\alpha_{0,\lambda} = 4\pi\kappa_{0,\lambda}/\lambda$. $Q_s = C_s/\pi a^2$ and $Q_a = C_{abs}/\pi a^2$ are respectively the scattering and absorption efficiencies of bubble of radius $a = a_{32}$. Note that in the presence of an absorbing host medium ($\alpha_0 > 0$), Q_a is negative for non-absorbing particles such as bubbles [73]. In the GO regime ($x = 2\pi a_{32}/\lambda >> 1$ here) and weakly absorbing host medium ($x\kappa_0 < 1$), $Q_a = -8x\kappa_{0,\lambda}/3$ gives a good approximation [49, 62]. In addition, for non-absorbing scatterers and without a diffraction contribution $Q_s = 1$ [20–22]. Therefore, the absorption coefficient is just $\alpha_0\ (1 - f_v)$ while the scattering coefficient is $0.75 f_v/a_{32}$. The scattering phase function at a typical wavelength $\lambda = 1.7\ \mu$m predicted by the GO theory [26] is reported in Fig. 8 versus the scattering angle. It tends to show that the scattering by an optically large bubble is essentially forward (i.e. $0 \le \Theta < \pi/2$).

In order to compare the above radiative properties with the experimental data from the identification method, which include the wave nature of radiation such as diffraction, we analyze the evolution of transport properties such as the transport extinction, denoted by β_{tr}, and the transport scattering albedo, denoted by ω_{tr}, defined by [49, 71]:

$$\beta_{tr} = \alpha + \sigma\,[1 - <\mu>]\ \text{and}\ \omega_{tr} = \frac{\sigma}{\beta_{tr}}\,[1 - <\mu>] \tag{29}$$

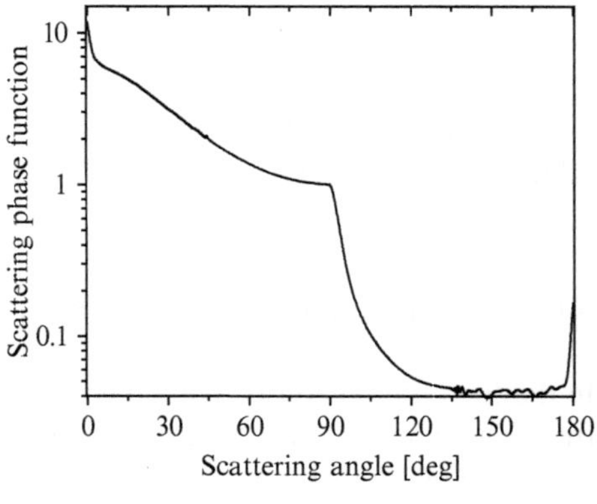

Fig. 8 Scattering phase function at wavelength $\lambda = 1.7\ \mu$m and $n_0/n_1 = 1.44$ versus scattering angle predicted by the GO theory

where $<\mu>$ is the scattering anisotropy factor given by [20–22]:

$$< \mu > = \frac{1}{2} \int_0^{\pi} \Phi(\Theta) \sin \Theta d\Theta \quad (30)$$

Figures 9 and 10 depict respectively the transport extinction coefficient and scattering albedo according to the radiation wavelength of current porous fused quartz. Lines are results from the prediction model while symbols are experimental results after introducing original data from reference [48] in (29) and (30). The abrupt increase of extinction coefficient and decrease of scattering albedo around 2.7 μm is due to absorption of hydroxyl contents in the fused quartz matrix. Above 3.5 μm, glasses have generally high absorption. This results in an increase of the extinction and a decrease of the scattering albedo. As can be noted, the predicted transport radiative properties match well the experimental data proving the adequacy of the Independent scattering theory under the GOA in the current case.

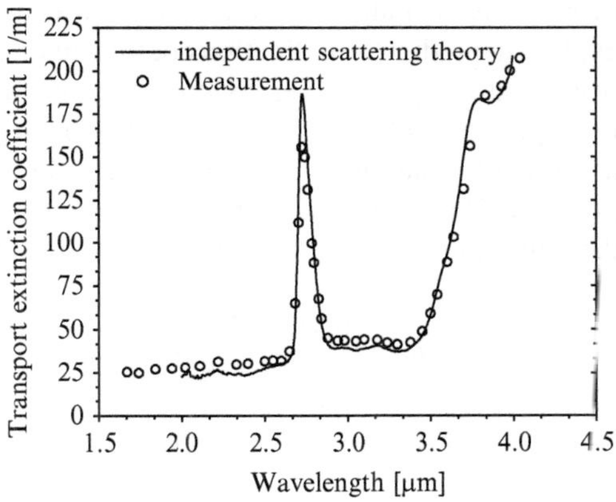

Fig. 9 Comparison between the prediction and experimental data of transport extinction coefficient of porous fused quartz

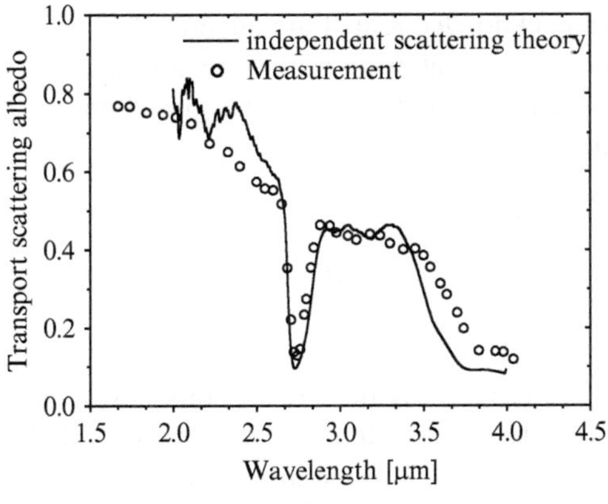

Fig. 10 Comparison between the prediction and experimental data of transport scattering albedo of porous fused quartz

3.1.3 Computation of Radiative Properties for MPA

The radiative properties of each phase constituting the two-phase materials are determined using (20)–(23). The absorption coefficient of the glass matrix α_0 is simply given by (20) in which values of κ_0 are reported in Fig. 7. The absorption coefficient of bubble contents is taken to be zero as consistent with the HPA.

The scattering coefficients require the knowledge of average paths between successive radiation interactions with boundaries $dist_{\mathrm{bd},i}$ in the substance i. For spherical and uniform scatterer size, the average paths in-between scattering of radiation $dist_{\mathrm{bd},i}$ (for $i = 0$ and 1) can be retrieved using the famous Mean-beam-length approach [42, 74]:

$$dist_{bd,0} = \frac{4V_0}{A_0} = \frac{a}{0.75}\frac{1-f_v}{f_v} \text{ and } dist_{bd,1} = \frac{4V_1}{A_1} = \frac{a}{0.75} \tag{31}$$

where V_0 and A_0 refer respectively to the volume and surface of the continuous phase while $V_1 = 4\pi a^3/3$ and $A_1 = 4\pi a^2$ are respectively the volume and surface of a sphere of radius a. Note that the Mean-beam-approach has been previously applied to predict the extinction coefficient of packed bed of opaque spheres and the result is exactly that given by the left equation in (31). The result for the radiation propagation inside bubbles is rather new but it has been confirmed by (1) theoretical analysis of radiation intensity crossing a thin layer of similarly dispersed medium [40] and (2) the Ray-tracing method carried out inside a sphere [41, 46]. Now, using the definitions (21) and (22), we obtain the scattering coefficients for the MPA:

$$\begin{aligned} \sigma_0 &= \frac{0.75}{a}\frac{f_v}{1-f_v}\rho_0, \quad \sigma_{01} = \frac{0.75}{a}\frac{f_v}{1-f_v}(1-\rho_0), \\ \sigma_1 &= \frac{0.75}{a}\rho_1, \quad \text{and} \quad \sigma_{10} = \frac{0.75}{a}(1-\rho_1) \end{aligned} \tag{32}$$

with f_v the bubble volume fraction (or porosity) and $a = a_{32} = 0.64$ mm the average bubble radius calculated from the particle size distribution (Fig. 5). We can acknowledge that the quantity $1/(1 - f_v)$ in (32) corresponds to the Brewster's non-point scattering scaling factor S_R in (7). ρ_0 and ρ_1 are respectively the hemi-spherical reflectivities at the glass/bubble and bubble/glass interfaces defined as follows:

$$\rho_i = 2\int_0^1 \rho'_i \cos\chi \, d\cos\chi \quad \text{for } i = 0 \text{ and } 1 \tag{33}$$

In (33), ρ_i' is the Fresnel reflectivity and depends on wavelength through the optical refractive indexes and on the angle between the incoming radiation and the outward normal to the interaction surface, denoted here by χ [20–22].

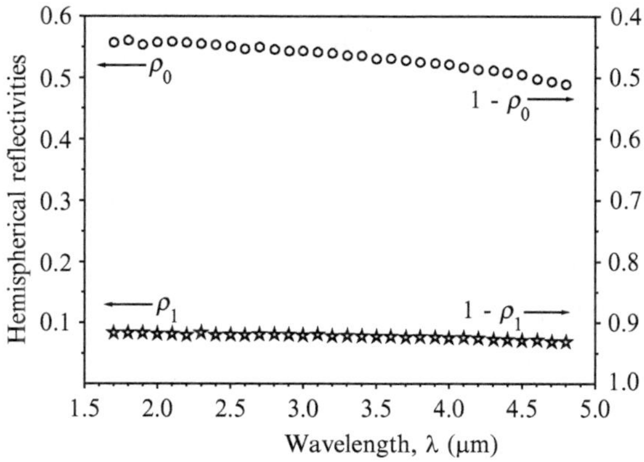

Fig. 11 Evolution of hemispherical reflectivities at glass/bubble and bubble/glass interfaces versus wavelength

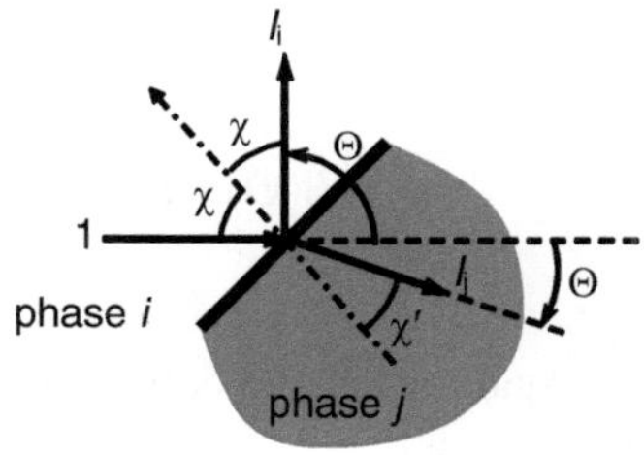

Fig. 12 Schematization of reflection and refraction of incident intensity unit on interface formed by the substances i and j with a refraction index ratio $n_j/n_i > 1$

For current fixed porosity and bubble size, the scattering coefficients follow the evolution of ρ_0, $1 - \rho_0$, ρ_1, or $1 - \rho_1$ with wavelength as in Fig. 11. For bubbles of radius a, the scattering phase functions can be obtained analytically. Let denote by $I_i(\Theta)$ the intensity of solid angle dΩ reflected to the substance i into the angle Θ from the incoming direction (see Fig. 12). For a specularly reflecting sphere, this scattered intensity can be expressed per incident intensity unit as:

$$I_i(\Theta) = \rho_i'(\chi) \tag{34}$$

where χ is related to the scattering angle Θ by the specular reflection law $2\chi = \pi - \Theta$. Integrating (34) over all scattering directions and using (33) gives:

$$\int_0^{4\pi} I_i(\Theta) d\Omega = 8\pi \int_0^{\pi/2} \rho'_i(\chi) \sin\chi \cos\chi \, d\chi = 4\pi\rho_i \tag{35}$$

In (35), the following relationship d$\Omega = \sin 2\chi$d(2χ)dφ was used. Now, the scattering phase functions in the substance i due to reflections is retrieved using the conventional definition (23):

$$\Phi_i(\Theta) = \frac{\rho'_i(\chi)}{\rho_i} \text{ for } i = 0 \text{ and } 1 \tag{36}$$

The derivation of scattering phase functions due to transmissions from one phase to other requires the analysis of refracted rays at interfaces using the reflection and refraction laws. The energy flux transmitted through an element area $\mathrm{d}A = a^2 \sin\chi$ $\mathrm{d}\chi$ $\mathrm{d}\varphi$ to the phase j per incident intensity unit is [26]:

$$[1 - \rho'] \cos \chi \, dA \tag{37}$$

The intensity transmitted at a large distance l (far from dA) in a solid angle $\sin\Theta$ $\mathrm{d}\Theta$ $\mathrm{d}\varphi$ [21, 26]:

$$I_j(\Theta) = a^2 \frac{[1 - \rho'] \cos \chi \, \sin \chi \, d\chi}{l^2 \sin \Theta d\Theta} \tag{38}$$

with $\chi = \Theta + \chi'$ and χ' the corresponding refraction angle. Integration of (38) over all scattering angles gives:

$$\int_0^{4\pi} I_j(\Theta) d\Omega = 2\pi \frac{a^2}{l^2} \int_0^{\pi/2} [1 - \rho_i'] \cos \chi \, \sin \chi d\chi = \pi \frac{a^2}{l^2} [1 - \rho_i] \tag{39}$$

Again using the definition (23), we obtain the scattering phase function due to refractions:

$$\Phi_{ij}(\Theta) = 4 \frac{1 - \rho_i'}{1 - \rho_i} \frac{\cos \chi \sin \chi \, d\chi}{\sin \Theta \, d\Theta} \text{ for } i = 0, j = 1 \text{ and } i = 1 \text{ and } j = 0 \tag{40}$$

In Fig. 13 is depicted the evolution of these scattering phase functions of porous fused quartz at the wavelength $\lambda = 1.7$ μm for which $n_1/n_0 \approx 1/1.44$. It can be observed that for spherical bubbles, the scattering phase functions due to transmissions are identical, i.e. $\Phi_{10} = \Phi_{01}$. The scattering phase function due to reflection in the glass matrix Φ_0 reaches a constant maximum value at scattering angles $\Theta < \pi - 2 \arcsin(n_1/n_0) = 92.3°$ for $n_0/n_1 = 1.44$ due to total internal reflection. The same phenomenon is responsible for zero values of $\Phi_{10} = \Phi_{01}$ at $\Theta > \pi/2 - \arcsin(n_1/n_0) = 46.01°$ for $n_0/n_1 = 1.44$.

3.1.4 Validity of Both Approaches

The radiative transfer in low porosity closed cell foams is illustrated here. The results of hemispherical transmittance and reflectance are used to study the competition between both prediction approaches in such material. The configuration is 1D sample exposed to a collimated radiation in one side. The experimental data

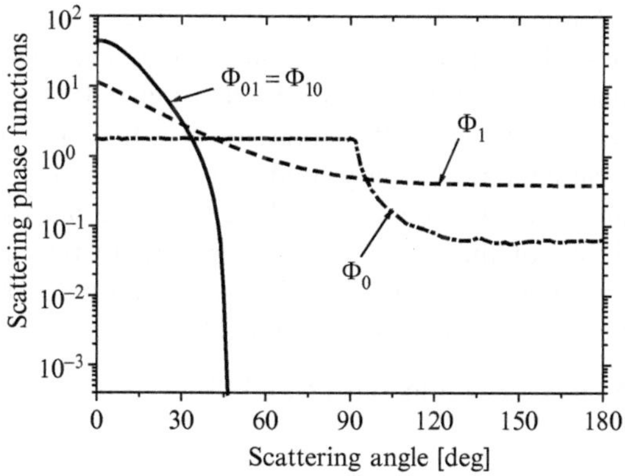

Fig. 13 Scattering phase functions of porous fused quartz for the MPA at wavelength λ = 1.7 μm

corresponding to the above porous fused quartz sample are obtained from a Fourier transform infrared (FTIR) spectrometer, which uses unpolarized radiation and disregards the sample emission thanks to the radiation modulation and the phase sensitive detection. The hemispherical reflectance and transmittance are respectively defined by [20]:

$$R_h = \frac{2\pi \int_{-1}^{0} I(0,\mu)\mu d\mu}{q_{inc}} \tag{41}$$

and

$$T_h = \frac{2\pi \int_{0}^{1} I(L,\mu)\mu d\mu}{q_{inc}} \tag{42}$$

The denominator $q_{\text{inc}} = I_{\text{inc}}\Delta\Omega_{\text{inc}}$ designates the incoming collimated radiation flux of intensity I_{inc} and a divergence solid angle $\Delta\Omega_{\text{inc}}$. It enters the sample perpendicularly to its surface at the abscise $z = 0$. The numerators correspond respectively to the radiation flux leaving the sample from the entering and opposite faces. In the prediction approaches, $I(z,\mu)$ corresponds to the intensity (solution of the transport equations without a self-emission contribution) leaving the sample at the abscise z with a direction of cosine μ measured from the normal to the sample surface. The values of $\mu > 0$ and $\mu < 0$ concern respectively the transmittance and reflectance.

To solve the transport equations in both approaches, the famous Discrete Ordinate Method (DOM) is adopted. It consists of subdividing (1) the angular domain into discretized angles following a quadrature (e.g. a Gauss quadrature [21, 22] of

12 orders is sufficient); and (2) the space into control volumes. This method is well-described in the radiation transfer textbooks [18, 20–22] and not detailed here. Note that in the MPA, the intensity $I(z,\mu)$ includes the radiation from both continuous and dispersed phases. Its expression as a function of $I_0(z,\mu)$ and $I_1(z,\mu)$ depends on the treatment of boundary conditions. For instance, we assume that the sample surfaces are optically smooth, which neglects the presence of bubble slices (which occupy 4–8% of the entire sample surface). In this case, only the continuous phase is considered present at the face exposed to the external radiation source. The boundary conditions of the MRTE are:

- For the continuous phase [48]

$$I_0(0,\mu) = \delta_{\mu,1}(1-\rho')I_{inc}\left(\frac{n_1}{n_0}\right)^2 + \rho' I_0(0,-\mu)\ for\ 0 < \mu \le 1 \tag{43}$$

$$I_0(L,\mu) = \rho' I_0(L,-\mu) for -1 \le \mu < 0 \tag{44}$$

- For the dispersed phase

$$I_1(0,\mu) = \rho' I_1(0,-\mu) for\, 0 < \mu \le 1 \tag{45}$$

$$I_1(L,\mu) = \rho' I_1(L,-\mu) for -1 \le \mu < 0 \tag{46}$$

where ρ' is the Fresnel reflectivity of sample boundaries. It depends on the angle between the current incident intensity and the inward normal to the boundary. $\delta_{\mu,\mu'}$ is the Kroneker delta function, which is equal to 1 if $\mu = \mu'$ and 0 elsewhere. According to (43), the continuous phase receives all transmitted radiation from the external source. Only after scattering, an amount of this radiation is transferred to the dispersed phase via (19). It is clear that the intensity that leaves a sample boundary is just the intensity leaving the continuous phase since we assume that no bubble cuts boundaries [48].

$$I(0,\mu) = \delta_{-1,\mu}\rho' I_{inc} + \left(\frac{n_0}{n_1}\right)^2 (1-\rho')I_0(0,\mu)\ for -1 \le \mu < 0 \tag{47}$$

$$I(L,\mu) = (1-\rho')I_0(L,\mu)\ for\ 0 < \mu \le 1 \tag{48}$$

In the HPA, $I(z,\mu)$ in (41)–(42) are the solutions of the usual RTE, calculated at each of the boundaries. The corresponding boundary conditions follow the (43), (44), (47) and (48) in which the intensities I_0 are substituted by the intensities in the single homogeneous and continuous medium.

In Figs. 14 and 15, the transmittance and reflectance computed from the HPA and MPA are compared with the experimental data of the porous fused quartz

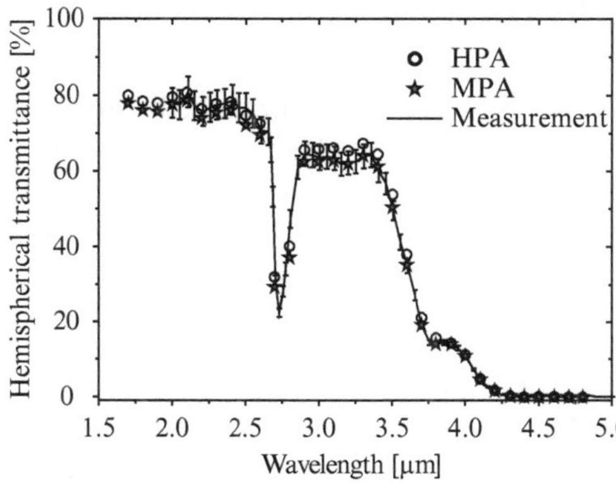

Fig. 14 Hemispherical transmittance from 1D sample of porous fused quartz of thickness 9.9 mm

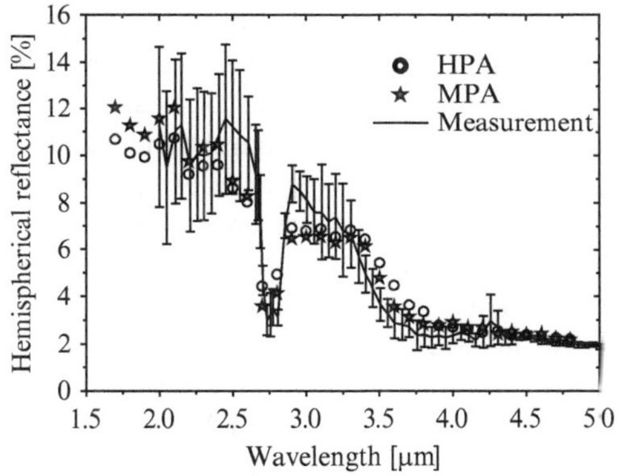

Fig. 15 Hemispherical reflectance from 1D sample of porous fused quartz of thickness 9.9 mm

sample. It can be observed that the predictions from both approaches are globally close to each other. In the near infrared region ($\lambda < 3.5$ μm), the transmittance (resp. reflectance) result from HPA is slightly lower (resp. higher) than the transmittance (resp. reflectance) result from MPA. It may be attributed to the dependent scattering effects neglected in the HPA (with which the Independent scattering theory is adopted here) while they are accounted for in the MPA. Moreover, in the spectral region where fused quartz becomes significantly absorbing (i.e. $\lambda >$ 3.5 μm), the predictions are overlapped. In this case, the evolution of transmittances and reflectances is mainly governed by the absorption of fused quartz because the scattering albedo approaches zero. The calculated transmittances and reflectances agree with the experimental data since they fall within the experimental uncertainty envelope. As a conclusion, the HPA (using Independent scattering theory) and MPA appear both suitable to model the radiative transfer in low porosity ($f_v \ll 1$) closed cell foams under the GO regime, especially concerning the hemispherical transmittance and reflectance.

3.2 Packed Beds

Usually, packed or fluidized beds are used as fuel particles in combustors or reacting particles in chemical reactors. In both applications, heat transfer within the particle bed becomes the limiting controlling mechanism. Moreover, the radiative transfer is generally the dominant heat transfer mode within the beds due to high temperature involved; therefore, it should be evaluated with accuracy to optimize the system efficiency. Depending on the application, the beds may be constituted of opaque particles, semitransparent particles, or a combination of these types of particles [75].

The radiative problem in such media has been widely studied. Currently, various prediction approaches are suggested. Most of them use the HPA in which the radiative properties are either derived from the Independent scattering theory [50–52], the correlated scattering theories [35, 36, 42], or the Ray-tracing techniques [41, 44, 46]. Each of these methods has its own weaknesses and advantages. For example, the Independent scattering theory, although practical, is restricted to very low volume fraction of scatterers. The correlated scattering approaches improve the prediction from the Independent scattering model. The current knowledge concerns however beds of spherical scatterers in non-absorbing substance. The Ray-tracing methods are suitable for beds of arbitrary geometry but restricted by the GOA. Moreover, they are more time consuming than the preceding predicting models.

In this section, the radiative transfer in typical packed bed media, and more precisely through samples of opaque or semitransparent particles, is analyzed in the framework of the HPA and MPA. Moreover, the prediction models of radiative properties in the HPA, presented above, are compared. In this aim, we limit our attention to transparent surrounding substances (i.e. $\alpha_0 = 0$). Analysis of radiative properties of packed bed in presence of a semitransparent host medium can be found elsewhere [41, 46].

3.2.1 Bed Microstructure and Optical Properties

As for porous fused quartz, the internal structure of packed (or fluidized) beds is relatively simple. They are generally represented as arrangements of spherical particles with constant or dispersed size in a fluid substance. The distance between two neighboring particles controls the porosity of the medium.

For simplicity, radiative transfer through packed beds of uniform particle size d and wavelength fulfilling the criterion $x = \pi d/\lambda = 100 \gg 1$ is analyzed. The sample thicknesses and the packing fractions are in the range $2 \leq L/d \leq 16$ and $0 < f_v < 0.45$, respectively. For beds of opaque particles, the optical properties can be represented by the particle reflectivities ρ_0, ρ_0', ρ_1, and ρ_1', the refraction index of the surrounding substance n_0, and the particle absorption index κ_1. For beds of

Table 1 Properties of opaque particle so that $x = 100$

ρ_0, ρ_1	0.6	0.9
n_0	1	
κ_1	0.05	
$\alpha_1 a$	10	

Table 2 Properties of semitransparent particle so that $x = 100$

n_0	1			
n_1	1.5			
κ_1	0	0.00025	0.001	0.0025
$\alpha_1 a$	0	0.05	0.2	0.5
Q_a	0	0.083	0.287	0.552
Q_s	1	0.917	0.713	0.448

semitransparent particles, the optical properties are represented by the refraction indexes of two substances, n_1 and n_0, and the particle absorption index κ_1. The Tables 1 and 2 summarize the properties of packed bed media analyzed hereafter. To be consistent with analysis reported in literature, we restrict our study to the case of specular reflecting particles. The directional reflectivities of opaque particles ($\rho'_1 = \rho_1$ and $\rho'_0 = \rho_0$) are assumed angle independent. Moreover, we assume that the external and internal reflectivities are identical ($\rho_1 = \rho_0$).

3.2.2 Computation of Radiative Properties for HPA

Bed of Opaque Particles

The Independent scattering theory (28) can be considered as the zeroth order approximation. In this model, the unknowns are the scattering Q_s and absorption efficiencies Q_a and the single scattering phase function ϕ. In the limit of optically large particles and without a diffraction contribution, when a radiation beam interacts with the particle, it is either reflected at the continuum-particle interface or totally absorbed in the particle. As a consequence, the efficiency factors are just $Q_s = \rho_0$ and $Q_a = 1 - \rho_0$. Since $a_{32} = a$ for monodispersed size, the extinction coefficient and scattering albedo are given by:

$$\beta = \frac{0.75 f_v}{a} \text{ and } \omega = \rho_0 \tag{49}$$

The scattering phase function is mainly governed by the angular distribution of the reflected radiation [20–22]. In the case of specular reflection with the reflectivity independent of the angle, the scattering phase function is isotropic, i.e.

$$\phi(\Theta) = \frac{\rho'_0(\Theta)}{\rho_0} = 1 \tag{50}$$

Remark: for diffuse reflecting particles, the extinction coefficient and scattering albedo are unchanged while scattering phase function is anisotropic and follows the well-known formula: [20–22]:

$$\phi(\Theta) = \frac{3}{8\pi}(\sin\Theta - \Theta\cos\Theta) \tag{51}$$

The correlated scattering theories scale the Independent scattering theory by introducing the dependent scattering corrections. For large opaque particles, the dependent scattering phenomena reduce to the non-point scattering and can be handled by multiplying the above extinction coefficient and/or scattering albedo in (49) by scaling factors while the scattering phase function is unchanged. The resulting radiative properties are summarized below according to the authors:

Singh and Kaviany [36]: $\beta = 0.75 f_v/a \times S_r$ and $\omega = \rho_0$ with S_{r} given by (5)

Brewster [56]: $\beta = 0.75 f_v/a \times S_r$ and $\omega = \rho_0$ with S_{r} given by (7)

Kamiuto [35]:$\beta = 0.75 f_v/a \times \gamma$ and $\omega = 1 - (1 - \rho_0)/\gamma$ with γ given by (6)

The RT method applied to opaque particle beds in the HPA was first developed by Coquard and Baillis [44]. Later, a simpler algorithm has been suggested by Randrianalisoa and Baillis [41, 46] and this last is adopted in this chapter. The RT method is performed on a packed bed medium reconstructed from X-ray tomography image analysis [76] or approached by an arrangement of particles [44]. In the last case, the particle shapes, sizes and arrangement must be chosen so as to reproduce faithfully the structural morphology of the real medium. Usually, a 3D random arrangement of spheres is sufficient to model packed beds of spherical particles. In practice, the particle arrangement is generated within a finite cubic box of dimensions much greater than the particle size. For instance, a ratio of box size to particle diameter of 10 is sufficient. In Fig. 16 we illustrate a typical packed bed

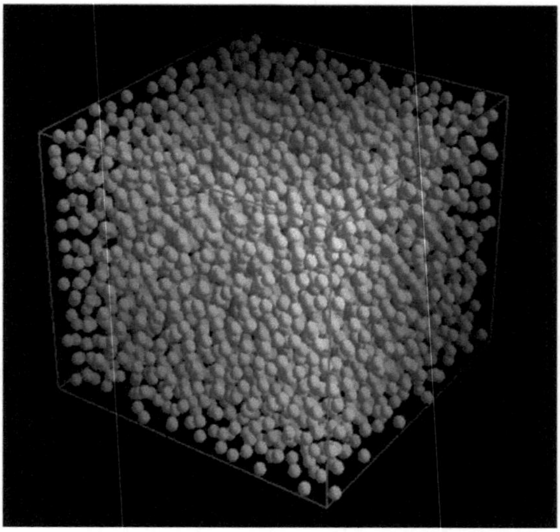

Fig. 16 Typical 3D bed of spherical particles

where the gray spheres represent the particles whereas the black font represents the surrounding substance. This particle arrangement is generated using the algorithm detailed in reference [44].

Recall that the aim of the RT algorithm is for analyzing the histories of a great number (N_{ray}) of radiation bundles propagating inside this medium and then to calculate the total extinction distance (*dist*), the fraction of bundles undergoing scattering (f_s), and the angular fraction of scattered bundles W(Θ). As previously described, the radiative properties are determined from these quantities through the formula (11), (14), and (15). Hereafter, the main steps in the RT method are summarized. For clarity purpose, the RT algorithm is illustrated in 2-D on Fig. 17. More details can be found in reference [46].

For a given radiation bundle, the RT process consists in:

(S1) – Select, in a random way, the initial location of the bundle on a particle surface (e.g. the point s_0 in Fig. 17) and the initial direction, oriented toward the surrounding medium as it has been scattered (e.g. the vector ds_0 in Fig. 17). The particle, from which the bundle path starts, is randomly chosen among the existing particles.

(S2) – Track the bundle path through the sample until it undergoes extinction. The type of the extinction event (i.e. scattering or absorption) is chosen by considering the most probable one through classical Monte Carlo tests [20–22]. The bundle may be absorbed in a particle (e.g. at the position s_1 in Fig. 17). Therefore, the absorption distance is given by the path traveled between the initial position s_0 and the absorption location s_1. For opaque particles, the scattering event is considered when the ray is reflected at a particle surface (e.g. at the position s_1 in Fig. 17).

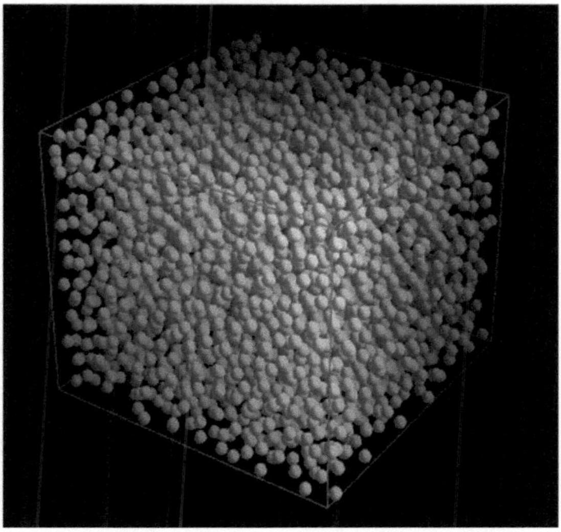

Fig. 17 Illustration of the Ray-tracing algorithm on a 2D dispersed medium [41, 46]

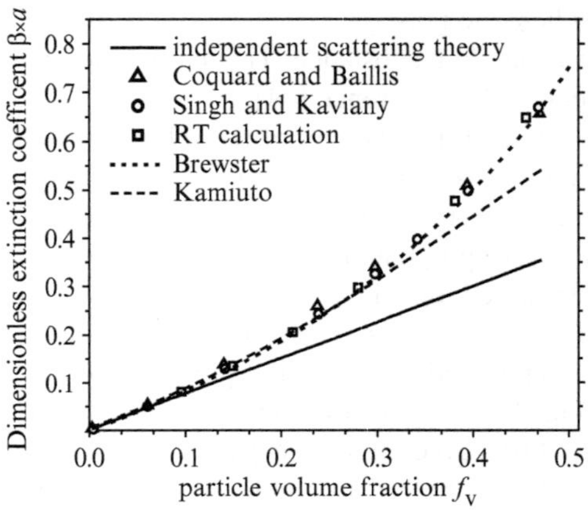

Fig. 18 Dimensionless extinction coefficient of opaque large spheres

In this case, the scattering distance is just the distance from the initial position s_0 to the location at which the ray interacts with the external particle surface, i.e. s_1. The extinction distance (either the scattering or absorption distance) is stored. When a scattering event occurs, the angle Θ between the incoming ray (i.e. ds_0) and the direction of the ray after scattering (i.e. ds_1 in Fig. 17) is also stored. The track of the path of the bundle is stopped after the extinction event.

The steps (S1) and (S2) are carried out for N_{ray} different bundles. Through this algorithm, *dist* is the sum of extinction distances traveled by the N_{ray} bundles; f_s is the ratio of the number of scattering events to the total number of extinction events, which is N_{ray}; therefore $f_a = 1 - f_s$ corresponds to the fraction absorption events. $W(\Theta)$ is the ratio between rays scattered into the angular interval Θ and $\Theta + d\Theta$ from the incoming direction and the total scattered rays, which is $f_s \times N_{ray}$. $d\Theta$ is the elementary angle.

Figure 18 summarizes the dimensionless extinction coefficient, $\beta \times a$, as a function of the packing fraction calculated from the above prediction models. As we can note, the RT method, the correlated theories of Singh and Kaviany and that of Brewster are in close agreement. This means that these models capture appropriately the non-point scattering phenomenon. On the other hand, the Independent scattering theory, which does not take into account the non-point scattering phenomenon, and the Kamiuto correlated theory fails when the particle concentration is greater than 0.1 and 0.3, respectively. Remember that the Kamiuto theory is based on comparisons of the extinction and scattering coefficients from the Independent scattering theory and those from the multiple scattering of waves under the Keller's approximation (i.e. by considering only the first and second scattering terms in the multiple scattering diagrams) [77]. The failure of the Kamiuto theory can be explained by the fact that, at large packing fractions ($f_v > 0.3$), high orders of scattering terms (third and higher) are significant but are not accounted for by the Kamiuto scaling factor γ. The scattering albedo from Kamiuto theory normalized by the scattering albedo from Independent scattering is reported in Fig. 19 for $\rho_0 = 0.9$ and 0.6. We observe that the lower the particle reflectivity and the higher

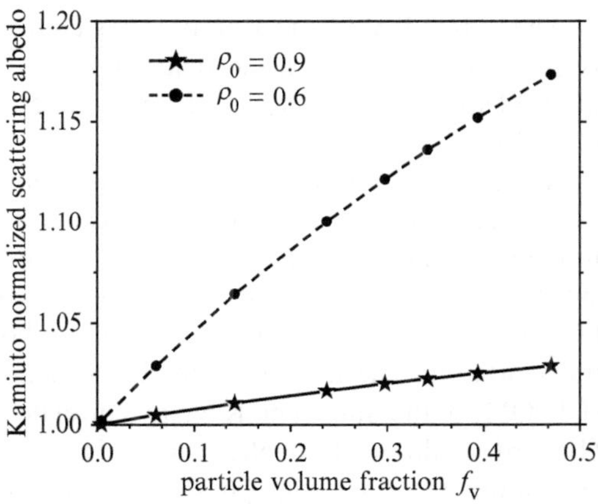

Fig. 19 Ratio of the scattering albedo from Kamiuto theory to the scattering albedo from Independent scattering theory of opaque large spheres

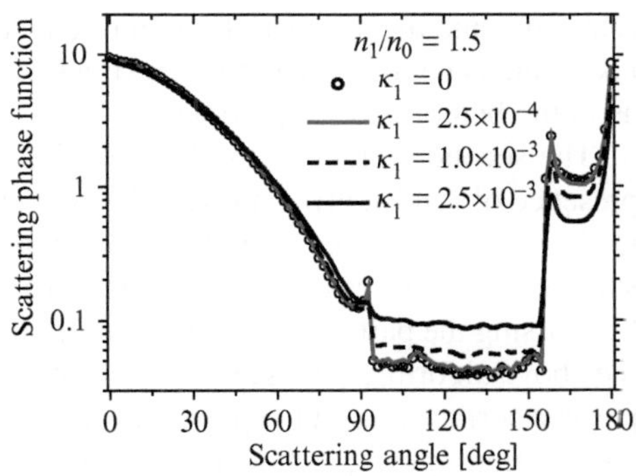

Fig. 20 Scattering phase functions of semitransparent particles with $n_1/n_0 = 1.5$

the packing fraction are, the greater the deviation of the Kamiuto theory. However, both theories converge at low packing fractions.

Bed of Semitransparent Particles

The radiative properties of semitransparent particle beds are first studied through the Independent scattering theory. The extinction coefficient is identical to that of opaque particle (i.e. $\beta = 0.75 f_v/a$) and the scattering albedo is equal to Q_s. The scattering and absorption efficiencies Q_s and Q_a are summarized in Table 2 for each particle optical properties. As expected, the higher the particle absorption index is, the greater the absorption efficiency and the smaller the scattering efficiency. The corresponding scattering phase functions are plotted in Fig. 20. The scattering phase function presents rainbow peaks in various scattering angles in the backward angular region. They correspond to rays scattered after undergoing internal reflections inside particles. When the particle becomes more absorbing, internal transversals are attenuated by absorption and the magnitude of rainbow peaks decreases.

The suitability of the Kamiuto correlated theory is also studied. The model is given by (6) in which the scattering albedo from Independent theory to scale is Q_s (Table 2). The scattering phase function is that from Independent theory (Fig. 20).

In beds of semitransparent or transparent particles, the dependent scattering phenomena include the non-point scattering and the radiation transportation. Remember that this latter is due to the transportation of radiation beams across substantial distances (through the particles) compared to the interparticle distance [10, 36]. Few methods enable to account for both non-point scattering and radiation transportation. The first one is the Dependence included discrete ordinate method (DIDOM) or the Singh and Kaviany correlated theory for semitransparent particles [10, 36]. It differs from the usual RTE (in the HPA) by the in-scattering term containing the scattering phase function. In fact, to account for the radiation transportation effect, the DIDOM employs an unusual scattering phase function, which depends not only on the scattering angle but also on the exit point of the scattering and the number of internal reflections. The non-point scattering has been taken into account by scaling the extinction coefficient from Independent scattering as in the bed of opaque particles.

The second method is the RT described in the above section. However, in the presence of semitransparent particles, the algorithm is slightly different, especially in the step (S2), since scattering does not occur uniquely at the external particle surface as was the case for opaque particles. In fact, the ray bundles can be scattered by crossing the particle after one or multiple internal transversals. In the step (S2) of the above algorithm, the scattering event is considered when the ray is reflected at a particle surface (e.g. at the position s_1 in Fig. 17) or when it crosses a particle (e.g. at the position s_2 or s_3 in Fig. 17 by representing only the two first interior transversals). Generally, the surface reflection and the two or three first internal reflections contain most of the scattered energy. When the ray is reflected at the particle surface (e.g. at s_1) or exits from the same side as it first enters the particle (e.g. at s_3), the scattering distance is just the distance from the initial position s_0 to the location at which the ray interacts with the external particle surface, i.e. s_1. Now, when the ray crosses the particle from one side (e.g. at s_1) to the opposite side (e.g. at s_2), the scattering distance is given by the distance $s_1 - s_0$ plus a transportation distance, denoted by d_{tr}. For specularly spherical particles, it was shown that d_{tr} only depends on the particle size d and on the ratio of the refraction indexes of the particles and the host medium, defined by $\eta = n_1/n_0$, according to (52) [41, 46]. According to these considerations, d_{tr} is about $0.72\,d$ with $n_1/n_0 = 1.5$.

$$d_{tr}/d = \begin{cases} 1.02 - \exp\left[-\left(\frac{K_1}{\eta} - K_2\right)\right] & \text{for} \quad 0.5 \leq \eta < 1 \text{ and } 1 < \eta \leq 8 \\ 1 & \text{elsewhere} \end{cases} \tag{52}$$

with $K_1 = 3.20$ and $K_2 = 1.47$ for $0.5 \leq \eta < 1$ and $K_1 = 2.89$ and $K_2 = 0.72$ while for $1 < \eta \leq 8$. Note that the result of the (52) reaches a minimal value when the contrast of refraction indexes is small (i.e. η tends to 1); it increases as long as η

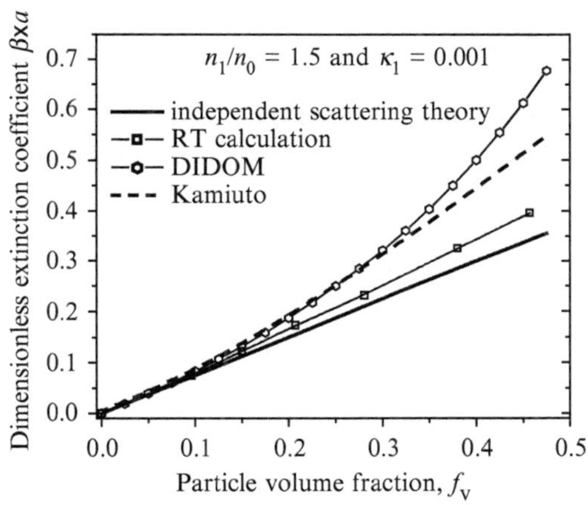

Fig. 21 Dimensionless extinction coefficient $\beta \times a$ of semitransparent particle bed with $n_1/n_0 = 1.5$ and $\kappa_1 = 0.001$

moves away from 1; and finally, it converges to the asymptotic value about 1 for $\eta > 8$ and $\eta < 0.5$. In fact, when $\eta > 8$ or $\eta < 0.5$, the rays striking a particle far from its center undergo a surface reflection because the local reflectivity approaches 1. Only the incident rays approximately normal to the particle surface have a significant probability to be refracted inside the particle; they travel at a distance about d before exiting this particle. For η approaching 1, the reflectivity approaches zero, thus almost all the rays interacting with the particle cross it without being reflected. In addition, the ray directions are not much altered. As a result, the calculated average distance traveled by these rays inside the particle tends to approach the theoretical value of the mean distance of scattering inside a particle, $2d/3$.

The evolution of the dimensionless extinction coefficient (β multiplied by the particle radius for convenience) according to the particle volume fraction is shown in Fig. 21 for an absorption index $\kappa_1 = 0.001$. At small packing fractions at which the dependent scattering effects are insignificant, we observe that both predictions converge to $0.75 \times f_v$. At high packing fractions: (1) the DIDOM gives extinction coefficients identical to that of opaque particles but much greater than the extinction coefficients from other predictions. Note that the DIDOM extinction coefficient has not the same significance as that predicted from other models. In fact, it takes into account only the non-point scattering effects while the RT extinction coefficient, for example, includes both non-point scattering and the ray transportation effects. (2) The RT results are slightly greater than the Independent scattering predictions. This means that the non-point scattering effect prevails but also that it is weaker for semitransparent particles than for opaque particles due to the ray transportation effect. In fact, this latter tends to increase the extinction mean-free-path and consequently, it gives opposite effects to the non-point scattering. The failure of the Independent scattering model is more noticeable at high packing fraction. The deviation between the Independent scattering and the current RT method is insignificant as soon as $f_v < 0.20$. In this case, the non-point scattering and ray transportation effects are either insignificant or cancelled out. (3) At moderate and high particle volume fractions, the extinction coefficient from the Kamiuto correlated theory is greater than that from the RT. Note that this extinction coefficient does not include the ray transportation correction. Finally, both prediction methods give identical scattering albedo, $\omega \approx 0.71$, which equals the scattering efficiency reported in Table 2. The scattering phase function from the RT method is identical to that of Independent scattering depicted in Fig. 20.

3.2.3 Computation of Radiative Properties for MPA

The formulae (32), (36) and (40) established in the above section are applicable to dispersed media constituted of spherical scatterers. In the following, they are applied to calculate the radiative properties of beds of opaque and semitransparent particles with parameters summarized in Tables 1 and 2.

For opaque particles with uniform reflectivities $\rho'_0 = \rho_0 = \rho'_1 = \rho_1$ (=0.6 or 0.9) and uniform size of radius a: (1) the scattering coefficients in the continuous phase, due to radiation reflections σ_0 and transmissions σ_{01}, vary proportionately to $0.75/a \times f_v/(1-f_v)$. It can be shown that the evolution of this factor versus packing fraction is identical to the scaled extinction coefficients of opaque particles in the HPA depicted in Fig. 18. The scattering coefficients in the dispersed phase are constant: $\sigma_1 = 0.75\rho_1/a$ and $\sigma_{10} = 0.75(1-\rho_1)/a$. (2) The scattering phase functions due to reflections are isotropic: $\Phi_0 = \Phi_1 = 1$ (Fig. 22). For highly reflecting particles, the refraction index ratio n_1/n_0 is generally much greater than 1. In this case, the refraction angle χ' is small, and the incidence χ and scattering angles Θ are quasi-identical. Therefore, the scattering phase functions due to transmissions vary approximately as $\Phi_{10} = \Phi_{01} \approx 4\cos\Theta$ for $0 \leq \Theta < \pi/2$ and are null elsewhere as plotted in Fig. 22. Moreover, the comparison of the scattering phase functions $\Phi_{01} = \Phi_{10}$ of porous fused quartz (in Fig. 13) and that of bed of semitransparent particles tends to show that the higher the ratio n_1/n_0, the larger the angular interval over which the scattering occurs ($\Phi_{01} = \Phi_{10} > 0$).

For semitransparent particle beds, the scattering coefficients are similar to that of opaque particles but with the following reflectivities $\rho_0 \approx 0.091$ and $\rho_1 \approx 0.596$ for

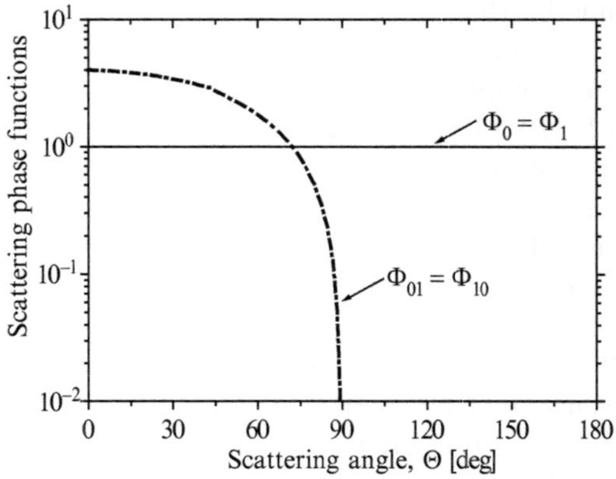

Fig. 22 Scattering phase functions of opaque particle bed in the MPA

$n_1/n_0 = 1.5$. In addition, the scattering phase functions are similar to that of porous fused quartz (Fig. 13) except that Φ_0 and Φ_1 are interchanged because $n_1/n_0 = 1.5$ here (instead of 1/1.44). The scattering phase function Φ_1 due to reflections within the particle reaches a constant maximum value at scattering angles $\Theta < \pi - 2\arcsin(n_0/n_1) = 96.4°$ while the scattering phase function due to transmissions $\Phi_{10} = \Phi_{01}$ decreases to zero at $\Theta > \pi/2 - \arcsin(n_0/n_1) = 48.2°$.

3.2.4 Validity of Both Approaches

Bed of Opaque Particles

The hemispherical transmittances through samples of specularly opaque particle beds are analyzed using the HPA and MPA. Consistent with data reported in literature, the configuration adopted is a plane parallel slab of packed bed exposed on one side to diffuse incident radiation. In both prediction approaches, we assumed that no particle crosses the sample surfaces; therefore, the boundary conditions (43)–(48) can be applied but with $\rho' = 0$ when $n_0 = 1$. In the HPA, the results based on radiative properties predicted by the correlated theories of Singh and Kaviany, and Kamiuto, and the Ray-tracing method are reported. The results for MPA are taken from reference [40]. In each case, the reference data are from the well-known direct MC simulation, similar to that described elsewhere [78, 79]. In Figs. 23 and 24, the transmittances versus the reduced sample thickness L/d are shown for particle reflectivities of 0.6 and 0.9 and a particle volume fraction of 0.278. When the slab thickness increases, the medium optical thickness increases and, as a result, the transmittances decreases. We can note that in all cases, the results of HPA based on the RT method and on the Singh and Kaviany correlated theory are close to the MC simulation whereas the results based on the Kamiuto correlated theory overestimate the MC solution noticeably. The significant deviation of the Kamiuto results may be attributed to the overestimation of the albedo

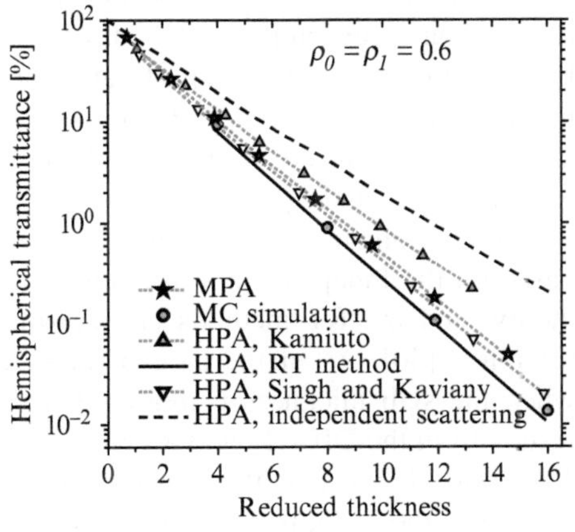

Fig. 23 Transmittance through beds of opaque spheres versus reduced bed thickness with $f_v = 0.278$, $\rho_0 = \rho_1 = 0.6$, and $x = \pi d/\lambda = 100$

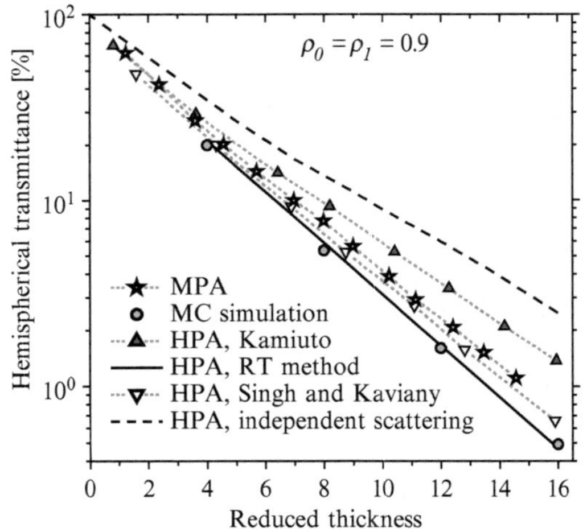

Fig. 24 Transmittance through beds of opaque spheres versus reduced bed thickness with $f_v = 0.278$, $\rho_0 = \rho_1 = 0.9$, and $x = \pi d/\lambda = 100$

(see Fig. 22). Concerning the MPA, it predicts satisfactorily the MC solutions. This can be explained as follows. For opaque particles, the RTE in the dispersed phase (i.e. within particles) of the MPA (18) becomes useless since the radiation intensity in this substance approximates zero due to the infinite absorptivity of opaque particles ($\alpha_1 a = 10 > 1$). The MPA reduces therefore to a single RTE with radiative properties identical to that from Brewster correlated theory. In other words, the MPA tends to reduce to the HPA in the case of dispersed media with opaque scatterers.

Bed of Semitransparent Particles

To illustrate the competition between the HPA and MPA for analyzing the radiative transfer calculation in beds of non-opaque particles, the hemispherical transmittances through samples of specularly reflecting particles with refraction index $n_1 = 1.5$ and concentration $f_v = 0.278$ in transparent host medium are shown in Figs. 25 and 26 as a function of the reduced sample thickness L/d. The results for transparent particles ($\kappa_1 = 0$ or $\alpha_1 a = 0$), weakly absorbing particles with $\kappa_1 = 0.00025$ (or $\alpha_1 a = 0.05$), and highly absorbing particles with $\kappa_1 = 0.001$ (or $\alpha_1 a = 0.2$) and $\kappa_1 = 0.0025$ (or $\alpha_1 a < 0.5$) are depicted. The data from direct MC simulation are still considered as reference. The results of the HPA calculation based on the Independent scattering theory, the DIDOM, the Kamiuto correlated theory, and the RT methods are shown. The MPA results are still taken from reference [40]. As in the case of opaque particles, the transmittances decrease when the slab thickness increases. Once again, we can note that generally, the results from the DIDOM and RT method are in close agreement with the MC data. This also shows that these two prediction methods are the most suitable for analyzing the radiative transfer in semitransparent medium embedding densely

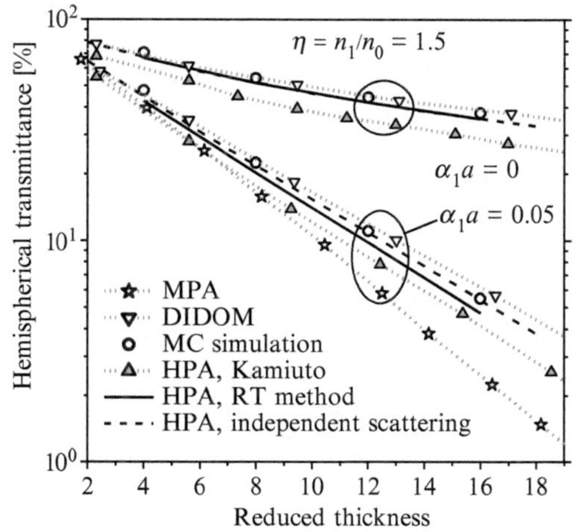

Fig. 25 Transmittance through beds of semitransparent spheres versus reduced bed thickness with $f_v = 0.278$, $n_1 = 1.5$, $n_0 = 1.0$, $\alpha_1 a = 0$ and 0.05

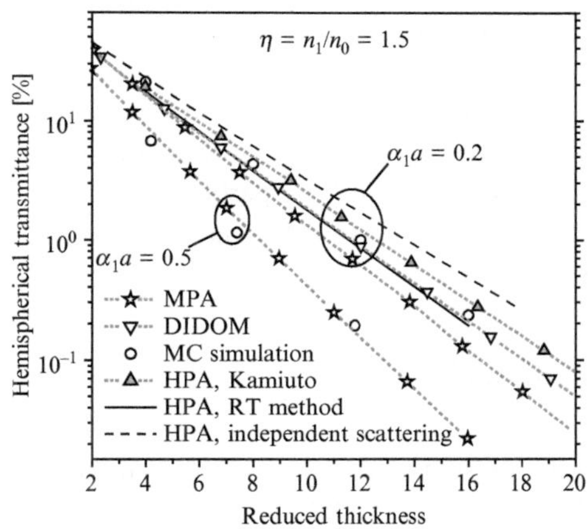

Fig. 26 Transmittance through beds of semitransparent spheres versus reduced bed thickness with $f_v = 0.278$, $n_1 = 1.5$, $n_0 = 1.0$, $\alpha_1 a = 0.2$ and 0.5

packed semitransparent particles. According to the considered simulation parameters (in particular $f_v = 0.278$ and $\eta = 1.5$), the Independent scattering theory model is suitable as soon as particles are slightly absorbing ($\alpha_1 a < 0.05$ here). These results tend to confirm previous experimental observations which pointed out (through analyzes of transmittances and reflectances of latex or glass particles in transparent water or air) that the dependent scattering effects are negligible even for $f_v = 0.7$ when dealing with optically large weakly absorbing or transparent particles. [32, 56] However, our analysis also points out that the Independent scattering theory breakdowns when particles are more absorbing ($\alpha_1 a \geq 0.2$ for example). In fact, given that the ray transportation effects become less significant, it overestimates the transmittance due to the lack of the non-point scattering correction. The Kamiuto model underestimates the transmittance for non-absorbing particles and overestimates the transmittance for absorbing particles. The failure of the Kamiuto model for absorbing particles can be explained in a similar manner as for opaque particles. However, when the particles are transparent, the transmittance from the Kamiuto theory is under-evaluated probably due to the overestimation of the extinction coefficient (see Fig. 21). The MPA underestimates the transmittances. The deviation is more pronounced for weakly absorbing particles (e.g. $\alpha_1 a < 0.2$ here) than for strongly absorbing particles (e.g. $\alpha_1 a = 0.5$). The models of scattering phase functions in the MPA are probably the main sources of inaccuracy because they do not reproduce correctly the scattering pattern of spherical semi-transparent particles. In fact, we showed that the scattering pattern of semitransparent particles presents rainbow peaks corresponding to the rays scattered after undergoing internal reflections inside particles. Therefore, there exists a strong correlation between the rays incident on the particle and those transmitted through the particle. This correlation between incident and scattered rays is captured in HPA prediction models because they use or predict rigorously the scattering phase function of a single particle. However, this is disregarded in the MPA, which splits the scattering pattern into four uncorrelated scattering phase functions [namely Φ_0, Φ_1, Φ_{01}, and Φ_{10}]. For refractive index ratios $n_1/n_0 > 1$ and transparent or slightly absorbing particles ($\alpha_1 a < 0.2$ here), the order of magnitude of these rainbow peaks is significant. Neglecting them leads to an underestimation of the scattering albedo ω [40] and, as a consequence, to an underestimation of the transmittances. Since the rainbow peaks are absent for refraction index ratios $n_1/n_0 < 1$ (see Fig. 8) and much attenuated for absorbing particles (see Fig. 20 for $\kappa_1 = 0.0025$), this explains why the MPA is better in these cases.

4 Conclusions

In the present study, we have shown the ability of the conventional Homogeneous Phase Approach (HPA) and the recently developed Multi-Phase Approach (MPA) to model the radiative transfer in two-phase dispersed media. For each approach, the simplest and realistic radiative properties models are identified. The competition between the HPA and MPA for predicting the radiative transfer through glass containing bubbles, packed beds of semitransparent or opaque spheres is highlighted through their comparison with experimental data or Monte Carlo simulation of hemispherical transmittance and reflectance. The following conclusions can be drawn.

Although exact solutions of radiative transfer in dispersed media constituted of closely spaced scatterers are not expected theoretically through the radiative transfer theory, prior investigations and the current one (including both experimental and numerical analysis) have shown its suitability by using the notion of effective radiative properties which differ from radiative properties of dilute media.

The traditional HPA is suitable for a wide range of two-phase materials provided that their effective radiative properties are accurately known and that the thermal equilibrium can be assumed. The simple and well-known independent scattering theory gives accurate prediction as long as the dispersed phase volume fraction is small or the scatterers are far from each other. Otherwise, more advanced models such as the correlated scattering theory or ray-tracing methods, enabling to approximate the complex dependent scattering phenomena, should be used. For dispersed media constituted of spherical opaque scatterers in non-absorbing environment, the Brewster or Singh and Kaviany correlated model are the most useful, otherwise the ray-tracing methods are an alternative solution. The practical one is the Randrianalisoa and Baillis' algorithm based on mean-free-path analysis.

In the MPA, the effective radiative properties can be predicted analytically for spherical scatterers; while the recourse to ray-tracing technique is indispensable for complex scatterer shape. The MPA is generally suitable for two-phase dispersed media as shown by the analysis on packed beds of opaque particles and glass embedding bubbles. It can however be inaccurate for the case of continuum less refracting than scatterers, and weakly absorbing scatterers with regular and specularly reflecting boundaries. In fact, these scatterers present rainbow peaks influencing the radiative transfer but not modeled by the current MPA. An example of these media is packed beds of semitransparent and specularly reflecting spheres. The MPA needs some improvements to be well competitive. Further analysis should be conducted on the modeling of radiative properties, notably to specify how to take into account, in the scattering phase functions, the correlations between radiations incident on the scatterer and those scattered.

References

1. Mujeebu, M.A., Abdullah, M.Z., Bakar, M.Z.A., Mohamad, A.A., Abdullah, M.K.: Applications of porous media combustion technology – a review. Appl. Energy **86**, 1365–1375 (2009)
2. Sathe, S.B., Peck, R.E., Tong, T.W.: A numerical analysis of heat transfer and combustion in porous radiant burners. Int. J. Heat Mass Transf. **33**, 1331–1338 (1990)
3. Damm, D.L., Fedorov, A.G.: Radiation heat transfer in SOFC materials and components. J. Power Sources **143**, 158–165 (2005)
4. Steinfeld, A.: Solar thermochemical production of hydrogen – a review. Sol. Energy **78**, 603–615 (2005)
5. Glicksmann, L.R., Arduini-Schuetz, M.C., Sinofsky, M.: Radiation heat transfer in foam Insulation. Int. J. Heat Mass Transf. **109**, 809–812 (1987)
6. Kuhn, J., Ebert, H.P., Arduini-Schuster, M.C., Buttner, D., Fricke, J.: Thermal transport in polystyrene and polyurethanes foam insulations. Int. J. Heat Mass Transf. **35**, 1795–1801 (1992)
7. Kaviany, M.: Principles of Heat Transfer in Porous Media, 2nd edn. Springer, New York (1995)
8. Quintard, M., Whitaker, S.: Fundamentals of transport equation formulation for two-phase flow in homogeneous and heterogeneous porous media. In: Parlange, M.B., Hopmans, J.W. (eds.) Vadose Zone Hydrology: Cutting Across Disciplines, pp. 3–57. Oxford University Press, New York (1999). Chapter 1
9. Howell, J.R.: Radiative transfer in porous media. In: Vafai, K. (ed.) Handbook of Porous Media, 2nd edn, pp. 663–698. Marcel Dekker, New York (2000). Chapter 15

10. Kaviany, M., Singh, B.P.: Radiative heat transfer in porous media. In: Hartnett, J.P., Irvine, T. (eds.) Advances in Heat Transfer, vol 23, pp. 133–186. Academic, San Diego (1993)
11. Adzerikho, K.S., Nogotov, E.F., Trofimov, V.P.: Radiation Heat Transfer in Two-Phase Media, 1st edn. CRC, Boca Raton (1993)
12. Foldy, L.L.: The multiple scattering of waves. I. General theory of isotropic scattering by randomly distributed scatterers. Phys. Rev. **67**, 107–119 (1945)
13. Goldberg, M.L., Watson, K.M.: Collision Theory. Wiley, New York (1964)
14. Ishimaru, A.: Wave Propagation and Scattering in Random Media. Academic, New York (1978)
15. Apresyan, L.A., Kravtsov, Y.A.: Radiative Transfer Theory. Nauka, Moscow (1983)
16. Papanicolaou, G.C., Burridge, R.: Transport equations for the Stokes parameters from Maxwell equations in a random media. J. Math. Phys. **16**, 2074–2082 (1975)
17. Mishchenko, M.I., Travis, L.D., Lacis, A.A.: Multiple Scattering of Light by Particles: Radiative Transfer and Coherent Backscattering. Cambridge University Press, Cambridge (2006)
18. Chandrasekhar, S.: Radiative Transfer. Oxford University Press, Oxford (1950)
19. Lenoble, J.: Radiative Transfer in Scattering and Absorbing Atmospheres: Standard Computational Procedures. Deepak, Hampton (1985)
20. Brewster, M.Q.: Thermal Radiative Transfer and Properties. Wiley, New York (1992)
21. Modest, M.F.: Radiative Heat Transfer. Academic, New York (2003)
22. Siegel, R., Howell, J.R.: Thermal Radiation Heat Transfer. Taylor & Francis, New York (2002)
23. Mishchenko, M.I.: Far-field approximation in electromagnetic scattering. J. Quant. Spectrosc. Radiat. Transf. **100**, 268–276 (2006)
24. Kerker, M.: The Scattering of Light and Other Electromagnetic Radiation. Academic, New York (1966)
25. Van de Hulst, H.C.: Light Scattering by Small Particles. Dover, New York (1981)
26. Bohren, C.F., Huffman, D.R.: Absorption and Scattering of Light by Small Particles. Wiley, New York (1983)
27. Durant, S., Calvo-Perez, O., Vukadinovic, N., Greffet, J.J.: Light scattering by a random distribution of particles embedded in absorbing media: full-wave Monte Carlo solutions of the extinction coefficient. J. Opt. Soc. Am. A **24**, 2953–2962 (2007)
28. Waterman, P.C., Truell, R.: Multiple scattering of waves. J. Math. Phys. **2**, 512–537 (1961)
29. Twersky, V.: On scattering of waves by random distributions. I. Free-space scatterer formalism. J. Math. Phys. **3**, 700–715 (1962)
30. Lax, M.: Multiple scattering of waves. II. The effective field in dense systems. Phys. Rev. **85**, 621–629 (1952)
31. Cartigny, J.D., Yamada, Y., Tien, C.L.: Radiative transfer with dependent scattering by particles: Part I – Theoretical investigation. J. Heat Transf. **108**, 608–613 (1986)
32. Yamada, Y., Cartigny, J.D., Tien, C.L.: Radiative transfer with dependent scattering by particles: Part II – Experimental investigation. J. Heat Transf. **108**, 614–618 (1986)
33. Drolen, B.L., Tien, C.L.: Independent and dependent scattering in packed-sphere systems. J. Thermophys. Heat Transf. **1**, 63–68 (1987)
34. Kumar, S., Tien, C.L.: Dependent absorption and extinction of radiation by small particles. J. Heat Transf. **112**, 178–185 (1990)
35. Kamiuto, K.: Correlated radiative transfer in packed-sphere systems. J. Quant. Spectrosc. Radiat. Transf. **43**, 39–43 (1990)
36. Singh, B.P., Kaviany, M.: Modelling radiative heat transfer in packed beds. Int. J. Heat Mass Transf. **35**, 1397–1405 (1992)
37. Dombrovsky, L.: Thermal radiation from nonisothermal spherical particles of a semitransparent material. Int. J. Heat Mass Transf. **43**, 1661–1672 (2000)
38. Dombrovsky, L.: Large-cell model of radiation heat transfer in multiphase flows typical for fuel–coolant interaction. Int. J. Heat Mass Transf. **50**, 3401–3410 (2007)

39. Zeghondy, B., Estelle, Y., Taine, J.: Determination of the anisotropic radiative properties of a porous material by radiative distribution function identification (RDFI). Int. J. Heat Mass Transf. **49**, 2810–2819 (2006)
40. Gusarov, A.V.: Homogenization of radiation transfer in two-phase media with irregular phase boundaries. Phys. Rev. B **77**, 144201 (2008)
41. Randrianalisoa, J., Baillis, D.: Radiative transfer in dispersed media: comparison between homogeneous phase and multiphase approaches. J. Heat Transf. **132**, 023405 (2010)
42. Brewster, M.Q.: Volume scattering of radiation in packed beds of large, opaque spheres. J. Heat Transf. **126**, 1048–1050 (2004)
43. Coquard, R., Baillis, D.: Radiative properties of dense fibrous medium containing fibers in the geometric limit. J. Heat Transf. **128**, 1022–1030 (2006)
44. Coquard, C., Baillis, D.: Radiative characteristics of opaque spherical particle beds: a new method of prediction. J. Thermophys. Heat Transf. **18**, 178–186 (2004)
45. Tancrez, M., Taine, J.: Direct identification of absorption and scattering coefficients and phase function of a porous medium by a Monte Carlo technique. Int. J. Heat Mass Transf. **47**, 373–383 (2004)
46. Randrianalisoa, J., Baillis, D.: Radiative properties of densely packed spheres in semitransparent media: a new geometric optics approach. J. Quant. Spectrosc. Rad. Transf. **111**, 1372–1388
47. Jones, P.D., McLeod, D.G., Dorai-Raj, D.E.: Correlation of measured and computed radiation intensity exiting a packed bed. J. Heat Transf. **118**, 94–102 (1996)
48. Randrianalisoa, J., Baillis, D., Pilon, L.: Improved inverse method for radiative characteristics of closed-cell absorbing porous media. J. Thermophys. Heat Transf. **20**, 871–883 (2006)
49. Dombrovsky, L.: The propagation of infrared radiation in a semitransparent liquid containing gas bubbles. High Temp. **42**, 133–139 (2004)
50. Mischler, D., Steinfeld, A.: Nonisothermal nongray absorbing–emitting–scattering suspension of Fe_3O_4 particles under concentrated solar irradiation. J. Heat Transf. **117**, 346–354 (1995)
51. Qiu, T.Q., Longtin, J.P., Tien, C.L.: Characteristics of radiation absorption in metallic particles. J. Heat Transf. **117**, 340–345 (1995)
52. Dembele, S., Delmas, A., Sacadura, J.F.: A method for modeling the mitigation of hazardous fire thermal radiation by water spray curtains. J. Heat Transf. **119**, 746–753 (1997)
53. Fedorov, A.G., Viskanta, R.: Radiation characteristics of glass foams. J. Am. Ceram. Soc. **83**, 2769–2776 (2000)
54. Hottel, H.C., Sarofim, A.F., Dalzell, W.H., Vasalos, I.A.: Optical properties of coatings, effect of pigment concentration. J. Thermophys. Heat Transf. **9**, 1895–1898 (1971)
55. Ishimaru, Kuga, Y.: Attenuation constant of a coherent field in a dense distribution of particles. J. Optic. Soc. Am. **72**, 1317–1320 (1982)
56. Brewster, M.Q., Tien, C.L.: Radiative transfer in packed fluidized beds: dependent versus independent scattering. J. Heat Transf. **104**, 573–579 (1982)
57. Ivezic, Z., Menguc, P.: An investigation of dependent/independent scattering regimes using a discrete dipole approximation. Int. J. Heat Mass Transf. **39**, 811–822 (1996)
58. Kaviany, M., Singh, B.P.: Radiative heat transfer in packed beds. In: Quintard, M., Todorovic, M. (eds.) Heat and Mass Transfer in Porous Media, pp. 191–202. Elsevier, Amsterdam (1992)
59. Zeghondy, B., Iacona, E., Taine, J.: Experimental and RDFI calculated radiative properties of a mullite foam. Int. J. Heat Mass Transf. **49**, 3702–3707 (2006)
60. Petrasch, J., Wyss, P., Steinfeld, A.: Tomography-based Monte Carlo determination of radiative properties of reticulate porous ceramics. J. Quant. Spectrosc. Radiat. Transf. **105**, 180–197 (2007)
61. Baillis, D., Sacadura, J.F.: Thermal radiation properties of dispersed media: theoretical prediction and experimental characterization. J. Quant. Spectrosc. Radiat. Transf. **67**, 327–363 (2000)
62. Dombrovsky, L., Randrianalisoa, J., Baillis, D., Pilon, L.: Use of Mie theory to analyze experimental data to identify infrared properties of fused quartz containing bubbles. Appl. Opt. **44**, 7021–7031 (2005)

63. Baillis, D., Sacadura, J.F.: Directional spectral emittance of a packed bed: influence of the temperature gradient in the medium. J. Heat Transf. **124**, 904–911 (2002)
64. Fedorov, A.G., Viskanta, R.: Radiative transfer in a semitransparent glass foam blanket. Phys. Chem. Glasses **41**, 127–135 (2000)
65. Fedorov, A.G., Pilon, L.: Glass foams: formation, transport properties, and heat, mass, and radiation transfer. J. Non-Cryst. Solids **311**, 154–173 (2002)
66. Malitson, I.H.: Interspecimen comparison of the refractive index of fused silica. J. Opt. Soc. Am. **55**, 1205–1209 (1965)
67. Beder, E.C., Bass, C.D., Shackleford, W.L.: Transmissivity and absorption of fused quartz between 0.2 μm and 3.5 μm from room temperature to 1500 degree C. Appl. Opt. **10**, 2263–2268 (1971)
68. Touloukian, Y.S., DeWitt, D.P.: Thermal radiative properties: nonmetallic solids. In: Touloukian, Y.S., DeWitt, D.P. (eds.) Thermophysical Properties of Matter, vol 8. Plenum, New York (1972)
69. Khashan, M.A., Nassif, A.Y.: Dispersion of the optical constants of quartz and polymethyl methacrylate glasses in a wide spectral range: 0.2–3 μm. Opt. Commun. **188**, 129–139 (2001)
70. Randrianalisoa, J., Baillis, D., Pilon, L.: Modeling radiation characteristics of semitransparent media containing bubbles or particles. J. Opt. Soc. Am. A **23**, 1645–1656 (2006)
71. Dombrovsky, L., Randrianalisoa, J., Baillis, D.: Modified two-flux approximation for identification of radiative properties of absorbing and scattering media from directional–hemispherical measurements. J. Opt. Soc. Am. A **23**, 91–98 (2006)
72. Dombrovsky, L., Randrianalisoa, J., Baillis, D.: Infrared radiative properties of polymer coatings containing hollow microspheres. Int. J. Heat Mass Transf. **50**, 1516–1527 (2007)
73. Mundy, W.C., Roux, J.A., Smith, A.M.: Mie scattering by spheres in an absorbing medium. J. Opt. Soc. Am. **64**, 1593–1597 (1974)
74. Mills, A.F.: Heat Transfer, 2nd edn. Prentice Hall, New Jersey (1999)
75. Jäger, K., Lipiński, W., Katzgraber, H.G., Steinfeld, A.: Determination of thermal radiative properties of packed-bed media containing a mixture of polydispersed particles. Int. J. Therm. Sci. **150**, 502–508 (2009)
76. Haussener, S., Lipiński, W., Petrasch, J., Wyss, P., Steinfeld, A.: Tomographic characterization of a semitransparent-particle packed bed and determination of its thermal radiative properties. J. Heat Transf. **131**, 072701 (2009)
77. Keller, J.B.: Stochastic equation and wave propagation in random media. Proc. Symp. Appl. Math. **13**, 145–170 (1964)
78. Yang, Y.S., Howell, J.R., Klein, D.E.: Radiative heat transfer through a randomly packed bed of spheres by the Monte Carlo method. J. Heat Transf. **105**, 325–332 (1983)
79. Rousseau, B., De Sousa Meneses, D., Echegut, P., Di Michiel, M., Thovert, J.F.: Prediction of the thermal radiative properties of an X-ray μ-tomographied porous silica glass. Appl. Opt. **46**, 4266–4276 (2007)

Predictions of Effective Thermal Conductivity of Complex Materials

Ramvir Singh

Abstract In this review, a comprehensive and systematic effort is made to incorporate the most significant and popular models for calculation of the effective thermal conductivity of complex materials and discuss their limitations. A brief review of the numerical techniques for prediction of the effective thermal conductivity of multi-phase materials is presented and discussed. The real structures and geometries around us are so vast and vivid, that one cannot use a single model to estimate effective thermal conductivity of complex materials in the whole range due to their inherent limitations.

1 Introduction

The approach of thermo-mechanical properties of conventional materials to their optimum value does not allow the fulfillment of the needs of new age requirements. Therefore, there is an urgent need to develop a new class of materials to fulfill requirements of the present millennium. Hence, we need tailored materials made up of a combination of two or more materials with controlled volume fractions. Such materials are treated as composite materials for analysis purposes. Accurate prediction of effective thermo-mechanical properties of these materials still remains a challenging task for engineers and scientists working in the field of materials science.

Multi-phase materials are made up of more than one phases which are rarely distributed uniformly, and each component has its own properties and proportion, and thus contributes towards the effective system. Once mixed together, the components of different types will more or less interact with each other and the

R. Singh
Department of Physics, University of Rajasthan, Jaipur 302 004, India
e-mail: singhrvs@rediffmail.com, rvs2020@gmail.com

A. Öchsner and G.E. Murch (eds.), *Heat Transfer in Multi-Phase Materials*,
Adv Struct Mater 2, DOI 10.1007/8611_2010_5,

properties at the interfacial region will exhibit a transition from one component to the other. Such effects usually turn even more complicated when the components are at different phase states, such as in a semi-frozen soil system. Multi-phase materials are increasingly used in various fields, but analysis and investigation efforts are severely lagging behind. The challenges in studying complex multi-phase materials come mainly from the inherent variety and randomness of their internal microstructures, and the coupling between the components of different phases. For systems with three or more phases, the microstructures or the phase distribution can become much more complicated. Even in a two-component alloy, if there are pores or small fractures inside, they should be treated as three-phase materials for analysis purposes. Yet other often-encountered multi-phase structure is that of unsaturated porous soils.

The basic parameters needed to be studied in the context of the heat transfer through complex materials are (1) thermal conductivity, (2) thermal diffusivity and (3) volumetric specific heat. These parameters are interrelated to each other. Complete thermophysical behaviour of a material can be obtained by knowledge of any two of the above parameters.

The thermal conductivity of a material is defined by Fourier's equation [1],

$$Q = -\lambda A\left(\frac{d\theta}{dx}\right) \tag{1}$$

where Q is the rate of heat flow, $(d\theta/dx)$ is the temperature gradient, Q and $(d\theta/dx)$ both are normal to the area A of the specimen.

For multi-phase materials the effective value of the thermophysical parameter lies in between that of the constituent phases and is denoted by adding a suffix 'e' throughout the chapter. The effective value of a parameter is found to depend upon following geometrical factors,

1. Thermal conductivities of the constituents
2. Fractional volumes of the constituents
3. Size and morphology of the constituent particles
4. Structure of the material
5. Inter phase interactions of the constituents

Beside these physical conditions e.g. pressure and temperature also affect the effective property of these complex materials.

The study of heat transfer in complex materials has inspired a large number of workers to derive theoretical expressions for evaluating effective thermo-physical properties. Further, due to the advancement of computing technology in recent times, there has been a rapid development of powerful numerical methods to evaluate effective parameters of such materials. Among all the thermo-physical properties, the thermal conductivity is mostly studied due to its significance and wide applications in heat transfer problems. So this chapter is devoted to the prediction of effective thermal conductivity (ETC) of multi-phase complex materials.

2 Complex Materials

Complex materials are multi-phase systems in which nearly all the phases are distributed randomly. As such, the behaviour of these materials are dictated by each and every component of the different phases, i.e. its overall macroscopic property is not equal to that of any single constituent, rather it is a collective one contributed by all components forming the system. Therefore, the effective property is actually the equivalent property. Inclusion of other components into a basic material can be beneficial, acting as reinforcement or supplements to improve the performance of the material and by combining them we can obtain a tailored behaviour as a final result. Alloys, cellular structures, and fiber reinforced polymer composites are just such examples. In the present era of emerging new technological materials most of them are complex multi-phase materials in terms of both physics and structures, thus increasing the urgency for more robust theoretical and computational tools.

3 Mathematical Modeling

Mathematical modeling of multi-phase materials for prediction of thermo-physical coefficients are still preferred because of low cost, easy and quick to use, and reasonable accuracy for certain specific cases, especially when the microstructure can be simplified. In the present chapter, more emphasis has been given to the significant and popular models and the author also reviews some recent progresses in numerical modeling of effective thermal properties for multi-phase materials.

3.1 Models for Two-Phase Materials

In this section, theoretical models for calculation of the thermal conductivity of two phase materials are discussed. In the following discussion, a comprehensive and systematic effort is made to incorporate most of the popular models and discuss their limitations.

The earliest model in this direction was given by Maxwell [2] for the prediction of the effective thermal conductivity by assuming random size spheres dispersed into a continuous medium. The effective thermal conductivity for such a system can be represented as

$$\lambda_e = \lambda_c \frac{[2\lambda_c + \lambda_d - 2\phi_d(\lambda_c - \lambda_d)]}{[2\lambda_c + \lambda_d + \phi_d(\lambda_c - \lambda_d)]} \tag{2}$$

where λ_c is the thermal conductivity of the continuous phase, λ_d is the thermal conductivity of dispersed phase and ϕ_d is the volume fraction of the dispersed phase respectively.

Fricke [3] and Burger [4] independently modified the Maxwell's equation for particles having ellipsoidal shape. The expression obtained by Fricke and Burger is

$$\lambda_e = \frac{\lambda_c \phi_c + \lambda_d (1 - \phi_c) F}{\phi_c + (1 - \phi_c) F} \tag{3}$$

where $F = \frac{1}{3} \sum_{i=1}^{3} \left[1 + \left\{ \left(\frac{\lambda_d}{\lambda_c} \right) - 1 \right\} g_i \right]^{-1}$ and $\sum_{i=1}^{3} g_i = 1$

They assumed that particles were noninteracting. F is the ratio of the average temperature gradients in the two phases and g_i are the semi principal axes of the ellipsoid.

De Vries [5] in his derivation of the effective thermal conductivity used $g_1 = g_2 = 1/8$ and $g_3 = 3/4$. This implies that the two minor axes of ellipsoid are the same and the major axis is three times that of the minor axis.

Bruggeman [6] used Maxwell's model for cylindrical particles. He obtained an expression for the effective thermal conductivity in the following form:

$$\lambda_e = \frac{\lambda_c \left[1 - \left(1 - \frac{\lambda_d}{\lambda_c} \right) \frac{2}{3} \phi_d \delta \right]}{[1 + (\delta - 1)\phi_d]} \tag{4}$$

Here ϕ_d is the fractional volume of inclusions and δ is determined from λ_c and λ_d. The value of δ is $\frac{3\lambda_c}{2\lambda_c + \lambda_d}$ for spherical particles, $\frac{5\lambda_c + \lambda_d}{3(\lambda_c + \lambda_d)}$ for cylindrical particles, and $\frac{\lambda_c + 2\lambda_d}{3(\lambda_d)}$ for plates and scales.

Lord Rayleigh [7] assumed that particles are spherical in shape and they are arranged in a cubical array. The expression for the effective thermal conductivity given by Rayleigh was:

$$\lambda_e = \frac{\lambda_c \left[1 - 2\phi.k - 1.65(\phi)^{10/3} Ak \right]}{\left[1 + \phi.k - 1.65(\phi)^{10/3} Ak \right]} \tag{5}$$

Here $k = \frac{(\lambda_c - \lambda_d)}{(2\lambda_c + \lambda_d)}$ and $A = \frac{(3\lambda_c - 3\lambda_d)}{(4\lambda c + 3\lambda_d)}$

Rayleigh's model was so rigid and artificial that it could not predict the λ_e of mixtures in practical cases.

Wiener [8] developed a model based on the resistors concept. This concept was a turning point in theoretical models for the effective thermal conductivity. Wiener considered the system to be made of alternate layers of solid and fluid in the form of slabs. These slabs can be arranged in different configurations with respect to the direction of heat flux (Fig. 1).

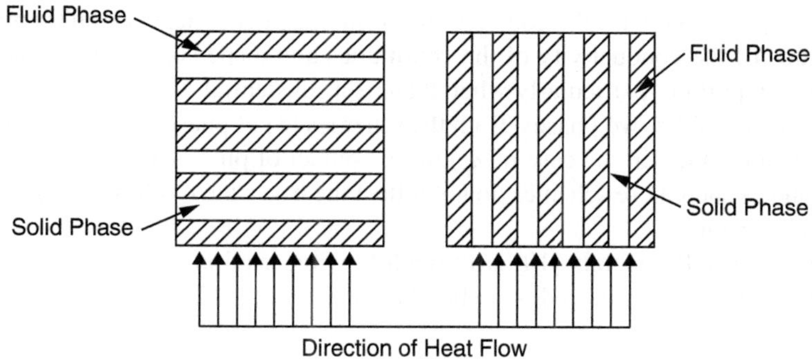

Fig. 1 Model of dispersed resistors giving extreme conductivity values (λ_{II} and $\lambda_{\perp}$)

Parallel configuration: In this configuration the plane of equivalent slabs is parallel to the direction of heat flow i.e. the two phases are thermally parallel to the heat flux. The effective thermal conductivity for parallel configuration $\lambda_{||}$ can be expressed as the weighted arithmetic mean of the conductivities of the solid and fluid phases and is written as

$$\lambda_{||} = [\phi\lambda_f + (1 - \phi)\lambda_s] \tag{6}$$

The parallel configuration offers minimum insulation resulting in a maximum value of the effective thermal conductivity.

Perpendicular configuration: When the direction of heat flow is perpendicular to the plane of slabs, then it offers maximum insulation and the value of the effective thermal conductivity is a minimum. In this configuration, the constituent phases are thermally in series with the direction of heat flow. Effective thermal conductivity in this case is represented as $\lambda_{\perp}$ and is given by the weighted harmonic mean of the conductivities of the constituent phases.

$$\lambda_{\perp} = \left[\frac{\phi}{\lambda_f} + \frac{(1 - \phi)}{\lambda_s}\right]^{-1} \tag{7}$$

The above equations for the effective thermal conductivity are the limiting formulae for all possible conductivities of phases for a given value of the porosity ϕ.

Later, Woodside and Messmer [9] analyzed these relations more critically and concluded that for both the distributions (parallel and perpendicular configurations), one should have

$$\left[\frac{d\lambda_e}{d\lambda_s}\right]_{\lambda_s = \lambda_f} = (1 - \phi) \tag{8}$$

Hence (8) should be satisfied by different relations for the effective thermal conductivity applicable to different kinds of distributions.

In a dispersed system, the conductivity of the aggregate depends on the relative magnitude of the conductivity of the continuous and dispersed phases. This information is important, particularly when there is a substantial difference between the conductivities of the two phases. Both the schemes [8] of phase distribution assume either ideal contact (in the case of $\lambda_{||}$) or no contact of phases at all (as in the case of $\lambda_{\perp}$) and as such these schemes do not reflect the true state of phase distribution in a natural system.

In this series, Bernshtein [10] gave the following relation assuming the material is in the form of plates placed in a chessboard-like order:

$$\lambda = \lambda_f \left[\frac{4(1-\phi)}{1+\frac{\lambda_f}{\lambda_s}} + 2(\phi - 1) \right] \quad \phi \geq 0.5$$

and

$$\lambda = \lambda_f \left[\frac{4(1-\phi)}{1+\frac{\lambda_f}{\lambda_s}} + \frac{\lambda_s}{\lambda_f}(1-2\phi) \right] \quad \phi \leq 0.5 \tag{9}$$

These schemes gave a considerable improvement over Wiener's model, yet are quite far from the realistic structure of granular materials.

Lichtenecker [11] has given an empirical relation to express the behavior of a two-phase system that has been named the "Logarithmic law of mixing". The expression for the effective thermal conductivity as per this law can be written as

$$\log(\lambda_e) = \phi_f \log(\lambda_f) + \phi_s \log(\lambda_s) \tag{10}$$

where the respective conductivities and corresponding volume fractions are represented by subscripts s and f. The effective thermal conductivity of a mixture should be found between upper and lower limiting values for distinctly dispersion type of systems. This equation can also be written in a different form as:

$$\lambda_e = (\lambda_f)^{\phi_f} . (\lambda_s)^{\phi_s} \tag{11}$$

For a mixture of n components its general form will be

$$\log \lambda_e = \sum_{i=1}^{n} \phi_i \log(\lambda_i) \tag{12}$$

Equation (11) is intended only for particles having two-directional randomization and oriented in the third direction. Bruggeman [6] has extended Lichtenecker's relation by randomizing the phases in three dimensions.

Russel [12] developed a model for predicting the effective thermal conductivity by assuming that the cubes of one phase are arranged in a cubic array into other

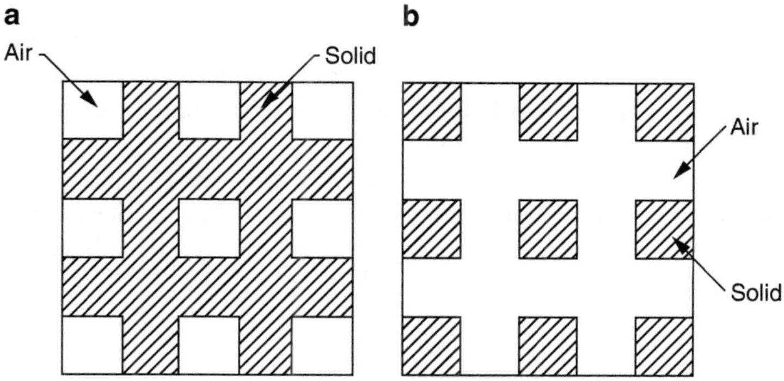

Fig. 2 Russel's model for the effective thermal conductivity of porous media: (**a**) Cubes of air with solid substance in between (**b**) Cubes of solid separated by air spaces

phase (Fig. 2). If the dispersed cubes are solid and the continuous phase is fluid, then

$$\lambda_e = \frac{\lambda_f\left[(\phi_s)^{\frac{2}{3}} + \left(\frac{\lambda_f}{\lambda_s}\right)\left(1 - \phi_s^{\frac{2}{3}}\right)\right]}{\left[(\phi_s)^{\frac{2}{3}} - \phi_s + \left(\frac{\lambda_f}{\lambda_s}\right)\left(1 + \phi_s - \phi_s^{\frac{2}{3}}\right)\right]} \tag{13}$$

In the opposite case, when the dispersed phase is fluid and the solid phase is continuous one, then (13) is modified to

$$\lambda_e = \frac{\lambda_s\left[\left(\phi_f\right)^{\frac{2}{3}} + \left(\frac{\lambda_s}{\lambda_f}\right)\left(1 - \phi_f^{\frac{2}{3}}\right)\right]}{\left[\phi_f^{\frac{2}{3}} - \phi_f + \left(\frac{\lambda_s}{\lambda_f}\right)\left(1 - \phi_f - \phi_f^{\frac{2}{3}}\right)\right]} \tag{14}$$

For porous materials, Ribaud [13] later proposed an equation by assuming that the pores are joined in a cubical manner resulting into an expression for λ_e as

$$\lambda_e = \lambda_s(\phi_s)^{\frac{2}{3}} + \lambda_f\left(\phi_f\right)^{\frac{1}{3}} \tag{15}$$

A natural system, however, is not as simple as predicted by these formulae, for there are always contacts among the particles. Ignoring this important fact restricts the applicability of these relations to natural systems.

Powers [14] has made a survey of methods for calculating the thermal conductivity of aggregates of almost any type. He has shown that when there is an increase in porosity of the dispersed phase to nearly 50%, the dispersed phase can no longer be regarded as discontinuous phase but it behaves as a continuous phase. In several cases having concentrations between 25 and 75%, both phases can be a dispersed phase. Such systems are termed as mixtures. Powers concluded (for mixtures) that the particles are no longer distributed systematically, but may be elongated in one or more directions and oriented randomly.

Bogomolov [15] developed a similar kind of expression by taking into account that solid spheres are packed into a tetrahedral packing. The resultant expression so obtained has been extensively used for determination of the effective thermal conductivity:

$$\lambda_e = 3\pi\lambda_f \ln\left[\frac{0.43 + 0.31\phi}{\phi - 0.26}\right] \tag{16}$$

Assad [16] gave an expression for effective thermal conductivity of sandstone rocks. His relation was an empirical one.

$$\lambda_e = \lambda_s\left[\frac{\lambda_f}{\lambda_s}\right]^{B\phi} \tag{17}$$

where B is a constant and is related to the characteristics of sandstone.

More developments in the expressions of heat conduction took place with the new models and formulations resulting in new kinds of application.

Kunni and Smith [17] took a practical approach and proposed a relation for effective thermal conductivity of loose granular materials as

$$\lambda_e = \lambda_f\left[\frac{\phi + \beta(1-\phi)}{e + (2\lambda_f/3\lambda_s)}\right] \tag{18}$$

where β is an adjustable parameter ranging between 0.9 and 1and

$$e = e_2 + \left[\frac{\phi - 0.259}{0.217}\right](e_1 - e_2) \tag{19}$$

Here e_1 and e_2 are dependent on values of ϕ for loose and compact packing. When $\phi < 0.259$ then $e = e_2$ and when $\phi > 0.476$ then $e = e_1$.

Woodside and Messmer [9] proposed three modes of heat conduction using the resistor approach. They assumed that there is solid to solid conduction, fluid to fluid conduction and solid to fluid conduction and vice-versa. Their expression for the effective thermal conductivity is

$$\lambda_e = \alpha\left[\frac{\lambda_s\lambda_f}{\lambda_s(1-\gamma) + \lambda_f\gamma}\right] + \beta\lambda_s + \delta\lambda_f \tag{20}$$

where $\alpha + \beta + \gamma = 1$ and $\alpha\gamma + \beta = (1 - \phi)$. α, β, γ are parameters for cube formation and δ is reciprocal of formation factor F which is equal to $\phi - 0.03$.

The estimation of the effective thermal conductivity using a curve-fitting technique has been presented by Sugawara et al. [18, 19]. The expression given by them for the measurement of effective thermal conductivity of soil, rocks and other granular materials is:

$$\lambda_e = \left[(1 - A)\lambda_s + A\lambda_f\right] \tag{21}$$

where

$$A = \frac{2^n}{(2^n - 1)}\left[1 - (1 + \phi)^{-n}\right]$$

and n represents an empirical number.

Chaudhary and Bhandari [20] extended the Lichtnecker model by considering the series and parallel resistors concept for a two-phase system. The random distribution of series and parallel resistors is represented by an empirical factor n, which denotes the probability of orientation of parallel resistors in the direction of heat flow. The resultant expression for the effective thermal conductivity is given by:

$$\lambda_e = \left(\lambda_{||}\right)^n \left(\lambda_{\perp}\right)^{1-n} \tag{22}$$

where

$$\lambda_{||} = \left[\phi\lambda_f + (1 - \phi)\lambda_s\right]$$

$$\lambda_{\perp} = \left[\frac{\phi}{\lambda_f} + \frac{(1 - \phi)}{\lambda_s}\right]^{-1}$$

and

$$n = \frac{k(1 - \log\phi)}{\log\left[\phi(1 - \phi)\left(\frac{\lambda_s}{\lambda_f}\right)\right]}$$

here k is an empirical constant.

In this vein Cheng and Vachon [21] proposed a model for randomly distributed particles in a continuous phase. Their model is represented by the equation for the effective thermal conductivity as:

$$\frac{1}{\lambda_e} = 2\int_0^{\frac{B}{2}} \left[\frac{dx}{\lambda_c + B(\lambda_d - \lambda_c) - Cx^2(\lambda_d - \lambda_c)}\right] + \frac{1 - B}{\lambda_c} \tag{23}$$

where $B = \left[\frac{3\phi}{2}\right]^{\frac{1}{2}}$, $C = 4\left[\frac{2\phi}{3}\right]^{\frac{1}{2}}$ and $\phi = \phi_d$

Here x is the dimension of dispersed phase along the x axis.

This expression was found to be suitable for values of $\phi \leq 0.667$.

Later on, a bound technique with lower and upper bound came due to the lack of a general expression for determination of the effective thermal conductivity. The technique estimates the closest optimum value of effective thermal conductivity. A general set of bounds were given by Hashin and Strikman [22] incorporating a variational principle for particles having cylindrical geometry as:

$$\lambda_f + \frac{\phi_s}{\left[\frac{1}{(\lambda_s - \lambda_f)} + \frac{\phi_f}{3\lambda_f}\right]} \langle \lambda_e \langle \lambda_s + \frac{\phi_f}{\left[\frac{1}{(\lambda_f - \lambda_s)} + \frac{\phi_s}{3\lambda_s}\right]} \tag{24}$$

Prager [23] presented a solution for the bounds for particles having cylindrical geometry using effective thermal conductivity values of some other materials whose conductivity ratios are the same but the constituent phases may be different. A set of bounds have been derived by Schulgasser [24] for fibrous reinforced materials applying symmetry considerations. His bounds are expressed as

$$\lambda_s\lambda_f \left[\lambda_{\mathrm{II}} - \frac{\phi\left\{\frac{(1-\phi)}{2}(\lambda_f - \lambda_s)^2\right\}}{\phi\frac{(1-\phi)}{2}\lambda_{\mathrm{II}} + (\lambda_s - \lambda_f)I_1}\right]^{-1} \langle \lambda_e \langle \lambda_s\lambda_f \left[\lambda_{\mathrm{II}} - \frac{\phi\left\{\frac{(1-\phi)}{2}(\lambda_s - \lambda_f)^2\right\}}{\phi\frac{(1-\phi)}{2}\lambda_{\mathrm{II}} + (\lambda_f - \lambda_s)I_2}\right] \tag{25}$$

where I_1 and I_2 are geometry dependent factors. These can be evaluated considering statistically three dimensional isotropic medium.

Using statistical properties of materials and generalized functions, Hori [25] developed a model for effective constants of a heterogeneous medium. Under specific conditions his theory yielded bounds obtained by Weiner [8] as well as by Hashin and Strikman [22].

Kumar and Chaudhary [26] developed a model by presenting an empirical bound for the effective thermal conductivity based on random distribution of series and parallel resistors.

Their expression for ETC with lower and upper bound is written as

$$\left[\lambda + \phi_f\left(1 - \phi_f\right)\left\{\lambda_f\lambda_s e^{\omega(2\phi_f - 1)}\right\}^{\frac{1}{2}}\right] < \lambda_e < \left[\lambda - \phi_f\left(1 - \phi_f\right)\left\{\lambda_f\lambda_s e^{\omega(2\phi_f - 1)}\right\}^{\frac{1}{2}}\right] \tag{26}$$

where

$$\lambda = \lambda_s\left(1 - \phi_f\right)e^{\omega\phi_f} + \lambda_f\phi_f e^{-\omega(1-\phi_f)}$$

and

$$\omega = \left[\frac{\lambda_f}{\lambda_s} - 1\right] \quad \text{for } \lambda_f < \lambda_s$$

$$\omega = \left[1 - \frac{\lambda_s}{\lambda_f}\right] \quad \text{for } \lambda_f > \lambda_s$$

This model was further modified by Pande et al. [27]. The proposed model of Pande is written mathematically in terms of bounds as

$$\frac{1}{(e^b - 1)}\left[(\lambda_f - \lambda_s)e^{b\phi} + \lambda_s e^b - \lambda_f\right] > \lambda_e > \frac{1}{(e^a - 1)}\left[(\lambda_f - \lambda_s)e^{a\phi} + \lambda_s e^a - \lambda_f\right] \quad (27)$$

where $a = \left[\frac{\lambda_f - \lambda_s}{(\lambda_s \lambda_f)^{1/2}}\right]$ and $b = \left[\frac{\lambda_f{}^2 - \lambda_s{}^2}{\lambda_f{}^2 + \lambda_s{}^2}\right]$

Zimmerman [28] modified the Fricke [3] relation for fluid saturated rocks having different types of pores. For very small porosity, Fricke [3] showed that the effective thermal conductivity is given as:

$$\frac{\lambda_e}{\lambda_s} = (1 - \beta\phi) \quad (28)$$

where

$$\beta = \frac{1 - r}{3}\left[\frac{4}{2 + (r - 1)M} + \frac{1}{1 + (r - 1)(1 - M)}\right]$$

Here $r = \lambda_f/\lambda_s$ and M is a factor that depends on the aspect ratio of the pore. Zimmerman [28] gave a more interesting and useful analytical expressions for β for three limiting cases as:

For thin cracks $\beta \rightarrow \frac{(1-r)(1+2r)}{3r}$

For spherical pores $\beta \rightarrow \frac{3(1-r)}{2+r}$

And for needle like pores $\beta \rightarrow \frac{(1-r)(5+r)}{3(1+r)}$

Torquato and Rintoul [29] developed rigorous bounds for the effective thermal conductivity of dispersions that are given in terms of the phase contrast between the inclusion and matrix, the interface strength, volume fraction, and higher order morphological information, including morphological information. Their bounds give very accurate predictions of effective thermal conductivity for dispersions of metallic particles in epoxy matrices. Their upper and lower bounds are:

$$\frac{\lambda_e}{\lambda_f} \leq D_U(C) = 1 + (\alpha + 3C - 1)\phi_s - \frac{E_U}{F_U} \quad (29)$$

where

$$E(U) = \phi_s\Big\{\phi_f\Big[C(5-5\alpha-6C)-(\alpha-1)^2\Big]-3C\Big\}^2$$

and

$$F(U) = 6C+(\alpha-1+2C)^2+\Big[3\phi_f+(\alpha-1)\Big(2\xi_s\phi_f+\phi_f^2\Big)\Big]$$
$$+C\phi_s(\alpha-1+2C)^2\times\left[\frac{16}{9}+3\phi_s(1+\phi_2)\right]+3C\Big[\phi_f(\alpha-1+2C)+1\Big]^2$$

$$\frac{\lambda_e}{\lambda_f} \geq D_L(C) = \left\{1+\left(\frac{1}{\alpha}-1\right)\phi_s-\frac{E_L}{F_L}\right\}^{-1} \quad (30)$$

where

$$E_L = \phi_s\Big[2\phi_f(\alpha-1+2C)(\alpha-1)+6C\Big]^2$$

and

$$F_L = 6\alpha^2\Big\{3C+6C^2+\phi_s\Big[(\alpha-1)^2-4C^2\Big]\Big\}+\left(\alpha-\alpha^2\right)$$
$$\times\left\{4\Big[\phi_f(\alpha-C-1)-3C\phi_s\Big]^2+2\xi_s\phi_f(\alpha-1+2C)^2\right\}$$

The parameters C, α and ξ are defined in Torquato and Rintoul [29].

Pande et al. [30] also gave an expression for the prediction of effective thermal conductivity of a granular system by considering regular geometry of dispersed phase.

$$\lambda_e = \lambda_f\left[1+3.7396\left(\frac{\lambda_s-\lambda_f}{\lambda_s+2\lambda_f}\right)\phi^{2/3}\right] \quad (31)$$

This represents the interaction between gas and solid particles up to sixth order for two-phase systems. The higher orders are negligible due to their very small contribution. According to the ratios of thermal conductivities of the constituents, the above relation is represented as:

$$\lambda_e = \lambda_f\left[1+3.844(\phi)^{\frac{2}{3}}\right] \quad \text{for } \lambda_s >> \lambda_f$$

$$\lambda_e = \lambda_f\left[1-1.154(\phi)^{\frac{2}{3}}\right] \quad \text{for } \lambda_s << \lambda_f$$

$$\lambda_e = \lambda_f\left[1+2.307\left(\frac{\lambda_s-\lambda_f}{\lambda_s+2\lambda_f}\right)(\phi)^{\frac{2}{3}}\right] \quad \text{for } \lambda_s = \lambda_f$$

Pande and Chaudhary [31] further modified these equations for effective continuous medium as:

$$\lambda_e = 0.6132(\lambda_f\lambda_s)^{\frac{1}{2}}\left\{1 - 1.154\left(\psi_f\right)^{\frac{2}{3}}\right\} \quad \text{for } \psi_f = |\phi_f - 0.5| \tag{32}$$

and

$$\lambda_e = 0.6132(\lambda_f\lambda_s)^{\frac{1}{2}}\left\{1 + 3.844(\psi_s)^{\frac{2}{3}}\right\} \quad \text{for } \psi_s = |\phi_s - 0.5|$$

Here ϕ_s and ϕ_f are the volume fractions of solid and fluid phases.

Hadley [32] in his contribution for the determination of effective thermal conductivity gave a model by taking into account the average of temperature field over different phases. The expression so obtained for effective thermal conductivity is written as

$$\lambda_e = \frac{\lambda_f\left[(1-\phi)F + \frac{\lambda_s}{\lambda_f}\{1-(1-\phi)F\}\right]}{\left[1 - \{(1-\phi)(1-F)\} + \frac{\lambda_s}{\lambda_f}(1-\phi)(1-F)\right]} \tag{33}$$

The value of F lies between 0 and 1.

The expression (33) for effective thermal conductivity is modified for packed metal powders as:

$$\frac{\lambda_e}{\lambda_f} = (1-\beta)\frac{\left[(1-\phi)F_0 + \frac{\lambda_s}{\lambda_f}\{1-(1-\phi)F_0\}\right]}{\left[1 - \{(1-\phi)(1-F_0)\} + \frac{\lambda_s}{\lambda_f}(1-\phi)(1-F_0)\right]} + \alpha\frac{\left[2\left(\frac{\lambda_s}{\lambda_f}\right)^2\phi + \left(\frac{\lambda_s}{\lambda_f}\right)(3-2\phi)\right]}{\left[(3-\phi)\left(\frac{\lambda_s}{\lambda_f}\right) + \phi\right]} \tag{34}$$

F_0 is a parameter like F above and β represents the degree of consolidation.

Verma et al. [33] concluded that the parameter F could be expressed as

$$F = \exp\left[-\psi\left(\frac{\lambda_f}{\lambda_s}\right)^{1/3}\right] \tag{35}$$

Here ψ is the sphericity of the particles. The value of F given by Verma et al. is 0.82 for granular systems and 0.75 for emulsion like systems. They have also applied a resistor model to obtain the expression for effective thermal conductivity of two-phase systems with spherical inclusions.

$$\lambda_e = \frac{\left[\lambda_f\left\{2.598\phi^{1/3}(\lambda_s - \lambda_f) + 3.224\phi^{-1/3}\lambda_f\right\}\right]}{\left[\left(1 - 1.2407\phi^{1/3}\right)\left\{2.5985\phi^{1/3}(\lambda_s - \lambda_f) + 3.224\phi^{-1/3}\lambda_f\right\} + \lambda_f\right]} \tag{36}$$

where ϕ is the volume fraction of the solid phase.

Later, Misra et al. [34] improved this relation by replacing ϕ by the porosity correction term F_p and provided a relation for λ_e as

$$\lambda_e = \frac{\left[\lambda_f\left\{2.598{F_p}^{1/3}(\lambda_s - \lambda_f) + 3.224{F_p}^{-1/3}\lambda_f\right\}\right]}{\left[\left(1 - 1.2407{F_p}^{1/3}\right)\left\{2.5985{F_p}^{1/3}(\lambda_s - \lambda_f) + 3.224{F_p}^{-1/3}\lambda_f\right\} + \lambda_f\right]} \tag{37}$$

where $F_p = \exp\left[-C_2(1-\phi)^{2/3}\right]$

The value of constant C_2 can be further expressed as

$$C_2 = 2.736e^{-0.004(\lambda_s/\lambda_f)}$$

Based on a lumped-parameter method, Hsu et al. [35] developed algebraic expressions for the stagnant thermal conductivity of two and three dimensional periodic media. Geometries considered were arrays of touching and non-touching in line square and circular cylinders. Their expression for stagnant thermal conductivity for two dimensional arrays of square cylinders having a square cross section $a \times a$ and a width of c as shown in Fig. 3a is given as:

$$\frac{\lambda_e}{\lambda_f} = \frac{\gamma_a\gamma_c}{\lambda} + \frac{\gamma_a(1-\gamma_c)}{1+(\lambda-1)\gamma_a} + \frac{(1-\gamma_a)}{1+(\lambda-1)\gamma_a\gamma_c} \tag{38a}$$

For in-line cubes the expression for thermal conductivity (Fig. 3b) is

$$\frac{\lambda_e}{\lambda_f} = \left(1 - \gamma_a^2 - 2\gamma_c\gamma_a + 2\gamma_c\gamma_a^2\right) + \frac{\gamma_a^2\gamma_c^2}{\lambda} + \frac{\left(\gamma_a^2 - \gamma_a^2\gamma_c^2\right)}{(1-\gamma_a+\gamma_a/\beta)} + \frac{2\left(\gamma_c\gamma_a - \gamma_c\gamma_a^2\right)}{(1-\gamma_a\gamma_c+\gamma_a\gamma_c/\beta)} \tag{38b}$$

where $\gamma_a = a/l$, $\gamma_c = c/a$ and β is the ratio of solid to fluid thermal conductivity respectively. A comparison of the results based on equation (38b) with existing experimental data shows that they are in excellent agreement with the experimental data if the contact resistance parameter $\gamma_c = 0.13$ is chosen at a porosity of 0.36. However, no comparison with experimental data at different porosities was given. In fact, the two parameters, γ_a and γ_c are related to the porosity through a nonlinear equation. It appears that the two parameters, γ_a and γ_c are a function of porosity, but no functional dependence was reported by them.

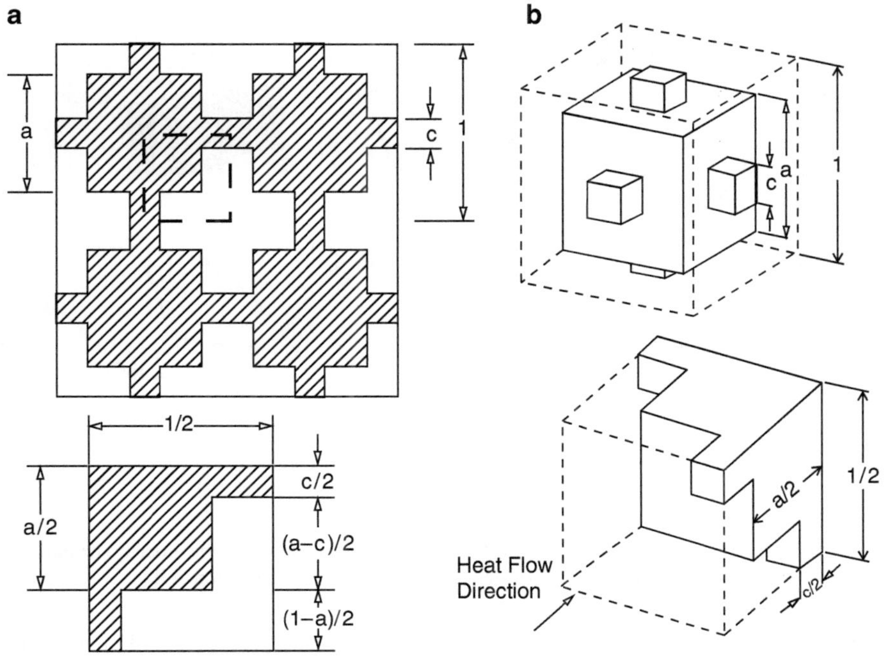

Fig. 3 (**a**) An array of touching square cylinders (*above*) and its unit cell (*below*). (**b**) In-line touching cubes (*above*) and their unit cell (*below*)

Singh et al. [36] presented a geometrical model for estimation of the effective thermal conductivity by using the resistors approach of two-phase systems with spherical inclusions. Their expressions for spherical and cubic particles are:

$$\lambda_e = \frac{\left[\lambda_f\left\{\lambda_f + 0.8060F^{2/3}\left(\lambda_s - \lambda_f\right)\right\}\right]}{\left[\lambda_f + F^{2/3}\left\{0.8060\left(\lambda_s - \lambda_f\right)\left(1 - 1.2407F^{1/3}\right)\right\}\right]} \tag{39}$$

$$\lambda_e = \frac{\left[\lambda_f\left\{\lambda_f + F^{2/3}\left(\lambda_s - \lambda_f\right)\right\}\right]}{\left[\lambda_f + F^{2/3}\left\{\left(\lambda_s - \lambda_f\right)\left(1 - F^{1/3}\right)\right\}\right]} \tag{40}$$

where F, the porosity correction, is written as

$$F = \left[1 - \exp\left\{-0.92\phi_s^{\ 2}\ln\left(\lambda_s/\lambda_f\right)\right\}\right]$$

Boomsma and Poulikakos [37] have developed a model for the effective thermal conductivity of saturated porous metal foams based on three-dimensional geometry for unit cell termed the tetrakaidecahedron (Fig. 4). The foam structure was

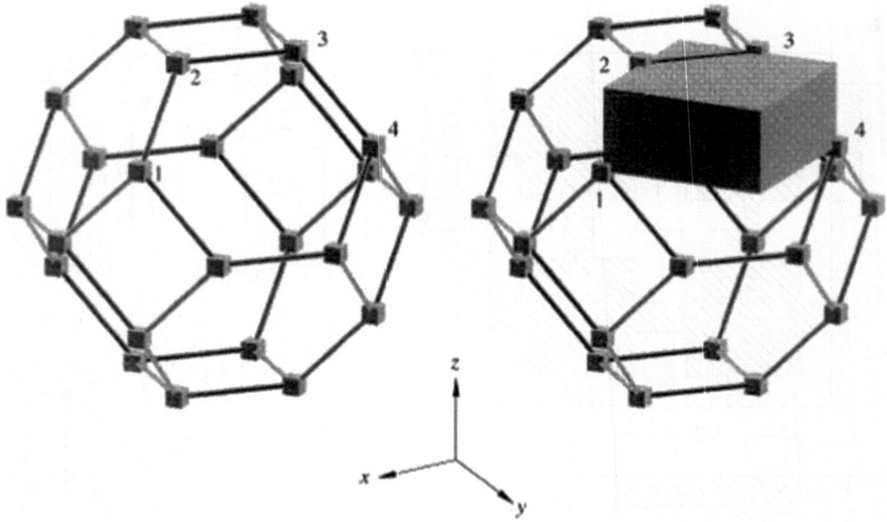

Fig. 4 The tetrakaidecahedron geometry with cylindrical ligaments and cubic nodes. The unit cell is shown on the right as a solid block located in a single tetrakaidecahedron cell

represented with cylindrical ligaments attached to cubical nodes at their centers. The resultant expression for effective thermal conductivity so obtained is

$$\lambda_e = \frac{\sqrt{2}}{2[R_A + R_B + R_C + R_D]} \tag{41}$$

where $R_A = \dfrac{4F}{[\{2e^2 + \pi F(1-e)\}\lambda_1 + \{4 - 2e^2 - \pi F(1-e)\}\lambda_2]}$

$$R_B = \frac{(e-2F)^2}{[(e-2F)e^2\lambda_1 + \{2e - 4F - (e-2F)e^2\}\lambda_2]}$$

$$R_C = \frac{\left(\sqrt{2} - 2e\right)^2}{\left[\left\{2\pi F^2\left(1 - 2\sqrt{2e}\right)\lambda_1\right\} + 2\left\{\sqrt{2} - 2e - \pi F^2\left(1 - 2\sqrt{2e}\right)\right\}\lambda_2\right]}$$

$$R_D = \frac{2e}{[e^2\lambda_1 + (4 - e^2)\lambda_2]}$$

$$F = \sqrt{\frac{\sqrt{2}\{2 - (5/8)e^3\sqrt{2} - 2(1-\phi)\}}{\left(3 - 4e\sqrt{2} - e\right)\pi}}$$

and $e = 0.339$ respectively.

Calmidi and Mahajan [38] developed a model for high porosity fibrous metal foams based on the structure of a metal foam matrix. They assumed that the

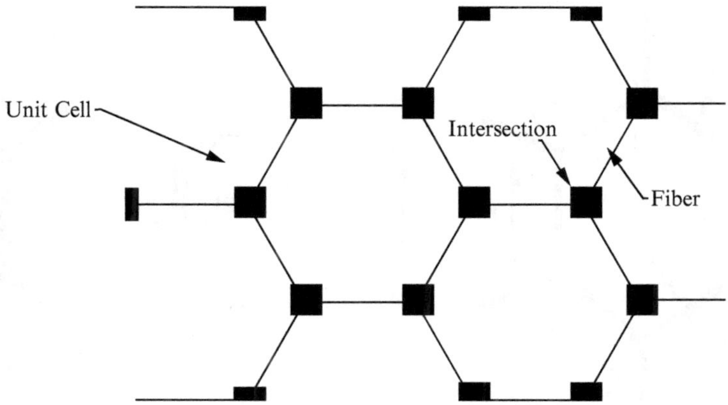

Fig. 5 Hexagonal structure of metal foam matrix

structure of a metal foam consisted of dodecahedron like cells with 12–14 hexagonal faces (Fig. 5). The lumping of material at the point of intersection of the fibers was taken into account as square. The expression for effective thermal conductivity is written as

$$\lambda_e = \left[\left(\frac{2}{\sqrt{3}}\right)\left\{\frac{r\left(\frac{b}{L}\right)}{\lambda_f + \left(1+\frac{b}{L}\right)\frac{(\lambda_s-\lambda_f)}{3}} + \frac{(1-r)\left(\frac{b}{L}\right)}{\lambda_f + \frac{2}{3}\left(\frac{b}{L}\right)(\lambda_s-\lambda_f)} + \frac{\frac{\sqrt{3}}{2}-\frac{b}{L}}{\lambda_f + \frac{4r}{3\sqrt{3}}\left(\frac{b}{L}\right)(\lambda_s-\lambda_f)}\right\}\right]^{-1} \quad (42)$$

where $$\frac{b}{L} = \frac{-r + \sqrt{r^2 + \frac{2}{\sqrt{3}}(1-\phi)\left\{2 - r\left(1+\frac{4}{\sqrt{3}}\right)\right\}}}{\frac{2}{3}\left\{2 - r\left(1+\frac{4}{\sqrt{3}}\right)\right\}}$$

Here r is defined as the area ratio.

An analytical model was given by Bhattacharya, Calmidi and Mahajan [39] for highly porous metal foams. With the hexagonal geometry and two-dimensional array of hexagonal cells, the expression for λ_e is

$$\lambda_e = \left[\left(\frac{2}{\sqrt{3}}\right)\left\{\frac{\left(\frac{t}{L}\right)}{\lambda_2 + \frac{(\lambda_1-\lambda_2)}{3}} + \frac{\frac{\sqrt{3}}{2}-\frac{t}{L}}{\lambda_2}\right\}\right]^{-1} \quad (43)$$

where $$\frac{t}{L} = \frac{-\sqrt{3} - \sqrt{3 + (1-\phi_1)(\sqrt{3}-5)}}{1 + \frac{1}{\sqrt{3}} - \frac{8}{3}}$$

and $$\lambda_e = F\{\phi_1\lambda_1 + (1-\phi_1)\lambda_2\} + \frac{(1-F)}{\left\{\frac{\phi_1}{\lambda_1} + \frac{(1-\phi_1)}{\lambda_2}\right\}}$$

where $F = 0.35$

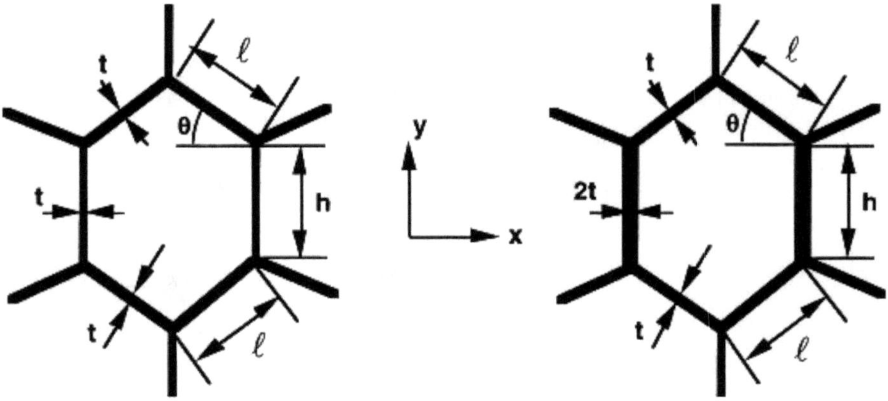

Fig. 6 Unit cell of a hexagonal honeycomb (**a**) with uniform thickness cell walls and (**b**) with non uniform thickness cell walls

Lu and Chen [40] presented a theoretical approach by considering a system made of hexagonal honeycombs (Fig. 6). The conductivities of a hexagonal honeycomb are in general anisotropic and can be represented by a second order tensor. The resulting conductivities expressed in the (x, y, z) directions are

$$\lambda_x = \rho\lambda_s \frac{Cos^2\theta}{(1 + h/l)} \tag{44}$$

$$\lambda_y = \rho\lambda_s \left(\frac{h/l + Sin\theta}{1 + h/l}\right)^2$$

$$\lambda_z = \rho\lambda_s$$

where $\rho = \dfrac{(1 + h/l)t/l}{Cos\theta(h/l + Sin\theta)}$

Recently, a generalized model was derived by Feng et al. [41] for the effective thermal conductivity of porous media based on the fact that statistical self-similarity exists in porous media. The proposed model assumes that porous media consist of two portions: randomly distributed non-touching particles and self-similarly distributed particles contacting each other with resistance. The latter are simulated by Sierpinski carpets with side length $L = 13$ and cutout size $C = 3, 5, 7$ and 9, respectively, depending upon the porosity concerned. Recursive formulae are presented and expressed as a function of porosity, ratio of areas, ratio of component thermal conductivities and contact resistance. There is no empirical constant and every parameter has a clear physical meaning. The model predictions are compared with the existing experimental data, and good agreement is found in a wide range of porosity of 0.14–0.80, and this verifies the validity of the proposed model. Their expression for the dimensionless effective thermal conductivity of porous media for randomly and self-similarly distributed non-touching particles is

$$\lambda_e^+ = \frac{\lambda_e}{\lambda_f} = \frac{A_{nt}}{A}\lambda_{e,nt}^+ + \left(1 - \frac{A_{nt}}{A}\right)\lambda_{e,sc}^+$$
$$= \frac{A_{nt}}{A}\left[\left(1 - \sqrt{1-\phi}\right) + \frac{\sqrt{1-\phi}}{1 + (1/\beta - 1)\sqrt{1-\phi}}\right] + \left(1 - \frac{A_{nt}}{A}\right)\lambda_{e,sc}^{+(n)} \quad (45)$$

Here A is the total area of a representative cross section and A_{nt} is an equivalent area of a cross section having the same porosity as the non-touching particles, with $0 \leq A_{nt}/A \leq 1$.

For non-touching particles, Hsu et al. [35] gave the following expression as:

$$\lambda_{e,nt}^+ \lambda_f = \lambda_{e,nt} = \lambda_f \left\{ \left(1 - \sqrt{1-\phi}\right) + \sqrt{1-\phi} \Big/ 1 + (1/\beta - 1)\sqrt{1-\phi} \right\}$$

For an n-stage carpet (Fig. 7), the dimensionless effective thermal conductivity $\lambda_{e,sc}^{+(n)}$ is given by $\lambda_{e,sc}^{+(n-1)}\left(\frac{1-C^n/L^n}{t^+(\beta^n-1)+1} + \frac{C^n}{C^n(\beta^n-1)+L^n}\right)^{-1}$, where $\beta^n = \beta^o/\lambda_{e,sc}^{+(n-1)}$ and the superscript $n = 1, 2, \ldots$. This equation represents a recursive algorithm for the thermal conductivity of self-similar porous media.

Singh et al. [42] developed an empirical relation or quick estimation of effective thermal conductivity of highly porous systems. The matrix is supposed to be made up of layers oriented parallel and perpendicular to the direction of heat flow.

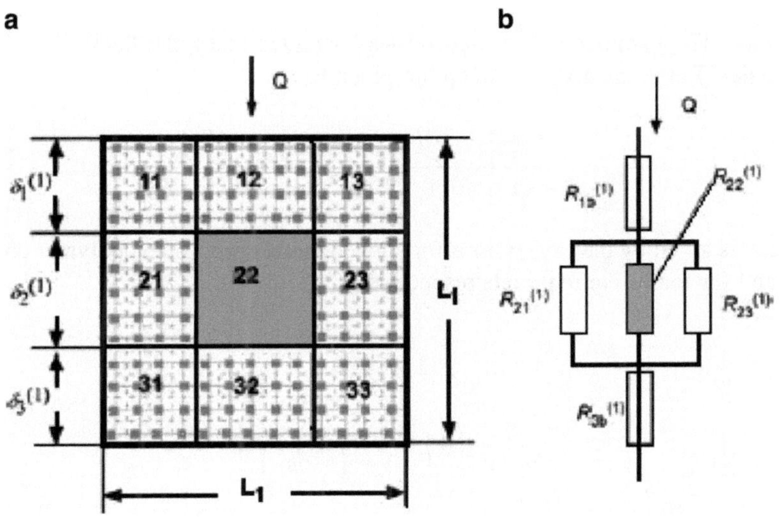

Fig. 7 The thermal conductivity model and the thermal–electrical analogy for a one-stage carpet with $L = 13$ and $C = 5$ (**a**) The thermal conductivity model for a one-stage carpet and (**b**) The network of the thermal–electrical analogy for the one-stage carpet

In order to incorporate varying individual geometries and non linear flow of heat flux lines generated by the difference in thermal conductivity of the constituent phases, a correlation term has been introduced. Their expression is:

$$\lambda_e = \lambda_{\text{II}}^F \cdot \lambda_{\perp}^{(1-F)} \quad F \geq 0, \quad 0 \leq F \leq 1 \tag{46}$$

where $\lambda_{\text{II}} = \phi\lambda_f + (1-\phi)\lambda_s$ and $\lambda_{\perp} = \frac{\lambda_s\lambda_f}{(1-\phi)\lambda_f+\phi\lambda_s}$ are upper and lower bounds on the effective thermal conductivity respectively and F is given by:

$$F = C\left\{0.3031 + 0.0623\ln\left(\phi\frac{\lambda_s}{\lambda_f}\right)\right\}$$

where C is a numerical constant and depends on the nature of the material.

Jagjiwanram et al. [43] generalized Singh and colleagues model assuming inclined slabs with the heat flux lines and derived a relation as:

$$\lambda_e = \left[\left(\phi\lambda_s + (1-\phi)\lambda_f\right)^2 Cos^2\theta + \frac{\lambda_s^2\lambda_f^2 Sin^2\theta}{\left(\phi\lambda_f + (1-\phi)\lambda_s\right)^2}\right]^{1/2} \tag{47}$$

where $Sin^2\theta = C_1\phi^{1/2}\ln\left(\lambda_s/\lambda_f\right) + C_2$ and constants C_1 and C_2 are different for each type of material.

Singh and Sharma [44] developed a model for predicting the effective thermal conductivity of particle filled polymer composites assuming spherical inclusions arranged in a three dimensional cylindrical array (Fig. 8). A Green's function technique has been applied to determine the value of effective thermal conductivity.

Recently, Singh et al. [45] extended Hadley's model to predict effective thermal conductivity of cellular and polymer composites considering random flow of heat flux lines. The parameter F introduced was obtained using the laws of statistical mechanics. Relations given by them are given below:

$$\lambda_e = \frac{\lambda_c\left((1-\phi)F + \frac{\lambda_d}{\lambda_c}[1-(1-\phi)F]\right)}{1-(1-\phi)(1-F) + \frac{\lambda_d}{\lambda_c}(1-\phi)(1-F)} \tag{48}$$

where F is given by the expressions for packed metal composites, polymer composites and for foam-like materials respectively:

$$F = e^{-\left(\frac{\lambda_c}{\lambda_d}\right)^{\frac{1}{3}}}$$

$$F = e^{-\left(\frac{\lambda_c}{\lambda_d}\right)^{\frac{1}{3}}}\left[1 + \left(\frac{\lambda_c}{\lambda_d}\phi_c\right)^{\frac{1}{3}}\right]$$

$$F = e^{-\left(\frac{\lambda_c}{\lambda_d}\right)^{\frac{2}{3}}}\left[1 + \left(\frac{\lambda_c}{\lambda_d}\right).^{\frac{2}{3}}\phi_c\right]$$

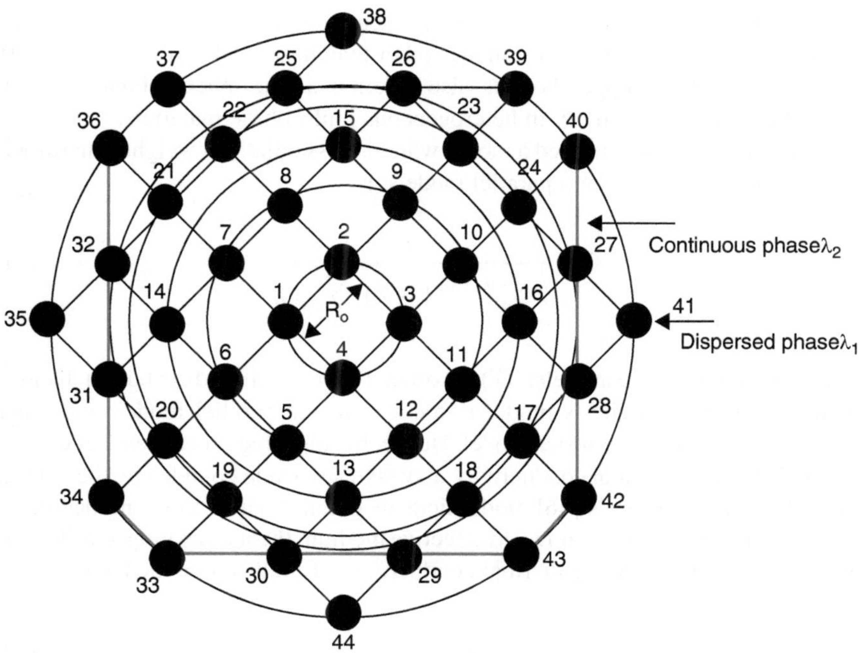

Fig. 8 Dispersion around source and sensor in an X–Y plane

3.2 *Models for Multi-Phase Materials*

A large number of theoretical models were developed for prediction of effective properties of multi-phase materials. Roughly, these models can be classified into three major categories:

- Models based on empirical equations
- Models based on mixing of phases
- Analytical solutions from physical laws

The most straightforward theoretical models for the calculation of effective thermal conductivity in heterogeneous multi-phase materials are the parallel and series models, which are formulated as

$$\lambda_e = \sum_{i=1}^{n} \phi_i \lambda_i \tag{49}$$

and

$$\lambda_e = \frac{1}{\sum\limits_{i=1}^{n} \frac{\phi_i}{\lambda_i}} \tag{50}$$

respectively. Note that $\sum \phi_i = 1$. Since the calculation in the present work does not take into account the heat transfer modes other than heat conduction, (49) and (50) define the lower and upper bounds also referred as the Wiener bounds of the effective thermal conductivity in heterogeneous multi-phase materials.

The Krischer [46] formulated a model which is, essentially, a weighted harmonic combination of the series and parallel models, as

$$\lambda_e = \frac{1}{Z\sum_{i=1}^{n} \frac{\phi_i}{\lambda_i} + (1-Z)/\sum_{i=1}^{n} \phi_i \lambda_i} \tag{51}$$

where, the weighting parameter (Z) is often called as the "distribution factor". When $Z = 0$, the Krischer's model is reduced to the parallel model, and when $Z = 1$, it is reduced to the series model. Hence, by adjusting Z between 0 and 1, the predicted effective thermal conductivity may take any value bounded by the Wiener bounds [8]. The Krischer's [46] model finds its greatest use in food engineering.

Carson et al. [47] using renowned effective medium theory, developed following formulation for the effective thermal conductivity of multi-phase media as

$$\sum_{i=1}^{n} \phi_i \frac{\lambda_i - \lambda_e}{\lambda_i + 2\lambda_e} = 0 \tag{52}$$

The derivation of (52) is based on the exact solution of the Laplace's heat transfer equation applied to a single sphere of thermal conductivity λ_1 being embedded in a larger continuous medium of conductivity λ_2, which is subjected to a steady temperature gradient in one Cartesian-coordinate direction. The only assumption made in the derivation procedure is that the distribution of the embedded components is completely random such that the resultant temperature gradient within the material is uniform, which is rather idealized.

Various empirical equations have been proposed to connect the effective property to the volume fractions of the multiple phase components, by means of fitting the experimental data [48–59]. One such empirical equation developed by Cosenza et al. [51] valid over the solids thermal conductivity (λ_s) ranging from 2 to 5 W m^{-1} K^{-1}, porosity (ϕ) from 0.4 to 0.6, and the volumetric water content (θ) from 0.1 to 0.4 is given as

$$\lambda_e = (0.8908 - 1.0959\phi)\lambda_s + (1.2236 - 0.3485\phi)\theta \tag{53}$$

Such equations have proved to be useful in targeted cases only.

Based on the two-phase fundamental models, two schemes have been implemented for the multiphase case. The first is to extend the two-phase basic models directly to multiphase cases. For example, the Parallel and Series equations were adopted to develop a simple linear relationship between the thermal conductivity and the degree of saturation, which provided a satisfactory correlation for the data measured as well as those reported recently for sandy and clay loams [60]. For three-phase media, Woodside and Messmer [9] proposed the 'quadratic parallel' model for the effective conductivity. Their expression is given as

$$\lambda_e = \left(\sum \lambda_i^{1/2}\phi_i\right)^2 \tag{54}$$

where λ_i with $i = \{1, 2, 3\}$ is the conductivity of each phase and ϕ_i the corresponding volume fraction. This model appears to be applicable when $i > 3$. Various weighted average models have also been proposed for such multiphase mixtures [61].

Maxwell's model has been extended by Brailsford and Major [62] for a wide range of dispersions. In this model, the constituent phases are mixed in a definite proportion for a two-phase system. This mixture is then embedded in a random mixture of the same two phases having conductivity equal to the average value of the conductivity of two phase system. Thus the effective thermal conductivity of such a three-phase system can be determined as

$$\lambda_e = \frac{\left[\lambda_c\phi_c + \lambda_{d_1}\phi_{d_1}\left(\frac{3\lambda_c}{2\lambda_c+\lambda_{d_1}}\right) + \lambda_{d_2}\phi_{d_2}\left(\frac{3\lambda_c}{2\lambda_c+\lambda_{d_1}}\right)\right]}{\left[\phi_c + \phi_{d_1}\left(\frac{3\lambda_c}{2\lambda_c+\lambda_{d_1}}\right) + \phi_{d_2}\left(\frac{3\lambda_c}{2\lambda_c+\lambda_{d_1}}\right)\right]} \tag{55}$$

where subscript c and d represent continuous and dispersed phases, respectively.

Chaudhary and Bhandari [63] extended Woodside and Messmer's model [9] for three-phase systems using the resistor approach. Their expression was given as:

$$\lambda_e = {\lambda_{\mathrm{II}}}^n {\lambda_\perp}^{(1-n)} \tag{56}$$

where n was obtained by the best fit technique using experimental values of calcareous sand stone [18]. When the three-phases are thermally in parallel or in series with the direction of heat flow the parallel and series effective thermal conductivity of the material are given as:

$$\lambda_{\mathrm{II}} = \psi\lambda_w + (\phi - \psi)\lambda_a + (1 - \phi)\lambda_s$$

$$\lambda_\perp = \left(\frac{\psi}{\lambda_w} + \frac{\phi - \psi}{\lambda_a} + \frac{1 - \phi}{\lambda_s}\right)^{-1}$$

This model was further extended by Singh et al. [64] by calculating n empirically.

The second route is to treat two of the multiple phases as one single phase with their own effective properties and then to mix this effective phase with another new phase, i.e. dividing a multiphase material into combination of several two-phase systems. The Maxwell models are the most suitable for developing such mixture models in general, and the Maxwell–De Loor model in particular is widely used, for it requires no geometrical parameters [65, 66]. Dobson et al. [67] rewrote this model for a four-phase system in to:

$$\lambda_e = \frac{[3\lambda_1 + 2(\phi_2 - \phi_3)(\lambda_2 - \lambda_1) + 2\phi_3(\lambda_3 - \lambda_1) + 2(\phi_4 - \phi_2)(\lambda_4 - \lambda_1)]}{[3 + 2(\phi_2 - \phi_3)(\lambda_1/\lambda_2 - 1) + \phi_3(\lambda_1/\lambda_3 - 1) + (\phi_4 - \phi_2)(\lambda_1/\lambda_4 - 1)]} \tag{57}$$

As the Maxwell models are based on the assumption that the dispersed phases are independently distributed in the continuous phase with negligible interactions with each other, these types of models fail to work properly if there exists any strong phase interactions inside the materials, unless some empirical parameters are introduced to account for those influences [67].

Verma et al. [68] has extended Hadley's model for three-phase materials by putting

$$\phi_1\langle\nabla T_1\rangle^1 + \phi_2\langle\nabla T_2\rangle^2 = (\phi_1 + \phi_2)\langle\nabla T_i\rangle^i$$

and

$$\lambda_1\phi_1\langle\nabla T_1\rangle^1 + \lambda_2\phi_2\langle\nabla T_2\rangle^2 = \lambda_i(\phi_1 + \phi_2)\langle\nabla T_i\rangle^i$$

in basic equations of two-phase materials, we obtain

$$\nabla\langle T\rangle = (\phi_1 + \phi_2)\langle\nabla T_i\rangle^i + \phi_3\langle\nabla T_3\rangle^3 \tag{58}$$

$$\lambda_e\nabla\langle T\rangle = \lambda_i(\phi_1 + \phi_2)\langle\nabla T_i\rangle^i + \lambda_3\phi_3\langle\nabla T_3\rangle^3 \tag{59}$$

Equations (58) and (59) show that a three-phase problem has been reduced to a two-phase problem, one phase being the intermediate phase having an average temperature gradient $\langle\nabla T_i\rangle^i$ and the second phase being the remainder with an average temperature gradient $\langle\nabla T_3\rangle^3$. If we compare (58) and (59) with the equations for two-phase materials, we see that the volume fraction and thermal conductivity of the intermediate phase should be given by

$$\phi_i = \phi_1 + \phi_2$$

and

$$\lambda_i = \frac{\lambda_1\phi_1}{\phi_1 + \phi_2}$$

The usefulness of this method depends upon the proper pairing of different phases. No selection rules for the pairing of different phases have been given by Hadley, but they suggest the following guidelines for composing the phases.

- When the amount of dispersed phase is small, the continuous phase should be treated as independent and the rest should be combined to form an intermediate phase.
- If the fractional volumes of phases are comparable, the intermediate phase should be formed in such a manner as to reduce the ratio of thermal conductivities of intermediate and independent phases.
- To define the intermediate thermal conductivity, the constituent of the pair playing a greater role in conduction should be considered. This has to be decided judiciously by considering the concentration and thermal conductivity of the constituents within the pair itself. It can be judged by comparing the products of their concentration and respective thermal conductivity. Thus the intermediate thermal conductivity can also be defined by

$$\lambda_i = \frac{\lambda_2 \phi_2}{\phi_1 + \phi_2}$$

Beniwal et al. [69] extended the work of Pande et al. for statistically homogeneous and regular multi-phase systems. The solution of Poisson's equation was used for effective neighboring interactions and modified field which thereafter yields the effective thermal conductivity of multi-phase systems as

$$\lambda_e = \lambda_c \left[1 + 3.844 \phi_c^{-\frac{1}{3}} \left\{ \left(\frac{\lambda_{d_1} - \lambda_c}{\lambda_{d_1} + 2\lambda_c} \right) \phi_{d_1} + \left(\frac{\lambda_{d_2} - \lambda_c}{\lambda_{d_2} + 2\lambda_c} \right) \phi_{d_2} \right\} \right] \tag{60}$$

when there is a very dilute dispersion of both the phases, the interactions may not spread over a large distance. Therefore, expression (59) has the form

$$\lambda_e = \lambda_c \left[1 + 3.489 \phi_c^{-\frac{1}{3}} \left\{ \left(\frac{\lambda_{d_1} - \lambda_c}{\lambda_{d_1} + 2\lambda_c} \right) \phi_{d_1} + \left(\frac{\lambda_{d_2} - \lambda_c}{\lambda_{d_2} + 2\lambda_c} \right) \phi_{d_2} \right\} \right] \tag{61}$$

Singh et al. [70] have developed a model for moist soil like materials. In the case of soil there are two possibilities. (1) The soil is dry or fully saturated, i.e. the system is a two-phase system or (2) the system is moist, i.e. the system is a three-phase system made of solid, liquid and gas phases. Let us express ψ_{ma} as the volume fraction of moisture in the pore space. In the case when the dispersion of the water in the air is small, i.e. $0 < \psi_{ma}{}^{2/3} < 0.4$, the thermal conductivity of moist air (ψ_{ma}) within the pore space is expressed by

$$\lambda_{ma} = \lambda_a \left(1 + 1.3884 \frac{\lambda_w - \lambda_a}{\lambda_w + 2\lambda_a} \psi_{ma}{}^{2/3} \right) \tag{62}$$

When $\psi_{ma}{}^{2/3}$ lies between 0.4 and 1.0, the λ_{ma} of moist soil air becomes

$$\lambda_{ma} = \lambda_w \left(1 + 1.3884 \frac{\lambda_w - \lambda_a}{\lambda_w + 2\lambda_a} \left(1 - \psi_{ma}{}^{2/3} \right) \right) \tag{63}$$

where λ_a and λ_w are the thermal conductivity of air and water respectively

If ψ_m be the volume fraction of moisture and ψ_a the volume fraction of air in the sample, then ψ_{ma}, the volume fraction of moisture in the pore space will be

$$\psi_{ma} = \frac{\psi_m}{\psi_a}$$

where $\psi_m = (m/M)\psi_a$. Here m and M represent the varying moisture content and moisture content at saturation by weight per cent respectively and ψ_a is the volume fraction of air. Then the effective thermal conductivity of moist soil is expressed, using (63), and putting $\xi_s = \psi_s - 0.5$, as

$$\lambda_e = \lambda_{ECM}\left(1 + 1.3884\frac{\lambda_w - \lambda_{ECM}}{\lambda_w + 2\lambda_{ECM}}\xi_s{}^{2/3}\right) \quad (64)$$

$$\lambda_e = \lambda_{ECM}\left(1 + 1.3884\frac{\lambda_w - \lambda_{ECM}}{\lambda_w + 2\lambda_{ECM}}\xi_{sa}{}^{2/3}\right) \quad (65)$$

where $\xi_{sa} = 0.5 - \psi_s$. Here ψ_s, ξ_s and ξ_{sa} denote the volume fraction of the solid phase, solid phase dispersion and air phase dispersion in the soil respectively. The effective thermal conductivity of the effective continuous medium is defined by Singh et al. [71] as

$$\lambda_{ECM} = P(\lambda_s\lambda_{ma})^{1/2}$$

where P is the coefficient which depends on the order of phase interaction.

Moosavi and Sarkomaa [72] presented a theoretical expression for estimating effective thermal conductivity of three-phase composite materials by incorporating circular cylindrical geometry. Their expression is written as

$$\lambda_e = 1 - \frac{2f_1}{(\lambda_1\lambda_2 - \xi_1\xi_2)/(\lambda_2 - \xi_2)} - \frac{2f_2}{(\lambda_1\lambda_2 - \xi_1\xi_2)/(\lambda_1 - \xi_1)} \quad (66)$$

where $$\lambda_i = \frac{1}{\gamma_i} + c_1 f_i - c_2\gamma_i f_i^4 - c_3\gamma_{2-\delta_{i2}} f_i f_{2-\delta_{i2}}^3$$

$$\xi_i = c_4 f_i - c_5\left(\gamma_i f_i^4 + \gamma_{2-\delta_{i2}} f_i f_{2-\delta_{i2}}^3\right)$$

$$\gamma_i = \frac{1 - k_i}{1 + k_i}$$

where λ_i is the thermal conductivity of the ith phase, f_i is the volume fraction of the ith phase and c_i (i = 1–5) are constants.

Recently, Gori and Corasaniti [73] built up a cubic cell model for the thermal conductivity of three-phase porous media. Water absorbed was considered to either cover the solid particles or form liquid-bridges between different particles. Analytical models were thus derived. Unfortunately, the critical water content for liquid-bridge formation was unknown and had to be determined empirically, hence causing gross errors in predictions. In a three-phase olivine system, the water distribution is different according to the amount of water content. If the water content W is lower than $W_a = 0.083$, which is an empirical value proposed in [74] the water is adsorbed around the solid particle and no water bridges are established among the adjacent particles (Fig. 9b). In this case the expression for the effective thermal conductivity given by them is

$$\frac{1}{\lambda_T} = \frac{\beta - 1 - \delta/3}{\beta\lambda_c} + \frac{\beta\delta}{3\left[\lambda_c\left(\beta^2 - 1\right) + \lambda_w\right]} + \frac{\beta}{\lambda_s + 2/3\delta\lambda_w + \lambda_c\left(\beta^2 - 1 - 2/3\delta\right)} \tag{67}$$

where

$$\delta = \frac{W}{1-\phi} = 6\frac{l_{wa}}{l_s}$$

If $W > W_a$, Fig. 9c, d, the amount of water accumulated among the solid particles is the funicular one, V_{wf}/V_s. In order to simplify the model, V_{wf}/V_s is assumed linearly proportional to the real porosity of the porous medium between 0.183, for $\phi = 0.4764$, and 0.226, for $\phi = 0.2595$. The resulting expression is

$$\frac{V_{wf}}{V_s} = \frac{V_{wf}}{V_v}\left(\beta^3 - 1\right) = \left[0.183 + \frac{0.226 - 0.183}{0.4764 - 0.2595}(0.4764 - \phi)\right]\left(\beta^3 - 1\right)$$

where the variables are given as

$$\gamma = \frac{l_w}{l_s} = \sqrt[3]{\frac{V_w}{V_s} - \frac{V_{wf}}{V_s} + 1}$$

$$\gamma_f = \frac{l_{wf}}{l_s} = \sqrt{\frac{V_{wf}/V_s}{3(\beta - \gamma)}}$$

In the configuration of Fig. 9c, where $\gamma_f < 1, \lambda_T$ is given by

$$\frac{1}{\lambda_T} = \frac{\beta^2 - \beta\gamma}{\lambda_c\left(\beta^2 - \gamma_f^2\right) + \lambda_w\gamma_f^2} + \frac{\beta\gamma - \beta}{\lambda_c\left(\beta^2 - \gamma^2\right) + \lambda_w\gamma^2} + \frac{\beta - \beta\gamma_f}{\lambda_c\left(\beta^2 - \gamma^2\right) + \lambda_w\left(\gamma^2 - 1\right) + \lambda_s} + \frac{\beta\gamma_f}{\lambda_s - \lambda_w\left(\gamma^2 - 1 + 2\beta\gamma_f - 2\gamma\gamma_f\right) + A} \tag{68}$$

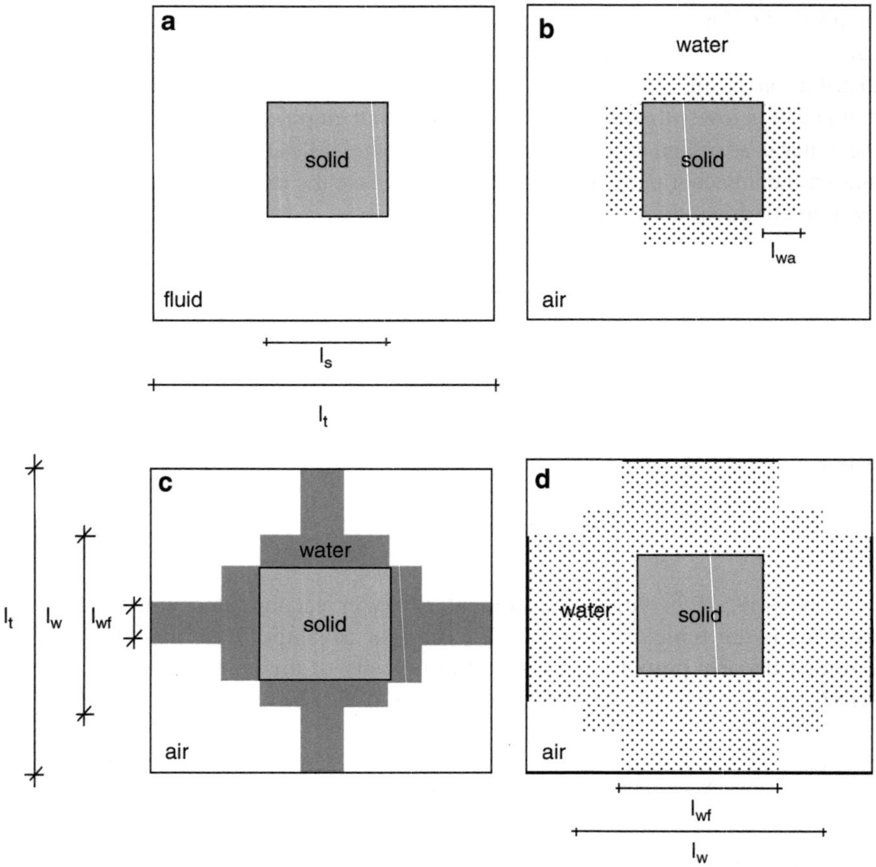

Fig. 9 (**a**) Unit cell (**b**) Water adsorbed around the solid particle (**c**) Cubic cell model with water around the solid particle, according to absorption and capillarity among the adjacent particles (**d**) Water adsorbed around the solid particle and disposed among adjacent particles

where $A = \lambda_c(\beta^2 - \gamma^2 + 2\beta\gamma_f + 2\gamma\gamma_f)$

For $\gamma > 1$, Fig. 9d, λ_T has the following expression

$$\begin{aligned}\frac{1}{\lambda_T} = {} & \frac{\beta^2 - \beta\gamma}{\lambda_c\left(\beta^2 - \gamma_f^2\right) + \lambda_w\gamma_f^2} + \frac{\beta\gamma - \beta\gamma_f}{\lambda_c(\beta^2 - \gamma^2) + \lambda_w\gamma^2} \\ & + \frac{\beta\gamma_f - \beta}{\lambda_c(\beta^2 - \gamma^2 - 2\beta\gamma_f + 2\gamma\gamma_f) + \lambda_w(\gamma^2 + 2\beta\gamma_f - 2\gamma\gamma_f)} \\ & + \frac{\beta}{\lambda_s + \lambda_w(\gamma^2 - 1 + 2\beta\gamma_f - 2\gamma\gamma_f) + \lambda_c(\beta^2 - \gamma^2 - 2\beta\gamma_f + 2\gamma\gamma_f)}\end{aligned} \tag{69}$$

Ma et al. [75, 76] have developed approximate fractal geometry models, for the effective thermal conductivity of saturated/unsaturated porous media based on the thermal–electrical analogy technique (Fig. 10) and on statistical self-similarity of porous media. The proposed thermal conductivity models were expressed as a function of porosity, ratio of areas, ratio of component thermal conductivities,

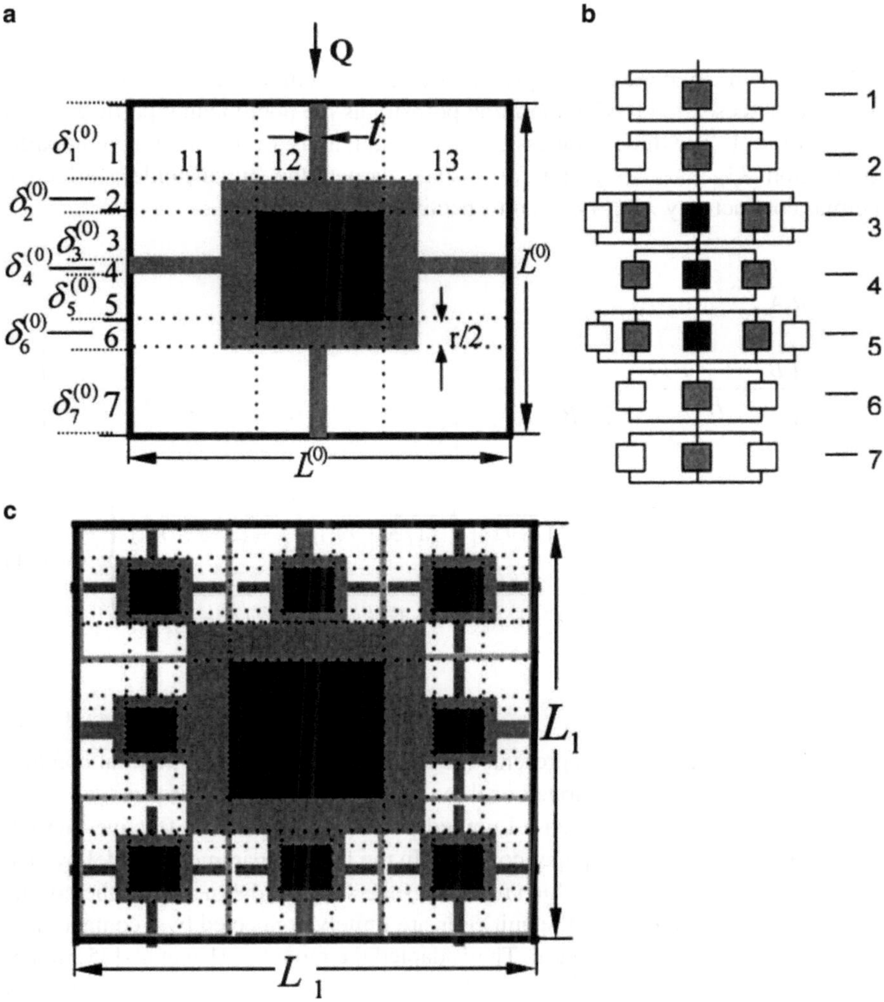

Fig. 10 The thermal conductivity model and the thermal–electrical analogy for a 0-stage and one-stage carpet in three phases (**a**) the thermal conductivity model for 0-stage carpet (**b**) the network of the thermal–electrical analogy for 0-stage carpet, and (**c**) the thermal conductivity model for one-stage carpet

and saturation. A quite simple recursive algorithm for the effective thermal conductivity for three-phase porous medium is thus presented as:

$$\lambda_e^+ = \frac{\lambda_e}{\lambda_g} = \frac{A_{nt}}{A}\lambda_{e,nt}^+ + \left(1 - \frac{A_{nt}}{A}\right)\lambda_{e,sc}^+$$

$$= \frac{\lambda_e}{\lambda_g} = \frac{A_{nt}}{A}\lambda_{e,nt}^+ + \left(1 - \frac{A_{nt}}{A}\right)\lambda_{e,sc}^{+(n)}, n = 1, 2, 3, \ldots \quad (70)$$

Here A is the total area of a representative cross section and A_{nt} is an equivalent area of a cross section having the same porosity as the nontouching particles, with $0 < A_{nt}/A < 1$. The dimensionless thermal conductivity for the nontouching particle portion $\lambda_{e,nt}^+ = \lambda_{e,nt}/\lambda_g$ and for n-stage carpet, the dimensionless effective thermal conductivity $\lambda_{e,sc}^{+(n)}$ is given respectively as:

$$\lambda_{e,nt}^+ = \frac{l_t}{R_t A_t \lambda_g}$$

$$= \left\{\frac{B - 1 - F/2}{B} + \frac{F}{2\left[(B-1) + \beta_{wg}\right]} + \frac{1}{\beta_{sg} + F\beta_{wg}/2 + (B - 1 - F/2)}\right\}^{-1}$$

$$\lambda_{e,sc}^{+(n)} = \lambda_{e,sc}^{+(n-1)} \left\{ \begin{array}{l} \dfrac{2/3 - r^+}{[t^+\beta_{wg}^n + (1 - t^+)]} + \dfrac{r^+}{(1/3 + r^+)\beta_{wg}^n + (2/3 - r^+)} \\ + \dfrac{1/2 - t^+}{2/3 - r^+ + \beta_{wg}^{(n)} r^+ + \beta_{sg}^{(n)}/3} + \dfrac{t^+}{2\beta_{wg}^{(n)}/3 + \beta_{sg}^{(n)}/3} \end{array} \right\}^{-1} \quad (71)$$

where
$A_t = l_t \times 1, \beta_{wg} = \lambda_w/\lambda_g, \beta_{sg} = \lambda_s/\lambda_g, \beta_{wg}^n = \beta_{wg}^0/\lambda_{e,sc}^{+(n-1)}$, and $\beta_{sg}^n = \beta_{sg}^0/\lambda_{e,sc}^{+(n-1)}$ respectively. Equation (71) represents the recursive algorithm for thermal conductivity of self-similarly unsaturated or three-phase porous media [76].

Gerstner et al. [77] proposed a Lumped parameter model for effective thermal conductivity of three-phase paper coatings. In the lumped parameter model by Hsu et al. [35], a unit cell is used which represents the simplified structure of the porous medium. The solid phases of the unit cells are finitely connected by a contact plate and surrounded by the fluid phase. They adapted the model of Hsu et al. [35] model (Fig. 3) using a square cylindrical solid phase (λ_{s1}), representing the pigment, and introduce a second solid phase (λ_{s2}) for the connecting plates, representing the binder, the fluid phase being air when representing a dry coating or a chosen liquid/ mix of liquids when considering either a wet coating or one that has absorbed ink vehicle, for example. The width of the binder bridge is c and can be related to the pigment size a as the contact parameter $\gamma = ca{-}1$. The unit cell can be split into

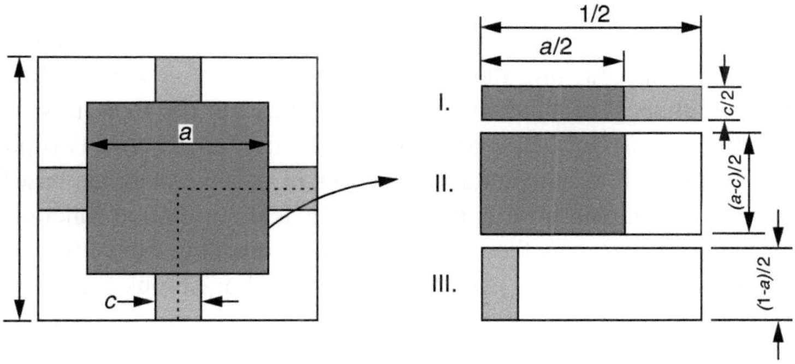

Fig. 11 Modified three-phase unit cell based on the lumped parameter model [35] and its decomposition into serial and parallel layers (I–III) of thermal resistors

three parallel layers (Fig. 11 I–III), consisting of the relative combinations of solid phases and each solid phase with the fluid phase, respectively. Their expression for the overall effective thermal conductivity of the unit cell is:

$$\frac{\lambda_e}{\lambda_f} = \frac{\gamma a}{(\lambda - \mu)a + \mu} + \frac{1-\gamma}{(\lambda - 1)a + 1} + \frac{1-a}{(\mu - 1)\gamma a + 1} \tag{72}$$

where λ_e is the effective thermal conductivity of the unit cell, λ_f the thermal conductivity of the fluid phase, $\lambda = \lambda_f/\lambda_{s1}$ the conductivity ratio of the fluid phase to the pigment phase, and $\mu = \lambda_f/\lambda_{s2}$ the conductivity ratio of the fluid phase to the binder phase.

3.3 Numerical Methods

Rapid development of computational techniques in the past decades and the limited power of the existing theoretical models in dealing with the complex materials have tremendously enhanced numerical capabilities in modeling the thermo-physical behaviors of multi-phase materials.

A complete numerical determination of the effective properties of a multi-phase material has to include two major steps:

- To reproduce properly the multi-scale microstructures using computer algorithms, and then
- To solve the relevant partial differential equations for effective thermal conductivity of a multi-phase material with acceptable efficiency and accuracy

As the rapid development of the microscopy techniques, several methods have been proposed to reconstruct the microstructures of multi-phase materials using computers [78–83]. However, none of them can deal well with the interactions between the grains, thus unsuitable for energy transfer problems that are extremely sensitive to contacts. As the microscopy instruments and the observation technologies are also well developed now, the image mapping has become a highly powerful tool for approaching the real structures in more geometric details such as the element shapes, orientations and connections, on the materials properties [84, 85]. Better reconstruction processes have been used to generate two-phase [86, 87] and multi-phase [88, 89] random structures of porous materials based on the digital micro-tomographic information and statistical correlation functions. More about these reconstruction methods based on image mapping can be found in some excellent review papers [90, 91] and books [92, 93]. For fibrous porous materials, Pan et al. [94, 95] have done a series of analyses to theoretically characterize the microstructure using statistical density distribution function approaches.

Moreover, for multi-phase materials, their microstructures can be influenced by the phase interactions. When such interactions are non-negligible, approaches have to be developed to reflect the influences during structure generation. Losic et al. [88] proposed a reconstruction process with given phase probabilities and an overall correlation function to form lamellar clay films on solid surfaces and dispersed clay dots on solid structures. Mohanty [96] adopted a Monte Carlo annealing algorithm to generate unsaturated porous media by using the law of lowest interfacial energy, but it failed to differentiate various liquid–solid interactions.

Wang et al. [97–99] have done excellent work in this direction. A three-dimensional mesoscopic method was developed and discussed by them for predicting the effective thermal conductivity of multiphase random porous media. The energy transport equations were solved using the Lattice Boltzmann method [100–102] for multi-phase conjugate heat transfer through a porous structure whose morphology was characterized by a random generation-growth algorithm. This method was designed to reproduce structural assembles of elements with random sizes, locations and orientations, and connections, each of which grows from randomly distributed seeds and the growth is guided by a few given probabilistic growth rates. They suggested that for different types of microstructures (granular, fibrous or netlike), the algorithms will be different but still bear the same principles. The temperature and the heat flux were calculated according to [103] as:

$$T = \sum_{\alpha} g_{\alpha}$$

$$q = \sum_{\alpha} e_{\alpha} g_{\alpha} \left(\frac{\tau_g - 0.5}{\tau_g} \right)$$

here g_{α} is the equilibrium distribution in each direction, e_{α} the discrete lattice velocity and τ_g the dimensionless relaxation time for each phase.

After solving the temperature field their expression for effective thermal conductivity is given as

$$\lambda_e = \frac{L. \int q.dA}{\Delta T \int dA} \tag{73}$$

where q is the steady heat flux through the media cross section area dA between the temperature difference ΔT with a distance L. All of these parameters can be theoretically determined, and there were no empirical factors existed in this model.

4 Limitations of the Theoretical Models and Discussion

We have briefly reviewed several existing models available in the literature for predicting effective thermal conductivity of complex materials. A number of different structures were considered. The real structures and geometries of materials around us are so vast and vivid, that one cannot use a single model to explain the thermal behaviour of various systems due to their inherent limitations. In real systems, the kind of structure we face does not match with the geometries discussed in various models. Therefore, results of these models vary with the experimental values. Particularly, when the ratio of thermal conductivities of solid to fluid phase was large. The limitations in the existing models, and thus challenges in developing new ones, are summarized below:

Almost all the models discussed here are based on simplified physics and are developed either using the concept of modified flux or considering the phases made up of different resistors with certain over-idealized assumptions. Therefore, they are incapable of dealing with the issue of phase interactions in real structures. This may make the models easy and quick to use, but the simplifications also restrict the applicability to simple structures and unable to tackle directly the effects of morphological changes in the microstructure on the material properties. Any efforts in combining the models for more complex structures will lead to escalating complexity in the model. Most modifications to improve the accuracy of the models will in turn narrow their applicability. In recently developed models, the empirical parameters have been introduced with no valid physical significance, whose values have to be determined case by case based on experimental data, thus their applicability is limited and become powerless for wider applications. Even for the relatively successful theoretical models, one can only calculate the properties of existing materials, rather than make predictions for new class of materials. Therefore, there are not of much value in optimizations or design for novel materials The resistor concept embodies linear flow of heat where Ohm's law is followed, while the flux concept is based on material field from external or internal sources and thus the flux density and its path depends upon the conductivity of the material through which the heat flow is maintained. These models do not describe the behavior of an actual system. The resistor model is an exact solution for a two-phase system arranged in the form of slabs while the flux model is an exact solution whose micro geometry and phase distribution is completely prescribed like spheres in cubic array with lattice type structures. In general, the materials around us do not belong to either category as their phase distribution and grain arrangement is completely undefined. Therefore, the models developed using these concepts cannot be directly applied to natural two phase materials. However, at the same time, the resistor model yields the maximum and minimum limits on the effective thermal conductivity. Similarly, the flux model gives an insight as to how the flux modification takes place in a periodic structure like beads which are statistically homogeneous but locally heterogeneous. The difference between real and assumed situations increases rapidly as the ratio of component conductivity increases or decreases from a moderate value. This led to a new idea i.e. randomization of phase distribution. The resulting expression for the effective thermal conductivity depends upon the method of randomization applied.

The idea of random phase distribution using the flux concept resulted in Maxwell's relation [2], Fricke's relation [3], the Bruggeman theory [6] of variable dispersion and Brailsford and Major's relation [62]. The relation established by Maxwell is suitable for dilute dispersion and moderate conductivity ratios. The results of Maxwell's [2] formula are satisfactory for $\phi = 0.5$. The concept of randomization used in resistor models with averaging techniques is better reflected in models of Woodside and Messmer [9], Chaudhary and Bhandari [20], Cheng and Vachon [21], and Kumar and Chaudhary [26].

The model developed by Lichtnecker [11] does not incorporate structure and mode of packing of a system. It is suitable at very low and very high dispersions. When the ratio of thermal conductivities of solid and fluid phase is more than 20, the value predicted for the effective thermal conductivity is overestimated. The expression of Rayleigh [7] is restricted to cellular materials and emulsions. Fricke [3] and Burger's [4] relation gives a good result for packed systems of quartz sand, glass beads in different fluids only. The expression of Kunni and Smith [17] provides a lower value of the effective thermal conductivity for lower ratio of thermal conductivities of solid and fluid phase. Sugawara and Yoshizawa [18] provided an empirical relation for effective thermal conductivity, which yields correct values of λ_e for soils. They used $n = 6.5$ in their equation (21).

The bound technique is really useful in estimating the closest optimum value of the effective thermal conductivity of two-phase systems. The Hashin-Strikman bound [22] is a most general bound which predicts the effective thermal conductivity with the least possible knowledge about the two phase system. The Hashin-Strikman bound is of wider use in the range $0.1 < \lambda_d/\lambda_c < 10$. Moreover, when the ratio (λ_d/λ_c) becomes too large or too low, the Hashin-Strikman bound becomes broader and turns out to be unrealistic. The Kumar and Chaudhary bound [26] is equally well in the region $0.1 < \lambda_d/\lambda_c < 10$ but at the same time it is narrower than the Hashin-Strikman bound when $\lambda_d/\lambda_c < 10^{-2}$ in the case of sands and soils.

The expressions developed on the basis of an averaging technique seem to be more relevant than the models, which are the outcome of rigorous mathematics. As an example, the results of Hadley's model [32] are better than Lichtnecker [11], Brailsford [62] and Pande [27]. Pande's model is better when the fluid phase is air in a two-phase system. The model developed by Boomsma and Poulikakos [37] is based on the idealized three-dimensional basic cell geometry of foam, the tetrakaidecahedron. This geometric shape results from filling a given space with cells of equal size yielding minimal surface energy. The foam structure was represented with cylindrical ligaments, which attach to cubic nodes at their centers. It was found that the model estimated the effective thermal conductivity very well for these experimental configurations. This three dimensional model fits the experimental data very well for the parameter range experienced in metal foams. Calmidi and Mahajan [38], Bhattacharya [39] in their models considered that the structure consists of a two dimensional array of hexagonal cells where the fibers form the sides of the hexagons and a circular blob of the metal at the intersection of the fibers. Their analysis reflected that the effective thermal conductivity depends strongly on the porosity and the ratio of the cross sections of the fiber and intersection. The Ma et al. models [75, 76] have only two parameters, A_{nt}/A and t^+ in which every parameter has a clear physical meaning whereas other models contain several empirical constants without physical interpretation. This model is analytically related to several variables, such as ϕ, β, t^+ and A_{nt}/A, whereas other numerical solutions cannot provide such an analytical relationship. The recursive algorithm for the thermal conductivity obtained using this model was also quite simple. The advantage of this model over the others is, thus, quite evident.

Now, due to rapid developments in computational techniques to reconstruct the microstructures of multi-phase complex materials, the effect of microstructure on thermo-mechanical properties of the material can be calculated more precisely. Therefore, numerical methods are more accurate and robust approaches in prediction and design of new class of materials to fulfill requirements of the present millennium. As a result, these techniques are becoming increasingly popular and more widely used in the materials science and engineering. In the present review, we presented only a glimpse of the technique developed for the prediction of the effective thermal conductivity of multi-phase complex materials and a lot of work remains to be done in this direction.

Acknowledgement I wish to sincerely thank Dr. V. S. Kulhar, Associate Professor, Department of Physics, University of Rajasthan, Jaipur, India for helpful suggestions, which have certainly improved both the content and quality of the material.

References

1. Fourier, J.B.: Theorie analytique de la chaleur, Paris, English translation by A. Freeman. Dover Publication Inc., New York (1955)
2. Maxwell, J.C.: A Treatise of Electricity and Magnetism, 3rd edn. Clarendon, Oxford (1904)
3. Fricke, H.: A mathematical treatment of the electric conductivity and capacity of disperse systems I. The electric conductivity of a suspension of homogeneous spheroids. Phys. Rev. **24**, 575 (1924)
4. Burgers, H.C.: Generalized Maxwell's approach to ellipsoidal particles. Phys. Z. **20**, 73 (1919)
5. De Vries, D.A.: The thermal conductivity of soil. Meddelingen Van de Landbouwhoge School Te Wageningen (English translation) **52**(1), 1–73 (1952)
6. Bruggeman, D.A.G.: Dielectric constant and conductivity of mixtures of isotropic materials. Ann. Phys. **24**, 636 (1935)

7. Rayleigh, L.: On the influence of obstacles arranged in rectangular order upon the properties of medium. Philos. Mag. **34**, 481–502 (1892)
8. Weiner, O.: Lamellare Doppelbrechung. Phys. Z. **5**, 332 (1904)
9. Woodside, W., Messmer, J.H.: Thermal conductivity of porous media. I. Unconsolidated sands. J. Appl. Phys. **32**, 1688 (1961)
10. Bernshtein, R.S.: Studies of Burning Processes of Natural Fuels. State Power Press, Moscow (1948)
11. Lichtnecker, K.: Logarithmic law of mixing. Phys. Z. (Germany) **27**, 115 (1926)
12. Russel, H.W.: Principles of heat flow in porous insulators. J. Am. Ceram. Soc. **18**, 1–5 (1935)
13. Ribaud, M.: Theoretical study of thermal conductivity of porous and pulverulent materials. Chaleur. Ind. **18**, 36 (1937)
14. Powers, A.E.: Conductivity in aggregates. Res. Dev. Rep. KAPL-2145 (1961)
15. Bogomolov, V.Z.: Heat transfer in dispersed body. Trans. Phys. Astron. Inst., Agriculture Press (1941)
16. Assad, Y.: A study of the thermal conductivity of fluid bearing porous rocks. Ph.D. thesis, University of California, USA (1955)
17. Kunni, M., Smith, J.M.: Heat transfer characteristics of porous rocks. AIChE J. **6**, 71 (1960)
18. Sugawara, A., Yoshizawa, Y.: An investigation on the thermal conductivity of porous materials and its application to porous rock. Aust. J. Phys. **14**, 469 (1961)
19. Sugawara, A.: Jpn. J. Appl. Phys. (Japan) **30**, 899 (1961)
20. Chaudhary, D.R., Bhandari, R.C.: Heat transfer through dispersed medium. Indian J. Pure Appl. Phys. **6**, 135 (1968)
21. Cheng, S.C., Vachon, R.I.: The prediction of thermal conductivity of two and three phase solid heterogeneous mixtures. Int. J. Heat Mass Transf. **12**, 249 (1969)
22. Hashin, Z., Strikman, S.: A variational approach to the theory of effective magnetic permeability of multiphase materials. J. Appl. Phys. **33**, 3125 (1962)
23. Prager, S.: Variational treatment of hydrodynamic interaction on polymers. J. Chem. Phys. **50**, 4305 (1969)
24. Schulgasser, K.: On the conductivity of fiber reinforced materials. J. Math. Phys. **17**, 382 (1976)
25. Hori, M.: Statistical theory of effective electrical, thermal, and magnetic properties of random heterogeneous materials. II. Bounds for the effective permittivity of statistically anisotropic materials. J. Math. **14**, 514 (1973)
26. Kumar, V., Chaudhary, D.R.: Prediction of effective thermal conductivity of two-phase porous materials using resistor model. Indian J. Pure Appl. Phys. **18**, 984 (1980)
27. Pande, R.N., Kumar, V., Chaudhary, D.R.: Bounds on the effective thermal conductivity of two-phase systems. Pramana J. Phys. **20**, 339–346 (1983)
28. Zimmerman, R.W.: Thermal conductivity of fluid-saturated rocks. J. Petrol. Sci. Eng. **3**, 219 (1989)
29. Torquato, S., Rintoul, M.D.: Effect of the interface on the properties of composite media. Phys. Rev. Lett. **75**, 4067 (1995)
30. Pande, R.N., Kumar, V., Chaudhary, D.R.: Thermal conduction in a homogeneous two phase system. Pramana J. Phys. **22**, 63–70 (1984)
31. Pande, R.N., Chaudhary, D.R.: Thermal conduction through loose and granular two-phase materials at normal pressure. Pramana J. Phys. **23**, 599 (1984)
32. Hadley, G.R.: Thermal conductivity of packed metal powders. Int. J. Heat Mass Transf. **20**, 909 (1986)
33. Verma, L.S., Shrotriya, A.K., Singh, R., Chaudhary, D.R.: Thermal conduction in two-phase materials with spherical and non-spherical inclusions. J. Phys. D Appl. Phys. **24**, 1729 (1991)
34. Misra, K., Shrotriya, A.K., Singh, R., Chaudhary, D.R.: Porosity correction for thermal conduction in real two-phase systems. J. Phys. D Appl. Phys. **27**, 732 (1994)
35. Hsu, C.T., Cheng, P., Wong, K.W.: A lumped parameter model for stagnant thermal conductivity of spatially periodic porous media. ASME J. Heat Transf. **117**, 264–269 (1995)

36. Singh, K.J., Singh, R., Chaudhary, D.R.: Heat conduction and a porosity correction term for spherical and cubic particles in a simple cubic packing. J. Phys. D Appl. Phys. **31**, 1681 (1998)
37. Boomsma, K., Poulikakos, D.: On the effective thermal conductivity of a three-dimensionally structured fluid-saturated metal foam. Int. J. Heat Mass Transf. **44**, 827 (2001)
38. Calmidi, V.V., Mahajan, R.L.: The effective thermal conductivity of high porosity fibrous metal foams. J. Heat Transf. **121**, 466 (1999)
39. Bhattacharya, A., Calmidi, V.V., Mahajan, R.L.: Thermophysical properties of high porosity metal foams. Int. J. Heat Mass Transf. **45**, 1017 (2002)
40. Lu, T.J., Chen, C.: Thermal transport and fire retardance properties of cellular aluminum alloys. Acta Mater **47**, 1469 (1999)
41. Feng, Y., Yu, B., Zou, M., Zhang, D.: A generalized model for the effective thermal conductivity of porous media based on self-similarity. J. Phys. D Appl. Phys. **37**, 3030–3040 (2004)
42. Singh, R., Kasana, H.S.: Computational aspects of effective thermal conductivity of highly porous metal foams. Appl. Therm. Eng. **24**, 1841 (2004)
43. Jagjiwanram Singh, R.: Effective thermal conductivity of highly porous two-phase systems. Appl. Therm. Eng. **24**, 2727 (2004)
44. Singh, R., Sharma, P.: Effective thermal conductivity of polymer composites. Adv. Eng. Mater. **10**, 366–370 (2008)
45. Singh, R., Sharma, P., Bhoopal, R.S., Verma, L.S.: Prediction of effective thermal conductivity of cellular and polymer composites, Indian J. Pure Appl. Phys. Submitted for publication (2010)
46. Krischer, O.: Trocknungstechnik Band I, Die wissenschaftlichen Grundlagen der Trocknungstechnik. Springer Verlag, Berlin (1978)
47. Carson, J.K., Lovatt, S.J., Tanner, D.J., Cleland, A.C.: Thermal conductivity bounds for isotropic, porous materials. Int. J. Heat Mass Transf. **48**, 2150–2158 (2005)
48. Alharthi, A., Lange, J.: Soil water saturation: dielectric determination. Water Resour. Res. **23**, 591–595 (1987)
49. Batchelor, G.K., O'Brien, R.W.: Thermal or electrical conduction through a granular material. Proc. R. Soc. Lond. A Math. Phys. Sci. **355**, 313–333 (1977)
50. Birchak, J.R., Gardner, C.G., Hipp, J.E., Victor, J.M.: High dielectric constant microwave probes for sensing soil moisture. Proc. IEEE **62**, 93–98 (1974)
51. Cosenza, P., Guerin, R., Tabbagh, A.: Relationship between thermal conductivity and water content of soils using numerical modelling. Eur. J. Soil Sci. **54**, 581–587 (2003)
52. Friedman, S.P.: Soil properties influencing apparent electrical conductivity: a review. Comput. Electron. Agr. **46**, 45–70 (2005)
53. Hallikainen, M.T., Ulaby, F.T., Dobson, M.C., Elrayes, M.A., Wu, L.K.: Microwave dielectric behavior of wet soil. 1. Empirical-models and experimental observations. IEEE Trans. Geosci. Rem. Sens. **23**, 25–34 (1985)
54. Jacobsen, O.H., Schjonning, P.: A laboratory calibration of time domain reflectometry for soil water measurement including effects of bulk density and texture. J. Hydrol. **151**, 147–157 (1993)
55. Jones, S.B., Or, D.: Modeled effects on permittivity measurements of water content in high surface area porous media. Physica B **338**, 284–290 (2003)
56. Miyamoto, T., Annaka, T., Chikushi, J.: Soil aggregate structure effects on dielectric permittivity of an andisol measured by time domain reflectometry. Vadose Zone J. **2**, 90–97 (2003)
57. Topp, G.C., Davis, J.L., Annan, A.P.: Electromagnetic determination of soil water content: measurements in coaxial transmission lines. Water Resour. Res. **16**, 574–582 (1980)
58. Wang, J.R., Schmugge, T.J.: An empirical model for the complex dielectric permittivity of soils as a function of water content. AIEEE Trans. Geosci. Rem. Sens. **18**, 288–295 (1980)

59. Whalley, W.R.: Considerations on the use of time domain reflectometry (TDR) for measuring soil water. J. Soil Sci. **44**, 1–9 (1993)
60. Chandrakanthi, M., Mehrotra, A.K., Hettiaratchi, J.P.A.: Thermal conductivity of leaf compost used in bio-filters: an experimental and theoretical investigation. Environ. Pollut. **136**, 167–174 (2005)
61. de Veris, D.A., Wijk, WRv: Physics of Plant Environment, pp. 210–235. North-Holland Publishing Company, Amsterdam (1963)
62. Brailsford, A.D., Major, K.G.: Thermal conductivity of aggregates of several phases, including porous materials. Br. J. Appl. Phys. **15**, 313 (1964)
63. Chaudhary, D.R., Bhandari, R.C.: Heat transfer through a three-phase medium. Br. J. Appl. Phys. Ser. 2 **1**, 815–817 (1968)
64. Singh, R., Beniwal, R.S., Kumar, V.: Effective thermal conductivity of three-phase systems. Indian J. Pure Appl. Phys. **22**, 556–558 (1984)
65. De Loor, G.P.: Dielectric behaviour of heterogeneous mixtures. Appl. Sci. Res. B Elect. Acoust. Optic. Math Methods **11**, 310 (1964)
66. De Loor, G.P.: The dielectric properties of wet soils. The Netherlands Remote Sensing Board: BCRS Report **90–13**, 1–39 (1990)
67. Dobson, M.C., Ulaby, F.T., Hallikainen, M.T., Elrayes, M.A.: Microwave dielectric behavior of wet soil, Part II: Dielectric mixing models. IEEE Trans. Geosci. Rem. Sens. **23**, 35–46 (1985)
68. Verma, L.S., Shrotriya, A.K., Singh, R., Chaudhary, D.R.: Prediction and measurement of effective thermal conductivity of three-phase systems. J. Phys. D. Appl. Phys. **24**, 1515–1526 (1991)
69. Beniwal, R.S., Singh, R., Pande, R.N., Kumar, V., Chaudhary, D.R.: Thermal conduction in heterogeneous multi-phase systems. Indian J. Pure Appl. Phys. **23**, 289–294 (1985)
70. Singh, A.K., Singh, R., Chaudhary, D.R.: Prediction of effective thermal conductivity of moist porous materials. J. Phys. D Appl. Phys. **23**, 698–702 (1990)
71. Singh, R., Beniwal, R.S., Chaudhary, D.R.: Thermal conduction of multi-phase systems at normal and different interstitial air pressures. J. Phys. D Appl. Phys. **20**, 917–922 (1987)
72. Moosavi, A., Sarkomaa, P.: The effective conductivity of three-phase composite materials with circular cylindrical inclusions. J. Phys. D Appl. Phys. **36**, 1644–1650 (2003)
73. Gori, F., Corasaniti, S.: Experimental measurement and theoretical prediction of thermal conductivity two and three-phase water/olivine systems. Int. J. Thermophys. **24**, 1339–1353 (2003)
74. Gori, F.: Theoretical model for predicting the effective thermal conductivity of unsaturated frozen soils. In: Proc. Fourth Int. Conf. on Permafrost, Fairbanks, Alaska, 1983, pp. 363–368.
75. Ma, Y., Yu, B., Zhang, D., Zou, M.: A self-similarity model for effective thermal conductivity of porous media. J. Phys D Appl. Phys. **36**, 2157–2164 (2003)
76. Ma, Y., Yu, B., Zhang, D., Zou, M.: Fractal geometry model for effective thermal conductivity of three-phase porous media. J. Appl. Phys. **95**, 6426–6433 (2004)
77. Gerstner, P., Paltakari, J., Gane, P.A.C.: A lumped parameter model for thermal conductivity of paper coatings. Transport Porous Med. **78**, 1–9 (2009)
78. Torquato, S., Kim, I.C., Cule, D.: Effective conductivity, dielectric constant and diffusion coefficient of digitized composite media. J. Appl. Phys. **85**, 1560–1571 (1999)
79. Zhang, H.F., Ge, X.S., Ye, H.: Randomly mixed model for predicting the effective thermal conductivity of moist porous media. J. Phys. D Appl. Phys. **39**, 220–226 (2006)
80. Tacher, L., Perrochet, P., Parriaux, A.: A generation of granular media. Transport Porous Med. **26**, 99–107 (1997)
81. Pilotti, M.: Generation of realistic porous media by grains sedimentation. Transport Porous Med. **33**, 257–278 (1998)
82. Yang, A., Miller, C.T., Turcoliver, L.D.: Simulation of correlated and uncorrelated packing of random size spheres. Phys. Rev. E **53**, 1516–1524 (1996)

83. Li, Y.S., LeBoeuf, E.J., Basu, P.K., Mahadeyan, S.: Stochastic modeling of the permeability of randomly generated porous media. Adv Water Resour. **28**, 835–844 (2005)
84. Mori, Y., Hopmans, J.W., Mortensen, A.P., Kluitenberg, G.J.: Estimation of vadose zone water flux from multi-functional heat pulse probe measurements. Soil Sci. Soc. Am. J. **69**, 599–606 (2005)
85. Mortensen, A.P., Hopmans, J.W., Mori, Y., Simunek, J.: Multi-functional heat pulse probe measurements of coupled vadose zone flow and transport. Adv. Water Resour. **29**, 250–267 (2006)
86. Li, D.S., Saheli, G., Khaleel, M., Garmestani, H.: Quantitative prediction of effective conductivity in anisotropic heterogeneous media using two-point correlation functions. Comput. Mater. Sci. **38**, 45–50 (2006)
87. Makrodimitris, K., Papadopoulos, G.K., Philippopoulos, C., Theodorou, D.N.: Parallel tempering method for reconstructing isotropic and anisotropic porous media. J. Chem. Phys. **117**, 5876–5884 (2002)
88. Losic, N., Thovert, J.F., Adler, P.M.: Reconstruction of porous media with multiple solid phases. J. Colloid Interface Sci. **186**, 420–433 (1997)
89. Torquato, S.: Modeling of physical properties of composite materials. Int. J. Solids Struct. **37**, 411–422 (1999)
90. Young, I.M., Crawford, J.W., Rappoldt, C.: New methods and models for characterising structural heterogeneity of soil. Soil Till. Res. **61**, 33–45 (2001)
91. Torquato, S.: Statistical description of microstructures. Annu. Rev. Mater. Res. **32**, 77–111 (2002)
92. Adler, P.M.: Porous Media: Geometry and Transports. Butterworth/Heinemann, Stoneham (1992)
93. Torquato, S.: Random Heterogeneous Materials: Microstructure and Macroscopic Properties. Springer, New York (2002)
94. Pan, N.: Physics of Fibrous Soft Matters. Higher Education Publishing, Inc., Beijing (2005)
95. Pan, N., Gibson, P.: Thermal and Moisture Transport in Fibrous Material. Woodhead Publishing Ltd, Cambridge, UK (2006)
96. Mohanty, S.: Effect of multiphase fluid saturation on the thermal conductivity of geologic media. J. Phys. D Appl. Phys. **30**, L80–L84 (1997)
97. Wang, M., Wang, J., Pan, N., Chen, S., He, J.: Three-dimensional effect on the effective thermal conductivity of porous media. J. Phys. D Appl. Phys. **40**, 260 (2007)
98. Wang, M., Pan, N.: Modeling and prediction of the effective thermal conductivity of random open-cell porous foams. Int. J. Heat Mass Transf. **51**, 1325 (2008)
99. Wang, M., Pan, N.: Predictions of effective physical properties of complex multiphase materials. Mater. Sci. Eng. R **63**, 1–30 (2008)
100. He, X.Y., Chen, S.Y., Doolen, G.D.: A novel thermal model for the lattice boltzmann method in incompressible limit. J. Comput. Phys. **146**, 282–300 (1998)
101. Peng, Y., Shu, C., Chew, Y.T.: Simplified thermal lattice Boltzmann model for incompressible thermal flows. Phys. Rev. E **68**, 026701 (2003)
102. Peng, Y., Shu, C., Chew, Y.T.: A 3D incompressible thermal lattice Boltzmann model and its application to simulate natural convection in a cubic cavity. J. Comput. Phys. **193**, 260–74 (2003)
103. D'Orazio, A., Succi, S.: Boundary conditions for thermal lattice boltzmann simulations. In: Computational Science – ICCS 2003, Pt. I Proceedings, Heidelberg, vol. 2657, pp. 977–986 (2003)

Lattice Monte Carlo Analysis of Thermal Diffusion in Multi-Phase Materials

T. Fiedler, I.V. Belova, A. Öchsner, and G.E. Murch

Abstract This Chapter addresses the numerical simulation of thermal diffusion in multi-phase materials. A Lattice Monte Carlo method is used in the analysis of two- and three-dimensional calculation models. The composites considered are assembled by two or three phases, each exhibiting different thermal conductivities. First, a random distribution of phases is considered and the dependence of the effective thermal conductivity on the phase composition is investigated. The second part of this analysis uses a random-growth algorithm that simulates the influence of surface energy on the formation of composite materials. The effective thermal conductivity of these structures is investigated and compared to random structures. The final part of the Chapter addresses percolation analyses. It is shown that the simulation of surface energy distinctly affects the percolation behavior and therefore the thermal properties of composite materials.

1 Introduction

The thermal properties of multi-phase materials (cf. Fig. 1) are of great importance in modern science. Next to common examples such as fiber-reinforced composites or thermal insulators, examples stretch from the characterization of frozen food [2, 3] to advanced phase change composites used for thermal energy storage [4, 5].

The thermal properties of multi-phase materials are of great importance in modern science. Next to common examples such as binary metallic mixtures or

T. Fiedler (✉), I.V. Belova, and G.E. Murch
The University of Newcastle, Callaghan, NSW 2308 Australia
e-mail: Thomas.Fiedler@newcastle.edu.au, Irina.Belova@newcastle.edu.au, Graeme.Murch@newcastle.edu.au

A. Öchsner
Technical University of Malaysia, 81310 UTM Skudai, Johor, Malaysia
The University of Newcastle, Callaghan, NSW 2308, Australia
e-mail: Andreas.Oechsner@gmail.com

A. Öchsner and G.E. Murch (eds.), *Heat Transfer in Multi-Phase Materials*,
Adv Struct Mater 2, DOI 10.1007/8611_2010_6,

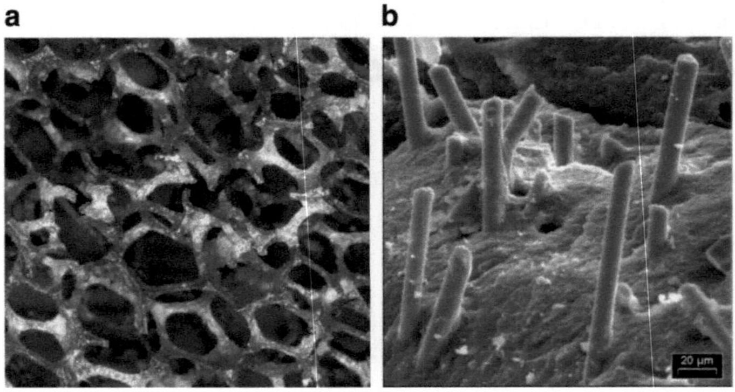

Fig. 1 Multi-phase materials: (**a**) cellular material, (**b**) fiber-reinforced composite [1]

thermal insulators, examples stretch from the characterization of frozen food [2, 3] to advanced phase change composites used for thermal energy storage [4, 5]. Due to the importance of this subject, conductivity properties of multi-phase materials have been the subject of intensive research. It should be highlighted here that although the current Chapter strictly addresses thermal properties, the mathematical models for electric conduction and mass diffusion are equivalent and with some caution can be used interchangeably. Landauer [6] addressed the electrical resistance of binary metallic mixtures using the effective medium theory. He investigated two-phase random mixtures similar to some of the models addressed in the current study. The comparison of his mathematical results with experimental data showed good agreement. Ben-Amoz [7] addressed the effective thermal properties of two-phase solids. He introduced improved Voigt and Reuss estimates as bounds for the effective thermal conductivities and diffusivities. Glatzmaier and Ramirez [8] conducted experimental measurements using the transient hot wire method. They suggested a mathematical volume averaging model in order to predict the effective thermal conductivity. Good agreement between their mathematical model and experimental measurements was observed for a glass beads packed bed, oil shale and coal. Rio et al. [9] introduced an approximate formula for the effective electrical conductivity of two-dimensional two-phase materials based on Keller's reciprocity theorem and observed good agreement with experimental data. Samantray et al. [10] addressed the effective thermal conductivity of two-phase materials. They suggested different models depending on the conductivity ratio of matrix and dispersed phase. Maxwell approaches were found to be accurate for phase fractions $\Phi < 0.1$ and $\Phi > 0.9$ where one phase is isolated and does not form an interconnected long range network. For the remaining phase fractions, three correlations for different conductivity ratios were introduced. Good agreement with extensive experimental data was found. Karthikeyan and Reddy [11] focused on composites with phase fractions between 0.1 and 0.9. They conducted a unit cell approach for conductivity ratios $r = \lambda_{\min}/\lambda_{\max} > 0.05$. Wang et al. [12] used random generation-growth methods to mimic real multi-phase structures. Using these models, they subsequently conducted Lattice Boltzmann analyses to solve the energy transport equations and obtain effective thermal properties. Good agreement with experimental data was found for copper-solder and water-sand composites. Their work was continued in [13] where also metallic foam materials and fibrous structures were analyzed.

The current work uses a modified Lattice Monte Carlo method towards thermal analysis of multi-phase materials. The Lattice Monte Carlo method for thermal analysis is elucidated in Sect. 2 of this Chapter. In Sect. 3, the effective thermal conductivities of materials with randomly dispersed species are obtained with high accuracy and compared to existing analytical relations. Furthermore, a random growth algorithm is introduced in order to simulate the effect of surface energy on thermal properties. Section 4 briefly addresses the effective thermal conductivity of three-phase systems. In Sect. 5, percolation analysis on the previously considered structures is performed and a close connection between percolation behavior and thermal properties is found.

2 Lattice Monte Carlo Method

The Lattice Monte Carlo method has proven to be an elegant approach to address a wide range of thermal or mass diffusion problems [14]. Previous publications dealing with thermal properties of materials have focused on the effective thermal conductivity of hollow sphere structures [15, 16], random shaped cellular aluminium [17], non-linear thermal material properties in composite structures [18] and transient heat transfer in phase change materials [19].

In the Lattice Monte Carlo (LMC) method, thermal diffusion is simulated by random walks of 'virtual' particles that represent very small – but finite – energy quantities. The random walks are directed by jump probabilities p_{j} which are in fact scaled thermal diffusivities D. In the case of a successful jump attempt, energy is transferred from a lattice node to a neighbor and the local temperatures T in the volumes represented by the nodes are changed. The arrangement of lattice nodes can be chosen arbitrarily and for reasons of simplicity a simple cubic arrangement with a constant jump length s is typically selected for this analysis.

A flow diagram of the Lattice Monte Carlo algorithm is shown in Fig. 2. At the beginning of a simulation, the initial conditions are implemented, i.e. a large population of 'virtual particles' is randomly distributed in the lattice model. Subsequently, in two random steps a particle is selected and its jump direction determined. Based on the thermal diffusivities of the origin and the destination nodes, a jump probability p_{j} is calculated. This value is then compared to a uniformly distributed random number $0 \leq \chi < 1$. If the jump probability is larger than the random number, the 'virtual' particle coordinates are updated. The calculation time t is directly incremented before the next particle is chosen. The accuracy of Lattice Monte Carlo analysis is governed by the number N_{P} of 'virtual' energy particles. The error can be estimated by $\varepsilon \approx \pm 1/\sqrt{N_{\mathrm{P}}}$. As an example, a population of 10^6 particles results in an error of approximately $\pm$ 0.1%. However, the

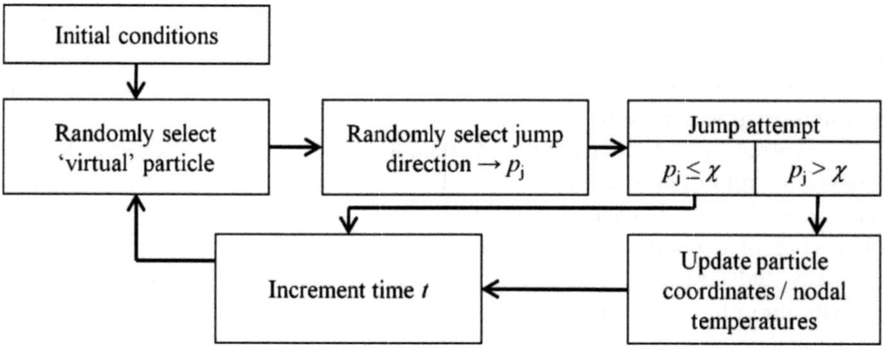

Fig. 2 Flow diagram for Lattice Monte Carlo analysis

calculation time increases with the number of particles N_p and therefore a balance between accuracy and computational load needs to be found. In addition, a sufficient long calculation time t must be simulated in order to ensure numerical convergence of the results. This can be guaranteed by a simple convergence analysis of the target value, i.e. the effective thermal conductivity.

At the end of a LMC random walk simulation, the particle displacements R are known for a large number N_p of particles. In order to determine the effective thermal diffusivity D_{eff} of a d-dimensional structure, the average mean square displacement $\langle R^2 \rangle$ is calculated and inserted into the Einstein equation:

$$D_{\text{eff}} = \frac{\langle R^2 \rangle}{2 \cdot d \cdot t}. \tag{1}$$

The effective thermal conductivity λ_{eff} is then obtained using the formula:

$$\lambda_{\text{eff}} = \frac{D_{\text{eff}}}{\rho_{\text{avg}} \cdot C_{\text{eff}}}, \tag{2}$$

where ρ_{avg} is the average density and C_{eff} the effective specific heat of the multi-phase material. The solution of thermal steady-state problems such as the current analysis is independent of the material parameters ρ and C and accordingly, these can be chosen arbitrarily for each phase i. For simplicity, $\rho_i = C_i = 1$ is selected and according to (2) the effective thermal conductivity is then identical to the effective diffusivity $\lambda_{eff} = D_{eff}$.

3 Two-Phase Systems

In this Section, systems assembled by two species A and B with corresponding thermal conductivities $\lambda_A > \lambda_B$ are considered. Several conductivity ratios $r = \lambda_B/\lambda_A < 1$ are addressed and calculation models are generated for a range of volume fractions $\Phi_B = 1 - \Phi_A$. Thereby, random distributions and modified distributions mimicking the effect of surface energy can be distinguished. A set of ten structures is generated for each configuration in order to monitor a possible scattering of results and ensure statistically representative results. If not mentioned differently, the displayed results are averaged values.

3.1 Random Distribution

First, random distributions are considered where two phases are dispersed in a simple cubic lattice model. During the model generation, no interaction occurs between members of the same or opposite species. Random distributions are generated according to the following procedure: initially, all lattice sites are assigned to species A ($\Phi_A = 1$, $\Phi_B = 1 - \Phi_A = 0$). A lattice site of A is randomly selected and assigned to species B. In the next step, the volume fraction Φ_B is updated. This procedure is repeated until the target volume fraction of species B is reached. Three examples of random distributions for the phase fractions $\Phi_B = 0.3$, 0.5, 0.7 are shown in Fig. 3. In the Figure, black pixels correspond to phase A and white pixels to phase B.

3.1.1 Two-Dimensional Analysis

System Size

The first series of calculations addresses the impact of the system size on the results of the numerical analysis. At the corners (two dimensions) or surfaces (three dimensions) of the calculation models, periodic boundary conditions are prescribed, i.e. the geometry is repeated. Small models that are repeated several times develop characteristic patterns and are unsuitable for representing *random* multi-phase

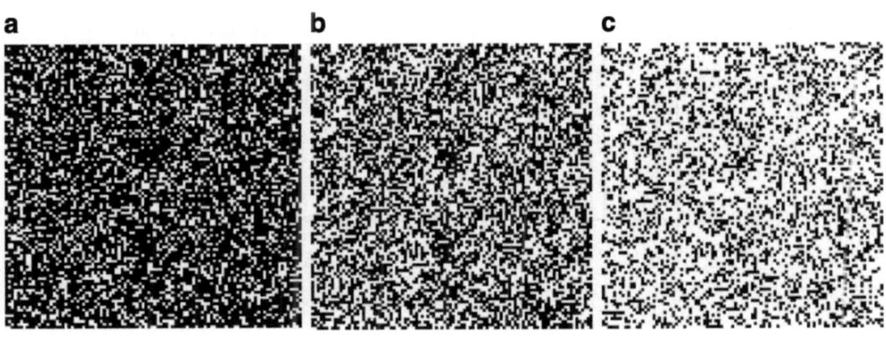

Fig. 3 Random distributions: (**a**) $\Phi_B = 0.3$, (**b**) $\Phi_B = 0.5$, (**c**) $\Phi_B = 0.7$

structures. An increase of the system size diminishes the influence of the periodic boundary conditions. Accordingly, preliminary investigations aim towards the determination of minimum system sizes to represent random distributions in the subsequent analysis.

Figure 4a shows the effective thermal conductivity of a two-phase system ($r = 0.1$) plotted versus the phase fraction Φ_B. The results obtained for the largest system (10,000^2 lattice sites) are drawn as a full line. Significant scattering of the results is only found for the smallest (10^2 lattice sites) of the considered models.

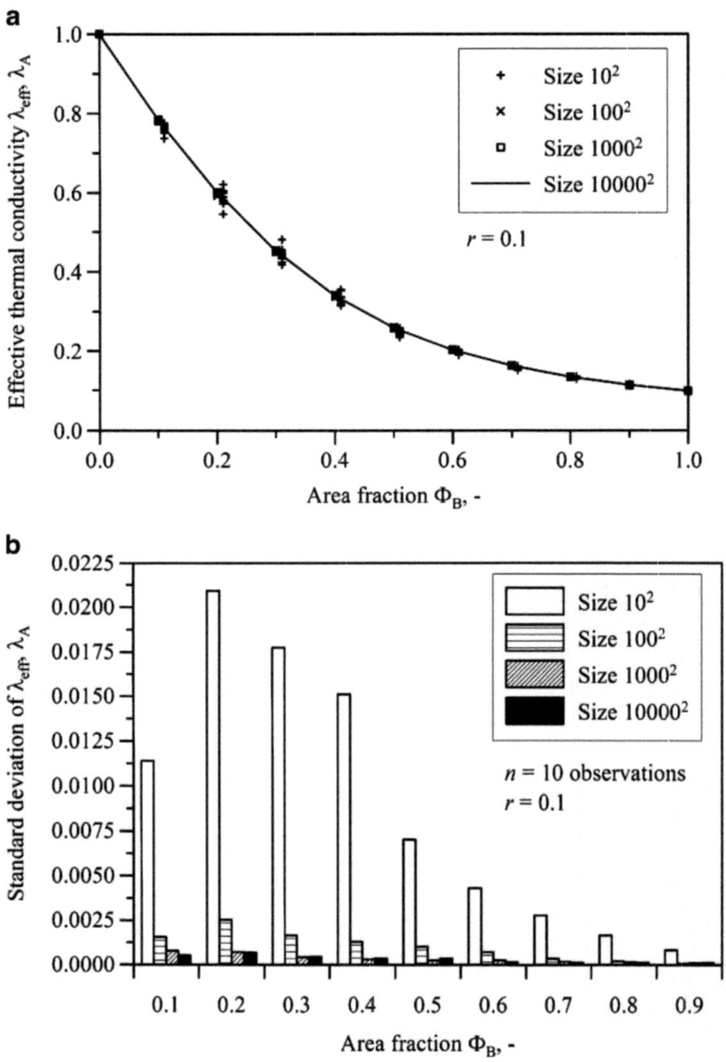

Fig. 4 Study of the effect of the system size: (**a**) Effective thermal conductivities plotted versus the area fraction Φ_B, (**b**) Standard deviation of the effective thermal conductivity

This observation can be quantified by comparing the standard deviation δ (calculated for $n = 10$ independent observations) of the effective thermal conductivity

$$\delta = \sqrt{\frac{\sum_{n} \left(\lambda_{\text{eff,n}} - \overline{\lambda_{\text{eff}}}\right)^2}{n-1}}. \tag{3}$$

The variable $\overline{\lambda_{\text{eff}}}$ is the average effective thermal conductivity for a given system size and phase fraction. The results are shown in Fig. 4b. It can clearly be seen that the standard deviation of the effective thermal conductivity δ decreases with increasing system size. The standard deviation also shows a dependency on the area fraction and maximum scattering is observed for $\Phi_B = 0.2$. Comparing the results of the two largest systems (1,000^2 and 10,000^2 nodes) no systematic change in the standard deviation can be observed. Accordingly, calculation models with 1,000^2 lattice nodes are chosen as the standard for the following two-dimensional analysis. A similar study was performed on three-dimensional models and 100^3 nodes was identified as a suitable system size.

Conductivity Ratio

In the following, the effective thermal conductivity of random two-phase systems is analysed in two dimensions. Four different conductivity ratios $r = 0.1, 0.01, 0.001$ and 0 are considered. Figure 5 shows the evolution of the effective thermal conductivity over the phase fraction Φ_B. The conductivity is given in terms of λ_A and accordingly for $\Phi_B = 0$ ($\Phi_A = 1$) the value 1 is obtained. An increase of the

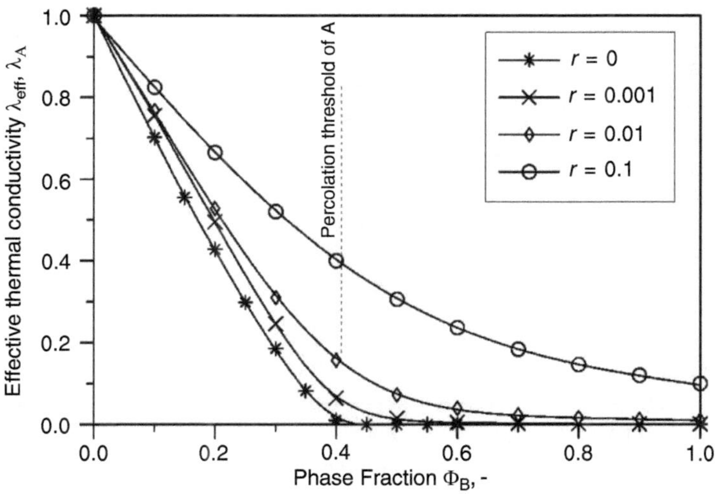

Fig. 5 Effective thermal conductivity of random two-phase systems (2D)

phase fraction Φ_B results in a decrease of λ_{eff} due to the lower thermal conductivity λ_B of phase B. The gradient of the curves is initially constant and its absolute value increases with the conductivity ratio r. The site percolation threshold for two-dimensional simple cubic structures ($\Phi_B = 1 - 0.59274621$ [20]) marks the transition to low effective thermal conductivities. Beyond this phase fraction, the conducting phase A fails to form interconnected long-range networks. Further increase of the phase fraction Φ_B shows fast convergence towards the effective thermal conductivity of species B: $\lambda_B = r\,\lambda_A$.

3.1.2 Three-Dimensional Analysis

Results of the three-dimensional analysis of randomly dispersed two-phase systems are shown in Fig. 6. A non-linear dependence of the effective thermal conductivity on the phase fraction Φ_B can be observed. Analogous to the two-dimensional analysis, the four conductivity ratios $r = 0.1$, 0.01, 0.001 and 0 are considered. The results for the minimum ratio $r = 0$ can be used to confirm the site percolation threshold of three-dimensional simple cubic structures: the best value has been estimated in the literature as 0.3116004 [21]. The effective thermal conductivity of two-phase materials becomes zero since the conducting phase A fails to form interconnected long-range networks.

For three-dimensional random structures, analytical models for the calculation of the effective thermal conductivity have been suggested in the literature. Figures 7 and 8 show a comparison of the LMC results with these analytical approaches for the conductivity rations $r = 0.1$ and 0.001. The effective thermal conductivity can

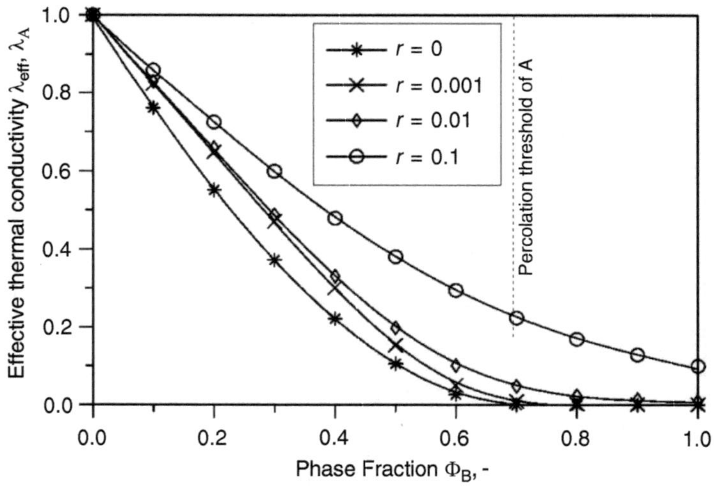

Fig. 6 Effective thermal conductivity of random two-phase systems (3D)

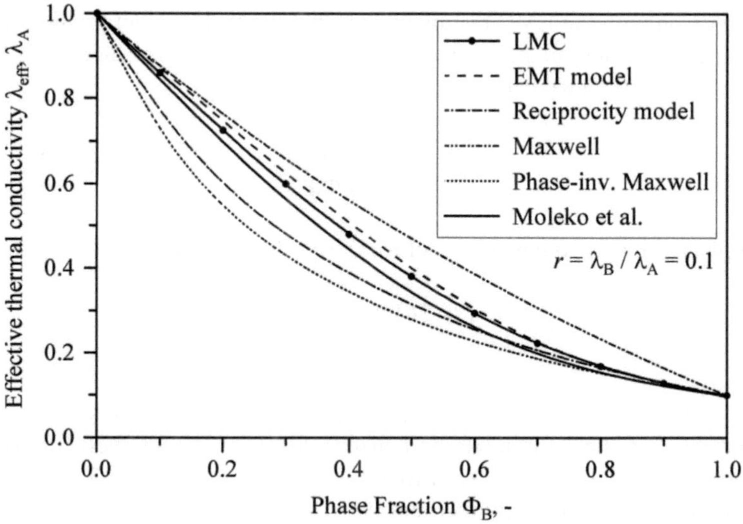

Fig. 7 Comparison between numerical simulation (LMC) and analytical models results for three-dimensional random distributions ($r = 0.1$)

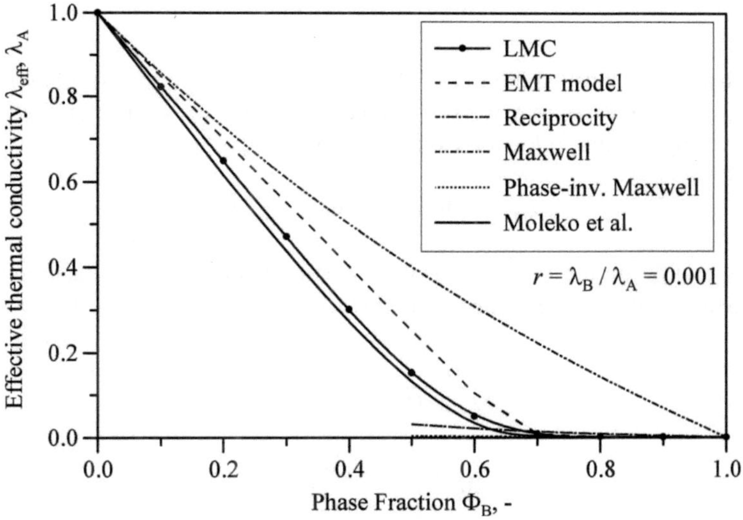

Fig. 8 Comparison between numerical simulation (LMC) and analytical models results for three-dimensional random distributions ($r = 0.001$)

be estimated using the effective medium theory (EMT) model [22] by the solution of the equation:

$$(1 - \Phi_B)\frac{\lambda_A - \lambda_{eff}}{\lambda_A + 2\lambda_{eff}} + \Phi_B \frac{\lambda_B - \lambda_{eff}}{\lambda_B + 2\lambda_{eff}} = 0. \tag{4}$$

Comparison with the LMC results reveals that the EMT model predicts systematically higher values, in particular for $r = 0.001$. A Maxwell approach for the range $0 \le \Phi_B \le 0.1$ [10] can be expressed as:

$$\lambda_{eff} = \frac{1 + 2\beta\Phi_B}{1 - \beta\Phi_B}\lambda_A \tag{5}$$

and the phase-inverse Maxwell model for $0.9 \le \Phi_B \le 1$ as:

$$\lambda_{eff} = \frac{(1 + 2\beta)(1 - \beta + 2\beta\Phi_B)}{(1 - \beta)(1 + 2\beta - \beta\Phi_B)}\lambda_A \tag{6}$$

with $\beta = (\alpha - 1)/(\alpha + 2)$ and $\alpha = \lambda_{dis}/\lambda_{con}$. The subscripts 'dis' and 'con' are abbreviations for the continuous and dispersed phase respectively. Reasonable agreement with the LMC results is found within these intervals. The reciprocity model based on the reciprocity theorem [9] is given by:

$$\lambda_{eff} = \left(\frac{1 + (\sqrt{\alpha} - 1)\Phi_B}{1 + \left(\sqrt{\frac{1}{\alpha}} - 1\right)\Phi_B}\right)\lambda_A. \tag{7}$$

Because of the close similarity of the current problem to the correlation effect problem during the random walks of tracer atoms on the corresponding crystal lattice, we suggest the following expression for the λ_{eff}:

$$\lambda_{eff} = (\Phi_A\lambda_A f_A + \Phi_B\lambda_B f_B)/f_0, \tag{8}$$

where f_A and f_B are the tracer correlation factors that can be calculated using the almost exact Moleko, Allnatt and Allnatt (MAA) diffusion kinetics theory [23]

$$f_A = \frac{H_A}{2\lambda_A + H_A}, f_B = \frac{H_B}{2\lambda_B + H_B}, \tag{9}$$

$$H_i = \lambda_i M_0 \frac{(\lambda_j + f_i\lambda_v)H + 2\lambda_i\lambda_j f_i}{(f_i\lambda_i + \lambda)H + 2\lambda_i\lambda f_i},$$

$$H_i = \lambda_i M_0 \frac{(\lambda_j + f_i\lambda_v)H + 2\lambda_i\lambda_j f_i}{(f_i\lambda_i + \lambda)H + 2\lambda_i\lambda f_i}, \tag{10}$$

where the function H is defined as a positive solution to the equation:

$$H^2 + 2\lambda H = M_0(\lambda_v H + 2\lambda_i\lambda_j), \tag{11}$$

where $\lambda = \Phi_B \lambda_A + \Phi_A \lambda_B$ and $\lambda_v = \Phi_A \lambda_A + \Phi_B \lambda_B$, $M_0 = 2f_0/(1 - f_0)$, and for the simple cubic lattice the geometric correlation factor $f_0 = 0.654....$ It can be shown that H/M_0, with H defined as a positive solution to (11) and with $f_0 = 2/3$ is identical to λ_{eff} given by the EMT (4).

In Figs. 7 and 8 a comparison between the LMC calculations and the analytical models is shown. Figure 7 shows results for the moderate value of $r = 0.1$, whereas Fig. 8 shows results for the small value of $r = 0.001$. It can be seen that for moderate r the EMT expression gives the best agreement and for small values of r the MAA model works the best. Overall, results shown in Figs. 7 and 8 indicate a very good *qualitative* agreement between the LMC data and the model derived from the MAA theory.

3.2 Modified Distribution

The creation of modified distributions is in principle similar to the procedure used for random distributions. However, a generation probability p is introduced as a function of the free surface area A_S. The general idea is to capture the generic structure of two-phase materials where a high surface energy causes 'clustering' of the growing phase. Figure 9 shows an example of normalized surface areas in a two-dimensional lattice. The matrix (initial phase) is represented by white squares and the growing phase with a high surface energy by black squares. Cursive numbers indicate that the surface area is altered due to periodic boundary conditions. Let us

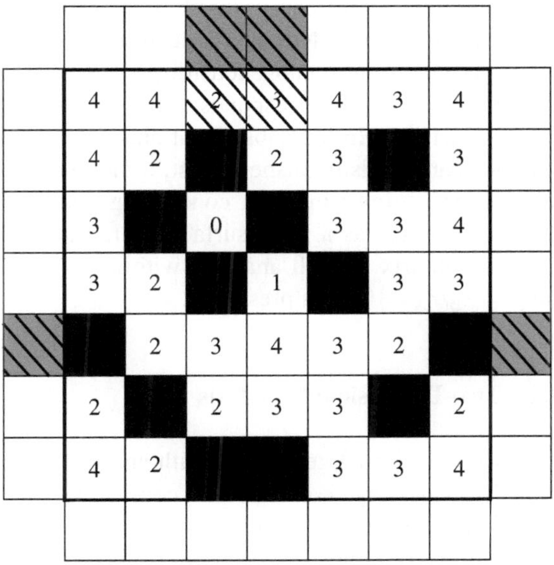

Fig. 9 Normalized surface areas A_S/s^2 in a two-dimensional lattice

consider a lattice site where all neighbors belong to the starting phase (i.e. are white squares). Thus, the surface area A_S exhibits its maximum value A_{max} which is 4 s^2 (two-dimensional) or 6 s^2 (three-dimensional), respectively. The symbol s is the distance between two adjoining lattice nodes (which are located in the centre of a square or cube). The generation probability $p(A_{max})$ assigned to the maximum surface area is called the initialization probability p_{init}.

In the compass of the present study, a linear probability function $p(A_S)$ is proposed. Considering the boundary conditions $p(0) = 1$ and $p(A_{max}) = p_{init}$ the following relation is obtained:

$$p(A_S) = \frac{p_{init} - 1}{A_{max}} \cdot A_S + 1. \tag{12}$$

Analogous to the algorithm used to generate random distributions (cf. Sect. 3.1), each lattice site is originally assigned to the matrix phase. A lattice site is then randomly chosen and the generation probability $p(A_S)$ is calculated in dependence on its surface energy. The result is compared to a uniformly distributed random number $0 \leq \chi < 1$. If the random number χ is smaller than the function value of p, the lattice site is assigned to the second phase. Otherwise, the site is rejected for the current attempt and a new lattice site is randomly chosen. The procedure is repeated until the target phase fraction is reached.

Figure 10 shows several examples of modified structures for different initialization properties p_{init}. A low value of p_{init} causes 'clustering' of the second species, since the phase is more likely to grow in the vicinity of seeds where the free surface area is reduced. In the limiting case, $p_{init} = 0$ a starting seed must be provided prior to the model generation and only one approximately circular (two-dimensional) or spherical (three-dimensional) cluster is formed. Deviations from this shape are caused by the finite system size used in the numerical analysis. An increase of p_{init} results in a more homogenous dispersion of the growing phase. The second limiting case $p_{init} = 1$ corresponds to the random distributions considered in the previous Sect. 3.1. In the subsequent analysis, two cases with different thermal properties can be distinguished. First, a thermally insulating phase with a large surface energy grows inside a conducting matrix. In the inverse problem, the conducting phase has a large surface energy and grows inside a thermally low conducting matrix. For all analyses within this Section, a constant conductivity ratio $r = \lambda_B/\lambda_A = 0.01$ is presumed.

3.2.1 Two-Dimensional Analysis

First, two-dimensional geometries with an insulating (λ_B) phase growing inside a thermally conducting (λ_A) matrix are considered. The results are shown in Fig. 11. In addition to the simulation of surface energy ($p_{init} < 1$), the results of Sect. 3.1.1 for random distributions ($p_{init} = 1$) are added for comparison. It can be observed that the random structures exhibit the smallest thermal conductivities among the

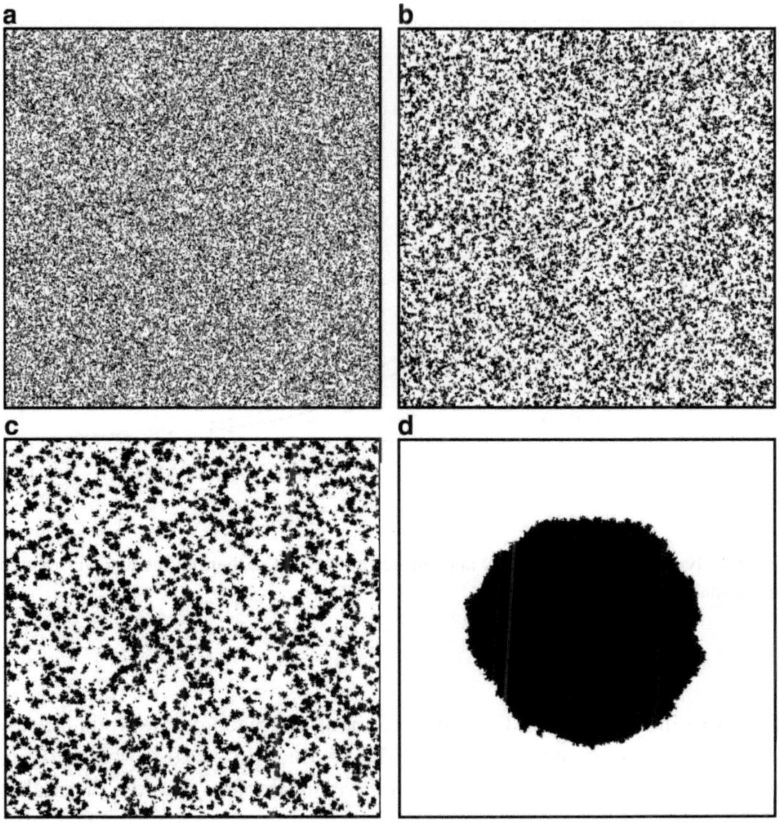

Fig. 10 Surface energy models ($\Phi_A = \Phi_B = 0.5$, resolution $1{,}000^2$ lattice nodes): (**a**) $p_{init} = 0.1$, (**b**) $p_{init} = 0.01$, (**c**) $p_{init} = 0.001$, (**d**) $p_{init} = 0$

considered geometries. A decrease of the initialization probability causes an increase of the effective thermal conductivity. A likely explanation is that the clustering of the thermal insulator B promotes the formation of long range networks of the conducting phase A. It will be shown in Sect. 5 that the percolation threshold is in fact affected by the initialization probability p_{init}. The minimum value $p_{init} = 0$ is considered as a special case: as indicated in Fig. 10, the insulating phase B (black pixels) forms a single approximately circular cluster inside the conducting matrix A (white pixels). In this case, the percolation threshold of phase B can be calculated by dividing the area of the circular inclusion (diameter d) by the area of the square formed by tangents of the circle: $\Phi_{perc,B} = \pi\,(d/2)^2/d^2 = \pi/4$. In a two-dimensional structure, only one percolating network can exist at any time. In other words, the conducting phase A cannot form any long-range networks beyond this threshold and as a consequence the effective thermal conductivity drops towards the lower values of the other structures. In addition to the LMC data, finite element results of two-dimensional models with spherical inclusions are added as square markers in

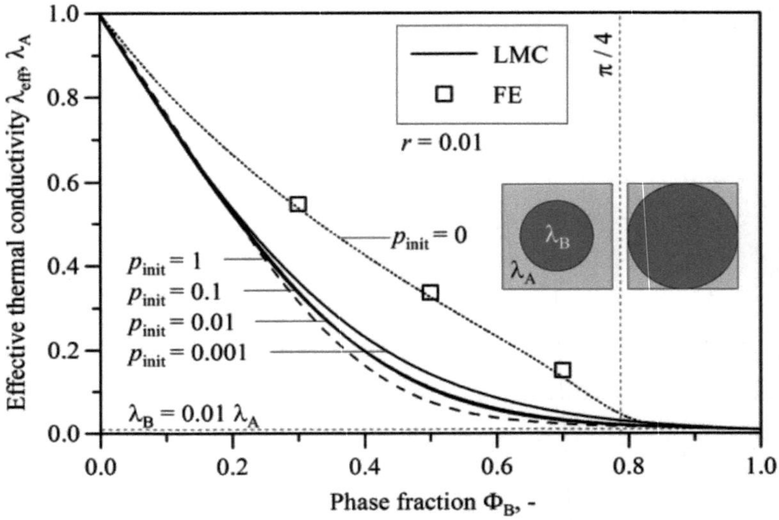

Fig. 11 Effective thermal conductivity and surface energy (2D): thermal insulator growing inside conducting matrix

Fig. 11. More information on this finite element approach can be found elsewhere [24]. An excellent agreement with the LMC results ($p_{\text{init}} = 0$) is obtained.

The second set of models addresses the *inverse* problem where a thermally high conducting phase (λ_A) grows within an insulating matrix (λ_B). The effective thermal conductivities are shown in Fig. 12. It is interesting to compare the results of the initialization probabilities $p_{\text{init}} = 0.1, 0.01, 0.001$ to random structures ($p_{\text{init}} = 1$): for phase fractions $\Phi_B > 0.31$ a small increase of the effective thermal conductivity can be observed. A likely explanation is that the surface energy causes clustering of the conducting phase which supports the formation of short-range networks. Interestingly, these structures exhibit slightly lower thermal conductivities than random structures for $\Phi_B < 0.31$. This phase fraction corresponds to the percolation threshold of the conducting phase in random distributions. However, it can also be noted that the impact of the parameter p_{init} (> 0) on the effective thermal conductivity is small and may be disregarded in most cases.

An interesting behavior is found for the limiting case $p_{\text{init}} = 0$. The conducting phase A is unable to form an interconnected long range network beyond $\Phi_B > 1 - \pi/4$, i.e. the circular clusters do not touch and the structures exhibit very low thermal conductivities. After percolation (i.e. $\Phi_B < 1 - \pi/4$), the thermal conductivity increases to the level of the other structures. Due to its importance for thermal conduction in composites, percolation will be considered separately in Sect. 5. Analogous to Fig. 11, finite element results (simulating geometries for $p_{\text{init}} = 0$) are indicated by square markers and a good agreement with the corresponding Lattice Monte Carlo results is found.

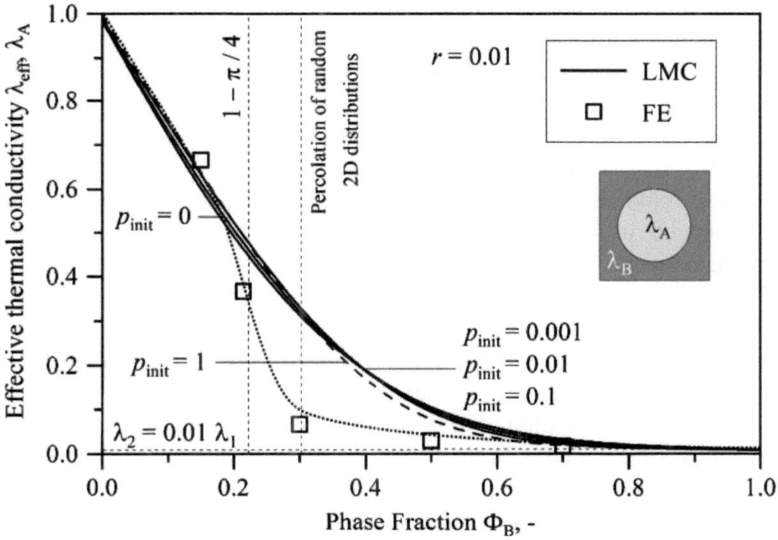

Fig. 12 Effective thermal conductivity and surface energy (2D): thermal conductor growing inside an insulating matrix

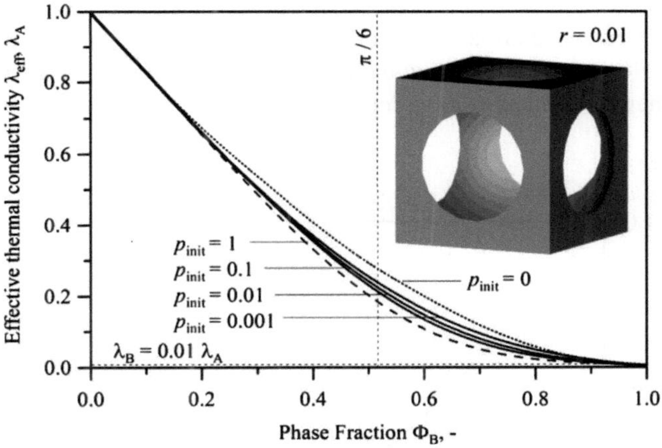

Fig. 13 Effective thermal conductivity and surface energy (3D): thermal insulator growing inside conducting matrix

3.2.2 Three-Dimensional Analysis

In this Subsection, three-dimensional analyses of the influence of surface energy on the effective thermal conductivity are performed. Analogous to the two-dimensional case, first a thermal insulator growing inside a conducting matrix is considered. The results are shown in Fig. 13. The random structure ($p_{\text{init}} = 1$)

exhibits the lowest conductivities at all phase fractions Φ_B. A decrease of the initialization probability causes clustering of the thermal insulator and as already observed in the two-dimensional structures promotes the formation of conducting long-range networks. As a consequence, the effective thermal conductivity increases. In the limiting case $p_{init} = 0$, the insulator forms spherical clusters inside the conducting matrix. At the phase fraction $\Phi_B = \pi/6$, the insulating phase B starts to percolate. Unlike the two-dimensional case no significant change in the effective thermal conductivity is observed. The explanation is that in three dimensions two percolating networks can coexist. An example for such a geometry is given in Fig. 13. Assuming a perfectly spherical growth of phase B, phase A percolates for all phase fractions $\Phi_B < 0.965$.

In Fig. 14, the *inverse* case (a thermal conductor growing inside an insulating matrix) is considered. As a reference, the effective thermal conductivities of random distributions are plotted as a dashed line. For phase fractions $\Phi_B < 0.4$, no strong dependence on the parameter p_{init} is found. In the range $0.4 < \Phi_B < 0.8$, an increase of the effective thermal conductivity with decreasing values of p_{init} (>0) can be observed. Similar to the two-dimensional structures, the two phases segregate due to the simulation of surface energy. As a result, the formation of conducting short-range networks is promoted resulting in a slightly increased conductivity. A different behavior is observed for the limiting case $p_{init} = 0$. At low phase fractions of the conductor $\Phi_A = 1 - \Phi_B$, clustering of phase A inhibits the connection of the conducting phase, i.e. neighboring spheres do not touch. Accordingly, the effective thermal conductivity is low in comparison to the other geometries. However, approaching the critical phase fraction $1 - \pi/6$, the (approximately) spherical clusters start to connect and the conductivity increases rapidly.

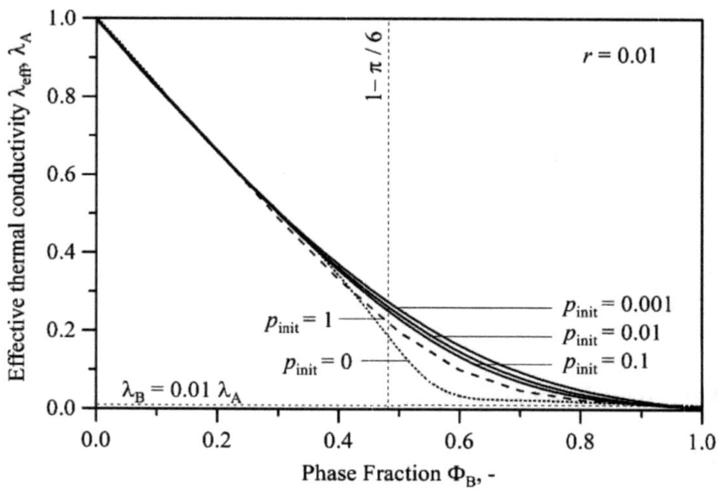

Fig. 14 Effective thermal conductivity and surface energy (3D): thermal conductor growing inside an insulating matrix

4 Three-Phase Systems

Due to the increased complexity of three-phase systems, only random distributions (cf. Sect. 3.1) are investigated. Three species with volume fractions Φ_A, Φ_B and Φ_C and corresponding thermal conductivities λ_A, λ_B and λ_C are considered. The thermal conductivities $\lambda_A > 0$, $\lambda_B = 0.5\ \lambda_A$ and $\lambda_C = 0.01\ \lambda_\varsigma$ are constant throughout the analysis. As in the previous Sections, values of the effective thermal conductivities are expressed as ratios of λ_ς.

4.1 *Two-Dimensional Three-Phase Systems*

First, two-dimensional models are considered. The results of the analysis are shown in the ternary plot Fig. 15a. Phase compositions of identical effective thermal conductivities (i.e. $\lambda_{eff} = 0.1\ \lambda_A$) are connected by lines. The circular markers in the graph correspond to results of the Lattice Monte Carlo calculations. It can be observed that all 'iso-conductivity' lines are in fact straight lines. Composites with a high fraction Φ_C exhibit low thermal conductivities due to the low value of λ_C. As required, the iso-line $\lambda_{eff} = 0.5\ \lambda_A$ intersects with 100% Φ_B. High effective thermal conductivities, i.e. $\lambda_{eff} > 0.9\ \lambda_A$ can only be observed for very high phase fractions Φ_A.

4.2 *Three-Dimensional Three-Phase Systems*

Next, three-dimensional three-phase systems are considered. The results are shown in the ternary plot Fig. 15b. Similar to Fig. 15a, phase compositions with identical effective thermal conductivities can be connected by straight lines. The comparison with the results of the two-dimensional analysis (cf. Fig. 15a) reveals distinctly higher effective thermal conductivities. A likely explanation is the formation of long-range networks (percolation) of the conducting phases (A and B) at relatively low phase fractions in comparison to the two-dimensional case. The percolation behavior of two-dimensional and three-dimensional structures is considered in the next Sect. 5 of this Chapter.

5 Percolation

The results of Sect. 3.2 indicate that the simulation of surface energy ($p_{init} < 1$) during the model generation distinctly changes the thermal properties of such structures. For identical phase compositions, different effective thermal conductivities are

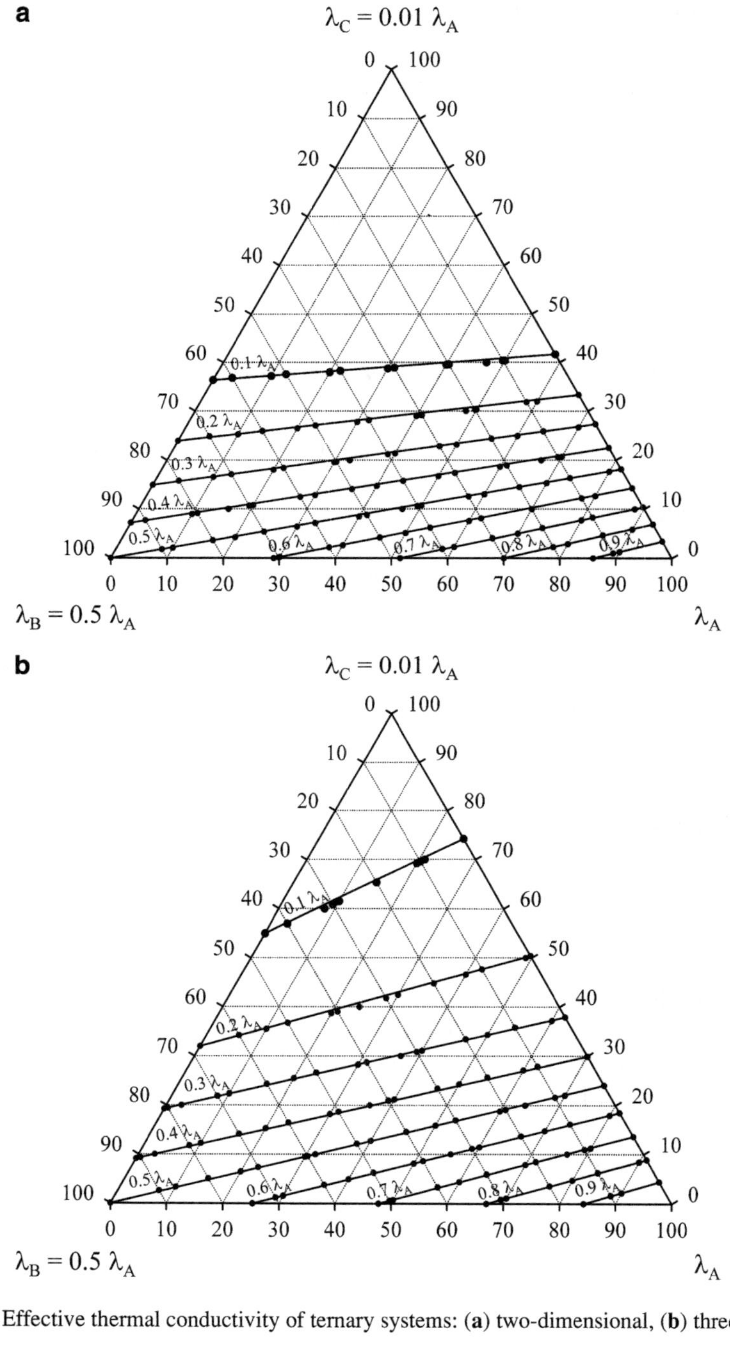

Fig. 15 Effective thermal conductivity of ternary systems: (**a**) two-dimensional, (**b**) three-dimensional

obtained. A likely explanation is a change of the percolation of the conducting phase threshold due to the surface energy. Let us consider a composite beyond this percolation threshold: no conducting long-range networks can be formed and the effective thermal conductivity exhibits low values in comparison to a percolating structure. This Section addresses percolation thresholds for random and modified two-phase structures. First, the algorithm used for the analysis is introduced. Second, the order of percolation and the influence of the system size are discussed. Afterwards, random distributions are considered and results are compared to values readily available in literature. In the final part of this Section, the dependence of the percolation threshold on the initialization probability p_{init} is investigated.

5.1 Percolation Algorithm

Figure 16 shows the algorithm used for percolation analysis on a simple two-dimensional lattice for one direction – the procedure for the orthogonal direction (s) is identical. The investigated phase, in the following referred to as species A, is represented by black boxes; all remaining phases are white boxes. In the first step, nodes of phase A inside the starting plane, i.e. $x = 1$, are identified (marked as light grey boxed in Fig. 16a. The selected nodes act as seeds for the following analysis. A seed is chosen and neighboring nodes of the same species A are marked as additional seeds (cf. Fig. 16b). It should be mentioned here that for the identification of neighbors, periodic boundary conditions are considered. In the next step, the

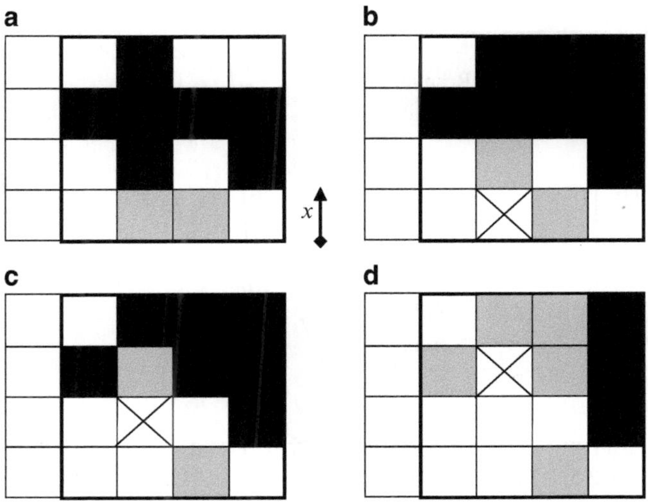

Fig. 16 Percolation algorithm on a two-dimensional lattice: investigated phase (*black*), other phases (*white*), active seeds (*light gray*), inactive seeds (*dark gray*)

previously used seed is erased, i.e. assigned to a different phase, so that it is not reused and thereby decreases the efficiency of the algorithm. Subsequently, a new seed is activated. In principal, any seed could be selected. However, in order to improve efficiency, the seed with the maximum distance to the starting plane is chosen. As before, neighboring nodes of phase A are registered as additional seeds and the old seed is erased (cf. Fig. 16c). This procedure is repeated until (1) the opposite plane is reached (percolation, cf. Fig. 16d) or (2) no more seeds are available (no percolation).

A range of different phase fractions is considered and for each fraction at least 100 different models are investigated. The amount of percolating structures divided by the total number of models is then used to calculate the percolation probability at the particular phase fraction. The obtained percolation probabilities are then plotted versus the phase fractions (for an example cf. Fig. 17).

5.2 *Order of Percolation*

In the case of *finite* systems a percolation order can be introduced. For simple cubic grid models, it is useful to distinguish three different orders (directions) of percolation. For first order, percolation is observed in at least one out of the three orthogonal x, y and z directions. Second and third order means percolation in two or three directions, respectively. In the context of thermal properties, first or second order percolation introduce anisotropic behavior. In Fig. 17, two different system sizes, namely 100^3 and 400^3 nodes are shown. Let us first consider the smaller

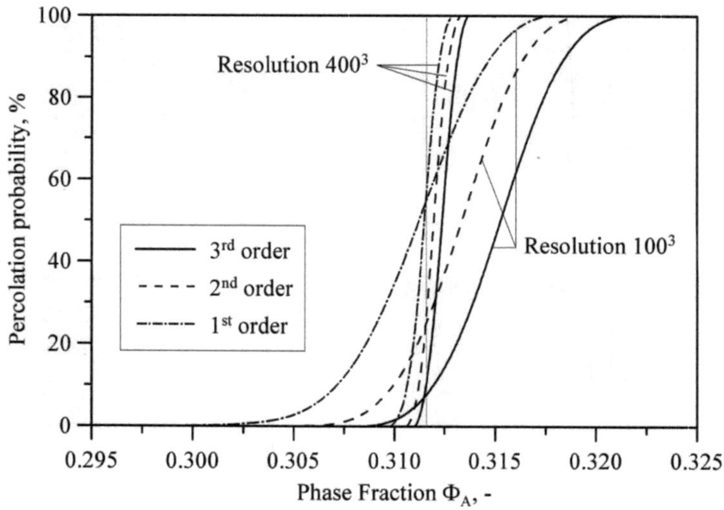

Fig. 17 Percolation order for system sizes 100^3 and 400^3

system. Percolation of the first order is already observed for relatively small phase fractions. The percolation probability exhibits values greater than zero for $\Phi_A > 0.302$. Second order percolation starts at $\Phi_A > 0.306$ and third order percolation only at $\Phi_A > 0.308$. The results for the larger system (400^3 nodes) indicate that the *n*th order ($n = 1, 2, 3$) percolation probabilities converge for increasing system size. In the case of an infinite system, the concept of percolation order becomes irrelevant and a singular percolation threshold of 0.3116004 [21] (dotted line) is obtained. As a consequence, only third order percolation on sufficiently large systems is considered in the following analysis.

5.3 System Size

The previous Subsection already indicates that the system size has a strong impact on the percolation behavior of finite systems. Therefore, the (third order) percolation behavior of models containing 100^3, 200^3, 400^3 and 800^3 nodes is now addressed. It can be observed that the percolation band, i.e. the range of phase fractions where the percolation probability is larger than 0 and smaller than 1, distinctly decreases with increasing system size. Convergence towards a step function at the reference solution (infinite system size) is observed. It is interesting to note that the solution of different system sizes intersects in close proximity to this reference solution. This behavior will be utilized in subsequent analysis: the percolation behavior of systems with the sizes 400^3 and 800^3 will be calculated and the phase fraction at the intersection of the probability curves will be assigned to the percolation threshold of the particular structure.

In principle, the results shown in Fig. 18 can also be interpreted as a ‘size effect’. Smaller structures exhibit a lower percolation probability (e.g. of a thermal conductor) at a particular phase fraction which will affect their thermal properties (e.g. decrease the effective conductivity).

5.4 Random Distributions

First, the percolation threshold of random distributions in simple cubic arrangements is considered. Figure 18 can be utilized to obtain the percolation threshold of three-dimensional structures. The probability curves for the system sizes 400^3 and 800^3 pixels intersect at $\Phi_A = 0.3116(4)$. The comparison with the reference solution given in literature with 0.3116004 [21] shows excellent agreement. The analysis is repeated for the two-dimensional case. Figure 19 shows the percolation probabilities for the system sizes 800^2, $1{,}600^2$, $3{,}200^2$ and $6{,}400^2$. Intersection of the curves assigned to the largest system sizes is observed at $\Phi_A = 0.5927(1)$ which corresponds to a deviation of less than 0.006% to the reference solution [20].

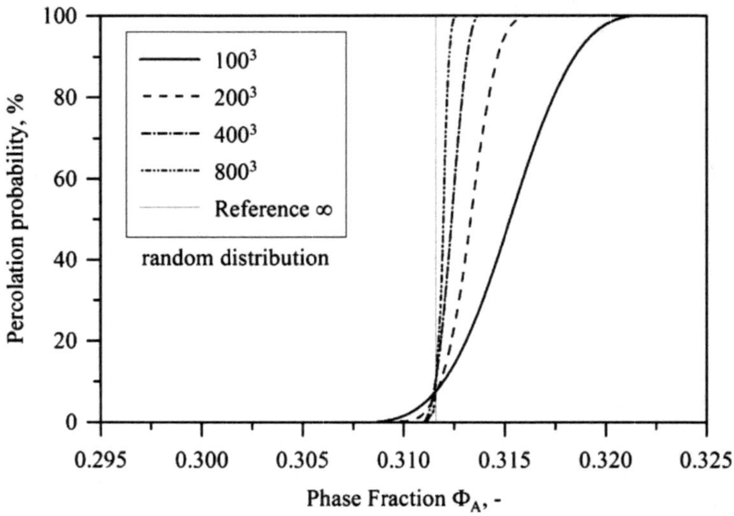

Fig. 18 Percolation behavior of random structures for different system sizes (3D)

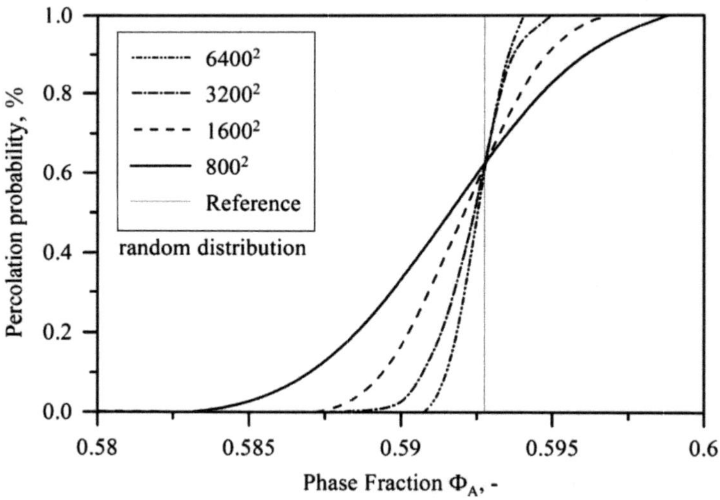

Fig. 19 Percolation behavior of random structures for different system sizes (2D)

The comparison of the two- and three-dimensional percolation thresholds reveals that three-dimensional structures percolate at lower phase fractions. This explains that three-dimensional structures exhibit higher thermal conductivities for similar phase compositions (for example cf. Fig. 15). Percolation of the conducting phase and therefore the formation of conducting long-range networks is achieved 'earlier' (i.e. at a lower phase fraction) than in two-dimensional structures resulting in higher effective conductivities.

5.5 *Modified Distributions*

In the following, the percolation behavior of modified distributions that simulate the influence of surface energy during the model generation is investigated.

5.5.1 Two-Dimensional Structures

First, the two-dimensional case is considered. In modified structures, a *new* phase (subscript 'N') with a high surface energy grows inside the *matrix* (subscript 'M'). The model generation is elucidated in Sect. 3.2 of this Chapter and controlled by the initialization probability p_{init}. Figure 10 shows the influence of the parameter on the geometry of the two-phase composites. Let us first consider the limiting cases: random structures ($p_{\text{init}} = 1$) were already addressed in the previous Sect. 5.4. The case $p_{\text{init}} = 0$ corresponds to a single circular cluster of the growing phase. Any deviation from the circular shape (cf. Fig. 10) is due to a limited model size of the calculation models. Accordingly, the percolation threshold of infinite structures can be calculated according to $\Phi_{\text{Perc}} = A_{\text{Circle}}/A_{\text{Square}} = \pi r^2/(2r)^2 = \pi/4$, which is the highest value among the considered composites (cf. Table 1). The percolation behavior of the structures with $p_{\text{init}} = 0.1$, 0.01 and 0.001 is obtained using the algorithm explained in Sect. 5.1. A decrease of the percolation threshold with decreasing initialization probabilities ($p_{\text{init}} > 0$) is observed. It is interesting to study the connection between percolation threshold and thermal properties: Fig. 12 shows results for a conducting phase growing inside a thermal insulating matrix. In the case of $p_{\text{init}} = 0$, a distinct decrease in the effective thermal conductivity is observed. Only after reaching the (high) percolation threshold of the conducting phase $\Phi_{\text{Perc,A}} = \pi/4$, the effective thermal conductivity increases towards the values of the random structures that percolate at a much lower fractions $\Phi_{\text{A}} = 0.3116$. Therefore, it is a likely conclusion that the decrease in thermal conductivity for $p_{\text{init}} = 0$ is caused by the high percolation threshold of the conducting phase.

The simulation of the surface energy of the growing phase also affects the percolation behaviour of the (remaining) matrix. Low values of p_{init} trigger strong clustering of the growing phase and cause the conservation of long range networks inside the matrix. This interdependence of clustering and percolation behaviour was already observed by Kikuchi [25]. This can be observed in Table 2 where the percolation threshold of the matrix continuously decreases with decreasing values of p_{init}. The percolation threshold of the limiting case $p_{\text{init}} = 0$ (circular clusters) is

Table 1 Percolation behaviour of the growing phase for different initialization probabilities p_{init} (2D)

p_{init}	Φ_{N}
0	$\pi/4 = 0.785$
0.001	0.5579
0.01	0.53273
0.1	0.54722
1	0.59271

Table 2 Percolation behaviour of the matrix phase for different initialization probabilities p_{init} (2D)

p_{init}	Φ_{M}
0	$1 - \pi/4 = 0.215$
0.001	0.4655
0.01	0.51700
0.1	0.55285
1	0.59271

Table 3 Percolation behaviour of the growing phase for different initialization probabilities p_{init} (3D)

p_{init}	Φ_{N}
0	$\pi/6 = 0.524$
0.001	0.1506
0.01	0.17930
0.1	0.24289
1	0.3116004 [21]

calculated according to $\Phi_{\text{Perc}} = (A_{\text{Square}} - A_{\text{Circle}})/A_{\text{Square}} = 1 - \pi/4$, all other values are obtained numerically. Again, it is of interest to study the interdependence of percolation behavior of the conducting phase and thermal properties. Figure 11 shows the effective thermal conductivities of structures where a thermal insulator grows inside a thermally conducting matrix. Comparison with Table 2 reveals that the effective thermal conductivity increases for low percolation thresholds of the conducting matrix, i.e. the maximum thermal conductivities are found for the minimum percolation threshold ($p_{\text{init}} = 0$, $\Phi_{\text{Perc}} = 0.215$).

5.5.2 Three-Dimensional Structures

Next, the three-dimensional case is considered. Table 3 shows the percolation behavior of the growing phase in modified structures. The largest percolation threshold is found for $p_{\text{init}} = 0$ where spherical clusters form inside the matrix phase. Assuming a perfectly spherical growth, geometrical analysis yields $\Phi_{\text{Perc}} = V_{\text{Sphere}}/V_{\text{Cube}} = 4/3\pi r^3/(2r)^3 = \pi/6$. For initialization probabilities larger than 0, a decrease of the percolation threshold in comparison to random structures is found. This behavior is very similar to the two-dimensional case (cf. Table 1). However, in three dimensions, distinctly lower percolation thresholds are obtained. The comparison of Fig. 14 and Table 3 reveals an increase of the effective thermal conductivity with decreasing values of the corresponding percolation threshold. This confirms the connection observed earlier for the two-dimensional case.

The percolation behaviour of the matrix phase is shown in Table 4. Analogous to the two-dimensional case (Table 2), the percolation threshold decreases with decreasing values of the initialization probability p_{init}. The smallest value is found for $p_{\text{init}} = 0$, where the growing phase forms spherical clusters inside the matrix phase. Assuming a perfect spherical growth (even after neighbouring clusters touch) geometrical analysis reveals a percolation threshold of 0.035. Again, comparison with the two-dimensional equivalent (cf. Table 2) reveals distinctly

Table 4 Percolation behaviour of the matrix phase for different initialization probabilities p_{init} (3D)

p_{init}	Φ_M
0	≈ 0.035
0.001	0.1248
0.01	0.18137
0.1	0.23818
1	0.3116004

lower percolation thresholds in three dimensions. The comparative analysis of Table 4 and Fig. 13 shows an increase of the effective thermal conductivity for decreasing percolation thresholds of the conducting matrix.

6 Conclusions

In this Chapter numerical simulations of thermal diffusion in multi-phase materials were performed. A Lattice Monte Carlo (LMC) method was used towards the analysis of two- and three-dimensional model structures. The considered composites are assembled by two or three phases, each exhibiting different thermal conductivities. First, a random distribution of phases was considered and the dependence of the effective thermal conductivity on the phase composition was investigated. The comparison between LMC results and analytical solutions showed good agreement, in particular for a model derived from the Moleko, Allnatt and Allnatt diffusion kinetics theory. The second part of this Chapter focused on a random-growth algorithm simulating the influence of surface energy on the formation of composite materials. The effective thermal conductivity of these structures was determined and distinct deviations from random structures (i.e. structures generated without the simulation of surface energy) were observed. In the final part of this Chapter, percolation analyses were performed on random and modified structures. A systematic connection between thermal properties and percolation behavior was found for two- and three-dimensional structures. A low percolation threshold of the conducting phase causes an increase in thermal conductivity and vice versa. The physical explanation is the existence of interconnected long-range networks that distinctly increase the thermal energy transfer. Three-dimensional structures exhibit lower percolation thresholds than of two-dimensional ones. Accordingly, they have higher effective thermal conductivities for similar phase compositions.

References

1. University of Magdeburg, Lab of Scanning Microscopy and Stereology
2. Tarnawski, V.R., Cleland, D.J., Corasaniti, S., et al.: Extension of soil thermal conductivity models to frozen meats with low and high fat content. Int. J. Refrig. **28**, 840–850 (2005)
3. Carson, J.K., Lovatt, S.J., Tanner, D.J., Cleland, A.C.: Predicting the effective thermal conductivity of unfrozen, porous foods. J. Food Eng. **75**, 297–307 (2006)

4. Sharma, A., Tyagi, V.V., Chen, C.R., Buddhi, D.: Review on thermal energy storage with phase change materials and applications. Renew. Sustain. Energy Rev. **13**, 318–345 (2009)
5. Fiedler, T., Öchsner, A., Belova, I.V., Murch, G.E.: Thermal conductivity enhancement of compact heat sinks using cellular metals. Defect Diffus. Forum **273–276**, 222–226 (2008)
6. Landauer, R.: The electrical resistance of binary metallic mixtures. J. Appl. Phys. **23**, 779–784 (1952)
7. Ben-Amoz, M.: The effective thermal properties of two phase solids. Int. J. Eng. Sci. **8**, 39–47 (1970)
8. Glatzmaier, G.C., Ramirez, W.F.: Use of volume averaging for the modelling of thermal properties of porous materials. Chem. Eng. Sci. **43**, 3157–3169 (1988)
9. Rio, J.A., Zimmerman, R.W., Dawe, R.A.: Formula for the conductivity of a two-component material based on the reciprocity theorem. Solid State Commun. **106**, 183–186 (1998)
10. Samantray, P.K., Karthikeyan, P., Reddy, K.S.: Estimating effective thermal conductivity of two-phase materials. Int. J. Heat Mass Transf. **49**, 4209–4219 (2006)
11. Karthikeyan, P., Reddy, K.S.: Effective conductivity estimation of binary metallic mixtures. Int. J. Therm. Sci. **46**, 419–425 (2006)
12. Wang, M., Pan, N., Wang, J., Chen, S.: Mesoscopic simulations of phase distribution effects on the effective thermal conductivity of microgranular porous media. J. Colloid Interface Sci. **311**, 562–570 (2007)
13. Wang, M., Pan, N.: Predictions of effective physical properties of complex multiphase materials. Mater. Sci. Eng. R **63**, 1–30 (2008)
14. Belova, I.V., Murch, G.E., Fiedler, T., Öchsner, A.: The Lattice Monte Carlo method for solving phenomenological mass and heat transport problems. Defect Diffus. Forum **279**, 13–22 (2008)
15. Fiedler, T., Öchsner, A., Belova, I.V., Murch, G.E.: Calculations of the effective thermal conductivity in a model of syntactic metallic hollow sphere structures using a Lattice Monte Carlo method. Defect Diffus. Forum **273–276**, 216–221 (2008)
16. Fiedler, T., Öchsner, A., Belova, I.V., Murch, G.E.: Recent advances in the prediction of the thermal properties of syntactic metallic hollow sphere structures. Adv. Eng. Mater. **10**, 269–273 (2008)
17. Fiedler, T., Solórzano, E., Garcia-Moreno, F., Öchsner, A., Belova, I.V., Murch, G.E.: Lattice Monte Carlo and experimental analyses of the thermal conductivity of random shaped cellular aluminium. Adv. Eng. Mater. (2008, submitted for publication)
18. Fiedler, T., Belova, I.V., Öchsner, A., Murch, G.E.: Non-linear calculations of transient thermal conduction in composite materials. Comput. Mater. Sci. **45**, 434–438 (2009)
19. Fiedler, T., Belova, I.V., Öchsner A., Murch G.E.: A Lattice Monte Carlo Analysis of Thermal Transport in Phase Change Materials. Defect Diffus. Forum **297–301**, 154–161 (2010)
20. Newman, M.E.J., Ziff, R.M., Zia, R.K.P.: Two-dimensional polymer networks near percolation. J. Phys. A Math. Theor. **41**, 1–7 (2008)
21. Jiří, Š., Nezbeda, I.: Percolation threshold parameters of fluids. Phys. Rev. E **79**, 041141–041147 (2009)
22. Bruggeman, D.A.G.: Berechnung verschiedener physikalischer Konstanten von heterogenen Substanzen. I. Dielektrizitätskonstanten und Leitfähigkeiten der Mischkörper aus isotropen Substanzen. Ann. Phys. (Berlin) **416**, 665–679 (1935)
23. Moleko, L.K., Allnatt, A.R., Allnatt, E.L.: A self-consistent theory of matter transport in a random lattice gas and some simulation results. Philos. Mag. A **59**, 141–160 (1989)
24. Fiedler, T., Pesetskaya, E., Öchsner, A., Grácio, J.: Numerical and analytical calculation of the orthotropic heat transfer properties of fibre-reinforced materials. Mater. Sci. Eng. **36**, 602–607 (2005)
25. Kikuchi, R.: Concept of the long-range order in percolation problems. J. Chem. Phys. **53**, 2713–2718 (1970)

Optimization of a Unit Periodic Cell in Lattice Block Materials Aimed at Thermo-Mechanical Applications

Pablo A. Muñoz-Rojas, Thiago A. Carniel, Emilio C.N. Silva and Andreas Öchsner

Abstract Lattice block materials (LBMs) are periodic cellular materials, made of truss-like unit cells, which usually present a significant enhancement in mechanical performance when compared to their parent material. This improvement is generally measured by their low weight to strength ratio but several other desirable properties can also be considered, including high capacity for kinetic energy absorption, enhanced vibrational and damping characteristics, acoustic noise attenuation, shear strength, fracture strength, and directional heat conduction or insulation. Using optimization techniques, it is possible to tailor LBMs for specific multifunctional needs, combining good performance in different, and sometimes competing, properties. This work presents a particular approach for a systematic design of unit periodic cells of LBMs aiming at enhanced simultaneous stiffness and heat transfer homogenized properties. The homogenization is developed using an asymptotic expansion in two scales, the unit cells are modeled using linear pin-jointed truss finite elements and the optimization algorithm employed is Sequential Linear Programming (SLP). Nodal coordinates and cross sectional areas might be adopted as design variables simultaneously and the necessary sensitivities are obtained analytically. Illustrative 2D and 3D examples are included.

P.A. Muñoz-Rojas (✉) and T.A. Carniel
Department of Mechanical Engineering, Center for Technological Sciences, Santa Catarina State University-UDESC, Campus Avelino Marcante, Joinville, SC 89223-100, Brazil
e-mail: pablo@joinville.udesc.br

E.C.N. Silva
Department of Mechatronics and Mechanical Systems Engineering Mechanical Engineering Building, University of São Paulo, São Paulo, SP 05508-900, Brazil

A. Öchsner
Faculty of Mechanical Engineering, Technical University of Malaysia, 81310 UTM Skudai, Johor, Malaysia

A. Öchsner and G.E. Murch (eds.), *Heat Transfer in Multi-Phase Materials*,
Adv Struct Mater 2, DOI 10.1007/8611_2010_32,

1 Introduction

Modern technological challenges increasingly demand new materials optimized for specific needs which oftentimes require competing properties. In this sense, for instance, a material can be required to be light and to show a high stiffness/conductivity ratio at the same time, or to support shear while being able to absorb impact energy efficiently. This kind of multifunctionality is best performed by porous or composite materials, which use the different properties of their constituents for an enhanced response to different types of simultaneous external loads [1].

Materials that present multifunctional response are observed in nature; for example, both complex and simple forms of life, like human bone and diatoms, exhibit cellular patterns which make them optimally suited for their tasks [2]. Examples of man-manufactured materials which try to exploit intelligently cellular structures for engineering applications are displayed in Fig. 1. Figure 1a shows the material used in a ceramic filter made of silicon carbide, alumina and zirconia. This application requires not only mechanical strength to withstand high temperature flow, but also to yield low pressure loss, erosion resistance, and chemical and thermal stability to avoid reaction with the molten metal being filtered. On the other hand, Figure 1b presents an open cell material, made up of truss-like unit cells which are distributed spatially in a periodic pattern. Materials formed by this kind of arrangement have been known as *lattice block materials* (LBMs), lattice-truss structures, lattice-block structures and cellular lattices.

LBMs are light and present good mechanical properties, particularly high stiffness and resistance to failure. Their greatest potential of use is foreseen in multifunctional applications ranging from automotive and aerospace components to biomedical, civil, sportive and domestic industry [3–6]. It is worth remarking that the development of high-precision manufacturing processes, such as rapid prototyping (e.g., Selective Laser Sintering, Digital Light Processing and Micro-stereolithography [7, 8]) has motivated growing industrial interest in truss microstructures aiming at high-performance applications. Practical high level engineering applications include sandwich beams and plates [9–11], photonic and phononic devices [12–14] and military equipment such as ship doors [15] and missiles [16]. At the same time, research on LBMs has gained attention from the scientific community, focusing on aspects such as their analysis at macro and microscopic levels [2, 17–21], further manufacturing possibilities [22–25], and their optimized design [6, 26–28].

Studies on well suited topologies for LBMs have a classical reference in the work of Gibson and Ashby [3]. More recently, Luxner et al. [17] have studied effective properties, elastoplastic behavior and localization phenomenon in LBMs resulting from different unit cell topologies, such as those shown in Fig. 2.

Manufacturing methods of LBMs are at present an active research field. Different technologies have been devised and continue to be proposed, so that an apparently complicated layout or topology of a particular unit cell should not be

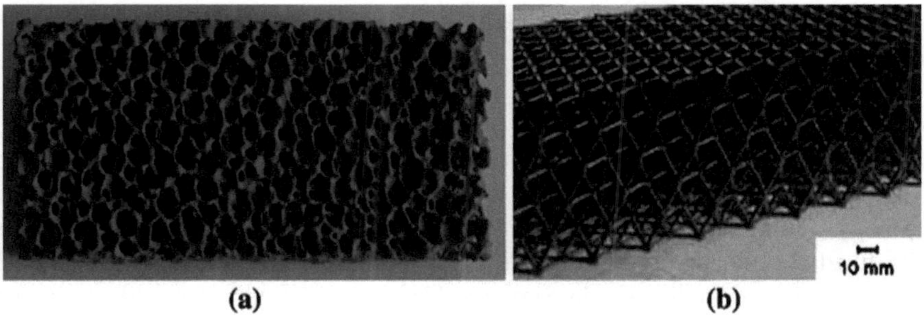

Fig. 1 Open cell structures. **a** ceramic foam filter used to remove impurities from liquid metals in casting [29], **b** Lattice Block Material made up of an aluminum alloy [26]

disregarded for production, although certainly an aspect of concern. As a consequence of these new manufacturing methods, efforts have been placed in the study of the structural behavior of LBMs resulting from particular fabrication procedures [31–33]. On the other hand, the possibility to manufacture a wide range of unit cell configurations opens the way for the application of optimization techniques to determine ideal layouts and topologies for LBMs withstanding multifunctional requirements. The optimization of a LBM can be defined as the determination of the best geometrical configuration of its unit cell, in order to maximize or minimize a given functional, or *multifunctional*, criterion. Techniques such as inverse analysis can be adopted for this purpose, although oftentimes the problem does not present solution unicity. In general, the functional criterion to be extremized corresponds to a macroscopic, average, or homogenized property. Homogenization for periodic microstructures is a well established theory, and classical references include Sanchez-Palencia [34] and Hassani and Hinton [35].

Hyun and Torquato [36] used standard topology optimization based on the Simple Isotropic Material with Penalization (SIMP) approach to obtain easily manufacturable two dimensional isotropic optimal cellular solids for effective bulk and shear moduli over the entire density range. Although their formulation was based in 2D elasticity, their results led to the conclusion that at intermediate densities, the optimal structures are given by Kagomé lattices. The authors suggest the possibilities of multifunctional applications of Kagomé structures based on

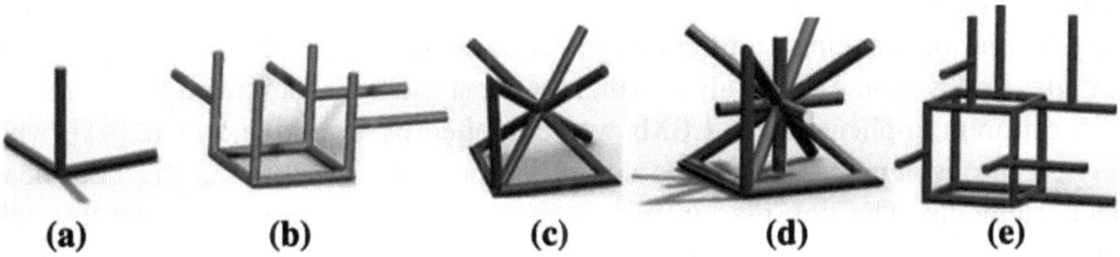

Fig. 2 Unit cells of periodic materials studied by [30] **a** Simple cubic (SC), **b** Translated simple cubic (TSC), **c** Body centered cubic (BCC), **d** Reinforced body centered cubic RBCC), **e** Gibson Ashby (GA)

their superior strength to elastic buckling when compared to triangular-like cell structures.

Sigmund [37] pioneered the use of homogenization methods to obtain and optimize macroscopic constitutive properties dealing directly with truss-like microstructures, i.e., employing truss and frame finite elements. His work intended to design materials with prescribed constitutive tensors. He adopted the usual topology optimization approach for trusses, in which a ground structure of bars with fixed coordinates is taken as the initial design, and only the cross sectional areas are allowed to change. In Sigmund [38] truss-like and frame-like cells are compared to optimize macroscopic constitutive properties. No apparent difference was observed in the topology of the optimized cells. Hence, it was assumed that truss elements can be used for the optimizing procedure with great computational advantage, especially in 3D cases, where the number of degrees of freedom per element in frames is twice that of the truss case. Further, Deshpande et al. [39] showed that stretching-dominated cellular solids are more weight efficient for structural applications than bending-dominated ones. In this regard, Evans et al. [40] and Suralvo et al. [24] mention that the absence of bending allows the stiffness and strength to vary linearly with relative density of the cell. In stretching-dominated structures the resultant strength-to-weight and stiffness-to-weight ratios are improved and the relative density can be reduced to as low as 2–3% [24]. Therefore, preliminary use of pin-jointed trusses in optimization is justified, although qualitative results obtained should be verified using cells formed by frame elements.

Yan et al. [26] presented an approach which optimized a LBM unit cell, having only nodal coordinates as design variables. They developed the formulation for 3D and 2D elasticity but presented results only for plane stress. The example shown concerned the maximization of the homogenized elastic tensor component associated to shear, that is, E_{1212}. The expressions for analytical sensitivities were detailed. More recently, Lippermann et al. [27] made an original contribution optimizing unit cells made of Euler–Bernoulli beam elements with respect to a resistance criterion at the microscale, which is known as stress localization. In this case there was no need for homogenized properties. They particularized the cross sections to be rectangular with one of the dimensions equal to unity. Thus, they adopted as design variables both the free dimension defining the cross section and nodal coordinates. 2D applications were presented showing the presence of local minima. Starting from different initial designs, the results converged systematically to one of the three configurations: squares with two diagonals, equilateral triangles and Kagomé grids, all of which being stretching-dominated structures.

A different application for LBMs was exploited by Prasad and Diaz [41], who devised the design of 2D bistable compliant periodic structures using nonlinear beam elements. The key idea in this case was to make use of the snap-through behavior of the slender beam elements in the unit cell. This concept had been presented before by Ohsaki and Nishiwaki [42], although restricted to the shape or layout design of regular compliant mechanisms, rather than to the design of a periodic material with bistable properties.

Dede and Hulbert [43, 44] introduced an original technique to optimize periodic lattice structures in order to obtain enhanced vibro-acoustic characteristics. The authors employed topology optimization aiming to design frame-made unit cells acting as compliant mechanisms to be integrated in large scale structures rendering improved dynamic response. To solve this problem, due to the existence of many local minima, the authors adopted a genetic algorithm. The possibility to obtain multifunctional properties exploiting lattice structures made of non-metallic materials, such as polymers, is highlighted in their work.

Gonella et al. [13] introduced a novel idea to design multifunctional LBMs combining ultralight characteristics, superior mechanical wave filtering properties and energy harvesting capabilities. The key issue is to introduce piezoelectric cantilevers in the unit lattice cells, so that, in addition to a bandgap desired behavior, the structure allows the conversion of localized kinetic energy into electrical energy. Foreseen applications include self-powered microelectromechanical systems (MEMS).

The present Chapter has its focus in the layout optimization of LBMs for enhanced response to simultaneous thermal and mechanical loading. Linear pin-jointed 3D bar elements are employed. Design variables include both, size (cross sectional areas) and shape (nodal coordinates) parameters simultaneously, thus reducing the number of bars needed in a ground structure. As explained above, one of the few research works to adopt this approach was presented by Lippermann et al. [27] but in a different context. All the sensitivities are developed analytically and the optimization is performed using Sequential Linear Programming (SLP). The formulation developed comprehends 3D and 2D structures. Examples of both cases are included.

The outline of the Chapter is as follows: Sect. 2 reviews briefly the concept of homogenization of materials with periodic microstructure using an asymptotic expansion; Sects. 3 and 4 particularize the application of this theory to obtain macroscopic thermal and mechanical properties of a continuous body based on a truss-like 3D unit cell; Sect. 5 shows the procedure adopted for applying periodic boundary conditions; Sect. 6 discusses the use of area and coordinate design variables at the same time; Sect. 7 defines the optimization problems, including the proposed multifunctional objective functions; Sect. 8 describes briefly the analytical sensitivity analysis approach adopted; Sect. 9 shows 2D and 3D numerical results; and Sect. 10 synthesizes the most important aspects of the work.

2 Homogenization of Materials with Periodic Microstructure

Among the wide range of existing materials, a particular group which exhibits regular and periodic structure can be identified. This means that any physical, mechanical or thermal property must obey the relation

$$F(\boldsymbol{x} + \boldsymbol{NY}) = F(\boldsymbol{x}) \tag{1}$$

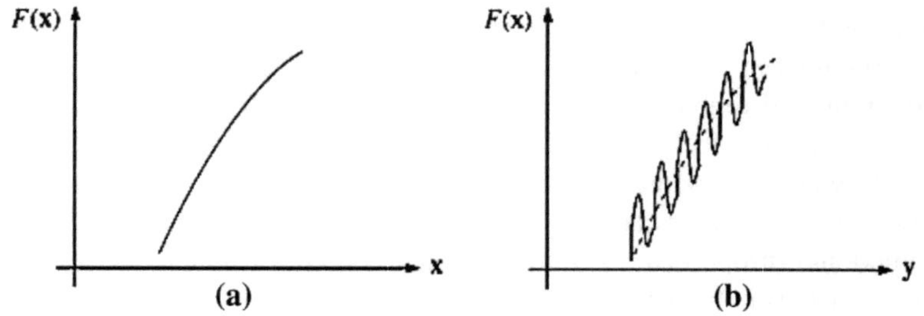

Fig. 3 Behavior of property *F(x)*: **a** global, macroscopic or average behavior and **b** local or microscopic behavior

where F is the property, $\boldsymbol{x} = [x_1, x_2, x_3]^T$ is the position vector of a given point, $\boldsymbol{N}$ is a 3 3 diagonal matrix of arbitrary integer numbers

$$\boldsymbol{N} = \begin{bmatrix} n_1 & 0 & 0 \\ 0 & n_2 & 0 \\ 0 & 0 & n_3 \end{bmatrix} \tag{2}$$

and $\boldsymbol{Y} = [Y_1, Y_2, Y_3]^T$ is a constant vector which determines the period of the structure, i.e., the dimension of the base cell. The period is usually very small when compared to the dimension of the global domain. Hence, the average physical, mechanical and thermal properties suffer rapid oscillations in the neighborhood of a given point $\boldsymbol{x}$, as shown in Fig. 3.

Figure 3 displays the same function in two scales: (a) a global, or macroscopic one, where the coordinates are given by $\boldsymbol{x}$; and a local, or microscopic one, where the coordinates are given by $\boldsymbol{y}$. Both scales are related by Eq. 3 through the parameter ε, which is typically very small.

$$\boldsymbol{y} = \frac{\boldsymbol{x}}{\varepsilon} \tag{3}$$

The macroscopic behavior can be obtained from the microscopic one using homogenization theory based on an asymptotic expansion in two (or more) scales.

3 Homogenization of Mechanical Properties

In this Section the basic homogenization equations for mechanical properties will be reviewed in the special perspective of a truss-like unit cell. The notation will follow Fonseca [45]. A similar, though more concise development can be found in Yan et al. [26].

Considering that the displacement field is expanded in two scales, one has

$$\boldsymbol{u}(\boldsymbol{x},\boldsymbol{y}) = \boldsymbol{u_0}(\boldsymbol{x}) + \varepsilon \boldsymbol{u_1}(\boldsymbol{x},\boldsymbol{y}) \tag{4}$$

where $\boldsymbol{u}_1$ is periodic in the dimension $\boldsymbol{Y}$ of the cell and the lower indexes "0" and "1" refer to the macro and microscale contributions, respectively.

If $\Phi = \Phi(\boldsymbol{x}, \boldsymbol{y})$ is a function where $\boldsymbol{y}$ depends implicitly of $\boldsymbol{x}$, it follows that

$$\frac{d\Phi}{d\boldsymbol{x}} = \frac{\partial \Phi}{\partial \boldsymbol{x}} + \frac{\partial \Phi}{\partial \boldsymbol{y}} \frac{d\boldsymbol{y}}{d\boldsymbol{x}} \tag{5}$$

and replacing Eq. 3 into Eq. 5 one gets

$$\frac{d\Phi}{d\boldsymbol{x}} = \frac{\partial \Phi}{\partial \boldsymbol{x}} + \frac{1}{\varepsilon} \frac{\partial \Phi}{\partial \boldsymbol{y}} \tag{6}$$

The stress–strain and strain–displacement relations are given respectively by

$$\sigma_{ij} = E_{ijkl} e_{kl} \tag{7}$$

$$\boldsymbol{e}_{kl} = \frac{1}{2} \left(\frac{\partial \boldsymbol{u}_k}{\partial \boldsymbol{x}_l} + \frac{\partial \boldsymbol{u}_l}{\partial \boldsymbol{x}_k} \right) \tag{8}$$

for which a more convenient notation is adopted, as given in Eqs. 9 and 10.

$$\sigma = \boldsymbol{E} \cdot \boldsymbol{e} \tag{9}$$

$$\partial_x = \frac{1}{2} \left(\frac{\partial (\cdot)_k}{\partial x_l} + \frac{\partial (\cdot)_l}{\partial x_k} \right) \tag{10}$$

The principle of virtual work is given by

$$\int_{\Omega} \sigma \cdot \delta e d\Omega - \int_{\Omega} \boldsymbol{b} \cdot \delta u d\Omega - \int_{\Gamma} \boldsymbol{t} \cdot \delta u d\Gamma = 0 \quad \forall \delta \boldsymbol{u} \in V_{\Omega} \tag{11}$$

where $\boldsymbol{b}$ is the body force, $\boldsymbol{t}$ is the surface force acting on Γ (boundaries of the domain Ω), $\boldsymbol{u}$ is the displacement suffered by the structure and V_{Ω} is the set of kinematically admissible displacements. It is assumed that there are no tractions acting on the internal borders of the base cell.

Using the same expansion applied in Eq. 4, it follows that $\boldsymbol{e}$ and $\delta \boldsymbol{e}$ can be written as

$$\boldsymbol{e} = \partial_x \boldsymbol{u}_0 + \varepsilon \partial_x \boldsymbol{u}_1 + \partial_y \boldsymbol{u}_1 \tag{12}$$

$$\delta \boldsymbol{e} = \partial_x \delta \boldsymbol{u}_0 + \varepsilon \partial_x \delta \boldsymbol{u}_1 + \partial_y \delta \boldsymbol{u}_1 \tag{13}$$

which, considering $\varepsilon \rightarrow 0$, reduce to

$$\boldsymbol{e} = \partial_x \boldsymbol{u}_0 + \partial_y \boldsymbol{u}_1 \tag{14}$$

$$\delta \boldsymbol{e} = \partial_x \delta \boldsymbol{u}_0 + \partial_y \delta \boldsymbol{u}_1 \tag{15}$$

Replacing Eqs. 9, 14 and 15 in the first term of Eq. 11, this part can be rewritten as

$$\int_{\Omega} \left(\partial_x \boldsymbol{u}_0 + \partial_y \boldsymbol{u}_1\right) \cdot \boldsymbol{E} \cdot \left(\partial_x \delta \boldsymbol{u}_0 + \partial_y \delta \boldsymbol{u}_1\right) d\Omega \tag{16}$$

which, considering the existence of periodic cells in the domain, becomes

$$\int_{\Omega} \frac{1}{|\boldsymbol{Y}|} \int_{Y} \left(\partial_x \boldsymbol{u}_0 + \partial_y \boldsymbol{u}_1\right) \cdot \boldsymbol{E} \cdot \left(\partial_x \delta \boldsymbol{u}_0 + \partial_y \delta \boldsymbol{u}_1\right) dY d\Omega \tag{17}$$

where Y is the cell period and $|\boldsymbol{Y}|$ is the total volume of the cell. In Eq. 17, the equality

$$\int_{\Omega} \Phi(\boldsymbol{x},\boldsymbol{y}) d\Omega = \frac{1}{|\boldsymbol{Y}|} \int_{\Omega} \int_{Y} \Phi(\boldsymbol{x},\boldsymbol{y})\, dY d\Omega \tag{18}$$

was used.

Since $\delta \boldsymbol{u}$ is arbitrary, one can choose $\delta \boldsymbol{u}_1 = \boldsymbol{0}$ and $\delta \boldsymbol{u}_0 \neq \boldsymbol{0}$, particularizing Eq. 17 to

$$\int_{\Omega} \frac{1}{|\boldsymbol{Y}|} \partial_x \delta \boldsymbol{u}_0 \int_{Y} \boldsymbol{E} \cdot \left(\partial_x \boldsymbol{u}_0 + \partial_y \boldsymbol{u}_1\right) dY d\Omega \tag{19}$$

which corresponds to the macroscopic term in Eq. 17.

On the other hand, choosing $\delta \boldsymbol{u}_0 = \boldsymbol{0}$ and $\delta \boldsymbol{u}_1 \neq \boldsymbol{0}$ in Eq. 17, it yields

$$\int_{\Omega} \frac{1}{|\boldsymbol{Y}|} \int_{Y} \partial_y \delta \boldsymbol{u}_1 \cdot \boldsymbol{E} \cdot \left(\partial_x \boldsymbol{u}_0 + \partial_y \boldsymbol{u}_1\right) dY d\Omega \quad \forall \delta \boldsymbol{u}_1 \quad \text{periodic } in\ Y \tag{20}$$

which corresponds to the microscopic term in Eq. 17.

In order to account for the external loading in the macroscopic part of the principle of virtual work contained in Eq. 11, it is considered that body forces obey the relation

$$\int_{\Omega} \boldsymbol{b} \cdot \delta \boldsymbol{u} d\Omega = \int_{\Omega} \langle \boldsymbol{b} \rangle \cdot \delta \boldsymbol{u}_0 d\Omega \tag{21}$$

where $\langle \cdot \rangle$ is the average of $(\cdot)$ in the cell, that is,

$$\langle \cdot \rangle = \frac{1}{|\boldsymbol{Y}|} \int_{Y} (\cdot) dY \tag{22}$$

At the same time, by hypothesis, no surface loads are assumed at the microscopic scale, resulting in

$$\int_{\Gamma} \boldsymbol{t} \cdot \delta \boldsymbol{u} d\Gamma = \int_{\Gamma} \boldsymbol{t} \cdot \delta \boldsymbol{u}_0 d\Gamma \tag{23}$$

Thus, the macroscopic equation can be expressed as

$$\int_{\Omega} \frac{1}{|\boldsymbol{Y}|} \partial_x \delta \boldsymbol{u}_0 \int_{Y} \boldsymbol{E} \cdot \left(\partial_x \boldsymbol{u}_0 + \partial_y \boldsymbol{u}_1\right) dY d\Omega - \int_{\Omega} \langle \boldsymbol{b} \rangle \cdot \delta \boldsymbol{u}_0 d\Omega - \int_{\Gamma} \boldsymbol{t} \cdot \delta \boldsymbol{u}_0 d\Gamma = 0 \tag{24}$$

while the microscopic equation is given by

$$\int_{\Omega} \frac{1}{|\boldsymbol{Y}|} \int_{Y} \partial_y \delta \boldsymbol{u}_1 \cdot \boldsymbol{E} \cdot \left(\partial_x \boldsymbol{u}_0 + \partial_y \boldsymbol{u}_1\right) dY d\Omega = 0 \tag{25}$$

At this point, it is possible to adopt the convenient separation of variables

$$\boldsymbol{u}_{1p} = -\chi_p^{kl}(\boldsymbol{x}, \boldsymbol{y}) \partial_x \boldsymbol{u}_0^{kl}(\boldsymbol{x}) \tag{26}$$

which, introduced in Eq. 14, provides

$$\boldsymbol{e} = \partial_x \boldsymbol{u}_0 + \partial_y \boldsymbol{u}_1 = \partial_x \boldsymbol{u}_0 - \partial_y \chi \, \partial_x \boldsymbol{u}_0 \tag{27}$$

where χ are the characteristic displacements of the unit cell, to be determined in a later stage.

Replacing Eq. 27 into Eq. 25, results in

$$\int_{\Omega} \frac{1}{|\boldsymbol{Y}|} \partial_x \boldsymbol{u}_0 \cdot \int_{Y} \partial_y \delta \boldsymbol{u}_1 \cdot \boldsymbol{E} \cdot \left(\boldsymbol{I} - \partial_y \chi\right) dY d\Omega = 0 \tag{28}$$

and the satisfaction of Eq. 28 implies that

$$\int_{Y} \partial_y \delta \boldsymbol{u}_1 \cdot \boldsymbol{E} \cdot \partial_y \chi dY = \int_{Y} \partial_y \delta \boldsymbol{u}_1 \cdot \boldsymbol{E} \cdot \boldsymbol{I} dY \tag{29}$$

where $\boldsymbol{I}$ is the 4th order identity tensor.

It is possible to find a family of functions χ (according to an additive constant) which satisfy Eq. 29. However, since the equations developed require only the derivatives of χ, given by $\partial_y \chi$, the value of the constant is indifferent. Therefore, by virtue of Eq. 27, the macroscopic equation given by Eq. 24 can be solved to yield

$$\int_{\Omega} \partial_x \delta \boldsymbol{u}_0 \cdot \boldsymbol{E}^H \cdot \partial_x \boldsymbol{u}_0 d\Omega - \int_{\Omega} \langle \boldsymbol{b} \rangle \cdot \delta \boldsymbol{u}_0 d\Omega - \int_{\Gamma} \boldsymbol{t} \cdot \delta \boldsymbol{u}_0 d\Gamma = 0 \tag{30}$$

where

$$\boldsymbol{E}^H = \frac{1}{|\boldsymbol{Y}|}\int\limits_Y \boldsymbol{E}\cdot(\boldsymbol{I} - \partial_y\chi)dY \tag{31}$$

is the homogenized constitutive tensor which, following Hassani et al. [35, 46], can be rewritten in indicial notation as

$$E^H_{ijkl}(\boldsymbol{x}) = \frac{1}{|\boldsymbol{Y}|}\int\limits_Y \left(E_{ijkl} - E_{ijpq}\frac{\partial \chi_p^{kl}}{\partial y_q}\right)dY \tag{32}$$

or in a compact form

$$\boldsymbol{D}^H(\boldsymbol{x}) = \frac{1}{|\boldsymbol{Y}|}\int\limits_Y \left(\boldsymbol{D} - \boldsymbol{D}\left[\varepsilon^{(1)} \quad \cdots \quad \varepsilon^{(NLC)}\right]\right)dY \tag{33}$$

that is,

$$D^{\boldsymbol{H}}_{ij}(\boldsymbol{x}) = \frac{1}{|\boldsymbol{Y}|}\int\limits_Y \left(D_{ij} - D_{ik}\varepsilon_k^{(j)}\right)dY \tag{34}$$

where D_{ij} is the constitutive tensor organized in compact *matrix* notation, $\boldsymbol{\varepsilon}^{(\beta)}$ is the strain field (organized in *vector* form) resulting from the *kl* "load case" of Eq. 29 (see Eqs. 32 and 36), and *NLC* stands for the number of columns of $\boldsymbol{D}$. The compact and full notations are related by the following replacements:

$$\begin{aligned} \sigma_1 = \sigma_{11}, \sigma_2 = \sigma_{22}, \sigma_3 = \sigma_{33}, \sigma_4 = \sigma_{12}, \sigma_5 = \sigma_{23}, \sigma_6 = \sigma_{31} \\ \varepsilon_1 = e_{11}, \varepsilon_2 = e_{22}, \varepsilon_3 = e_{33}, \\ \varepsilon_4 = 2e_{12} = \gamma_{12}, \varepsilon_5 = 2e_{23} = \gamma_{23}, \varepsilon_6 = 2e_{31} = \gamma_{31} \end{aligned} \tag{35}$$

and

$$D_{(\alpha)(\beta)} = E_{(ij)(kl)} \tag{36}$$

where the correspondence between Greek and Latin subscripts is given in Table 1.

It is important to notice that the resulting micro and macroscopic problems are uncoupled and their solution can be summarized as follows:

1. Find $\boldsymbol{\chi}$ solving Eq. 29 and determine $\partial_y\boldsymbol{\chi}$;
2. Calculate the homogenized tensor using Eq. 31;
3. Build the macroscopic equation given by Eq. 30.

Table 1 Correspondence of indexes in tensorial and compact notations

ij or *kl*	11	22	33	12	23	31
α or β	1	2	3	4	5	6

3.1 Numerical Determination of χ

In order to find χ, Eq. 29 may be solved numerically using the Finite Element Method, as first proposed by Guedes and Kikuchi [47]. The present particular approach considers truss-like unit cells, so that the discretization employs 3D linear bar elements. Following the usual procedure yields the system of equations below, also described by Yan et al. [26],

$$\boldsymbol{K}\chi = \boldsymbol{P} \tag{37}$$

where

$$\boldsymbol{K} = \sum_{e} \int_{\Omega^e} \boldsymbol{B}^T \boldsymbol{D} \boldsymbol{B} dY^e \tag{38}$$

and

$$\boldsymbol{P} = \sum_{e} \int_{\Omega^e} \boldsymbol{B}^T \boldsymbol{D} dY^e \tag{39}$$

$\boldsymbol{B}$ is the strain–displacement matrix and $\boldsymbol{D}$ is the constitutive matrix of the bar element referred to its global reference system of coordinates. $\boldsymbol{K}$ is the homogenized global stiffness matrix and $\boldsymbol{P}$ is a matrix containing global load cases resulting from the homogenization development. Each column of matrix $\boldsymbol{P}$ is a force vector related to an imposed unit strain given in a particular direction over the unit cell, as can be seen by inspection of Eq. 29.

The solution of Eq. 37 requires the application of periodic boundary conditions on the boundaries of the cell, which can be accomplished numerically by different means. The particular approach adopted in the present work will be discussed in Sect. 5.

3.2 Determination of the Mechanical Matrix B

Matrix $\boldsymbol{B}$ contains the derivatives of the interpolation functions that describe the global displacement field in the finite elements adopted. For the 3D bar elements employed, the isoparametric approach approximates both displacements and geometry using the same linear interpolation functions, parameterized by ξ in the range $-1 \leq \xi \leq 1$, as shown in Fig. 4.

The local displacement and geometry approximations are thus given by

$$u(\xi) = \frac{1}{2}(1 - \xi)u_1 + \frac{1}{2}(1 + \xi)u_2 \tag{40}$$

Fig. 4 Truss element showing its local parameterized reference system

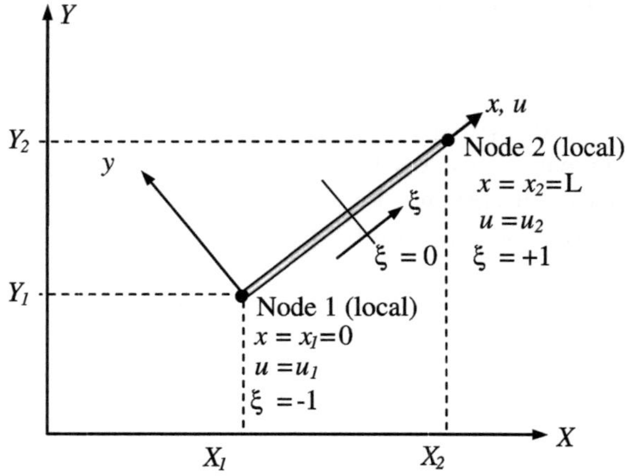

$$x(\xi) = \frac{1}{2}(1 - \xi)x_1 + \frac{1}{2}(1 + \xi)x_2 \tag{41}$$

and the global counterparts are given by

$$U(\xi) = \frac{1}{2}(1 - \xi)U_1 + \frac{1}{2}(1 + \xi)U_2 \tag{42}$$

$$X(\xi) = \frac{1}{2}(1 - \xi)X_1 + \frac{1}{2}(1 + \xi)X_2 \tag{43}$$

$$V(\xi) = \frac{1}{2}(1 - \xi)V_1 + \frac{1}{2}(1 + \xi)V_2 \tag{44}$$

$$Y(\xi) = \frac{1}{2}(1 - \xi)Y_1 + \frac{1}{2}(1 + \xi)Y_2 \tag{45}$$

$$W(\xi) = \frac{1}{2}(1 - \xi)W_1 + \frac{1}{2}(1 + \xi)W_2 \tag{46}$$

$$Z(\xi) = \frac{1}{2}(1 - \xi)Z_1 + \frac{1}{2}(1 + \xi)Z_2 \tag{47}$$

where $U(\xi)$ describes the displacement field in the global direction X, while U_1 and U_2 are the nodal displacements in this direction. Analogous descriptions are applied to the displacements $V(\xi)$ and $W(\xi)$, associated to the directions Y and Z, respectively. The same approach is applied to geometry parameterization. The interpolation functions used are identified as

$$N_1(\xi) = \frac{1}{2}(1 - \xi) \text{ and } N_2(\xi) = \frac{1}{2}(1 + \xi) \tag{48}$$

The Jacobian of the local coordinate transformation is given by

$$J = \frac{dx(\xi)}{d\xi} = -\frac{1}{2}x_1 + \frac{1}{2}x_2 = \frac{1}{2}(x_2 - x_1) = \frac{L}{2} \tag{49}$$

where L is the element length.

The relation between local and global coordinates in the bar elements is given by

$$x = Xl_1 + Ym_1 + Zn_1 \tag{50}$$

where l_1, m_1 and l_1 are the direction cosines of the angles α, β and γ between the local x axis and the global X, Y and Z axes, as shown in Fig. 5.

From the definition of the strain tensor (organized in a column vector), and replacing the analytical displacements by the interpolated fields, one has

$$\varepsilon = \begin{Bmatrix} \varepsilon_x \\ \varepsilon_y \\ \varepsilon_z \\ \gamma_{xy} \\ \gamma_{yz} \\ \gamma_{zx} \end{Bmatrix} = \begin{Bmatrix} \partial U/\partial X \\ \partial V/\partial Y \\ \partial W/\partial Z \\ \partial U/\partial Y + \partial V/\partial X \\ \partial V/\partial Z + \partial W/\partial Y \\ \partial W/\partial X + \partial U/\partial Z \end{Bmatrix} = \boldsymbol{BU} \tag{51}$$

where

$$\boldsymbol{B} = \begin{bmatrix} \frac{\partial N_1(\xi)}{\partial X} & 0 & 0 & \frac{\partial N_2(\xi)}{\partial X} & 0 & 0 \\ 0 & \frac{\partial N_1(\xi)}{\partial Y} & 0 & 0 & \frac{\partial N_2(\xi)}{\partial Y} & 0 \\ 0 & 0 & \frac{\partial N_1(\xi)}{\partial Z} & 0 & 0 & \frac{\partial N_2(\xi)}{\partial Z} \\ \frac{\partial N_1(\xi)}{\partial Y} & \frac{\partial N_1(\xi)}{\partial X} & 0 & \frac{\partial N_2(\xi)}{\partial Y} & \frac{\partial N_2(\xi)}{\partial X} & 0 \\ 0 & \frac{\partial N_1(\xi)}{\partial Z} & \frac{\partial N_1(\xi)}{\partial Y} & 0 & \frac{\partial N_2(\xi)}{\partial Z} & \frac{\partial N_2(\xi)}{\partial Y} \\ \frac{\partial N_1(\xi)}{\partial Z} & 0 & \frac{\partial N_1(\xi)}{\partial X} & \frac{\partial N_2(\xi)}{\partial Z} & 0 & \frac{\partial N_2(\xi)}{\partial X} \end{bmatrix} \tag{52}$$

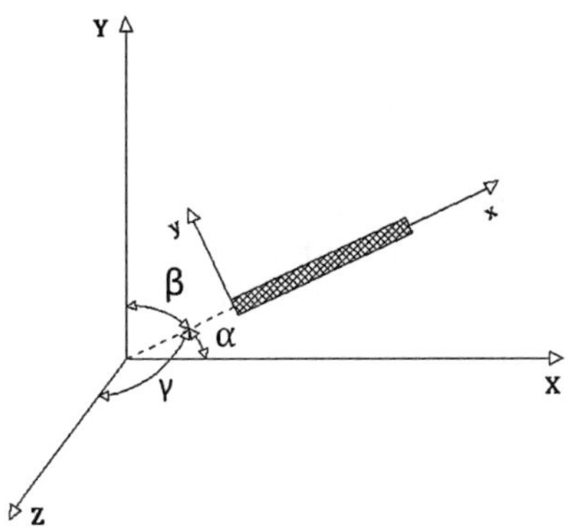

Fig. 5 Local and global coordinates of an arbitrary bar element

and $\boldsymbol{U} = [U_1 \quad V_1 \quad W_1 \quad U_2 \quad V_2 \quad W_2]^T$ are the element nodal global displacements.

By using Eqs. 48–50 and the chain rule of differentiation, the components of the matrix displayed in Eq. 52 can be found as follows

$$\frac{\partial N_i(\xi)}{\partial X} = \frac{\partial N_i(\xi)}{\partial \xi} \frac{\partial \xi}{\partial x} \frac{\partial x}{\partial X} \quad \text{if } i = 1 \rightarrow -\frac{l_1}{L}; \text{ if } i = 2 \rightarrow \frac{l_1}{L} \tag{53}$$

$$\frac{\partial N_i(\xi)}{\partial Y} = \frac{\partial N_i(\xi)}{\partial \xi} \frac{\partial \xi}{\partial x} \frac{\partial x}{\partial Y} \quad \text{if } i = 1 \rightarrow -\frac{m_1}{L}; \text{ if } i = 2 \rightarrow \frac{m_1}{L} \tag{54}$$

$$\frac{\partial N_i(\xi)}{\partial Z} = \frac{\partial N_i(\xi)}{\partial \xi} \frac{\partial \xi}{\partial x} \frac{\partial x}{\partial Z} \quad \text{if } i = 1 \rightarrow -\frac{n_1}{L}; \text{ if } i = 2 \rightarrow \frac{n_1}{L} \tag{55}$$

3.3 Determination of Matrix D

The one-dimensional elastic constitutive relation for a bar element in the local reference system can be written as

$$\sigma = \boldsymbol{D}'\varepsilon \tag{56}$$

or

$$\begin{Bmatrix} \sigma_x \\ \sigma_y \\ \sigma_z \\ \sigma_{xy} \\ \sigma_{yz} \\ \sigma_{zx} \end{Bmatrix} = \begin{bmatrix} E & 0 & 0 & 0 & 0 & 0 \\ 0 & 0 & 0 & 0 & 0 & 0 \\ 0 & 0 & 0 & 0 & 0 & 0 \\ 0 & 0 & 0 & 0 & 0 & 0 \\ 0 & 0 & 0 & 0 & 0 & 0 \\ 0 & 0 & 0 & 0 & 0 & 0 \end{bmatrix} \begin{Bmatrix} \varepsilon_x \\ \varepsilon_y \\ \varepsilon_z \\ \gamma_{xy} \\ \gamma_{yz} \\ \gamma_{zx} \end{Bmatrix} \tag{57}$$

where $\boldsymbol{D}'$ is the constitutive matrix of the bar element in its local reference system, as displayed in Fig. 5. Note that the same boldface notation σ is used for the stress tensor when organized in tensor or *vector* form. However, Eqs. 38 and 39 need the constitutive matrix to be referred to the global reference system. Since $\boldsymbol{D}'$ is a 4th order tensor, it is rotated as

$$\boldsymbol{D} = \boldsymbol{T}_{3D}^T \boldsymbol{D}' \boldsymbol{T}_{3D} \tag{58}$$

where T_{3D} is the transformation matrix given by

$$\boldsymbol{T}_{3D} = \begin{bmatrix} l_1^2 & m_1^2 & n_1^2 & l_1 m_1 & m_1 n_1 & n_1 l_1 \\ l_2^2 & m_2^2 & n_2^2 & l_2 m_2 & m_2 n_2 & n_2 l_2 \\ l_3^2 & m_3^2 & n_3^2 & l_2 m_2 & m_3 n_3 & n_3 l_3 \\ 2l_1 l_2 & 2m_1 m_2 & 2n_1 n_2 & l_1 m_2 + l_2 m_1 & m_1 n_2 + m_2 n_1 & n_1 l_2 + n_2 l_1 \\ 2l_2 l_3 & 2m_2 m_3 & 2n_2 n_3 & l_2 m_3 + l_3 m_2 & m_2 n_3 + m_3 n_2 & n_2 l_3 + n_3 l_2 \\ 2l_3 l_1 & 2m_3 m_1 & 2n_3 n_1 & l_3 m_1 + l_1 m_3 & m_3 n_1 + m_1 n_3 & n_3 l_1 + n_1 l_3 \end{bmatrix} \tag{59}$$

in which l_i, m_i and n_i with $i \in \{1, 2, 3\}$ are the direction cosines between the local x, y and z axes and the global X, Y and Z axes, respectively.

4 Homogenization of Thermal Properties

This Section shows the development of homogenization equations for heat transfer following the main guidelines described in Sect. 2. This time, as a deliberate equivalent option, an energetic approach is used instead of the principle of virtual work [48] adopted in Sect. 3. Once more, the homogenization equations are followed by their numerical counterparts via the finite element method particularized to truss-like unit cells.

The starting point is the definition of the thermal conductivity $\boldsymbol{Kt}$ as a periodic property in the domain

$$\boldsymbol{Kt}(\boldsymbol{x},\boldsymbol{y}) = \boldsymbol{Kt}(\boldsymbol{x},\boldsymbol{y}+\boldsymbol{Y}) \text{ and } \boldsymbol{y} = \boldsymbol{x}/\varepsilon,\ \varepsilon > 0 \tag{60}$$

The temperature field is expanded asymptotically in two scales, yielding

$$T = T(\boldsymbol{x},\boldsymbol{y}) = T_0(\boldsymbol{x}) + \varepsilon T_1(\boldsymbol{x},\boldsymbol{y}) \tag{61}$$

where T_1 is periodic in the dimension $\boldsymbol{Y}$ of the cell.

The temperature gradient is given by

$$\nabla_x T = \nabla_x T_0(\boldsymbol{x}) + \varepsilon \nabla_x T_1(\boldsymbol{x},\boldsymbol{y}) + \nabla_y T_1(\boldsymbol{x},\boldsymbol{y}) \tag{62}$$

where ∇_x and ∇_y are defined by

$$(\nabla_x)_i(\cdot) = \frac{\partial(\cdot)}{\partial x_i} \text{and} (\nabla_y)_i(\cdot) = \frac{\partial(\cdot)}{\partial y_i} \tag{63}$$

so that an energy functional for the average thermal conductivity can be written as

$$G(T) = -\frac{1}{2}\int_{\Omega} [\nabla_x T]^T \boldsymbol{Kt} \nabla_x T d\Omega + \int_{\Gamma} qT d\Gamma \tag{64}$$

where q is the heat flux on the surface, which is assumed to be independent of the scale ε, and Ω and Γ are the domain and the boundary of the macroscopic structure, respectively.

Substituting Eqs. 60–62 in Eq. 64,

$$\begin{aligned} G(T) = &-\frac{1}{2}\int_{\Omega} \left[\nabla_x T_0(\boldsymbol{x}) + \nabla_y T_1(\boldsymbol{x},\boldsymbol{y})\right]^T \boldsymbol{Kt}(\boldsymbol{x},\boldsymbol{y})\left[\nabla_x T_0(\boldsymbol{x}) + \nabla_y T_1(\boldsymbol{x},\boldsymbol{y})\right] d\Omega \\ &- \varepsilon \int_{\Omega} \left[\nabla_x T_1(\boldsymbol{x},\boldsymbol{y})\right]^T \boldsymbol{Kt}(\boldsymbol{x},\boldsymbol{y})\left[\nabla_x T_0(\boldsymbol{x}) + \nabla_y T_1(\boldsymbol{x},\boldsymbol{y})\right] d\Omega \\ &- \frac{\varepsilon^2}{2}\int_{\Omega} \left[\nabla_x T_1(\boldsymbol{x},\boldsymbol{y})\right]^T \boldsymbol{Kt}(\boldsymbol{x},\boldsymbol{y})\left[\nabla_x T_1(\boldsymbol{x},\boldsymbol{y})\right] d\Omega \\ &+ \int_{\Gamma} q T_0(\boldsymbol{x}) d\Gamma + \varepsilon \int_{\Gamma} q T_1(\boldsymbol{x},\boldsymbol{y}) d\Gamma \end{aligned} \tag{65}$$

Differentiating $G(T)$ and passing to the limit $\varepsilon \rightarrow 0$,

$$\begin{aligned} \lim_{\varepsilon\rightarrow 0}\{\delta G(T^{\varepsilon})\} = &-\int_{\Omega} \left[\nabla_x \delta T_0(\boldsymbol{x}) + \nabla_y \delta T_1(\boldsymbol{x},\boldsymbol{y})\right]^T \boldsymbol{Kt}(\boldsymbol{x},\boldsymbol{y})\left[\nabla_x T_0(\boldsymbol{x}) + \nabla_y T_1(\boldsymbol{x},\boldsymbol{y})\right] \\ & d\Omega + \int_{\Gamma} q\delta T_0(\boldsymbol{x}) d\Gamma \end{aligned} \tag{66}$$

where T_1 and δT_1 are periodic in the dimension Y of the cell. Thus, in order to satisfy Eq. 54, and using the relation given by Eq. 18, it follows that

$$-\int_{\Omega} \frac{1}{|\boldsymbol{Y}|}\int_{Y} \left[\nabla_x T_0(\boldsymbol{x}) + \nabla_y T_1(\boldsymbol{x},\boldsymbol{y})\right]^T \boldsymbol{Kt}(\boldsymbol{x},\boldsymbol{y})\left[\nabla_x \delta T_0(\boldsymbol{x})\right] dY d\Omega + \int_{\Gamma} q\delta T_0(\boldsymbol{x}) d\Gamma = 0 \tag{67}$$

for every admissible $\delta T_0(\boldsymbol{x})$, and

$$-\int_{\Omega} \frac{1}{|\boldsymbol{Y}|}\int_{Y} \left[\nabla_x T_0(\boldsymbol{x}) + \nabla_y T_1(\boldsymbol{x},\boldsymbol{y})\right]^T \boldsymbol{Kt}(\boldsymbol{x},\boldsymbol{y})\left[\nabla_x \delta T_1(\boldsymbol{x},\boldsymbol{y})\right] dY d\Omega = 0 \tag{68}$$

for every admissible $\delta T_1(\boldsymbol{x}, \boldsymbol{y})$, periodic in the dimension Y of the cell. Equations 67 and 68 stand for macro and microscopic terms, respectively.

Assuming a convenient separation of variables for $T_1(\boldsymbol{x}, \boldsymbol{y})$, one obtains

$$T_1(\boldsymbol{x},\boldsymbol{y}) = -R^{(m)}(\boldsymbol{x},\boldsymbol{y})\frac{\partial T_0}{\partial x_m} \Rightarrow \nabla_y T_1(\boldsymbol{x},\boldsymbol{y}) = -\nabla_y \boldsymbol{R}(\boldsymbol{x},\boldsymbol{y})\nabla_x T_0(\boldsymbol{x}) \quad (69)$$

where $\boldsymbol{R}(\boldsymbol{x}, \boldsymbol{y})$ are the characteristic temperatures of the unit cell.

Replacing Eq. 69 into the microscopic Eq. 68,

$$-\int_\Omega \frac{1}{|\boldsymbol{Y}|}\int_Y \left[\nabla_x T_0(\boldsymbol{x}) - \nabla_y \boldsymbol{R}(\boldsymbol{x},\boldsymbol{y})\nabla_x T_0(\boldsymbol{x})\right]^T \boldsymbol{Kt}(\boldsymbol{x},\boldsymbol{y})[\nabla_x \delta T_1(\boldsymbol{x},\boldsymbol{y})]dYd\Omega = 0$$
$$\forall \delta T_1 \text{periodic in } Y \quad (70)$$

which is satisfied if $\boldsymbol{R}(\boldsymbol{x}, \boldsymbol{y})$ is the solution of

$$\frac{1}{|\boldsymbol{Y}|}\int_Y \left[\boldsymbol{I} - \nabla_y \boldsymbol{R}(\boldsymbol{x},\boldsymbol{y})\right]^T \boldsymbol{Kt}(\boldsymbol{x},\boldsymbol{y})\left[\nabla_y \delta \boldsymbol{T_1}(\boldsymbol{x},\boldsymbol{y})\right]\boldsymbol{dY} = 0 \quad (71)$$

On the other hand, substituting Eq. 69 into the macroscopic Eq. 67, one has

$$-\int_\Omega \frac{1}{|\boldsymbol{Y}|}\int_Y \left[\nabla_x T_0(\boldsymbol{x}) - \nabla_y \boldsymbol{R}(\boldsymbol{x},\boldsymbol{y})\nabla_{\boldsymbol{x}} \boldsymbol{T_0}(\boldsymbol{x})\right]^T \boldsymbol{Kt}(\boldsymbol{x},\boldsymbol{y})[\nabla_{\boldsymbol{x}} \delta \boldsymbol{T_0}(\boldsymbol{x})]\boldsymbol{dYd\Omega}$$
$$+\int_\Gamma q\delta T_0(\boldsymbol{x})d\Gamma = 0 \quad (72)$$

which induces that, in analogy with Eqs. 30 and 31, the homogenized thermal conductivity tensor is defined by

$$\boldsymbol{Kt^H} = \frac{\boldsymbol{1}}{|\boldsymbol{Y}|}\int_{\boldsymbol{Y}} \boldsymbol{Kt}(\boldsymbol{x},\boldsymbol{y})\left[\boldsymbol{I} - \nabla_y \boldsymbol{R}(\boldsymbol{x},\boldsymbol{y})\right]\boldsymbol{dY} \quad (73)$$

or in indicial notation,

$$Kt^H_{ij} = \frac{1}{|\boldsymbol{Y}|}\int_Y \left(Kt_{ij} - Kt_{ip}\frac{\partial R^j}{\partial y_p}\right)dY \quad (74)$$

4.1 Numerical Determination of R

The determination of $\boldsymbol{R}$ follows a development completely analogous to the one presented in Sects. 3.1–3.3 for $\boldsymbol{\chi}$. Hence, in order to find $\boldsymbol{R}$, Eq. 71 must be solved. Again, the finite element method is employed to discretize a truss-like unit cell using linear 3D bar elements. The procedure leads to the following linear system, similar to the one described by Eq. 37,

$$\boldsymbol{CR} = \boldsymbol{Q} \tag{75}$$

where

$$\boldsymbol{C} = \sum_e \int_{\Omega^e} \boldsymbol{B}^T \boldsymbol{KtB} d\boldsymbol{Y}^e \tag{76}$$

and

$$\boldsymbol{Q} = \sum_e \int_{\Omega^e} \boldsymbol{B}^T \boldsymbol{Kt} d\boldsymbol{Y}^e \tag{77}$$

$\boldsymbol{B}$ is the matrix containing the derivatives of the interpolation functions and $\boldsymbol{Kt}$ is the thermal conductivity matrix of the bar element in its global reference system, $\boldsymbol{C}$ is the homogenized global thermal conductivity matrix and $\boldsymbol{Q}$ is a matrix containing load cases resulting from the homogenization development. Each column of matrix $\boldsymbol{Q}$ is related to an imposed unit thermal gradient vector given in a particular direction over the unit cell, as can be seen inspecting Eq. 71.

The solution of Eq. 75 requires the application of periodic boundary conditions, which will be discussed in Sect. 5.

4.2 Determination of the Thermal Matrix B

The three-dimensional Fourier Law for heat conduction is

$$-\boldsymbol{q}(\boldsymbol{X}) = \boldsymbol{Kt}\frac{d\boldsymbol{T}(\boldsymbol{X})}{d\boldsymbol{X}} \tag{78}$$

so that in the global domain of the unit cell it can be written as

$$-\begin{Bmatrix} q_X \\ q_Y \\ q_Z \end{Bmatrix} = \begin{bmatrix} kt_{11} & kt_{12} & kt_{13} \\ kt_{21} & kt_{22} & kt_{23} \\ kt_{31} & kt_{32} & kt_{33} \end{bmatrix} \begin{Bmatrix} \frac{dT}{dX} \\ \frac{dT}{dY} \\ \frac{dT}{dZ} \end{Bmatrix} \tag{79}$$

Using the linear interpolation functions given by Eq. 48 to approximate the temperature field, which is invariant for a change of frame, one gets

$$\begin{Bmatrix} \frac{dT}{dX} \\ \frac{dT}{dY} \\ \frac{dT}{dZ} \end{Bmatrix} = \begin{bmatrix} \frac{\partial N_1(\xi)}{\partial X} & \frac{\partial N_2(\xi)}{\partial X} \\ \frac{\partial N_1(\xi)}{\partial Y} & \frac{\partial N_2(\xi)}{\partial Y} \\ \frac{\partial N_1(\xi)}{\partial Z} & \frac{\partial N_2(\xi)}{\partial Z} \end{bmatrix} \begin{Bmatrix} T_1 \\ T_2 \end{Bmatrix} \tag{80}$$

where T_1 and T_2 are the nodal temperatures and the thermal matrix $\boldsymbol{B}$ is given by

$$\boldsymbol{B} = \begin{bmatrix} \frac{\partial N_1(\xi)}{\partial X} & \frac{\partial N_2(\xi)}{\partial X} \\ \frac{\partial N_1(\xi)}{\partial Y} & \frac{\partial N_2(\xi)}{\partial Y} \\ \frac{\partial N_1(\xi)}{\partial Z} & \frac{\partial N_2(\xi)}{\partial Z} \end{bmatrix} \tag{81}$$

with its terms given by Eqs. 53–55.

4.3 *Determination of Matrix Kt*

Equation 78 particularized for an element bar with reference to its local system of coordinates is given by

$$-\begin{Bmatrix} q_x \\ q_y \\ q_z \end{Bmatrix} = \begin{bmatrix} k & 0 & 0 \\ 0 & 0 & 0 \\ 0 & 0 & 0 \end{bmatrix} \begin{Bmatrix} \frac{dT}{dx} \\ \frac{dT}{dy} \\ \frac{dT}{dz} \end{Bmatrix} \tag{82}$$

where

$$\boldsymbol{Kt}' = \begin{bmatrix} k & 0 & 0 \\ 0 & 0 & 0 \\ 0 & 0 & 0 \end{bmatrix} \tag{83}$$

and $\boldsymbol{Kt}'$ is the thermal conductivity matrix with reference to the bar's local system of coordinates. Equations 76 and 77 need $\boldsymbol{Kt}'$ to be rotated to the global reference system. This is done through Eq. 84

$$\boldsymbol{Kt} = \boldsymbol{T}_{2D}^{T}\boldsymbol{Kt}'\boldsymbol{T}_{2D} \tag{84}$$

where $\boldsymbol{T}_{2D}$ is the transformation matrix for a 2nd order tensor, given by

$$\boldsymbol{T}_{2D} = \begin{bmatrix} l_1 & m_1 & n_1 \\ l_2 & m_2 & n_2 \\ l_3 & m_3 & n_3 \end{bmatrix} \tag{85}$$

in which l_i, m_i and n_i with $i \in \{1, 2, 3\}$ are the direction cosines between the local x, y and z axes and the global X, Y and Z axes, respectively.

5 Periodic Boundary Conditions

The linear systems resulting from the mechanical and thermal problems defined by Eqs. 37 and 75 must respect the periodicity condition over the whole domain,

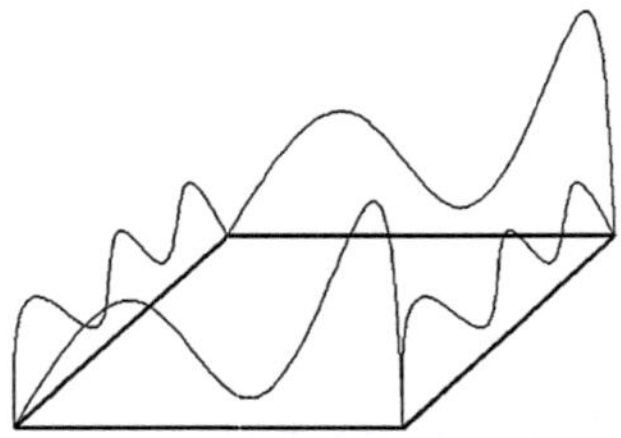

Fig. 6 Periodic boundary conditions on the base cell

which introduces a kinematic constraint on the unit cell. The imposed constraint enforces that displacement or temperature fields must display equal values on opposite borders of the cell, as depicted in Fig. 6.

After discretization of the base cell, several methods can be used to ensure the aforementioned constraint, including Lagrange multipliers, penalization [49] and condensation or reduction. This work employs the method of condensation as described by Yang et al. [50]. Initially, any prescribed displacement or temperature value is applied to one of the cell vertices (at least one *must* be constrained). The application of equality constraints between the proper degrees of freedom is accomplished by the use of a transformation matrix $\boldsymbol{T}_{\text{trans}}$ relating all the degrees of freedom ($\boldsymbol{\chi}$ or $\boldsymbol{R}$), represented generically by $\boldsymbol{u}$, to the unconstrained ones ($\tilde{\chi}$ or $\tilde{\boldsymbol{R}}$), represented generically by $\tilde{\boldsymbol{u}}$.

In Fig. 7 it is assumed that there are 2p nodes on the upper and lower faces, 2q nodes on the other two sides. The multi-point constrains for the unit cell can be expressed by

$$u_i = u_{i+p} \quad (i = 1, 2, \ldots, p) \tag{86}$$

on the upper and lower sides, and

$$u_j = u_{j+q} \quad (j = 1, 2, \ldots, q) \tag{87}$$

on the right and left sides.

This relation can be written as

$$\boldsymbol{u} = \boldsymbol{T}_{\text{trans}}\tilde{\boldsymbol{u}} \tag{88}$$

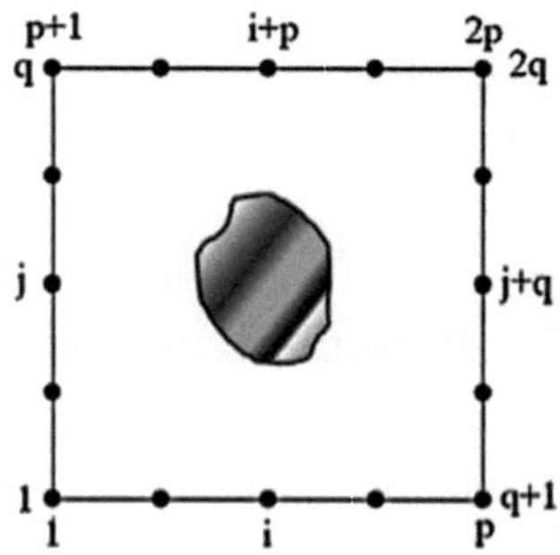

Fig. 7 Application of periodic boundary conditions on degrees of freedom of opposite sides in the base cell

For instance, in a system containing "n" degrees of freedom and the single constraint $u_J = u_I$, the transformation matrix will have dimension "$n\ x\ (n - 1)$" and will look like Eq. 89.

$$
T_{\text{trans}} = \begin{bmatrix}
1 & 2 & \cdot & \mathrm{I} & \cdot & \mathrm{J} & \cdot & n-1 & & \\
- & - & - & - & - & - & - & - & & \\
1 & 0 & \cdot & 0 & \cdot & 0 & \cdot & 0 & | & 1 \\
0 & 1 & \cdot & 0 & \cdot & 0 & \cdot & 0 & | & 2 \\
.. & .. & \cdot & .. & \cdot & .. & \cdot & .. & | & \cdot \\
0 & 0 & \cdot & 1 & \cdot & 0 & \cdot & 0 & | & \mathrm{I} \\
.. & .. & \cdot & .. & \cdot & .. & \cdot & .. & | & \cdot \\
0 & 0 & \cdot & 1 & \cdot & 0 & \cdot & 0 & | & \mathrm{J} \\
.. & .. & \cdot & .. & \cdot & .. & \cdot & .. & | & \cdot \\
0 & 0 & \cdot & 0 & \cdot & 0 & \cdot & 1 & | & n
\end{bmatrix} \tag{89}
$$

Thus, the global systems to be solved (Eqs. 37 and 75) become

$$\tilde{K}\tilde{u} = \tilde{P} \tag{90}$$

where, for the mechanical problem,

$$\tilde{K} = T^T_{\text{trans}} K T_{\text{trans}} \tag{91}$$

$$\tilde{P} = T^T_{\text{trans}} P \tag{92}$$

and after solving Eq. 90 for $\tilde{u}$, the value of u is recovered by means of Eq. 88. The thermal problem is treated by analogy with Eqs. 91 and 92, through condensation of C and Q.

The 2D asymmetric problem shown in Fig. 8 is taken from Yang et al. [50]. It consists of a 3 2 units cell containing an asymmetric hole in which the dimensions are given by "a". Figures 9 and 10 display the characteristic displacements and temperatures χ and R resulting from the respective 2D load cases. Each column of matrix P in Eqs. 37 and 39 corresponds to a load vector resulting from a particular imposed unit strain field. Similarly, each column of matrix Q in Eqs. 75 and 77 corresponds to a load vector resulting from a particular imposed unit thermal gradient field. The scale ratio in the figures is 1:4.

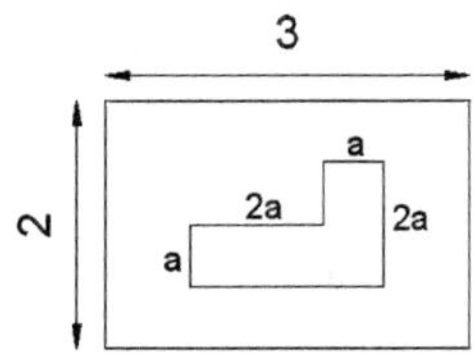

Fig. 8 Geometry and dimensions of an asymmetric cell adopted to show the application of periodic boundary conditions

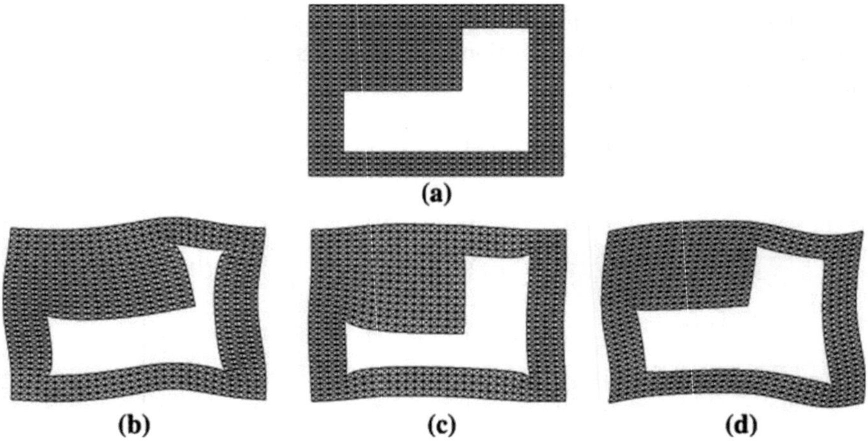

Fig. 9 Characteristic displacements **a** initial mesh, **b** unit ε_x, **c** unit ε_y, **d** unit γ_{xy}

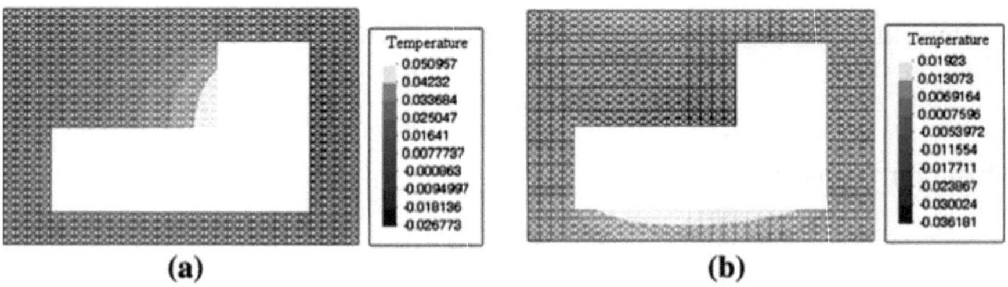

Fig. 10 Characteristic temperatures **a** unit dT/dX, **b** unit dT/dY

6 Design Variables: Why Areas *and* Coordinates Together?

The most common approach for dealing with trusses in structural optimization (usually topology optimization) uses densely populated ground structures [51, 52], and allows only cross sectional areas to be modified. Hence, a large amount of bars are needed to obtain realistic results. On the other hand, presumably, a combination of nodal coordinates and cross sectional areas would require much less design variables to lead to an acceptable optimized structure.

In order to show the point, consider the square design displayed in Fig. 11, in which the left edge has symmetry conditions. The structure is simply supported and subject to a single prescribed force. 42 bars with the same cross section and material are employed in the model. The objective is to minimize the weight while respecting the allowable stresses in every bar.

It is intuitive that starting from overestimated cross sectional areas and using only nodal coordinates as design variables, only a very limited improvement on weight can be attained. This case is characterized by Fig. 12a. On the other hand, if only areas are adopted as design variables, the weight decrease is dramatic,

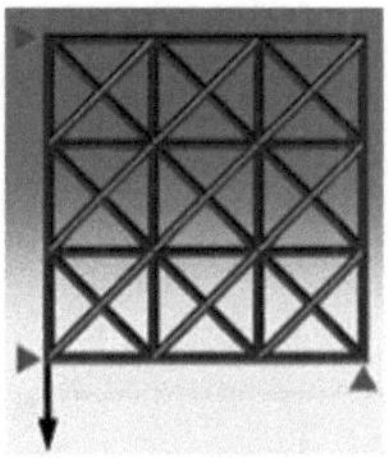

Fig. 11 Initial configuration for the weight minimization problem

although the geometry becomes complex, as in Fig. 12b. It can be seen in Figs. 12c–e that the introduction of coordinates provides a further enhancement in weight while rendering much simpler geometrical configurations.

Classical research works point out convergence difficulties when a combination of areas and coordinates is adopted as design variables, stress constraints are imposed, and SLP is adopted for weight optimization. Nevertheless, this problem has been successfully solved by Vanderplaats [53] and, in the different application considered in this Chapter no stress constraints are imposed. Hence, the joint adoption of areas and coordinates as design variables was shown to be effective.

7 Definition of the Optimization Problems

The study developed in this work considered 4 different optimization problems, i.e., 4 different objective functions. They correspond to maximize or minimize given components of the homogenized mechanical or thermal constitutive tensors, or their combinations to account for multifunctional responses. The design variables adopted can be any combination of cross sectional areas and nodal coordinates. Hence, the optimization problems can be defined as

Optimization problem 1: For given $i, j, k, l,$ maximize the corresponding component of the mechanical homogenized constitutive tensor E^H_{ijkl}

Optimization problem 2: For given $i, j,$ maximize or minimize the corresponding component of the thermal homogenized constitutive tensor K^H_{ij}.

Optimization problem 3: For given $i, j, k, l, m, n,$ maximize the ratio of the corresponding components of the mechanical and thermal homogenized constitutive tensors, $E^H_{ijkl}(\boldsymbol{x})/Kt^H_{mn}(\boldsymbol{x})$.

Optimization problem 4: For given $i, j, k, l, m, n,$ maximize the product of the corresponding components of the mechanical and thermal homogenized constitutive tensors, $E^H_{ijkl}(\boldsymbol{x}) \cdot Kt^H_{mn}(\boldsymbol{x})$.

In the optimization problems 1–4 the following constraints are applied:

$$f_L \cdot V_0 \leq V \leq f_U \cdot V_0 \tag{93}$$

$$A^i_{\min} \leq A^i \leq A^i_{\max} \quad m = 1, \ldots, nap \tag{94}$$

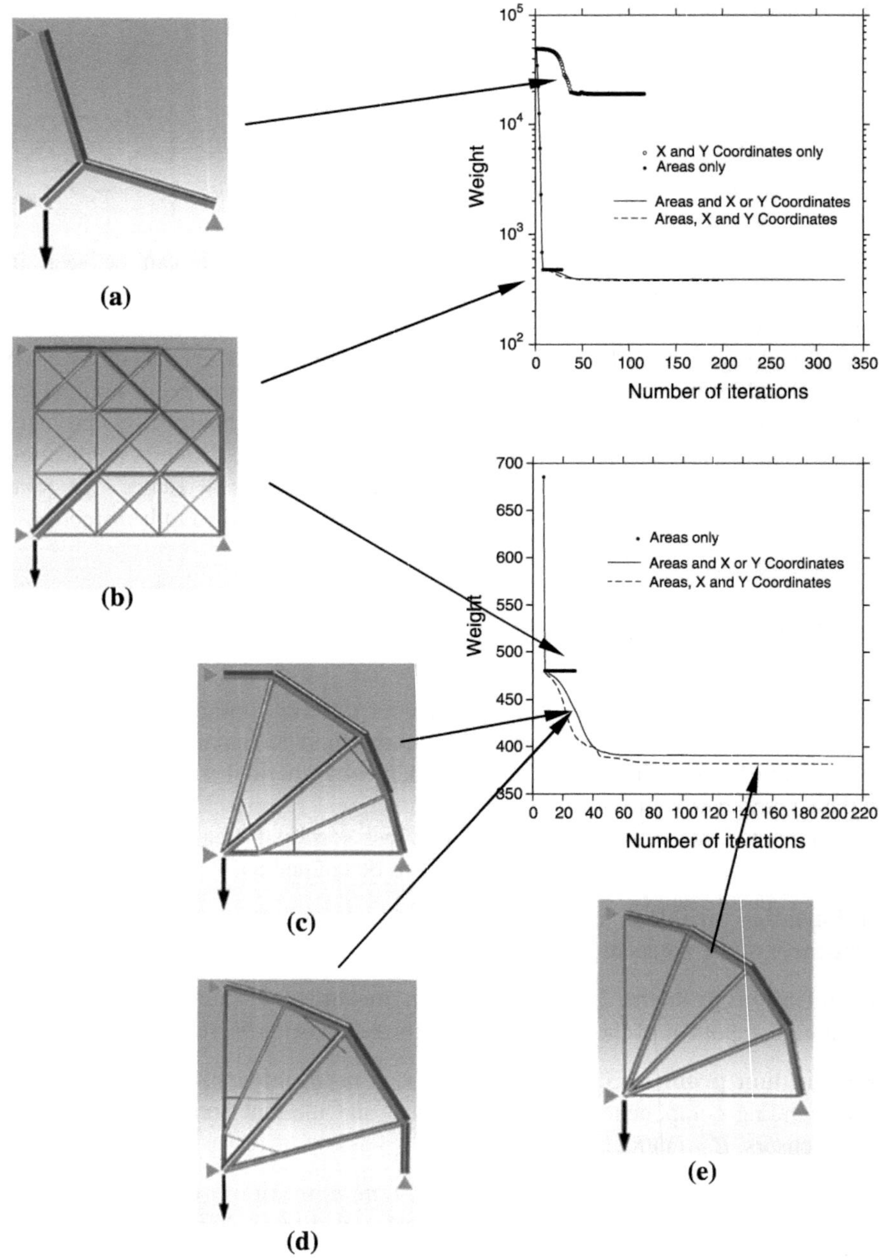

Fig. 12 Effect of using different combinations of areas and nodal coordinates as design variables in a weight minimization problem. **a** only *X* and *Y* nodal coordinates, **b** only areas, **c** areas and *X* nodal coordinates, **d** areas and *Y* nodal coordinates, **e** areas and *X* and *Y* nodal coordinates

$$X^i_{\min} \leq X^i \leq X^i_{\max} \quad m = 1, \ldots, ncopx \tag{95}$$

$$Y^i_{\min} \leq Y^i \leq Y^i_{\max} \quad m = 1, \ldots, ncopy \tag{96}$$

$$Z^i_{\min} \leq Z^i \leq Z^i_{\max} \quad m = 1, \ldots, ncopz \tag{97}$$

where *nap*, *ncopx*, *ncopy* and *ncopz* are the number of design variables associated to areas, and *X*, *Y* and *Z* coordinates, respectively. f_L and f_U are lower and upper bounding factors on volume. In all the examples discussed in Sect. 9, these factors were chosen in order to preserve a constant volume.

In this work, the optimization task is performed using SLP as previously done by Pedersen [54] and others for truss layout optimization. In this method the objective function and the constraints are linearized with respect to the current design point, defining an associated linear programming sub-problem. The optimum point for this sub-problem is adopted as the new current design project and the updated objective function and constraints are linearized again, now with respect to this point. The procedure continues iteratively until convergence. Hence a non-linear optimization problem is replaced by a sequence of linear programming problems. As a termination criterion, it can be established that the relative difference between two iterations should be less than a prescribed tolerance, for both the objective function and the design variables.

7.1 Linearization of the Objective Function and Volume Constraint

In order to apply SLP it is necessary to linearize the expressions for the homogenized mechanical and thermal coefficients. This is accomplished by performing a Taylor series expansion truncated in first order terms. As shown in Eqs. 98–100, the gradients with respect to the design variables are needed and their analytical determination will be detailed in Sect. 8.

$$E^H_{ijkl} = E^H_{ijkl0} + \sum_{m=1}^{nap} \left.\frac{\partial E^H_{ijkl}}{\partial A_m}\right|_0 \Delta A_m + \sum_{m=1}^{ncopx} \left.\frac{\partial E^H_{ijkl}}{\partial X_m}\right|_0 \Delta X_m + \sum_{m=1}^{ncopy} \left.\frac{\partial E^H_{ijkl}}{\partial Y_m}\right|_0 \Delta Y_m + \sum_{m=1}^{ncopz} \left.\frac{\partial E^H_{ijkl}}{\partial Z_m}\right|_0 \Delta Z_m \tag{98}$$

$$Kt^H_{ij} = Kt^H_{ij0} + \sum_{m=1}^{nap} \left.\frac{\partial Kt^H_{ij}}{\partial A_m}\right|_0 \Delta A_m + \sum_{m=1}^{ncopx} \left.\frac{\partial Kt^H_{ij}}{\partial X_m}\right|_0 \Delta X_m + \sum_{m=1}^{ncopy} \left.\frac{\partial Kt^H_{ij}}{\partial Y_m}\right|_0 \Delta Y_m + \sum_{m=1}^{ncopz} \left.\frac{\partial Kt^H_{ij}}{\partial Z_m}\right|_0 \Delta Z_m \tag{99}$$

$$V=V_0+\sum_{m=1}^{nap}\left.\frac{\partial V}{\partial A_m}\right|_0\Delta A_m+\sum_{m=1}^{ncopx}\left.\frac{\partial V}{\partial X_m}\right|_0\Delta X_m+\sum_{m=1}^{ncopy}\left.\frac{\partial V}{\partial Y_m}\right|_0\Delta Y_m+\sum_{m=1}^{ncopz}\left.\frac{\partial V}{\partial Z_m}\right|_0\Delta Z_m \quad (100)$$

where

$$\Delta A_m = A^m - A_0^m \quad (101)$$

$$\Delta X_m = X^m - X_0^m \quad (102)$$

$$\Delta Y_m = Y^m - Y_0^m \quad (103)$$

$$\Delta Z_m = Z^m - Z_0^m \quad (104)$$

In Eqs. 98–104 the lower index zero stands for values in the current design point.

7.2 Move Limits

The adequate and stable behavior of SLP demands introduction of move limits on the design variables [54, 55]. The move limits update scheme is external to the mathematical programming algorithm and affects strongly the process convergence and efficiency. The problems presented herein showed instabilities (convergence difficulties) with respect to different update strategies for move limits, especially regarding area design variables. Hence, initial move limits were made very small for all the design variables, and a different treatment was given for their update strategy, according to the design variable nature. Area move limits were kept constant, while coordinate move limits were updated in a very conservative fashion, with different update factors for each example presented. This strategy hindered efficiency on behalf of effectivity, as will become clear in the numerical examples presented in Sect. 9. It is worth remarking that Sigmund [37] mentions similar difficulties and recommends the use of small move limits to ensure final convergence. For the purpose of this work, efficiency was not a major concern and other optimization algorithms would probably result in faster convergence. This kind of study is left for future development.

8 Sensitivity Analysis

Sensitivity analysis deals with the determination of gradients. The focus in the present case are the derivatives resulting from the linearizations present in Eqs. 98–100. The gradients are found by means of analytical differentiation and validated via central finite differences. The main guidelines for the operational work involved is sketched below, but full development is not included in this text due to space limitations. All the detailed steps for the task can be found in Carniel et al. 2009 [56].

8.1 Gradients Related to the Mechanical Problem

8.1.1 Computation of $\frac{\partial E^H_{ijkl}}{\partial A_m}$

Particularization of Eq. 33 to a cell discretized using bar elements, yields

$$\boldsymbol{D^H}(\boldsymbol{x}) = \frac{1}{|\boldsymbol{Y}|}\sum_{n=1}^{nel}\left(\boldsymbol{D} - \boldsymbol{D}\left[\varepsilon^{(1)} \quad \ldots \quad \varepsilon^{(NLC)}\right]\right)^n A^n L^n \tag{105}$$

$$\boldsymbol{D^H}(\boldsymbol{x}) = \frac{\boldsymbol{1}}{|\boldsymbol{Y}|}\sum_{\boldsymbol{n=1}}^{\boldsymbol{nel}}\left(\boldsymbol{D} - \boldsymbol{DB}\left[\chi^{(\boldsymbol{1})} \quad \ldots \quad \chi^{(\boldsymbol{NLC})}\right]\right)^{\boldsymbol{n}} \boldsymbol{A^n L^n} \tag{106}$$

$$D^{\boldsymbol{H}}_{ij}(\boldsymbol{x}) = \frac{\boldsymbol{1}}{|\boldsymbol{Y}|}\sum_{\boldsymbol{n=1}}^{\boldsymbol{nel}}\left(\boldsymbol{D}_{ij} - \boldsymbol{D}_{ik}\boldsymbol{B}_{kl}\chi_l^{(j)}\right)^{\boldsymbol{n}} \boldsymbol{A^n L^n} \tag{107}$$

and by differentiating analytically with respect to the m-th design area,

$$\frac{\partial D^{\boldsymbol{H}}_{ij}}{\partial A_m} = \frac{1}{|\boldsymbol{Y}|}\left[\sum_{n=1}^{nel} D^n_{ij}L^n\delta_{nm} - \sum_{n=1}^{nel} D^n_{ik}B^n_{kl}\chi_l^{(j)n}L^n\delta_{nm} - \sum_{n=1}^{nel} A^nL^nD^n_{ik}B^n_{kl}\frac{\partial\chi_l^{(j)n}}{\partial A_m}\right] \tag{108}$$

where δ_{nm} is the Kronecker delta,

$$\delta_{nm} = \begin{cases} 1 & if \quad n = m \\ 0 & if \quad n \neq m \end{cases} \tag{109}$$

Table 1 provides the correspondence between $\frac{\partial D^H_{\alpha\beta}}{\partial A_m}$ and $\frac{\partial E^H_{ijkl}}{\partial A_m}$

8.1.2 Computation of $\frac{\partial \chi}{\partial A_m}$

In order to obtain the characteristic displacements gradient, Eq. 37 is differentiated with respect to the design areas, resulting in

$$\frac{\partial\chi}{\partial A_m} = \boldsymbol{K}^{-\boldsymbol{1}}\left(\frac{\partial\boldsymbol{P}}{\partial\boldsymbol{A_m}} - \frac{\partial\boldsymbol{K}}{\partial\boldsymbol{A_m}}\chi\right) \tag{110}$$

which is obtained easily with the aid of Eqs. 38 and 39.

8.1.3 Computation of $\frac{\partial E^H_{ijkl}}{\partial X_m}, \frac{\partial \mathbf{E}^{\mathbf{H}}_{\mathbf{ijkl}}}{\partial \mathbf{Y}_{\mathbf{m}}}$, and $\frac{\partial E^H_{ijkl}}{\partial Z_m}$

The development of expressions for sensitivities with respect to nodal coordinates is sketched in this section only for X components. The extension to Y and Z coordinates is straightforward by analogy. The development is done using compact notation and Table 1 provides the correspondence between $\boldsymbol{D}$ and $\boldsymbol{E}$.

After some manipulation, the expression for $\partial \boldsymbol{D}^H/\partial X_m$ yields

$$\frac{\partial D^H_{ij}}{\partial X_m} = \frac{1}{|\boldsymbol{Y}|}\left[\sum_{n=1}^{nel}\frac{\partial D^n_{ij}}{\partial X_m}A^nL^n + \sum_{n=1}^{nel}D^n_{ij}A^n\frac{\partial L^n}{\partial X_m} - \sum_{n=1}^{nel}A^nL^n\frac{\partial D^n_{ik}}{\partial X_m}B^n_{kl}\chi^{(j)n}_l \right.$$
$$\left. - \sum_{n=1}^{nel}A^nD^n_{ik}\frac{\partial\left(L^nB^n_{kl}\right)}{\partial X_m}\chi^{(j)n}_l - \sum_{n=1}^{nel}A^nL^nD^n_{ik}B^n_{kl}\frac{\partial\chi^{(j)n}_l}{\partial X_m}\right] \tag{111}$$

Notice that the complete determination of the desired sensitivity requires the evaluation of $\partial\boldsymbol{D}^n/\partial X_m$, $\partial L^n/\partial X_m$, $\partial(L^n\boldsymbol{B}^n)/\partial X_m$ and $\partial\boldsymbol{\chi}^n/\partial X_m$. The sensitivities $\partial\boldsymbol{D}^n/\partial X_m$ and $\partial(L^n\boldsymbol{B}^n)/\partial X_m$ can be obtained by differentiation of Eq. 58 and Eq. 52, respectively. On the other hand, the sensitivity $\partial\boldsymbol{\chi}^n/\partial X_m$ can be obtained by analogy with the procedure presented in Sect. 8.1.1.1. The sensitivity $\partial L^n/\partial X_m$ can be evaluated with the aid of the expression

$$L^n = \sqrt{\left(X^n_2 - X^n_1\right)^2 + \left(Y^n_2 - Y^n_1\right)^2 + \left(Z^n_2 - Z^n_1\right)^2} \tag{112}$$

where X_i^n, Y_i^n and Z_i^n are the global coordinates of the i-th local node belonging to the n-th element.

8.2 Gradients Related to the Thermal Problem

8.2.1 Computation of $\frac{\partial Kt^H_{ij}}{\partial A_m}$

Particularization of Eq. 74 to a cell discretized using bar elements, yields

$$\boldsymbol{Kt}^{\boldsymbol{H}}(\boldsymbol{x}) = \frac{1}{|\boldsymbol{Y}|}\sum_{n=1}^{nel}\left(\boldsymbol{Kt} - \boldsymbol{Kt}\,\boldsymbol{B}\left[\boldsymbol{R}^{(1)} \quad \dots \quad \boldsymbol{R}^{(NLC)}\right]\right)^nA^nL^n \tag{113}$$

$$Kt^{\boldsymbol{H}}_{ij}(\boldsymbol{x}) = \frac{1}{|\boldsymbol{Y}|}\sum_{n=1}^{nel}\left(Kt_{ij} - Kt_{ik}B_{kl}R^{(j)}_l\right)^nA^nL^n \tag{114}$$

where NLC in Eq. 113 stands for the number of load cases. Differentiating analytically with respect to the m-th design area, results

$$\frac{\partial Kt_{ij}^{H}}{\partial A_m} = \frac{1}{|\boldsymbol{Y}|}\left[\sum_{n=1}^{nel} Kt_{ij}^{n} L^{n} \delta_{nm} - \sum_{n=1}^{nel} Kt_{ik}^{n} B_{kl}^{n} R_{l}^{(j)n} L^{n} \delta_{nm} - \sum_{n=1}^{nel} A^{n} L^{n} Kt_{ik}^{n} B_{kl}^{n} \frac{\partial R_{l}^{(j)n}}{\partial A_m}\right] \tag{115}$$

where δ_{nm} is the Kronecker delta, defined by Eq. 109.

8.2.2 Computation of $\frac{\partial \boldsymbol{R}}{\partial A_m}$

Starting from Eq. 75, that is., $\boldsymbol{CR} = \boldsymbol{Q}$, the determination of the characteristic temperatures gradient is obtained by analogy with the equations presented in Sect. 8.1.2.

8.2.3 Computation of $\frac{\partial Kt_{ij}^{H}}{\partial X_m}, \frac{\partial \mathbf{Kt}_{\mathbf{ij}}^{\mathbf{H}}}{\partial \mathbf{Y}_{\mathbf{m}}}$ and $\frac{\partial Kt_{ij}^{H}}{\partial Z_m}$

As in Sect. 8.1.2, only the development of expressions for sensitivities with respect to nodal X coordinates is sketched in this Section. The extension to Y and Z coordinates is straightforward by analogy.

After some manipulation, the expression for $\partial \boldsymbol{Kt}^{H}/\partial X_m$ yields

$$\begin{aligned}\frac{\partial Kt_{ij}^{\boldsymbol{H}}}{\partial X_m} = \frac{1}{|Y|}\Bigg[&\sum_{n=1}^{nel} \frac{\partial Kt_{ij}^{n}}{\partial X_m} A^{n} L^{n} + \sum_{n=1}^{nel} Kt_{ij}^{n} A^{n} \frac{\partial L^{n}}{\partial X_m} - \sum_{n=1}^{nel} A^{n} L^{n} \frac{\partial Kt_{ik}^{n}}{\partial X_m} B_{kl}^{n} R_{l}^{(j)n} - \\ &- \sum_{n=1}^{nel} A^{n} Kt_{ik}^{n} \frac{\partial \left(L^{n} B_{kl}^{n}\right)}{\partial X_m} R_{l}^{(j)n} - \sum_{n=1}^{nel} A^{n} L^{n} Kt_{ik}^{n} B_{kl}^{n} \frac{\partial R_{l}^{(j)n}}{\partial X_m}\Bigg]\end{aligned} \tag{116}$$

Notice that the complete determination of the desired sensitivity requires the evaluation of $\partial \boldsymbol{Kt}^{n}/\partial X_m$, $\partial L^{n}/\partial X_m$, $\partial(L^{n}\boldsymbol{B}^{n})/\partial X_m$ and $\partial \boldsymbol{R}^{n}/\partial X_m$. The partial derivatives $\partial \boldsymbol{Kt}^{n}/\partial X_m$ and $\partial(L^{n}\boldsymbol{B}^{n})/\partial X_m$ can be obtained by differentiation of Eqs. 84 and 81, respectively. On the other hand, the sensitivity $\partial \boldsymbol{R}^{n}/\partial X_m$ can be obtained by analogy with the procedure presented in Sect. 8.1.1.1. The sensitivity $\partial L^{n}/\partial X_m$ can be evaluated with the aid of Eq. 112.

8.3 Gradients of the Volume Constraint

Having in mind that the total volume occupied by the bars in the cell is given by

$$V = \sum_{n=1}^{nel} A_n L_n, \tag{117}$$

the derivatives with respect to the m-th area or coordinate are easily identified as

$$\frac{\partial V}{\partial A_m} = \sum_{n=1}^{nel} \frac{\partial A_n}{\partial A_m} L_n = \sum_{n=1}^{nel} \delta_{nm} L_n = L_m \tag{118}$$

and

$$\frac{\partial V}{\partial X_m} = \sum_{n=1}^{nel} A_n \frac{\partial L_n}{\partial X_m}, \tag{119}$$

$$\frac{\partial V}{\partial Y_m} = \sum_{n=1}^{nel} A_n \frac{\partial L_n}{\partial Y_m}, \tag{120}$$

$$\frac{\partial V}{\partial Z_m} = \sum_{n=1}^{nel} A_n \frac{\partial L_n}{\partial Z_m} \tag{121}$$

where $\partial L^n/\partial X_m$, $\partial L^n/\partial Y_m$ and $\partial L^n/\partial Z_m$ can be evaluated with the aid of Eq. 112.

9 Numerical Applications

Numerical results obtained using the formulation proposed are presented in this Section. Although these results can be considered preliminary, one differential with respect to previous developments reported in the literature is the inclusion of a 3D example. All the four objective functions proposed were tested and some of the results allowed qualitative validation by visual inspection or comparison to previously published work. The initial cells are formed by truss elements in which Young's modulus, thermal conductivity and cross sectional areas are given by $E = 210$ [GPa], $k = 50$ [W/m°C] and $A = 10^{-5}$ [m^2], respectively. During the optimization process, lower bounds on areas and element lengths are prescribed to be equal to 10^{-7} [m^2] and 10^{-4} [m], respectively. In all the examples, the total volume of the unit cell is constrained to remain unchanged. The cells have equal dimension sides: in 2D cases each side equals 0.1 [m] and in 3D cases they equal 0.075 [m], resulting in relative densities of 49 and 33.5% respectively.

9.1 2D Examples

The initial 2D unit cell chosen for optimization is displayed in Fig. 13. In order to apply the method described, one component from the mechanical and one

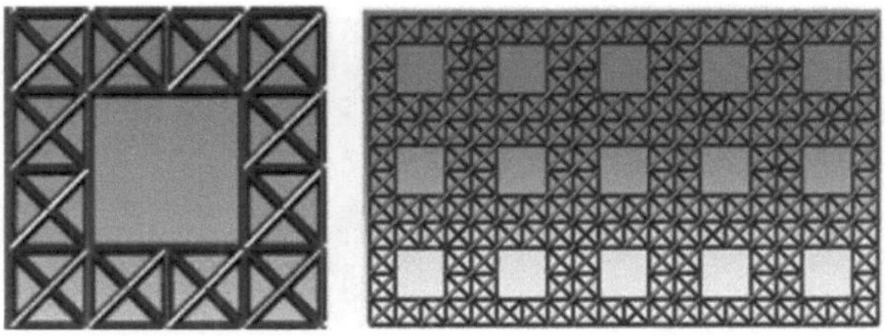

Fig. 13 Initial cell and corresponding periodic material

component from thermal homogenized tensors were selected to be considered in the objective functions. These were the shear (E^H_{1212}) component and the thermal conductivity component in direction X (Kt^H_{11}).

In the cell external vertical (right and left) edges, only the Y coordinates are chosen as design variables. Accordingly, in the cell external horizontal (upper and lower) edges, only the X coordinates are adopted as design variables. Opposite nodes in these edges are constrained to move equally (same displacements) and the four vertices are prescribed to keep unchanged.

Normalized diagrams showing the effect of rotations on E^H_{1111} and Kt^H_{11} are displayed in all the examples. The reference for normalization is the largest value of the tensorial component in a 360° rotation. These diagrams help to interpret the optimization results obtained. For the initial cell, they are displayed in Fig. 14. It is apparent that the largest normal and shear stiffness of the periodic structure material are oriented at 0° and 45 ° degrees with respect to the horizontal (X) axis, respectively. The heat conduction behavior is isotropic.

The homogenized mechanical and thermal constitutive tensors for this initial cell are given, respectively, by

$$\boldsymbol{E}^H = \begin{bmatrix} 0.1150 & 0.02706 & 0 \\ 0.02706 & 0.1150 & 0 \\ 0 & 0 & \boxed{0.02405} \end{bmatrix} \text{[GPa] and}$$

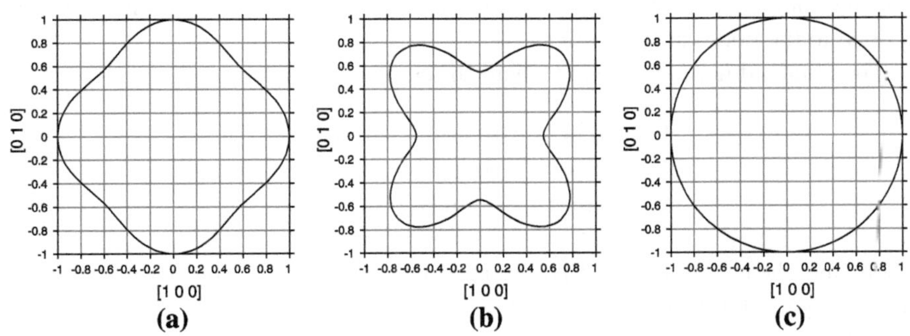

Fig. 14 Rotated tensorial components **a**E^H_{1111}, **b**E^H_{1212} and **c**Kt^H_{11} normalized with respect to $E^H_{1111} = 0.1150$ [GPa], $E^H_{1212} = 0.04397$ [GPa] and $Kt^H_{11} = 0.03897$ [W/m°C]

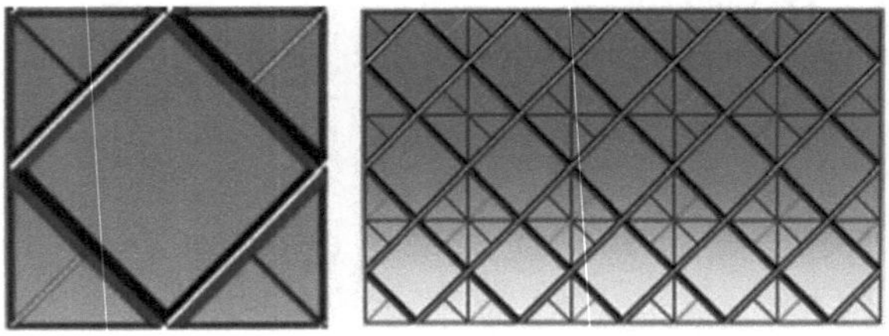

Fig. 15 Optimized cell and corresponding periodic material

$$Kt^{H} = \begin{bmatrix} 0.03897 & 0 \\ 0 & 0.03897 \end{bmatrix} \text{[W/m°C]}$$

where the values of the components E^H_{1212} and Kt^H_{11} are indicated.

For comparative purpose in the optimization cases to be studied, it is interesting to quantify the values of the four objective functions proposed for this initial cell:

- Objective function 1: E^H_{1212}= 0.02405[GPa];
- Objective function 2: $Kt^H_{11} = 0.03897$ [W/m°C];
- Objective function 3: $E^H_{1212}/Kt^H_{11} = 0.617$ [GPa m °C/W]
- Objective function 4: $E^H_{1212} \cdot Kt^H_{11} = 0.937$ [MPa W/m°C]

9.1.1 Maximization of Shear Stiffness (E^H_{1212})

In this case the shear component of the homogenized mechanical tensor E^H_{1212} was maximized keeping the overall volume of the cell unchanged ($f_U = 1$). All the element cross sectional areas and the coordinates *X* and *Y* of the internal nodes were adopted as design variables. Figure 15 depicts the optimized unit cell and the periodic material obtained. The cell is formed by a square rotated 45° with respect to the horizontal axis, and linked to the cell vertices by thin bars. This linking resulted due to the combination of the initial cell adopted (Fig. 13) and the fact that the homogeneous prescribed displacements (which avoid rigid motion) were applied to the nodes at the vertices.

The obtained homogenized mechanical tensor is given by

$$E^{H} = \begin{bmatrix} 0.09124 & 0.08886 & -0.001027 \\ 0.08886 & 0.09124 & -0.001026 \\ -0.001027 & -0.001026 & 0.09060 \end{bmatrix} \text{[GPa]}$$

where the shear component E^H_{1212} is indicated. Hence, the objective function 1 equals $E^H_{1212} = 0.09060$ [GPa], which is 3.76 times higher than the initial value.

The negative components in the homogenized tensor are caused small asymmetries in the unit cell, as will be discussed later. Two orthogonal planes of symmetry yield orthotropy and allow the negative components to be eliminated.

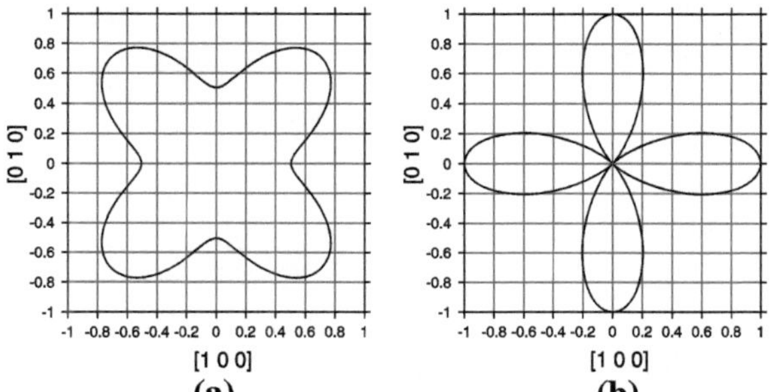

Fig. 16 Diagrams of the rotated values of **a** E^H_{1111} and **b** E^H_{1212} normalized with respect to $E^H_{1111} = 0.1806$ [GPa] and $E^H_{1212} = 0.0906$ [GPa]

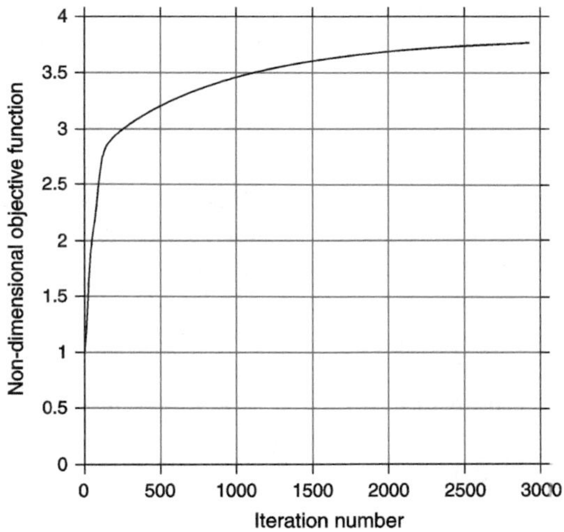

Fig. 17 Convergence of the non-dimensional objective function

Figure 16 shows the normalized diagrams for the rotated E^H_{1111} and E^H_{1212} components in the interpreted orthotropic cell.

Figure 17 presents the evolution of the objective function through the optimization process. The large amount of iterations is due to a very small value prescribed for move limits, as discussed in Sect. 6. Since the same qualitative convergence behavior is repeated in all the forthcoming examples, and due to conciseness requirements, the evolution of the objective function is presented only here and in the last example, for a 3D problem.

9.1.2 Maximization of Heat Conduction in Direction *X* (Kt^H_{11})

In this example, the homogenized thermal conductivity tensor component associated to conduction in direction *X* (Kt^H_{11}) is optimized for constant cell volume. All the cross sectional areas and the *X* and *Y* coordinates of the internal nodes in the cell are adopted as design variables. The optimal cell found is displayed in Fig. 18

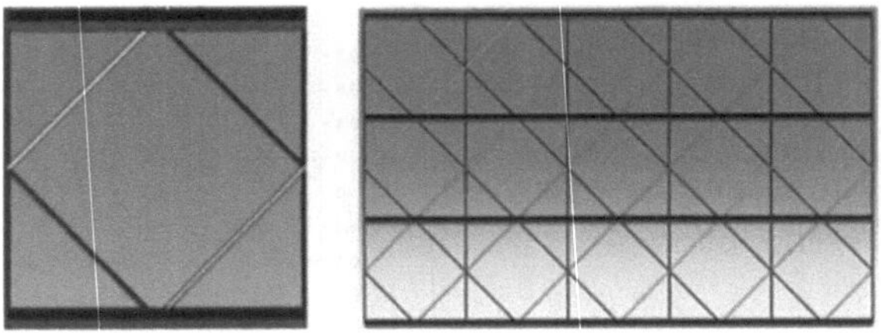

Fig. 18 Optimized cell and corresponding periodic material

as well as the corresponding periodic material. In this example, the ratio between the thicker and thinner cross sections is approximately 25.3, however thin bars should have vanished. Hence, the importance of results interpretation should not be underestimated.

The homogenized thermal tensor is given by

$$\boldsymbol{Kt}^H = \begin{bmatrix} 0.0845 & 0 \\ 0 & 0.002921 \end{bmatrix} \ [\mathrm{W/m^\circ C}]$$

where the value of the objective function 2 is recognized as $Kt_{11}^H = 0.0845$ [W/m°C], which is 2.16 times higher than the initial value.

Figure 19 displays the normalized diagram for the rotated Kt_{11}^H component.

Analyzing Fig. 18, it can be noticed that the initially internal bars were displaced to meet the upper and lower edges of the cell. At the same time, their cross sectional areas became larger, which is clearly in accordance with maximization of heat transfer in direction X. Figure 19 displays that, as a consequence, a negligible conductivity results in the vertical direction (aligned with the Y axis), thus favoring insulation.

9.1.3 Maximization of the Ratio Between Shear Stiffness and Thermal Conductivity in Direction X (E_{1212}^H/Kt_{11}^H)

This example aims to maximize the ratio between the shear E_{1212}^H and Kt_{11}^H thermal conductivity components, keeping a constant cell volume. Therefore, the

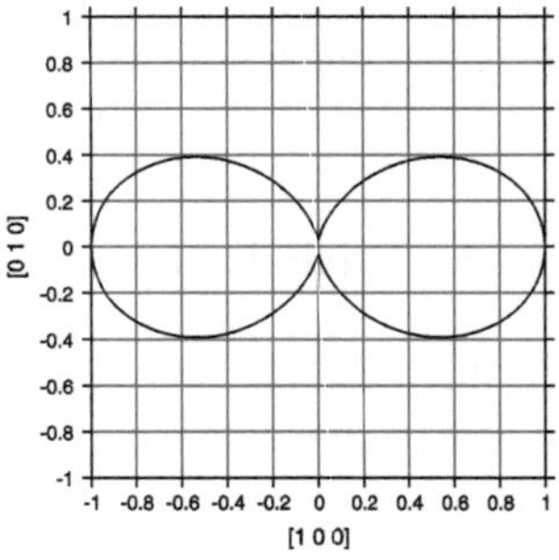

Fig. 19 Rotated tensorial component Kt_{11}^H normalized with respect to $Kt_{11}^H = 0.0845$ [W/m°C]

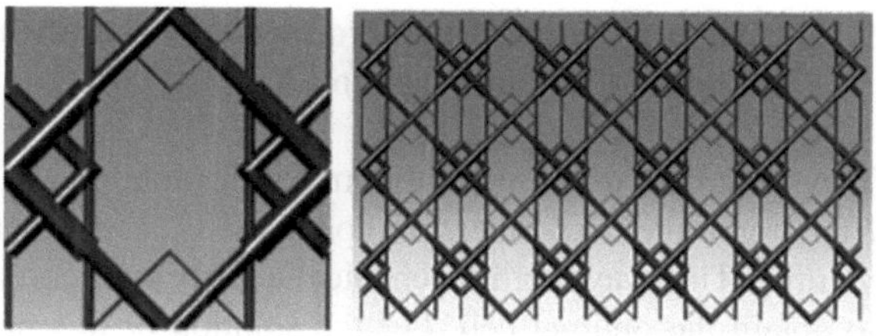

Fig. 20 Optimized cell for maximization of the ratio between the mechanical and thermal homogenized tensorial components shown, and corresponding periodic structure

procedure tends to maximize the mechanical component and minimize the thermal one. This problem is more complex than the previous ones since the two different properties considered may compete for the determination of the optimal cell. Only cross sectional areas are adopted as design variables and the optimized cell is displayed in Fig. 20.

The homogenized mechanical and thermal tensors are, respectively

$$E^H = \begin{bmatrix} 0.05910 & 0.05910 & 0 \\ 0.05910 & 0.0889 & 0 \\ 0 & 0 & 0.05910 \end{bmatrix} \text{[GPa] and}$$

$$Kt^H = \begin{bmatrix} 0.02933 & 0 \\ 0 & 0.04083 \end{bmatrix} \text{[W/m°C]}$$

In this case, the objective function 3 adopts the value $E^H_{1212}/Kt^H_{11} = 2.015$ [GPa·m·°C/W]. Comparing to the values of the initial cell, the objective function is 3.26 times higher, while the ratios $E^H_{1212}/E^H_{1212\_initial} = 2.46$ and $K^H_{11\_initial}/K^H_{11} = 1.33$. Therefore, there is a considerable gain both, in the ratio and in each individual component.

Rotations of the mechanical and thermal homogenized tensors provide the normalized plotted values for E^H_{1111} and E^H_{11} in Fig. 21, which evince an orthotropic behavior.

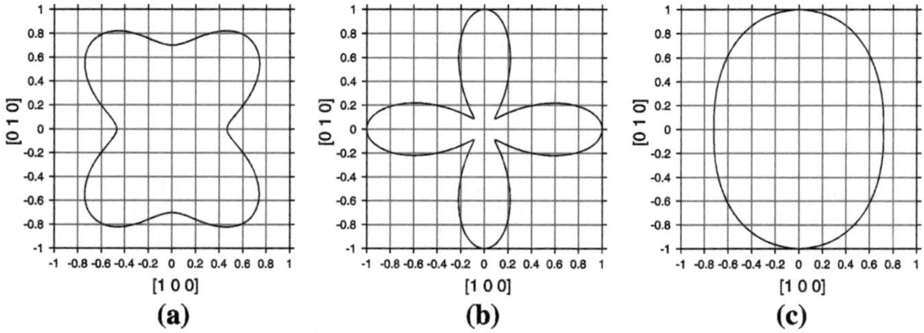

Fig. 21 Rotated tensorial components **a** E^H_{1111}, **b** E^H_{1212} and **c** Kt^H_{11} normalized with respect to $E^H_{1111} = 0.1266$ [GPa], $E^H_{1212} = 0.0591$ [GPa] and $Kt^H_{11} = 0.04083$ [W/m°C]

9.1.4 Maximization of the Product Between Shear Stiffness and Thermal Conductivity in Direction X ($E^H_{1212} \cdot Kt^H_{11}$)

This example deals with the maximization of the product between the shear E^H_{1212} and the Kt^H_{11} thermal conductivity components, keeping a constant cell volume. This way, both mechanical and thermal components will compete in order to obtain the optimal cell. For the solution of this problem, all the cross sectional areas and the X and Y coordinates of the internal nodes in the cell are adopted as design variables. Figure 22 shows the optimized cell and the corresponding periodic material.

The homogenized mechanical and thermal tensors are, respectively

$$E^H = \begin{bmatrix} 0.089540 & 0.07867 & 0 \\ 0.07867 & 0.09282 & 0 \\ 0 & 0 & \boxed{0.08451} \end{bmatrix} \text{[GPa] and}$$

$$Kt^H = \begin{bmatrix} \boxed{0.04439} & 0 \\ 0 & 0.04288 \end{bmatrix} \text{[W/m°C]}$$

and in this example, the final value of objective function 4 is given by $E^H_{1212} \cdot Kt^H_{11} = 3.751$ [MPa W/m°C], which is 4 times higher than the initial value. Furthermore, the ratios $E^H_{1212}/E^H_{1212\_initial} = 3.51$ and $K^H_{11}/K^H_{11\_initial} = 1.13$, showing gain for the product and for each component independently.

Figure 23 shows the normalized diagrams for rotated E^H_{1111}, E^H_{1212} and Kt^H_{11}.

The same problem was solved, this time adopting only cross sectional areas as design variables. The optimal cell found is displayed in Fig. 24 together with the corresponding periodic material.

The homogenized mechanical and thermal tensors are, respectively

$$E^H = \begin{bmatrix} 0.09859 & 0.09447 & 0 \\ 0.09447 & 0.09447 & 0 \\ 0 & 0 & \boxed{0.09447} \end{bmatrix} \text{[GPa] and}$$

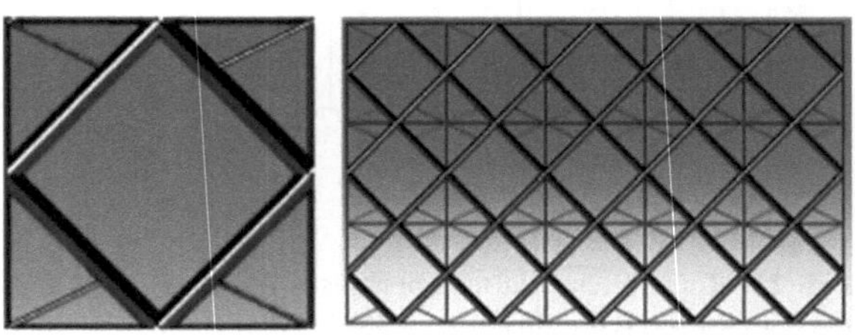

Fig. 22 Optimized cell for maximization of the ratio between the mechanical and thermal homogenized tensorial components shown, and corresponding periodic structure

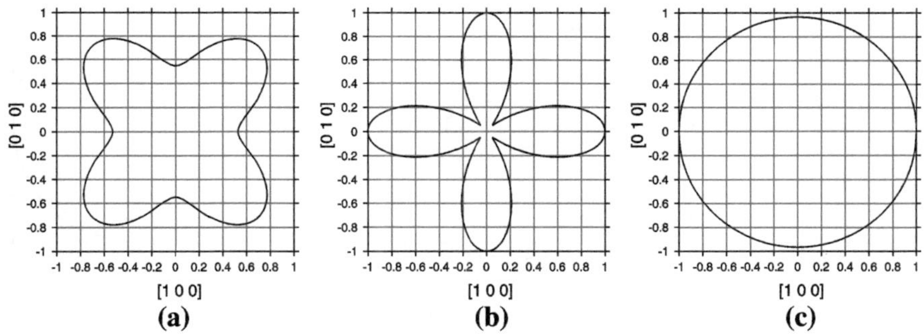

Fig. 23 Rotated tensorial components **a** E^H_{1111}, **b** E^H_{1212} and **c** Kt^H_{11} normalized with respect to $E^H_{1111} = 0.1694$ [GPa], $E^H_{1212} = 0.08451$ [GPa] and $Kt^H_{11} = 0.04439$ [W/m°C]

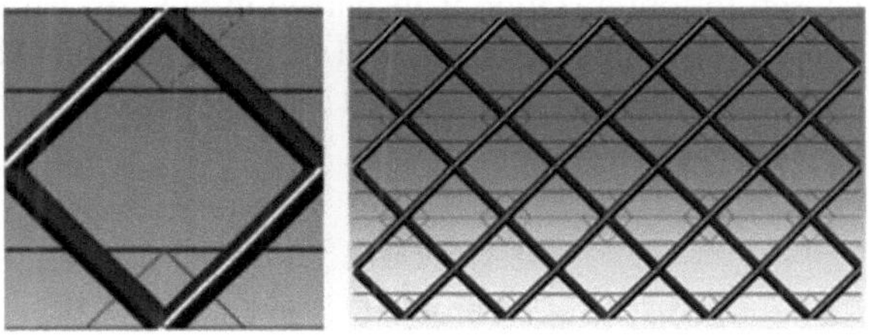

Fig. 24 Optimized cell for maximization of the ratio between the mechanical and thermal homogenized tensorial components shown, and corresponding periodic structure

$$\boldsymbol{Kt}^H = \begin{bmatrix} 0.04602 & 0 \\ 0 & 0.04504 \end{bmatrix} \text{ [W/m°C]}$$

and the objective function 4 is given by $E^H_{1212} \cdot Kt^H_{11} = 4.347$ [MPa W/m°C], which is 4.64 times higher than the initial value. The ratios $E^H_{1212}/E^H_{1212\_initial} = 3.93$ and $K^H_{11}/K^H_{11\_initial} = 1.16$. Note that this case, which employs less design variables than the former, led to a better result. This could possibly be attributed to the move limits adopted or to non unicity of the solution.

Rotations of the mechanical and thermal homogenized tensors provide the normalized plotted values for E^H_{1111}, E^H_{1212} and Kt^H_{11} in Fig. 25 which evince an orthotropic behavior.

9.2 *3D Example*

9.2.1 Maximization of the Product Between Shear Stiffness and Thermal Conductivity in Direction X ($E^H_{1212} \cdot Kt^H_{11}$)

A 3D example is presented, which aims to maximize the product between the homogenized constitutive component related to shear (E^H_{1212}) and heat conduction in X (Kt^H_{11}) direction, keeping a constant cell volume. The initial 3D unit cell

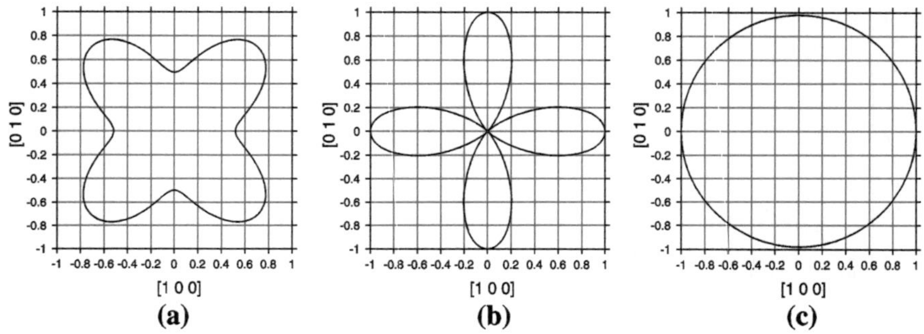

Fig. 25 Rotated tensorial components **a** E^H_{1111}, **b** E^H_{1212} and **c**Kt^H_{11} normalized with respect to $E^H_{1111} = 0.189$ [GPa], $E^H_{1212} = 0.09447$ [GPa] and $Kt^H_{11} = 0.04602$ [W/m°C]

chosen for optimization is displayed in Fig. 26. All the cross sectional areas were adopted as design variables, and their initial value was set to $A = 2\ \ 10^{-5}$ [m^2].

The mechanical and thermal homogenized tensors of the initial cell are identified, respectively, as

$$E^H = \begin{bmatrix} 15.41 & 3.466 & 3.466 & 0 & 0 & 0 \\ 3.466 & 15.41 & 3.466 & 0 & 0 & 0 \\ 3.466 & 3.466 & 15.41 & 0 & 0 & 0 \\ 0 & 0 & 0 & \boxed{2.702} & 0 & 0 \\ 0 & 0 & 0 & 0 & 2.702 & 0 \\ 0 & 0 & 0 & 0 & 0 & 2.702 \end{bmatrix} \text{[GPa] and}$$

$$Kt^H = \begin{bmatrix} \boxed{5.421} & 0 & 0 \\ 0 & 5.421 & 0 \\ 0 & 0 & 5.421 \end{bmatrix} \text{[W/m°C]}$$

From the tensors above, the initial value of the objective function 4 provides $E^H_{1212} \cdot Kt^H_{11} = 14.647 \quad [\text{GPa} \cdot \text{W/m}^{\text{o}}\text{C}]$.

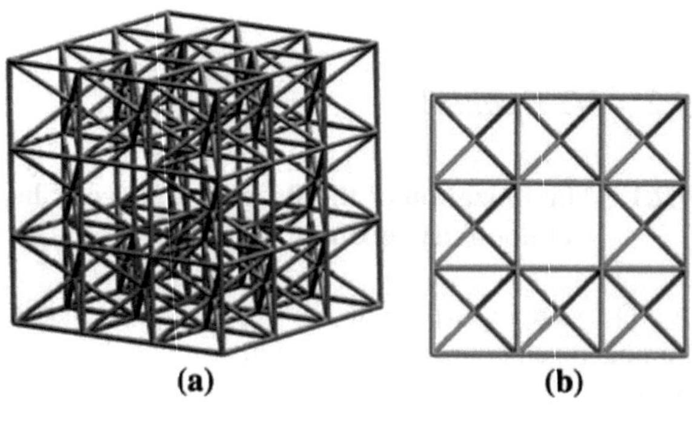

Fig. 26 **a** Initial 3D cell. **b** View of planes *XoY*, *XoZ* and *YoZ*

The homogenized mechanical and thermal homogenized tensors for the optimized material are, respectively

$$E^H = \begin{bmatrix} 23.32 & 5.828 & 5.841 & 0.02793 & -0.008291 & -0.04669 \\ & 5.909 & 5.840 & 0.02793 & -0.008292 & -0.04669 \\ & & 5.945 & \boxed{0.02793} & -0.008292 & -0.04669 \\ & & & 5.838 & -0.04669 & -0.008293 \\ & Sym & & & 5.788 & 0.02792 \\ & & & & & 5.790 \end{bmatrix} \text{[GPa] and}$$

$$Kt^H = \begin{bmatrix} \boxed{8.343} & 0.01996 & -0.03335 \\ & 4.197 & -0.05920 \\ Sym & & 4.203 \end{bmatrix} \text{[W/m°C]}.$$

The final value of the objective function 4 is given by $E^H_{1212} \cdot Kt^H_{11} = 48.71$ [GPa W/m°C], a value 3.32 times higher than for the initial cell. In addition, the ratios $E^H_{1212}/E^H_{1212\_initial} = 2.16$ and $K^H_{11}/K^H_{11\_initial} = 1.54$.

The optimized cell obtained is depicted in Fig. 27 and the normalized graphs in Figs. 28, 29 and 30 display the material properties behavior through rotations of the homogenized mechanical and thermal constitutive tensors. The components analyzed are E^H_{1111}, E^H_{2222}, E^H_{1212}, Kt^H_{11} and Kt^H_{22}, and the rotations are displayed in polar coordinates with respect to planes XoY ([1 0 0] and [0 1 0]), XoZ ([1 0 0] and [0 0 1]) and YoZ ([0 1 0] and [0 0 1]).

It should be noted that slight asymmetries in the cell render anisotropic terms in the homogenized tensors. Since no variable linking was imposed in the optimization process, minor asymmetries are expected even if the proper result is to be perfectly symmetric. Hence, an interpretation of the optimized cell result becomes necessary. In the homogenized optimum tensors obtained, many terms should be zero, including the negative ones. This would lead to orthotropic and isotropic

Fig. 27 Optimized 3D cell

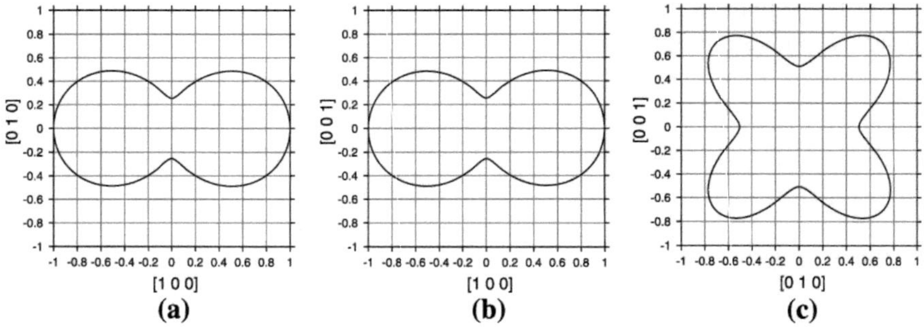

Fig. 28 Rotated E^H_{1111} (normalization value). **a** Plane *XoY* (23.32026 [GPa]), **b** Plane *XoZ* (23.32073 [GPa]). Rotated E^H_{2222} (normalization value). **c** Plane *YoZ* (11.6715 [GPa])

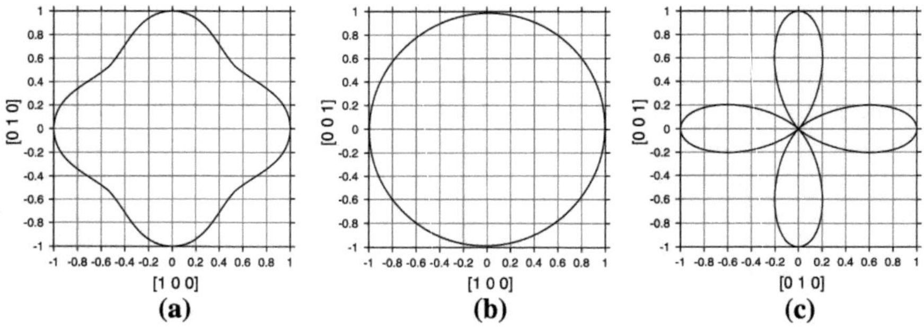

Fig. 29 Rotated E^H_{1212} (normalization value). **a** Plane *XoY* (5.8380 [GPa]), **b** Plane *XoZ* (5.8380 [GPa]). Rotated E^H_{2323} (normalization value). **c** Plane *YoZ* (5.788 [GPa])

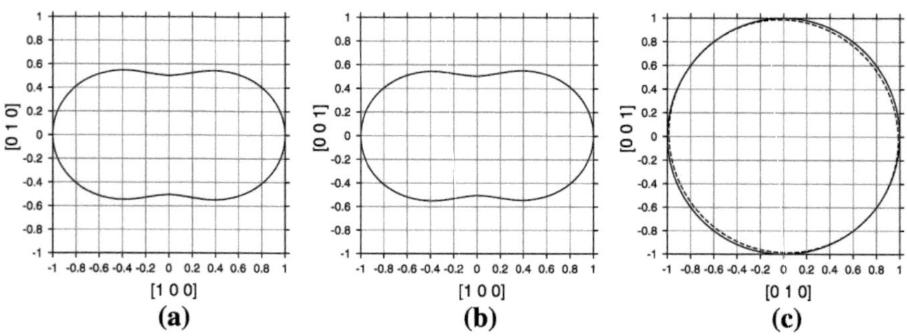

Fig. 30 Rotated Kt^H_{11} (normalization value). **a** Plane *XoY* (8.343 [W/m°C]), **b** Plane *XoZ* (8.343 [W/m°C]). Rotated Kt^H_{22} (normalization value). **c** Plane *YoZ* isotropic—*thick line* (4.203 [W/m°C]) and Plane *YoZ* anisotropic—*dashed line* (4.2592 [W/m°C])

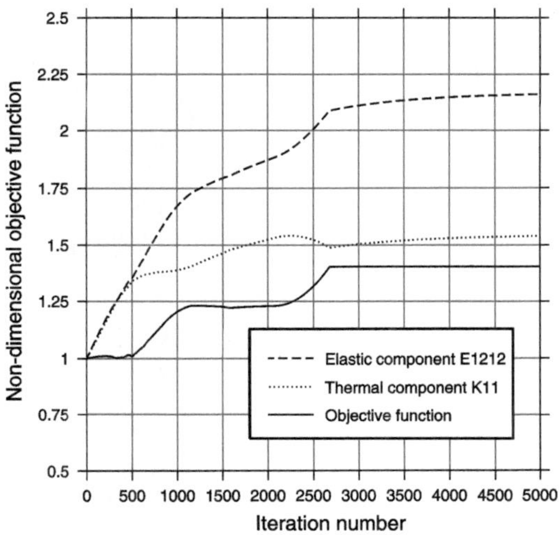

Fig. 31 Convergence of the non-dimensional objective function

tensors for the mechanical and thermal cases, respectively. Among all the Figs. 28, 29 and 30, only Fig. 30c reveals graphically a subtle effect of the numerical litter that causes anisotropy of the thermal tensor.

Hence, if it is known beforehand that for the application desired, the optimized material should be orthotropic, corresponding geometric symmetry conditions (in two orthogonal planes) and variable linking should be imposed in the unit cell. This would avoid spurious terms in the homogenized constitutive tensor.

Figure 31 shows the objective function evolution through the optimization process. Notice that approximately 2500 iterations are necessary for convergence. However, it should be remarked that the convergence criterion applied (in all the cases studied) considers oscillations in the objective function and also in the design variables.

Figure 32 displays the periodic structure formed by the optimized cell. Notice that horizontal bars favoring heat transfer appeared in the *X* direction, and no analogous bars showed up aligned to *Y* or *Z* directions. Thus, the structure is effectively tailored for the multifunctional task to which it was designed.

10 Concluding Remarks

The Chapter initially describes the homogenization theory applied to periodic materials made of truss-like unit cells. These materials, frequently referred to as Lattice Block Materials (LBMs), have been the focus of recent research especially for high performance engineering applications. A finite element code was implemented for the determination of their homogenized mechanical and thermal properties in either 2D or 3D situations.

In the sequence, a layout optimization procedure was proposed for obtaining tailored LBMs for multifunctional thermal and mechanical requirements. The main

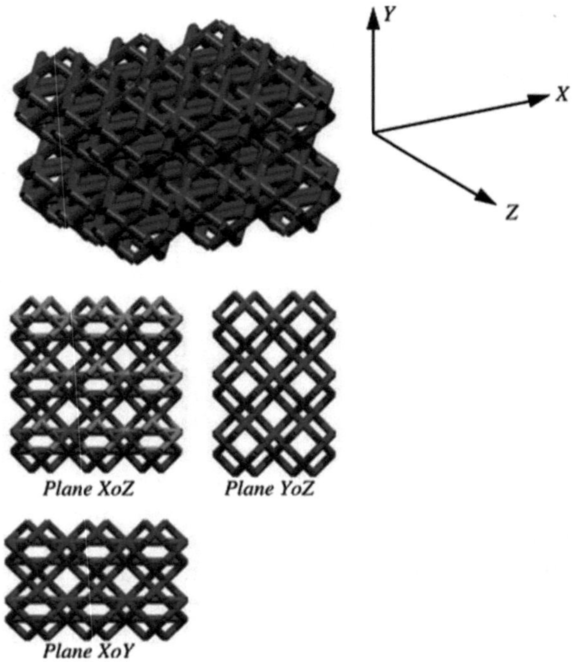

Fig. 32 Material resulting from the periodic arrangement of cells displayed in Fig. 27

particularities of the proposed approach are (i) the simultaneous use of size and shape design variables, thus reducing the size usually required for a ground structure; (ii) the definition of a multifunctional objective function; (iii) the development of analytical sensitivities with respect to the two types of design variables considered and (iv) the presentation of 3D results.

SLP was employed as the optimization algorithm. Although limitations were encountered with respect to stability for automatic schemes of move limits update, the method rendered good results provided very small move limits were adopted (constant for area variables). However, a more elaborate mathematical programming method is recommended for achieving better efficiency.

Promising preliminary results were obtained and the code implementation gives generality for further investigations. One interesting issue to be evaluated in the future is the effect of flexural and torsional stiffness in the material response, since the use of bar elements does not account for this aspect.

References

1. Torquato, S., Hyun, S., Donev, A.: Multifunctional composites: optimizing microstructures for simultaneous transport of heat and electricity. Phys. Rev. Lett. **89**(26), 266601-1–266601-4 (2002)
2. Luxner, M.H., Woesz, A., Stampfl, J., Fratzl, P., Pettermann, H.E.: A finite element study on the effects of disorder in cellular structures. Acta Biomater. **5**, 381–390 (2009)

3. Gibson, L.J., Ashby, M.F.: Cellular solids: structure and properties, 2nd edn. Cambridge University Press, Cambridge (1997)
4. Banhart, J.: Manufacture, characterisation and application of cellular metals and metal foams. Prog. Mater. Sci. **46**, 559–632 (2001)
5. Zhao, C.Y., Lu, W., Tassou, S.A.: Thermal analysis on metal-foam filled heat exchangers. Part II: tube heat exchangers. Int. J. Heat Mass Transf. **49**(15–16), 2762–2770 (2006)
6. Niu, B., Yan, J., Cheng, G.: Optimum structure with homogeneous optimum cellular material for maximum fundamental frequency. Struct. Multidiscip. Optim. **39**, 115–132 (2009)
7. Luxner, M.H., Stampfl, J., Pettermann, H.E.: Finite element modeling concepts and linear analyses of 3D regular open cell structures. J. Mater. Sci. **40**, 5859–5866 (2005)
8. McKown, S., Shen, Y., Brookes, W.K., Sutcliffe, C.J., Cantwell, W.J., Langdon, G.S., Nurick, G.N., Theobald, M.D.: The quasi-static and blast loading response of lattice structures. Int. J. Impact Eng. **35**, 795–810 (2008)
9. Ziegler, E., Accorsi, M., Bennet, M.: Continuum plate model for lattice block material. Mech. Mater. **36**, 753–766 (2004)
10. Liu, T., Deng, C., Lu, T.J.: Design optimization of truss-cored sandwiches with homogenization. Int. J. Solids Struct. **43**, 7891–7918 (2006)
11. Wallach, J.C., Gibson, L.J.: Mechanical behaviour of a three-dimensional truss material. Int. J. Solids Struct. **38**, 7181–7196 (2001)
12. Diaz, A.R., Haddow, A.G., Ma, L.: Design of band-gap grid structures. Struct. Multidiscip. Optim. **29**, 418–431 (2005)
13. Gonella, S., Albert, C.T., Liu, W.K.: Interplay between phononic bandgaps and piezoelectric microstructures for energy harvesting. J. Mech. Phys. Solids **57**, 621–633 (2009)
14. Martinsson, P.G., Movchan, A.B.: Vibrations of lattice structures and phononic band gaps. Q. J. Mech. Appl. Math. **56**(1), 45–64 (2003)
15. Brown, R.T., Zureick, A.: Lightweight composite truss section decking. Mar. Struct. **14**, 115–132 (2001)
16. Reynier, P., Reisch, U., Longo, J.M., Radespiel, R.: Flow predictions around a missile with lattice wings using the actuator disc concept. Aerosp. Sci. Technol. **8**, 377–388 (2004)
17. Luxner, M.H., Stampfl, J., Pettermann, H.E.: Numerical simulations of 3D open cell structures—influence of structural irregularities on elasto-plasticity and deformation localization. Int. J. Solids Struct. **44**, 2990–3003 (2007)
18. Yan, J., Cheng, G., Liu, L., Liu, S.: Concurrent material and structural optimization of hollow plate with truss-like material. Struct. Multidiscip. Optim. **35**, 153–163 (2008)
19. Fan, H.L., Fang, D.N., Jing, F.N.: Yield surfaces and micro-failure mechanism of block lattice truss materials. Mater. Des. **29**, 2038–2042 (2008)
20. Fan, H.L., Jin, F.N., Fang, D.N.: Nonlinear mechanical properties of lattice truss materials. Mater. Des. **30**, 511–517 (2009)
21. Lipperman, F., Fuchs, M.B., Ryvkin, M.: Stress localization and strength optimization of frame material with periodic microstructure. Comput. Methods Appl. Mech. Eng. **197**, 4016–4026 (2008)
22. Lim, J., Kang, K.: Mechanical behavior of sandwich panels with tetrahedral and Kagome truss cores fabricated from wires. Int. J. Solids Struct. **43**, 5528–5246 (2006)
23. Bele E, Bouwhuis BA, Hibbard GD (2008) Work hardening as a strengthening mechanism in periodic cellular materials. Mat. Sci. Eng. A Struct. 489: 29–37
24. Suralvo, M., Bouwhuis, B., McCrea, J.L., Palumbo, G., Hibbard, G.D.: Hybrid nanocrystalline periodic cellular materials. Scripta Mater. **58**, 247–250 (2008)
25. Kooistra, G.W., Wadley, H.N.G.: Lattice truss structures from expanded metal sheet. Mater. Des. **28**, 507–514 (2007)
26. Yan, J., Cheng, G., Liu, S., Liu, L.: Comparison of prediction on effective elastic property and shape optimization of truss material with periodic microstructure. Int. J. Mech. Sci. **48**, 400–413 (2006)
27. Lipperman, F., Ryvkin, M., Fuchs, M.B.: Fracture toughness of two-dimensional cellular material with periodic microstructure. Int. J. Fract. **146**, 279–290 (2007)

28. Liu, L., Yan, J., Cheng, G.: Optimum structure with homogeneous optimum truss-like material. Comput. Struct. **86**, 1417–1425 (2008)
29. Scheffler, M., Colombo, P.: Cellular ceramics: structure, manufacturing, properties and applications. Wiley–VCH, Weinheim (2005)
30. Luxner, M.H., Stampfl, J., Pettermann, H.E.: Linear and nonlinear investigations of regular open cell structures. In: Proceedings of IMECE'04, 2004 ASME International Mechanical Engineering Congress, Anaheim, California, USA (2004)
31. Lee, B., Kang, K.: A parametric study on compressive characteristics of wire-woven bulk Kagome truss cores. Compos. Struct. **92**, 445–453 (2010)
32. Hyun, S., Choi, J., Kang, K.: Effect of imperfections on the mechanical behavior of wire-woven bulk kagome truss PCMs under shear loading. J. Mech. Sci. Technol. **23**, 1270–1277 (2009)
33. Hyun, S., Choi, J., Kang, K.: Effects of imperfections on the mechanical behavior of a wire-woven bulk Kagome cellular metal under compression. Comp. Mater. Sci. **46**, 73–82 (2009)
34. Sanchez-Palencia, E.: Non-homogeneous media and vibration theory. Lecture Notes in Physics, vol. 127. Springer Verlag, Berlin (1980)
35. Hassani, B., Hinton, E.: Homogenization and structural topology optimization, Springer-Verlag, London (1999)
36. Hyun, S., Torquato, S.: Optimal and manufacturable two-dimensional, Kagomé-like cellular solids. J. Mater. Res. **17**(1), 137–144 (2002)
37. Sigmund, O.: Materials with prescribed constitutive parameters: an inverse homogenization problem. Int. J. Solids Struct. **31**, 2313–2329 (1994)
38. Sigmund, O.: Tailoring materials with prescribed elastic properties. Mech. Mater. **20**, 351–368 (1995)
39. Deshpande, V.S., Ashby, M.F., Fleck, N.A.: Foam topology bending versus stretching dominated architectures. Acta Mater. **49**, 1035–1040 (2001)
40. Evans, A.G., Hutchinson, J.W., Fleck, N.A., Ashby, M.F., Wadley, H.N.G.: The topological design of multifunctional cellular metals. Prog. Mater. Sci. **46**, 309–327 (2001)
41. Prasad, J., Diaz, A.R.: Synthesis of bistable periodic structures using topology optimization and a genetic algorithm. J. Mech. Des. **128**, 1298–1306 (2006)
42. Ohsaki, M., Nishiwaki, S.: Shape design of pin-jointed multistable compliant mechanisms using snapthrough behavior. Struct. Multidiscip. Optim. **30**, 327–334 (2005)
43. Dede, E.M., Hulbert, G.M.: Computational analysis and design of lattice structures with integral compliant mechanisms. Finite Elem. Anal. Des. **44**, 819–830 (2008)
44. Dede, E.M., Hulbert, G.M.: Topology optimization of structures with integral compliant mechanisms for mid-frequency response. Struct. Multidiscip. Optim. **39**, 29–45 (2009)
45. Fonseca, J.S.O.: Design of microstructures of periodic composite materials. Ph.D. Thesis, University of Michigan, Ann Arbor, Michigan, USA (1997)
46. Hassani, B., Hinton, E.: A review of homogenization and topology optimization I —homogenization theory for media with periodic structure. Comput. Struct. **69**, 707–717 (1998)
47. Guedes, J.M., Kikuchi, N.: Preprocessing and postprocessing for materials based on the homogenization method with adaptive finite element methods. Comput. Methods Appl. Mech. **83**(2), 143–198 (1990)
48. Muñoz-Rojas, P.A., Silva, E.C.N., Cardoso, E.L., Vaz, M., Jr.: On the application of optimization techniques to heat transfer in cellular materials. In: Öchsner, A., Murch, G.E., Lemos, M.J.S. (eds.) Cellular and porous materials: thermal properties simulation and prediction. Wiley-VCH Verlag GmbH & Co. KGaA, Weinheim (2008)
49. Cook, R.D., Malkus, D.S., Plesha, M.E.: Concepts and applications of finite element analysis. Wiley, New York (2002)
50. Yang, Q., Becker, W.: Numerical investigation for stress, strain and energy homogenization of orthotropic composite with periodic microstructure and non-symmetric inclusions. Comput. Mater. Sci. **31**, 169–180 (2004)

51. Bendsoe, M.P., Sigmund, O.: Topology optimization: theory, methods, and applications, 2nd edn. Springer-Verlag, Berlin, Germany (2004)
52. Beckers, M.: Optimisation de structures en variables discrètes. Ph.D. Thesis, Universite de Liege, Belgium (1997)
53. Hansen, S., Vanderplaats, G.N.: Approximation method for configuration optimization of trusses. AIAA J. **28**, 161–168 (1990)
54. Pedersen, P.: Optimal joint positions for space trusses. J. Struct. Div. ST12 (1973)
55. Vanderplaats, G.N.: Numerical optimization techniques for engineering design. McGraw-Hill, New York (1984)
56. Carniel, T.A.,.Muñoz-Rojas, P.A.: An´lise e otimização de células periódicas de "Lattice Block Materials" visando aplicações termo-mecânicas. Internal Report (in Portuguese). Computational Mechanics Laboratory, Department of Mechanical Engineering, State University of Santa Catarina, Brazil (2009)

Computational Model of Porous Media Using True 3-D Images

Khairul Alam, Mihnea S. Anghelescu, and Adrian Bradu

Abstract Thermally conductive foams are being developed for many engineering applications; and there is a need to develop analytical models to predict the thermal properties of such porous media. Most of the current models are based on volume averaging techniques, and often assume simple, ideal shapes for the pore geometry. The method described in this chapter focuses on modeling the thermal and flow properties of foams on the basis of its true microstructure. The approach is to take a three dimensional solid model of a real foam, obtained by imaging techniques, and use it as the basis for the numerical solution of the transport phenomena. This is a micro-model, in which the thermal phenomena are modeled at the pore level of the foam. The model is computationally intensive, as can be expected; but it does not require semi-empirical or experimentally derived constants such as permeability to derive a solution. By incorporating the effect of the true pore geometry on the thermal transport and fluid flow in the foam, this model is able to determine the thermal conductivity, permeability, friction factor and heat transfer coefficients. Graphitic carbon and silicon carbide foams are used in this study, but the approach that is described is quite general and can be applied to other porous media; it may also be applied to composites that contain phases with distinct boundaries at the micro-level.

1 Introduction

Foams are porous materials which are attractive for many engineering applications because their properties can be customized by varying the manufacturing process. Polymeric foams and metal foams are quite common in many engineering

K. Alam (✉), M.S. Anghelescu, and A. Bradu
Department of Mechanical Engineering, Ohio University, Athens, OH 45701, USA
e-mail: alam@ohio.edu; ma138903@ohio.edu; ab315905@ohio.edu

A. Öchsner and G.E. Murch (eds.), *Heat Transfer in Multi-Phase Materials*,
Adv Struct Mater 2, DOI 10.1007/8611_2010_7,

applications. Recent developments in foams have generated a new class of thermally conductive foams which have potential applications in thermal management and heat exchangers. For example, there are a variety of graphitic foams that can be produced in a range of densities by varying the porosity. The thermal conductivity of the foam can be also altered by changing the intrinsic conductivity of the solid phase through variations in heat treatment. By controlling these two dominant parameters, it is possible to produce foams of desired thermal conductivity over a wide range of values. Another example of conductive foam for thermal applications is the silicon carbide foam, which can be used at higher temperatures. The porosity of the silicon carbide foam and the intrinsic conductivity of the silicon carbide can also be varied to produce a desired bulk thermal property.

Conductive foams can greatly enhance heat transfer to a coolant flowing through the foam pores. In traditional heat sinks, a heat spreader with finned structures has been used to improve convective heat flux. Foams are similar to finned surfaces, but can support much higher heat flux because of the higher surface area. With the emergence of thermally conductive non-metallic foams, the design of thermal management strategies will require the development of new models that incorporate the unique characteristics of these new foams. It is important to note that these foams can be very different from the traditional metal foams that have been used in heat sinks. For example, the microstructure of the graphitic carbon foam varies due to the manufacturing process and heat treatments. This can lead to very different thermal transport and flow behavior within the foams. The pressure drop for flow through foam is a critical parameter for heat transfer applications; and it depends on the complex geometry of the foam at the pore level. The nature of the fluid flow and the thermal properties of the foam will determine the heat transfer effectiveness of the system. However, a highly random geometry at pore level makes it very difficult to analyze the properties and the behavior of this material by simple empirical formula. Because of the complex geometry, the transport properties of the foam are dependent on the three-dimensional geometry. It is difficult to replicate the thermal transport behavior by using a two-dimensional model, especially when a fluid flow in the foam is to be analyzed.

In this chapter, the approach focuses on modeling the thermal and flow properties of foams on the basis of its true microstructure, without using an averaging process or a simple pore geometry. The approach is to take a three dimensional solid model, obtained by imaging techniques, based on an accurate representation of the real geometry of carbon foam at the pore level. This is a micro-model, in which the thermal phenomena are modeled at the pore level of the foam. The solid model developed from the imaging process is then meshed and a numerical solution is obtained to predict the thermal behavior of the foam. The model is computationally intensive, as can be expected; but it forms a theoretical basis for the determination of the thermal conductivity, diffusivity, fluid friction and convective heat transfer coefficients. Such a model incorporates the effect of the pore geometry on the thermal transport and the fluid flow. As a result, the pore level model can provide the surface area of the foam (per unit volume) and the theoretical value of the permeability; in contrast, the traditional models require permeability as a parameter for the solution, determined either experimentally, or by semi-empirical methods.

Graphitic carbon foams and silicon carbide foams are used in this study, but the approach that is described is quite general and can be applied to other porous media; it can also be applied to composites that contain phases with distinct boundaries at the micro-level. Special attention will be given to the modeling of the carbon foam because of its complex microstructure that presents unique challenges to solid modeling, and numerical solution. The silicon carbide foam has a simpler microstructure, and the 3-D model for silicon carbide was developed to evaluate the success of this approach. It is important to note that a unique aspect of this approach is the development of the accurate solid model representation of 3D carbon foam microstructure, which can be used to investigate the bulk thermal, mechanical or other properties of porous media and to study the fluid flow and heat transfer phenomena through the porous media without the use of experimental or empirical parameters.

2 Silicon Carbide and Carbon Foams

The thermal conductivity of metals makes them a good candidate for enhancing heat transfer; while, non-metallic foams, such as polymeric foams tend to be insulators. Open cell conductive foams can be used as extended surfaces to improve the heat flux in a flow system and such foams will be the focus of this chapter. Open cell silicon carbide (SiC) foams have been used in high temperature systems. Recent developments include SiC coated foams that are made by chemical vapor deposition on a carbon foam skeleton [1]. The coating is thick enough that the foam properties are virtually as if the carbon skeleton was absent, and the deposition parameters can be controlled to improve the thermal properties of the foam. Figure 1a shows an SEM picture of a silicon carbide foam. The pores of this

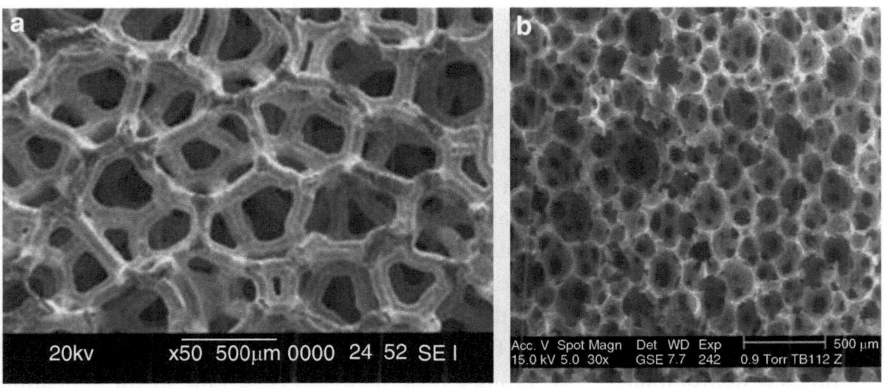

Fig. 1 (**a**) Silicon carbide foam (*Source*: [1]) and (**b**) SEM picture of graphitic carbon foam (*Source*: AFRL, Dayton, OH, USA)

foam are bigger than the pores of the carbon foam next to it; however, it should be noted that carbon foam comes in many different pore sizes.

Carbon foams are very versatile in that they can work as insulators as well as excellent conductors, and the cell structure can be open or closed. For example, reticulated vitreous carbon foam (RVC), which is produced by pyrolyzing a polymer, is an insulator. On the other hand, carbon foams made from mesophase pitches will have low thermal conductivity after carbonization, but can be graphitized by heat treatment to produce a highly conductive graphitic foam. The graphitic foams are also distinct from the RVC foams and metal foams in their microstructure. In RVC and metal foams, the pores are contained within ligaments that are similar to struts [2] and the size of the pore is similar to the size of the openings that interconnect the pores.

The graphitic carbon foam in Fig. 1b was made at the Air Force Research Laboratory (AFRL, Dayton, OH, USA). It can be seen that the microstructure of the two foams are very different. Graphitic foams tend to have bubble-shaped pores, and the pore wall structure is similar to stretched membranes, with thicker cross section at the junctions between the cells. The pore window in the wall membrane is typically smaller than the pore size. Another level of complexity is added when the anisotropy of the graphitic microstructure is considered. This is due to the higher degree of alignment of the graphene planes in the walls (ligaments) of the pores; while the junctions (nodes) have a disrupted, less graphitic structure. As a result, the thermal conductivity is anisotropic and inhomogeneous at the pore level of graphitic foams. Even at the bulk level, there is often a significant difference in the thermal conductivity between the vertical and the horizontal directions of the processing chamber where the foam is produced [3].

The microstructure of the carbon foam can be tailored by changing the precursor, the foaming process and the heat treatment conditions [4]. In addition, the low density and low coefficient of thermal expansion (CTE) of carbon foams make them suitable for utilization in many engineering applications. However, the wide range of microstructures of the carbon foam makes it a complex media that is harder to model and analyze. The following paragraphs describe the processing steps that produce the unique characteristics of the graphitic carbon foam.

The first carbon foams were developed by W. Ford in the 1960s as reticulated vitreous carbon foams by carbonizing thermosetting polymer foams [5]. In the 1990s, scientists at Air Force Research Laboratory (AFRL, Dayton, OH) developed graphitic carbon foams by heating a mesophase pitch precursor to the melting point while being pressurized in an inert atmosphere; this is then followed by "blowing", a process of releasing the gas pressure [6, 7]. The blowing process produces pores from bubbles which grow to form either a closed cell or an open cell structure; in the latter case the walls between the cells have openings that can allow a flow through the structure (Fig. 1). After the foaming process, the carbon foam is usually stabilized at approximately 170°C and a heat treatment consisting of carbonization at 1,000°C and graphitization at 2,700°C is applied [7]. Thermal conductivity of graphitic carbon foams generally range between 1 and 250 W/m°C depending on the microstructure, porosity and process parameters. From the SEM picture of a graphitic carbon foam produced at AFRL in Fig. 1, it can be seen that the pores of this foam are spherical or elliptical with diameter ranging between 100 and 350 μm.

Current producers of carbon foams in USA includes Graftech International (Parma, OH), Koppers (Pittsburg, PA), Poco Graphite (Decatur, TX), and Touchstone Research Laboratories (Triadelphia, WV). Graphitic foam development at the Oak Ridge National Laboratory has been described in several publications [5]. A method to manufacture carbon foams from precursors such as coal, petroleum pitch, coal tar pitch has been reported by researchers at West Virginia University [8]. Potential utilizations of non-graphitic carbon foams include structural applications such as tooling for composite materials manufacturing [9], stiffener inserts and core materials for composite sandwich structures. Non-graphitic carbon foams are also attractive as thermal protection materials, as they can be produced with very low thermal conductivity [10].

3 Model of Foams: The Macro-Model

Studies on the analysis of foams have generally employed two approaches to analyze porous media. The first is the classical macro-approach, in which the effect of the pores on the foam properties are averaged over a representative volume, or unit cell. A further simplification in the macro approach that is often adopted is to assume thermal equilibrium between the foam and the fluid in the pores which makes it easier to develop analytical solutions. The second approach, developed more recently, uses an idealized foam model in which the pore microstructure is built up by defining ligaments of simple geometry, which are then joined together, or selecting pores of specific shape, distributed uniformly throughout a volume; this may be termed the idealized micro-approach. These two approaches will be discussed is the following sections.

3.1 Macro-Model for Thermal Conductivity

The classical approach to modeling porous media is based on averaging the effect of randomly distributed pores, typically spherical in shape. An averaging method was used by Bauer [11] to determine the bulk (or effective) thermal conductivity of foams. The most important parameter for the conductivity is the porosity (void content) of a porous material, defined as [12]:

$$\varphi(\%) = \frac{V_f}{V} \cdot 100 = \frac{V_f}{V_f + V_s} \cdot 100 \tag{1}$$

where V_f is the volume of the pores and V_s is the volume of the solid phase.

Experimental determination of porosity is generally done by measuring the bulk density and then using the following relation:

$$\varphi(\%) = \left(1 - \frac{\rho_b}{\rho_s}\right) \cdot 100 \tag{2}$$

where ρ_b is the bulk (or effective) density and ρ_s is the solid phase density.

The surface area per unit volume of a porous material is defined as:

$$a_{sf} = \frac{A_{sf}}{V} = \frac{A_{sf}}{V_f + V_s} \tag{3}$$

where A_{sf} is the total pore surface area. It should be mentioned that experimental determination of surface area per unit volume of a porous material can be quite difficult. However, the 3-D model that will be described in a later section can be used to determine this value quite readily.

The traditional model for the bulk thermal conductivity of a foam as a function of the density was presented by Bauer [11] and also by Gibson and Ashby [2]:

$$k_{eff} = \frac{k_b}{k_s} = \left(\frac{\rho_b}{\rho_s}\right)^q \tag{4}$$

In the above equation, k_b is the effective thermal conductivity, k_s is the intrinsic solid phase thermal conductivity (also called the ligament conductivity), and k_{eff} is the non-dimensional ratio of the two values. The exponent q is usually between 1 and 2; this exponent represents the effect of the microstructure on the path of heat flow. A preferred representation is to use the exponent $1/n$ as shown below [3]:

$$k_{eff} = \frac{k_b}{k_s} = \left(\frac{\rho_b}{\rho_s}\right)^{\frac{1}{n}} = \left(1 - \frac{\varphi}{100}\right)^{\frac{1}{n}} \tag{5}$$

In this form, the bulk thermal conductivity k_b is higher as n increases; with the maximum possible value of k_b at $n = 1$. Therefore n represents the effectiveness of the pore geometry at different relative densities for heat flow through the foam and it can be called the "pore conduction shape factor" [13]. The value of n is generally significantly less than unity for most foams, and the lower limit is typically about 0.5.

When the porosity is very low, the value of n has been theoretically shown to be 2/3 (i.e., 0.667) for spherical pores [14]. A study by Bauer [11] suggested a value of $n = 0.77$ for liquid foams with porosity in the range of 60–95%. According to Ashby et al. [15], the exponent $(1/n)$ is generally between 1.8 and 1.65 for metal foams, so that:

$$\left(1 - \frac{\varphi}{100}\right)^{\frac{1}{0.556}} < k_{eff} < \left(1 - \frac{\varphi}{100}\right)^{\frac{1}{0.606}} \tag{6}$$

On the basis of experimental results of graphitic carbon foams, Klett et al. [16] modified (2) by adding a multiplier $\beta = 0.734$ to account for pore shape, and used the pore conduction shape factor $n = 0.701$:

$$k_{eff} = \frac{k_b}{k_s} = \beta\left(\frac{\rho_b}{\rho_s}\right)^{\frac{1}{n}} = 0.734\left(1 - \frac{\varphi}{100}\right)^{\frac{1}{0.701}} \tag{7}$$

It should be noted that the above equation is a volume-averaged result, in which the inhomogeneous graphene alignment is assumed to be isotropic at the bulk level. As discussed earlier, the graphitic foams may not be isotropic even at the bulk level; therefore, (2) must be used with an appropriate modifier to account for bulk anisotropy.

3.2 Volume-Averaged Model

The classical macro-approach of volume averaging has been used extensively when the thermal transport takes place in the solid and fluid phases. Under this approach, the quantities associated with the thermal transport and fluid flow phenomena are averaged over a representative elementary volume consisting of both the interstitial fluid and porous material [17]. For highly conductive foams, the two energy equation approach must be used (one for each phase) and the temperatures should be averaged separately for solid phase and fluid phase [18]. However, a thermal equilibrium assumption is often made in order to obtain analytical solutions. This approach has been described by Nield and Bejan [19] and Kaviany [20] and has been used in several studies for modeling fluid flow and convection through foams.

This averaging approach reduces the complexity of the general problem, but information about the transport phenomena at pore level and the influence on the overall transport phenomena are lost [17, 18, 46]. Such models typically need experimental values or semi-empirical parameters to compensate for the loss of information. In the case of fluid flow, the parameters are the permeability and the inertial coefficient of the porous media. The results are often presented as non-dimensional numbers such as friction factor, Reynolds number and Nusselt number which are based on the geometric parameters of the foam.

3.3 Volume-Averaged Fluid Flow Model

The steady state mass and momentum conservation equations for incompressible flow through porous media are [21]:

$$\nabla \cdot \langle \mathbf{u} \rangle = 0 \tag{8}$$

$$\frac{\rho}{\varphi^2}\langle \mathbf{u} \cdot \nabla \mathbf{u} \rangle = -\nabla \langle p \rangle^f + \frac{\mu}{\varphi}\nabla^2 \mathbf{u} - \frac{\mu}{K}\langle \mathbf{u} \rangle - \frac{\rho\, c_f}{\sqrt{K}} |\langle \mathbf{u} \rangle| \langle \mathbf{u} \rangle \tag{9}$$

where $\langle\rangle$ is the volume average symbol, $\boldsymbol{u}$ is the fluid velocity vector, p is the fluid pressure, ρ is the fluid density, μ is the fluid viscosity, φ is the porous media porosity, K is the porous media permeability and c_f is the porous media inertial coefficient. The mass and momentum conservation equations can be solved numerically for pressure and velocity fields provided that φ, K and c_f are known.

It should be mentioned that in comparison to Navier-Stokes equations, the momentum conservation equation for flow through porous media contains two extra terms, which are the last two terms on the right hand side of (9). They account for additional pressure loss due to the presence of the porous media. Since this equation is not amenable to analytical solution, it is often simplified to the case of a steady state, fully developed flow, without the effect of boundary walls. This leads to the Darcy-Forchheimer equation, which is the one dimensional momentum conservation equation for flow through porous media. It gives the following expression for the pressure drop:

$$\frac{\Delta p}{\Delta x} = \frac{\mu}{K} u_D + \frac{\rho\, c_f}{\sqrt{K}} u_D^2 \tag{10}$$

where the Darcy velocity (uniform filter velocity) in the porous media is given by:

$$u_D = \frac{\dot{m}}{\rho A_{ch}} \tag{10a}$$

and Δx is the length of the porous region in the direction of the flow, $\dot{m}$ is the mass flow rate and A_{ch} is the cross sectional area of the channel before it is filled with porous material.

The fluid flow regime in pipes and channels with constant flow area is established as either laminar or turbulent using the Reynolds number based on the equivalent hydraulic diameter as the length scale of the flow. In porous channels it is difficult to define an equivalent hydraulic diameter because the geometry is complex and the flow area is continuously changing. The square root of the permeability $\left(\sqrt{K}\right)$ can be used as the length scale of flow through porous media, and has been incorporated in the definition of the "modified" Reynolds number [22–24]:

$$\mathrm{Re}_K = \frac{\rho u_D \sqrt{K}}{\mu} \tag{11}$$

The fluid flow in porous media is in the "Darcy regime" when the velocities are small enough so that the inertial effects are insignificant and the quadratic term in the Darcy-Forchheimer equation is negligible. At high velocities, the quadratic term in the Darcy-Forchheimer equation cannot be neglected and the flow regime is "non-Darcy".

Equation (10) has been used to determine K and c_f from experimental measurements of pressure drop and uniform filter velocity for a large variety of porous materials: aluminum, nickel and carbon foams [21], compressed aluminum foam [25] and carbon foam [26]. Experimental testing on aluminum foams showed that permeability of a porous material is a strong function of porosity and pore size and inertial coefficient is influenced by the solid phase shape and pore structure [23].

In an attempt to unify the pressure loss characteristics for various porous media, a dimensionless friction coefficient has been used [22, 23]

$$f = \frac{\frac{\Delta p}{\Delta x}\sqrt{K}}{\rho u_D^2} \tag{12}$$

Equation (12) was used to determine f from experimental measurements of pressure drop and the Darcy velocity for aluminum foams [23] and Foametal [27].

Beavers and Sparrow [22] combined together (10)–(12) and obtained the following relation for friction coefficient:

$$f = c_f + \frac{1}{\mathrm{Re}_K} \tag{13}$$

Equation (10) describes the flow behavior of porous media and is widely used by researchers. It must be noticed that in the Darcy flow regime, the quadratic term in (10) can be neglected, so that the friction coefficient becomes equal to the inverse of the Reynolds number. The inertial coefficient and the permeability have been determined experimentally for aluminum foams by Paek et al. [23], and the results are described by the following equation:

$$f = 0.105 + \frac{1}{\mathrm{Re}_K} \tag{14}$$

For a high porosity foam (Foametal), the experimental results by Vafai and Tien [27] produced the following relationship:

$$f = 0.057 + \frac{1}{\mathrm{Re}_K} \tag{15}$$

The analysis of heat transfer due to flow through the foam is quite complex, and results from the volume-averaged equations and experiments are usually correlated in terms of the Nusselt number, Reynolds number and the Prandtl number [28]. Many forced convection studies for simple flows use the general Nusselt number correlation of the form:

$$Nu = CPe^m \tag{16}$$

where C is a constant, Pe is the Peclet number (product of the Reynolds and Prandtl numbers) and m is an exponent. In general, the solution for convection heat transfer requires the determination of the permeability (typically experimentally or semi-empirically) which determines the fluid flow. In addition, an effective conductivity and a representative length dimension have to be determined or estimated so that the Reynolds number and the Prandtl number can be defined. The complex nature of thermo-fluid transport in conductive foam has led to the development of numerical models that are based on pores of ideal shape. This is discussed in the following section.

4 Idealized Pore-Level Model

The development of pore-level micro-models for analyzing porous media is relatively recent, and has been made possible by the combination of advances in computational fluid dynamics (CFD), and enhancement in computer capabilities. Due to its complexity and randomness in the pore shape, dimensions and distribution in the solid matrix, the accurate representation of 3-D microstructure of foams as a solid model is computationally intensive. Therefore, studies have been carried out in which foam microstructure is often approximated by idealized geometry. Metal foams used for heat exchanger, (such as Duocel® produced by ERG Aerospace) have an open reticulated structure. It is generally reasonable to model reticulated structures as a set of ligaments with a geometric relationship [18]. Silicon carbide foam can also be modeled as a set of ligaments. Carbon foam, however, presents a special challenge because of the complex microstructure; and many different models have been proposed that are based on simple pore geometries.

Some of the carbon foam geometries that have been proposed for an idealized pore model are shown in Fig. 2, along with the SEM picture of a graphitic carbon foam. Sihn and Roy [29] approximated the carbon foam with a unit cell which is obtained by subtracting four identical spheres from a regular tetrahedron (Fig. 2a). The porosity of the unit cell is a function of the sphere diameter. Druma et al. [13, 30] proposed a body centered cubic (BCC) type structure for the unit cell of carbon foam, as shown in Fig. 2b. They also used ellipses (horizontal and vertical) to create the pores (Fig. 2c, d). Yu et al. [31] used a pore at the center of a cubic cell to form the model. A tetrakaidecahedral unit cell for carbon foam was used by Li et al. [32].

Using these idealized pore models, it is possible to determine the surface area and the thermal conductivity analytically [31, 48], or by numerical analysis.

The advantage of using a numerical micro-model that simulates the process at the pore level is that it can predict the experimental or semi-empirical parameter such as the "density exponent" (5). Druma et al. [13] showed that the plot for (5) for $n = 0.67$ matches the numerical solution only at very low porosities (below 10%); this is to be expected since this exponent was derived for low porosities. In general, a constant value of this exponent is not appropriate for a full range of porosity.

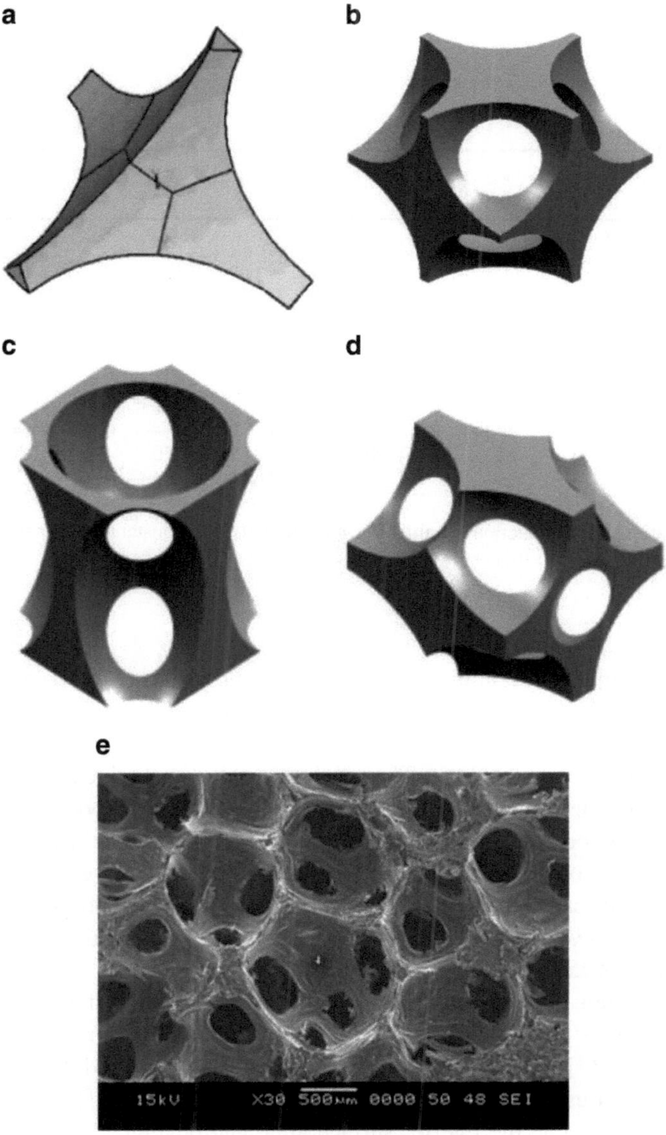

Fig. 2 Ideal pore models used to approximate carbon foam microstructure: (**a**) tetrahedron (*Source*: [29]), (**b**) BCC type cube (*Source*: [13]), (**c**) and (**d**) BCC type ellipse, vertical and horizontal (*Source*: [30]), (**e**) SEM picture of graphitic carbon foam (Source: Graftech International)

A plot of the numerical value of the exponent n has been presented by Druma [6] by writing (5) in the following form:

$$n = \frac{\ln\left(1 - \frac{\varphi}{100}\right)}{\ln k_{eff}} \tag{17}$$

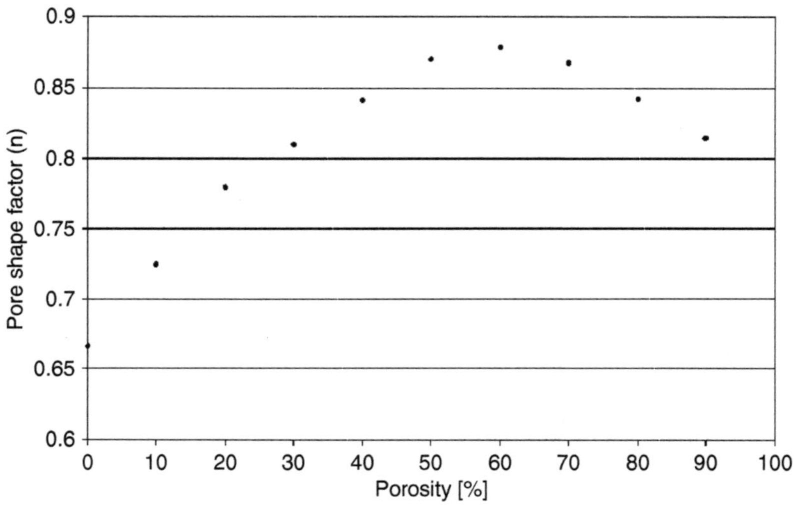

Fig. 3 Variation of "conduction parameter" with porosity for constant pore size for pores of 100 μm diameter (*Source*: [6])

The plot is reproduced in Fig. 3. It is obvious that for this case of a porous media with uniformly distributed 100 μm pores, the density exponent *n* is not constant, but varies between 0.65 and 0.88. A constant (average) density exponent can only be used for a limited porosity range. The maximum value is obtained before the pores become interconnected, since interconnected pores result in pockets of material that are not in good contact, and therefore not very effective in heat transfer. The exponent is not strongly affected by the size of the pores. However, the shape of the pores (e.g., spherical vs. elliptical) and the arrangement of the pores (face centered vs. body entered) will affect the density exponent. These conclusions demonstrate the usefulness of the numerical, pore level model.

A natural extension of the numerical analysis of a pore level model would be the study of the fluid flow and heat transfer in a porous media. Yu et al. [31], and Krishnan et al. [33] have developed numerical solutions for fluid flow and convection heat transfer using idealized pore level models. Yu et al. [31] used a cubic cell geometry to determine the interior surface area, and developed analytical expressions for the effective thermal conductivity and convective heat transfer. Karimian and Straatman[34, 35] followed up this study by numerically solving the fluid flow and heat transfer for the same body centered cubic cell geometry. Krishnan et al. [33] and Anghelescu [36] used the numerical model to determine the permeability, and friction factor due to fluid flow.

Anghelescu et al. [12] extended the pore level modeling approach to true three-dimensional structures of carbon foams. It is important to recognize that development of a solid three-dimensional image involves several steps before the numerical simulation can be undertaken. This is addressed in the following section.

5 Development of the True 3-D Image

The development of the 3-D microstructure of two different foams will be discussed in this section. The first is a silicon carbide foam made at Ultramet Inc. (Pacoima, CA, USA), and the second is a graphitic carbon foam manufactured at the Air Force Research Laboratory (AFRL, Dayton, OH, USA). The silicon carbide foam, as shown in Fig. 1 has a simpler microstructure. The carbon foam microstructure is more complex. Two different imaging processes have been used for the two foams. One approach [37] is to combine a light microscopy system with a highly accurate serial sectioning of the sample. A 90% porous graphitic carbon foam (similar to that shown in Fig. 1) which was produced at the Air Force Research Laboratory (AFRL, WPAFB, Dayton, Ohio) was first infiltrated with a polymer (epoxy) under vacuum. The 3 mm diameter sample was then analyzed by a novel instrument (Robo-Met.3D) which uses a diamond lapping film to polish the sample at about 3.5 μm per slice. The sample is imaged by bright field light microscopy after each slicing operation.

A second method is to do scanning by X-ray tomography to produce the images of the slices; this was the procedure used with the silicon carbide foam. This method has also been used on aluminum foams by Fiedler et al. [38], and by Hugo et al. [39]. The slices are separated by at least one voxel size of the specific instrument; this is the limit of the image resolution for this technology. The images produced either by Robo-Met.3D or by X-ray tomography, are then processed in a number of steps to provide the solid model. At the beginning, the image files of the slices are subject to a segmentation operation, which consists of defining the boundaries between the solid (the foam) and background in each slice, and is done manually and/or automatically. This can be done by customized or commercial software, such as the medical imaging software Amira [40]. Amira can also be used for the next step, which is the triangulation step. This step defines the entire surface of the foam by using a large set of triangles. After this step, the results are saved as STL files.

At this stage of the surface reconstruction, the point cloud is wrapped in a polygonal surface made of triangles and there are some choices for the software which can define the surface by fitting appropriate surfaces through the points defined by the triangulation. In the Geomagic Studio software, the regions of high curvature are identified and marked with curvature lines. The triangulated surface is divided into a number of areas, each of them bounded by curvature lines. Each of these areas is then divided into a number of patches. A patch is a convex quadrilateral which is manipulated in order to obtain acceptable warpage, skewness and aspect ratio. A geometric grid is generated for each quadrilateral patch and a Non-Uniform Rational B-Spline (NURBS) mathematical representation is calculated for each patch, using the grid as control points. The result is a file containing a closed surface, made of patches, that resembles the geometry of the foam microstructure. The NURBS patches are stitched together and the solid 3D carbon foam model is generated.

After NURBS fitting, the resulting surface can be saved as an IGES, STEP or other file format and transferred to the solid modeling software. This is the preferred mode if the foam and the fluid in the pores are to be meshed for fluid flow through the pores. Using Boolean operations such as "subtract" and "unite", the solid model of the foam and fluid can be combined using a software, such as SolidEdge. For example, in order to produce the solid model of the foam and the fluid in Fig. 4, a Boolean operation was carried out to subtract the shape of the foam from a parallelepiped representing the fluid. The foam is subsequently fitted into this empty space and aligned to the sides to produce the assembly consisting of carbon foam saturated with fluid.

The solid models were then used to generate the mesh using the software product HyperMesh. Another way to generate the solid mesh is to use the NURBS only (without constructing the solid model). The NURBS are loaded in HyperMesh where a volumetric mesh procedure is applied to the surface. The solid meshes for the two foams are shown in Fig. 5. The silicon carbide foam model has the dimensions of $1.50 \times 1.44 \times 1.40$; the graphitic carbon foam dimensions are $1.50 \times 1.48 \times 1.54$ (dimensions in mm).

The computational model with the mesh for the porous channel and the foam is discussed later in the development of the fluid flow model. The combination of the foam structure and the fluid makes the numerical model significantly bigger. As the number of elements and nodes increase far above one million, the issue of computational capabilities becomes important. One of the limiting factors of the modeling effort is the computational effort required to set up and run a model that reflects accurate details at the pore level of the microstructure. In order to have a model that

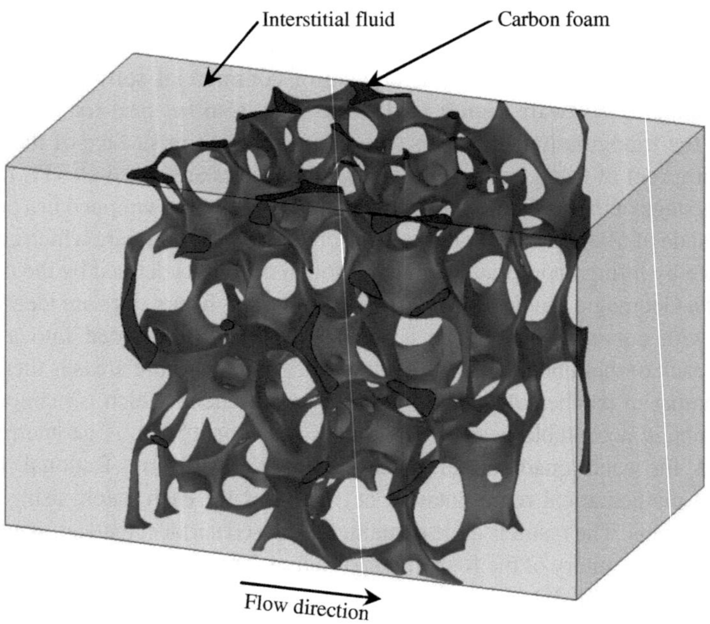

Fig. 4 3D solid model of porous carbon foam and fluid [36]

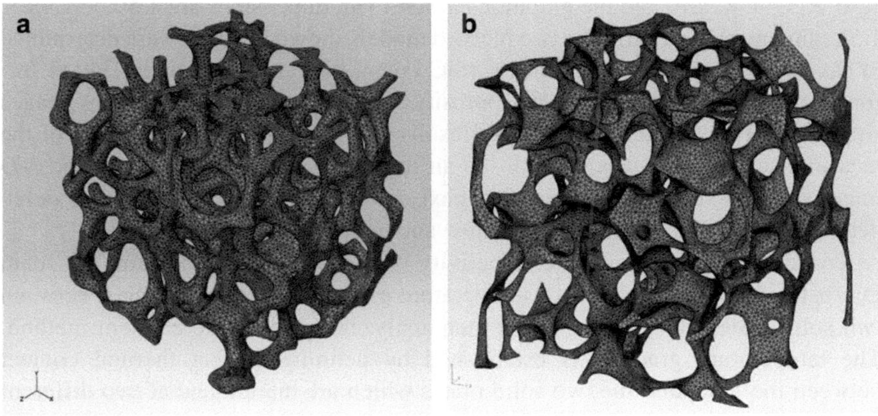

Fig. 5 Meshed solid models of 3D foam microstructure (**a**) Silicon carbide foam from Ultramet, 83% porous, with 6,611 m^2/m^3 of surface area (**b**) Graphitic carbon foam from AFRL, 90% porous, with 4,340 m^2/m^3 of surface area

is representative of the bulk material, the model should be large enough to have the same properties (density, thermal and mechanical properties, etc.) as the bulk material. It is important to note that the bulk properties of the foam depend on the pore geometry as well as the intrinsic (solid phase) properties of the foam ligaments. But the properties of the ligaments can vary from point to point, and are very difficult to measure. To keep the computational effort reasonable, the small volume of the model in this study was selected on the basis of similarity to the bulk porosity, which can be measured by the bulk density. The typical model size is in the order of cubic millimeters. This can present some difficulty in the simulation of fluid flow and heat transfer, which is discussed later.

A high performance computer (or super-computer) is generally needed for solid modeling of the foam. The number of elements in the model is typically one million or more, especially when the fluid flow is included. The numerical solution must be checked to make sure that it is mesh-independent, which also increases the computational requirements. With the continuous improvement in the capabilities of computational systems, it is expected that larger volume models would be handled more easily in future work and this would result in greater confidence in the results of the numerical model.

6 Thermal Conductivity Modeling with True 3-D Image

After the solid model is produced, it is quite straight forward for the solid modeling software to determine the porosity and the surface area per unit volume of the porous media. As mentioned earlier, the silicon carbide foam model has the dimensions of

1.50 × 1.44 × 1.40 and the graphitic carbon foam dimensions are 1.50 × 1.48 × 1.54 (dimensions in mm). The two meshed models shown in Fig. 5 were determined to have surface areas of 6,611 m^2/m^3 (SiC Foam, 83% porosity), and 4,340 m^2/m^3 for the graphitic carbon foam (90% porosity). The porosity of the real foam images were compared with the bulk density results and found to be within 1–2% of the measured value. This was taken to be an indication of the similarity of the 3-D image sample to the bulk material. The next step is to set up an appropriate model to determine thermal conduction, fluid flow and convective heat transfer.

To determine the thermal conductivity of the bulk (carbon) foam, the foam microstructure is subjected to a temperature gradient by sandwiching it between two solid plates, and the system is then analyzed by the finite element method. The temperature gradient is established by defining perfect thermal contact between the foam and the two solid plates which are maintained at two different temperatures. The temperature field in the system (carbon foam microstructure and plates) is governed by the steady state conduction heat transfer equation given by:

$$\frac{\partial}{\partial x_i}\left(k\frac{\partial T}{\partial x_i}\right) = 0 \qquad i = 1,\,2,\,3 \quad \text{(index notation)} \tag{18}$$

where T is the temperature and k is the thermal conductivity of the domain analyzed.

The following assumptions have been made in the analysis:

- The foam properties are constant
- The fluid in the pores has negligible effect on the bulk conductivity
- Convection and radiation are neglected.

The problem definition is completed by defining the boundary conditions as follows:

- A uniform heat flux is applied on the top surface of the upper plate
- A uniform temperature is specified on the bottom surface of lower plate
- Temperature and heat flux are continuous at the interface of the plate and carbon foam microstructure
- No heat transfer takes place at all on the other surfaces

The carbon foam model was analyzed with 519,397 elements. To evaluate mesh independence, the model was solved again with 1,357,568 elements; and it was determined that the two thermal conductivity results varied by less than 0.1%. For additional details of the carbon foam simulation, the reader is referred to the article by Anghelescu et al. [12] or the dissertation by Anghelescu [36]. The SiC foam in Fig. 5 was analyzed in the same manner. The model of the SiC foam contains 1,015,839 quadratic tetrahedral elements, which corresponds to 1,588,867 nodes. The plates, due to their simple geometry, have only 3,600 brick elements, with 5,766 nodes.

From the solution of the temperature field in the foam, the thermal conductivity is calculated by applying Fourier's law:

$$q_x = -k_b \frac{dT}{dx} = k_b \frac{\Delta T}{\Delta x} \tag{19}$$

where q_x is the heat flux applied on the top surface of the upper plate, ΔT is the temperature difference between the two solid plates and Δx is the thickness of the carbon foam in the heat flux direction.

The intrinsic conductivity of the solid phase in the foam may not be known accurately, or can be very difficult to measure. Therefore, the non-dimensional bulk thermal conductivity is determined in the solution; this is defined as:

$$k_{eff} = \frac{k_b}{k_s} \tag{20}$$

Three separate analyses are carried out in the x- y- and z-directions of the 3D solid model of foams to account for randomness in pore shape, dimensions and distribution. Figure 6a, b show the temperature distributions in two foam microstructures when the heat flux is applied on the top plate. Because of the boundary conditions applied, the heat transfer is predominantly one dimensional, in the heat flux direction. The heat flux distributions are shown in Fig. 7a, b. The heat flux has higher values in the thin ligaments of the two foams; these areas tend to be bottle necks in the thermal transport.

The thermal analysis results are summarized in Table 1 along with analytical and experimental results from other studies. It can be seen that the numerical models based on real foam geometry have much lower thermal conductivity k_{eff} as compared to the analytical or numerical values (based on idealized pores). In comparison to the values obtained from the pore level model, the value predicted by Bauer

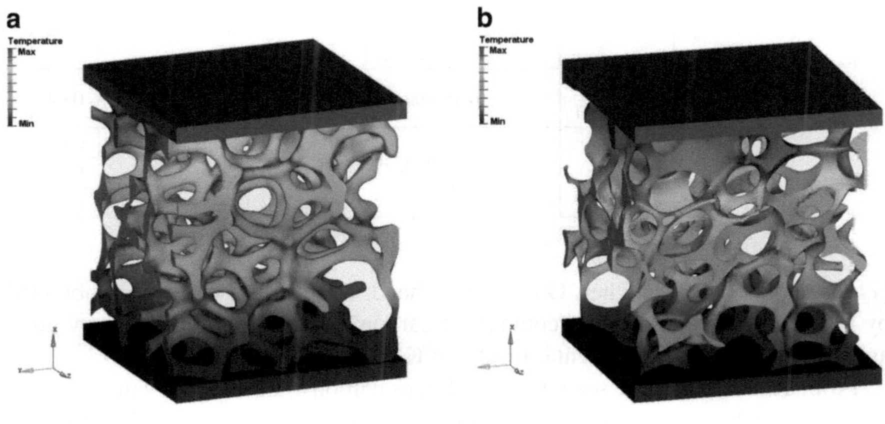

Fig. 6 Temperature distribution in SiC (**a**) and carbon foam (**b**)

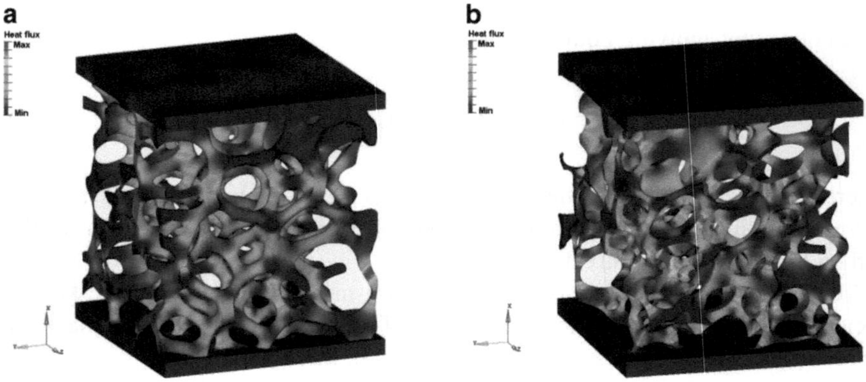

Fig. 7 Heat flux distribution in SiC (**a**) and carbon foam (**b**)

Table 1 Thermal conductivity results from different foam models

-	Source	Type of analysis		k_{eff} (% of solid phase)			n
83% SiC foam	Bauer (1993)	Analytical (spherical pores)		9.89			0.77
83% SiC foam	Present analysis	FEM on true foam geometry		x	5.55	6.30 (average)	0.61
				y	7.39		0.68
				z	5.95		0.63
90% carbon foam	Bauer (1993)	Analytical (spherical pores)		4.86			0.77
90% carbon foam	Druma et al. (2004)	FEM on ideal geometry	spherical pores	6.25			0.84
			ellipsoidal pores – horizontal	2.20			0.61
			ellipsoidal pores – vertical	9.20			0.98
90% carbon foam	Klett et al. (2004)	Experimental		2.62			0.64
90% carbon foam	Anghelescu et al. (2009)	FEM on true foam geometry		x	2.78	2.72 (average)	0.65
				y	2.93		0.66
				z	2.44		0.62

[11] is about 60–80% higher. On the other hand, the conductivity values obtained by FEM analysis on the true geometry of carbon foam microstructure show good agreement with the experimental results in Klett et al. [16].

From Table 1, it can be seen that both foams demonstrate a higher value of k_{eff} in the "y" direction compared to the other two directions. In particular, the "y" direction for the two foams has significantly higher conductivity (about 20%)

than the other two directions. These differences show that this particular solid model of carbon foam microstructure is not geometrically isotropic. This may be due to bulk geometric anisotropy, or unique sample characteristics. Graphitic carbon foams are known to exhibit higher bulk conductivity [3] in the vertical direction of the reactor; which can be due to geometric anisotropy and/or due to directional anisotropy of graphene alignment within the solid phase. The numerical results can only reflect the geometrical anisotropy in the 3-D image; therefore it is difficult to resolve this issue of anisotropy without examining images of multiple samples.

On the basis of the results shown in Table 1, it can be concluded that the accuracy of semi-empirical models depends on the adjustable parameter (such as the "density exponent", n); while ideal geometries (which do not have an adjustable parameter) may significantly over-predict the bulk thermal conductivity due to the geometric regularity of the model. The lower value of the non-dimensional conductivity in the true geometry is probably due to the higher tortuosity of the heat flux vectors caused by the randomness in pore shape, dimensions and distribution.

Fiedler et al. [38] obtained similar results in their study of thermal conductivity of aluminum foams. They analyzed the thermal conductivity of the foam by applying a lattice Monte Carlo method to a true foam image obtained by X-ray tomography. Their results showed that analytical or semi-empirical formulas can significantly over predict the experimental measurements as well as the numerical values obtained by the Monte Carlo simulation. On the other hand, the experimental and numerical simulation results were in good agreement.

Therefore, the numerical model of the true 3-D image can be expected to provide an accurate value of the thermal conductivity ratio k_{eff}. This is a useful quantity for heat transfer analysis since it relates the bulk conductivity value to the average solid phase conductivity (also known as the ligament conductivity). Once k_{eff} is calculated, the value of the average ligament conductivity can be calculated from known (measured) bulk conductivity; and (5) can be used to determine the relative density exponent (n). This exponent is a measure of the effectiveness of the pore geometry in thermal transport.

7 Fluid Flow and Heat Transfer in Real 3-D Foam

The interconnected pore structure of porous materials can be used as channels for fluid flow which offers a significant increase in surface area available for convective heat transfer. Highly porous aluminum foams have been investigated as possible solutions for thermal management of electronics [41, 42]. Hugo et al. [39] used a true 3-D image of an aluminum foam for a heat exchanger application. They used the iMorph software for developing the image output from X-ray tomography and carried out heat transfer and fluid flow calculations using StarCCM+ software. The aluminum foam used by Hugo et al. [39] is an ERG foam, with pore diameters of 4.5 mm (20 PPI) and a porosity of 92%; the larger pore size and the simpler microstructure can reduce the computational effort for constructing a solid model for numerical solution.

The small pore size and the complex microstructure make carbon foam modeling a challenging task. However, because of its light weight and very high thermal conductivity, graphitic carbon foam is an attractive candidate for heat sinks and heat exchanger core. Silicon carbide foam has potential for high temperature heat exchanger applications. This section addresses some of the issues in developing the model for a porous media heat exchange system consisting of a channel filled with an open cell porous material and saturated with a coolant flowing through the pores.

The analysis of fluid flow at the pore level of a real 3-D image of porous channel has the advantage of using basic fluid flow and heat transfer equations; additional terms (such as permeability, inertial coefficient) are not needed. Therefore, the model can be implemented with the governing equations of a Newtonian fluid by using one of several computational fluid dynamics (CFD) software products. The numerical simulations used in the following studies were obtained by the finite volume method as implemented in the FLUENT software [43]. It is important to note that the solid model of carbon foam used in the following simulations represents a very short flow length; and this may not represent the flow properties of the bulk foam. Even if periodic boundary conditions [33, 34] are used, the issue of a representative volume should be addressed because the true 3-D image has irregular, randomly distributed pores. It was discussed earlier that the volume of the model has the same porosity as the bulk; but the geometrical features of a small solid model may not be representative and may not provide accurate permeability and inertial coefficients. Therefore, in the following fluid flow analysis, a longer SiC solid model is used. The flow equations are solved for both foams by using symmetric boundary conditions on the four sides so that the effect of channel walls is avoided. The convective heat transfer study in a later section is carried out with flow through a carbon foam with emphasis on the effect of the foam ligament conductivity on the convective heat transfer.

The velocity and pressure fields of an incompressible, steady state, Newtonian fluid flow in laminar regime are governed by the mass conservation equation

$$\frac{\partial u_i}{\partial x_i} = 0 \qquad i = 1, 2, 3 \quad \text{(index notation)} \tag{21}$$

and the Navier-Stokes equations for momentum conservation, given (in index notation) by

$$\rho u_j \frac{\partial u_i}{\partial x_j} = -\frac{\partial p}{\partial x_i} + \mu \frac{\partial}{\partial x_j}\left(\frac{\partial u_i}{\partial x_j}\right) \quad i, j = 1, 2, 3 \tag{22}$$

where u_i are the fluid velocities, p is the fluid pressure, μ is the fluid viscosity and ρ is the fluid density. The fluid flow is assumed viscous and the body forces on the fluid are neglected. Due to the complexity of the geometry, a large number of finite

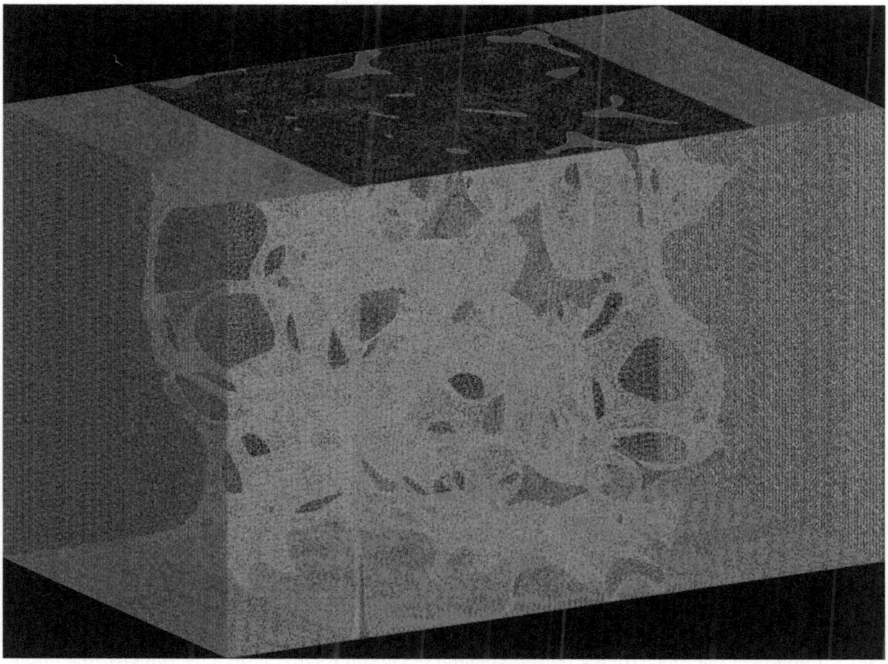

Fig. 8 Computational model for carbon foam in the flow channel

volume elements have to be used. Therefore, to reduce computational effort, the properties of the fluid are assumed constant with temperature.

The solid model assembly of porous channel (carbon foam microstructure and interstitial fluid) is discretized using unstructured tetrahedral mesh in commercial software HyperMesh [44] as shown in Fig. 8. The mesh created is imported into computational fluid dynamics (CFD) software FLUENT for fluid flow and heat transfer calculations. The unstructured tetrahedral mesh is first converted to polyhedral mesh in order to improve the mesh quality (eliminate bad elements) and reduce the computational time. In the polyhedral mesh 367,462 elements are generated for the solid phase and 1,148,766 elements for the fluid. The second order upwind scheme is used for discretizing the momentum and fluid energy conservation equations. The SIMPLE algorithm is used for pressure-velocity coupling.

Mesh independence of the finite volume solution was established by solving the problem using a different mesh density: 355,088 elements for solid and 756,387 for fluid. The difference between the two discretizations in terms of permeability and inertial coefficient of carbon foam microstructure is 1.6%, and 1.1%, respectively. The results presented in this chapter are obtained using the finer mesh.

The fluid flow equations (21) and (22) are for laminar flow regime because the Reynolds number calculated based on the equivalent hydraulic diameter of the flow channel is smaller than critical Reynolds number $Re_{cr} = 2{,}320$ for the entire range of velocities used in simulations. In order to account for the possibility of

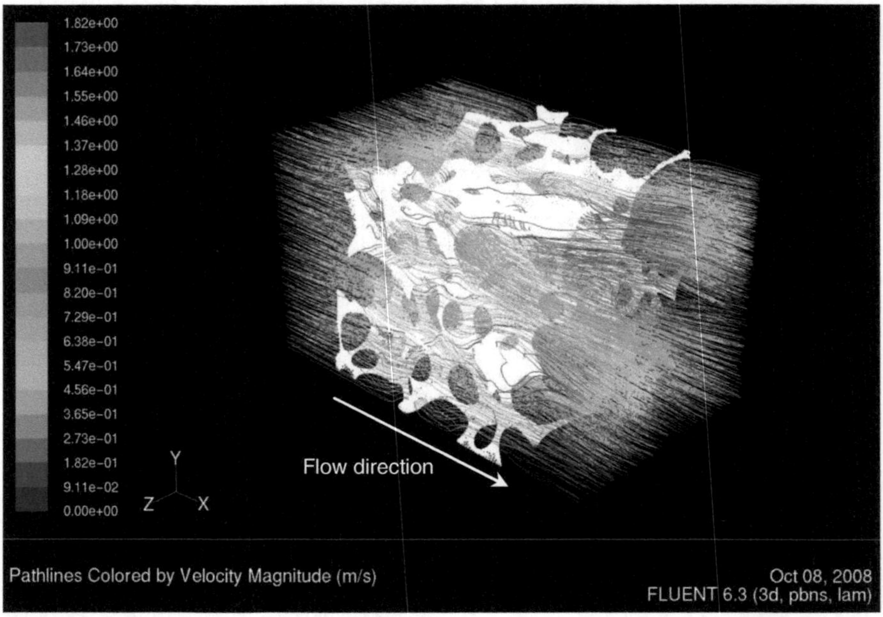

Fig. 9 Fluid flow pathlines (in the porous channel of carbon foam) colored by velocity magnitude for a Darcy velocity of 0.5 m/s [6]

turbulence due to the tortuous geometry, the simulations were also run by including turbulence in the flow model. The fluid flow turbulence was modeled using the Reynolds-Averaged Navier-Stokes (RANS) method. The turbulent viscosity and turbulence kinetic energy were modeled by employing the k-ε method as implemented in the commercial code FLUENT [43]. The difference between the laminar and turbulent flow simulation results in terms of permeability and inertial coefficient of carbon foam microstructure is 0.7% and 5.3%, respectively.

Fluid flow simulations for inlet (Darcy) velocities ranging between 0.01 and 1.5 m/s are run in order to determine the pressure drop as the flow moves across the solid model. Fluid (air) thermo-physical properties at 20°C are used in the simulations. Figure 9 shows the fluid flow pathlines in the carbon foam with velocity magnitudes shown in color; this simulation is for a Darcy fluid velocity of 0.5 m/s. The fluid flows along the positive x-axis to the right of the figure. The presence of the carbon foam in the channel reduces the cross-sectional area available for fluid flow and increases the fluid velocity. It can be noticed that the maximum fluid velocity in the porous channel is about 3.6 times higher than free stream velocity. This, in turn increases the pressure drop and the heat transfer. The tortuosity of the fluid flow pathlines around the foam ligaments can be observed in the simulations.

For the silicon carbide foam, the meshing process was also done in HyperMesh. Two models were created, with different element densities in order to check the mesh independence of the solution. The solid model for the SiC foam has the following dimensions: 2.9 mm length, 1.44 mm width and 1.40 mm height. The flow

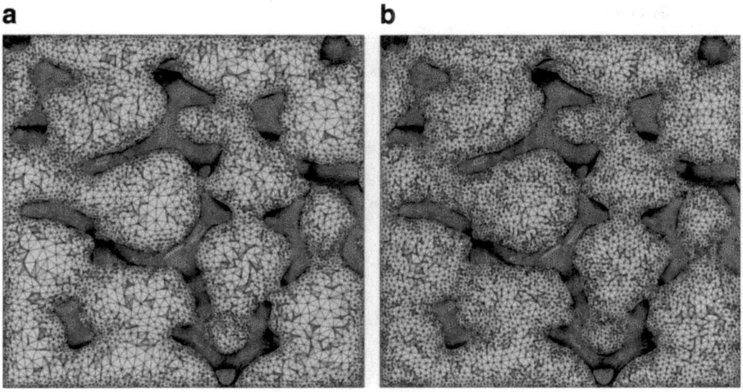

Fig. 10 Cross-sectional view through the HyperMesh models (**a**) 9,319,575 elements and (**b**) 14,012,640 elements

length in the fluid flow simulation model is approximately double the length of what was used in the thermal conductivity model. This longer model was used so that the flow is not significantly affected by end effects, such as penetration depth. The first model has the fluid domain discretized into 9,319,575 tetrahedral elements, while the second one has 14,012,640 elements. This model is bigger than what was used for analysis of thermal conductivity. Cross sectional views of the two models are shown in Fig. 10a, b.

After the models were loaded into FLUENT, the domain was converted into a polyhedral mesh to give a better mesh quality and improve the memory utilization. After the polyhedral conversion, the first SiC foam model was reduced to 1,179,862 polyhedral cells and the second one was reduced to 2,150,609 cells. The problem was then solved by the FLUENT code, and the fluid flow pathlines in the SiC foam are shown in Fig. 11. It can be seen that the pathlines are qualitatively more tortuous than higher porosity carbon foam.

The simulation results of pressure drop per unit length across the two different foams as a function of the Darcy velocity are shown in Fig. 12; along with the quadratic curves fitted through the data points. By comparing the plots in Fig. 12 with the Darcy-Forchheimer equation (10), the values of permeability and the inertial coefficient for the SiC foam can be determined. The permeability for the SiC foam was determined to be 2.1×10^{-9} m^2 and the inertial coefficient was calculated to be 0.14.

The value of permeability, and the heat transfer results obtained from the 3-D image were determined to be very similar to the values obtained by experimental measurements of the bulk SiC foam [45], which provides additional validation of the numerical model. Even though the foam model is only 2.9 mm long, the penetration depth is of the order of the square root of the permeability ($\sqrt{K}$), which is much smaller than the flow length. Therefore, the flow model from a small 3-D image can be used to predict the bulk flow behavior in a porous media.

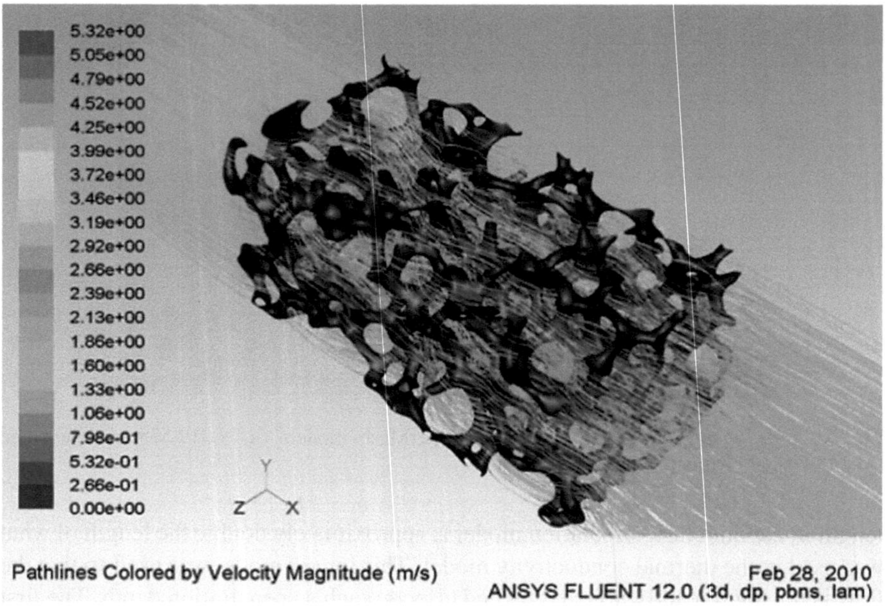

Fig. 11 Fluid flow pathlines through the SiC foam for a Darcy velocity of 1.5 m/s

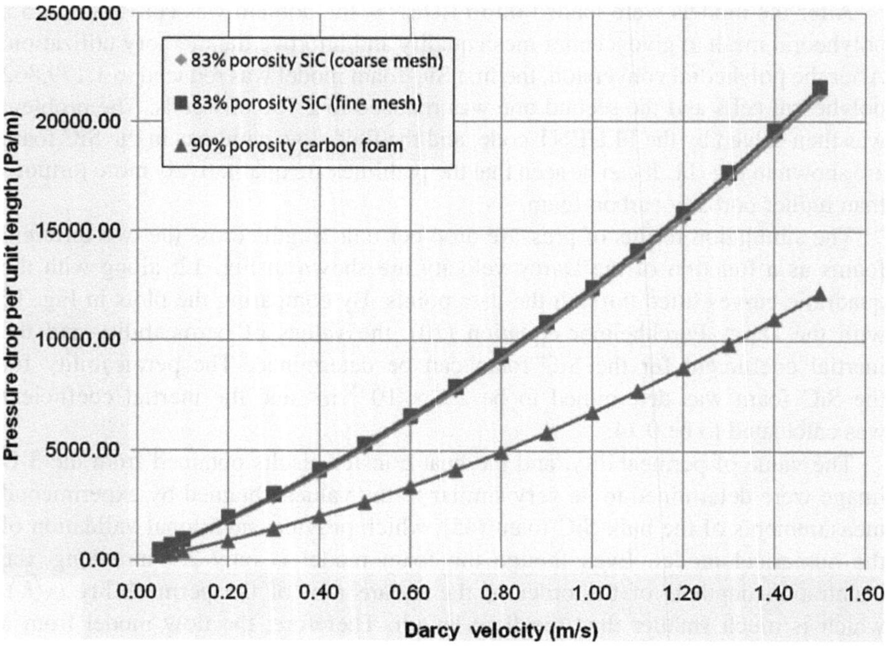

Fig. 12 Pressure drop in the two foams as a function of Darcy velocity

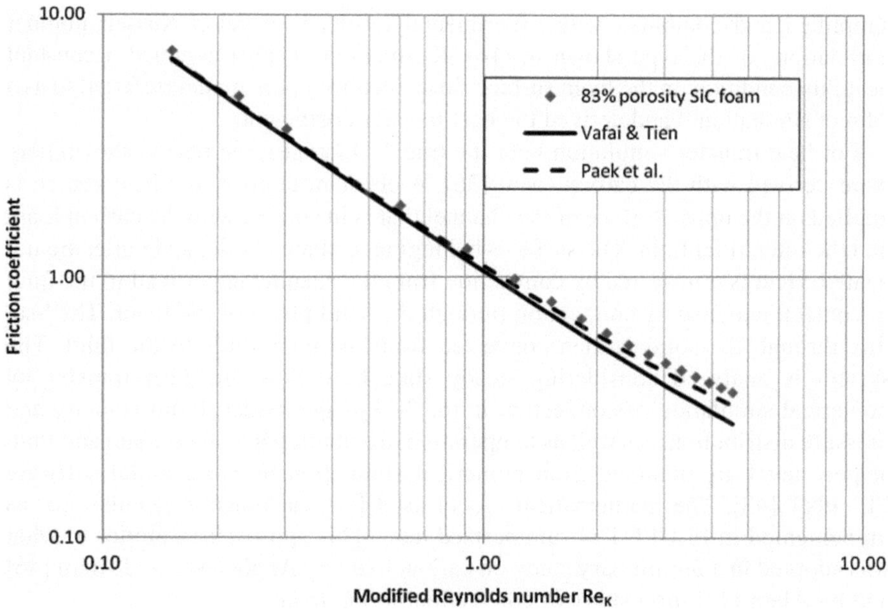

Fig. 13 Friction coefficient for flow in the SiC foam as a function of the modified Reynolds number

It is also important to note that the permeability was obtained from the 3-D image model without any adjustable parameters.

Figure 12 shows an unexpected result – the fluid pressure drop in the carbon foam is less than the pressure drop in the SiC foam. In general, carbon foam structure tends to be more tortuous then this silicon carbide foam. Therefore, the pressure drop in the carbon foam could have been expected to be higher than in the silicon carbide foam. However, in this case, the carbon foam has only 10% solid vs. 17% for the SiC foam. At very low densities, the pore structure and the resulting fluid flow can be significantly different from higher density foams of the same material. Figure 13 is a plot of the friction coefficient calculated from (12), versus the modified Reynolds number (11), for the SiC foam. This curve for the SiC foam is based on the permeability and inertial coefficient calculated from Fig. 12. It can be seen that the flow is in the Darcy regime for these simulations. For the purpose of comparison, the experimental correlations from Paek et al. [23] (14) and Vafai and Tien [27] (15) are also plotted. The plots are quite comparable over the range of the modified Reynolds number values.

7.1 Heat Transfer Calculation

Heat transfer calculations in a highly conductive porous medium are complex because of non-equilibrium thermal conditions. Therefore the equations are often simplified and combined with experimental results to develop Nusselt number correlations of the type shown in (16). Krishnan et al. [33] assumed a constant heat flux condition on the foam surfaces in an idealized pore model (referred to as a "direct simulation") and derived the heat transfer coefficient.

For heat transfer simulation with the true 3-D images, the results shown here were derived with the carbon foam [36]. A constant temperature heat source is applied on the upper surface of the channel that is in contact with the carbon foam and the interstitial fluid. The surfaces belonging to channel inlet and outlet are not heated. Heat is transferred by convection from the channel upper wall to the fluid in contact, and also by conduction through the solid phase of the foam. The heat flux through the solid is then convected from the pore walls to the fluid. The system is analyzed considering steady state fluid flow and heat transfer by numerical simulation of convection in the 3-D image model. Fluid velocity and pressure distributions, as well as temperature distribution in carbon foam and fluid at pore level are obtained from numerical simulation by commercial software FLUENT [43]. The mathematical model used for heat transfer calculations, as implemented in FLUENT is summarized here. This approach is similar to what was adopted in a preliminary study on carbon foam by Anghelescu and Alam [46] and by Alam [47] for a study of convection in SiC foam.

The steady state fluid temperature distribution in laminar flow regime is governed by the energy conservation equation

$$\rho c_p \frac{\partial}{\partial x_i}\left(u_i T_f\right) = \frac{\partial}{\partial x_j}\left(k_f \frac{\partial T_f}{\partial x_j}\right) \qquad i = 1,2,3 \quad \text{(index notation)} \tag{23}$$

where k_f is the fluid thermal conductivity.

The steady state temperature distribution in the carbon foam is governed by the conduction equation

$$\frac{\partial}{\partial x_j}\left(k_s \frac{\partial T_s}{\partial x_j}\right) = 0 \qquad i = 1,2,3 \quad \text{(index notation)} \tag{24}$$

where k_s is the thermal conductivity of the solid phase.

The following boundary conditions are applied for heat transfer analysis:

- Constant temperature on the upper surface of the porous channel (carbon foam and fluid)
- Constant fluid temperature at the inlet of the channel
- Symmetry on the sides and lower surface of the channel
- Temperature and heat flux are continuous at the solid–fluid interface

A wall boundary condition is used on the top surface of the porous channel in order to represent the constant temperature heat source. The two energy equations solved together yield the temperature distributions in carbon foam and fluid as a function of space coordinates x, y and z. The fluid flow computational model

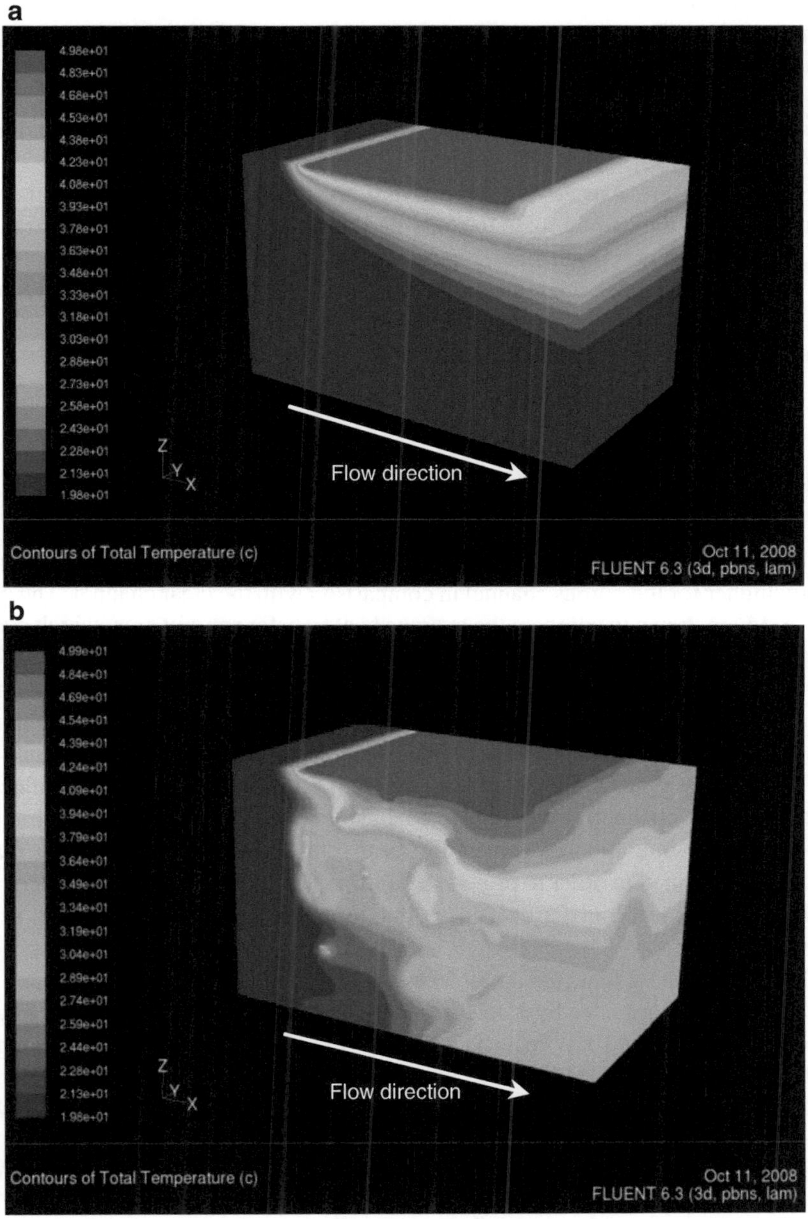

Fig. 14 Fluid temperature distribution in (**a**) clear channel (no foam) and in (**b**) porous channel with carbon foam for a Darcy velocity of 0.5 m/s

provides the velocity distributions necessary for solving fluid energy conservation equation. The thermo-physical properties of the fluid (air) are assumed constant with temperature during the convection heat transfer analysis so that the fluid flow governing equations and the energy conservation equations can be decoupled and solved independently. For additional details, readers are referred to Anghelescu [36].

Simulations are carried out by varying the solid phase (ligament) thermal conductivity in the carbon foam to study its influence on the effective heat transfer coefficient in the porous channel. Figure 14a shows the temperature distribution in the fluid in a channel without any foam, and Fig. 14b shows the temperature distribution in the porous channel (containing carbon foam of solid phase thermal conductivity of 50 W/m°C) for an inlet fluid velocity of 0.5 m/s.

7.2 *Effect of Foam Ligament Conductivity*

As can be expected, the fluid average temperature at the channel outlet is significantly higher for the porous channel in comparison with the clear channel. The plots in Fig. 15 are drawn to compare the average heat transfer per unit area, which can be represented by the "effective heat transfer coefficient" based on the heater area. The heat transfer is plotted as a function of Darcy flow velocity and the solid phase (ligament) thermal conductivity of the foam. Three different values for thermal conductivity of the solid phase in carbon foam are used in simulations, which are 10, 50 and 100 W/m°C. The results, as shown in Fig. 15, demonstrate the influence of the thermal conductivity of the foam ligaments on the effective heat transfer coefficient. A simulation of convection heat transfer in clear channel (no foam) was also performed in order to provide a baseline that would show the enhancement in the effective heat transfer coefficient due to the presence of carbon foam in the channel.

The results show that, for an air velocity of 1 m/s, the thin layer of foam can increase the heat transfer by a factor of three even if the foam ligaments have a conductivity of only 10 W/mK. The reason is that the ligaments are working as highly efficient fins [26]. The heat transfer in the porous channel is enhanced by the higher velocity due to restrictions in flow area and the thermal conductivity of the foam ligaments; but in this case the primary cause for the enhancement is the extended surface area available for convection. Beyond a thermal conductivity value of 100 W/mK, there is very little change in the heat transfer because the foam ligaments are almost isothermal across the small height of the thin channel. It should be noted that graphitic foam ligaments can have thermal conductivity exceeding 500 W/mK. If the flow channel height is extended further, using such high thermal conductivity foam would be highly effective in enhancing the heat transfer.

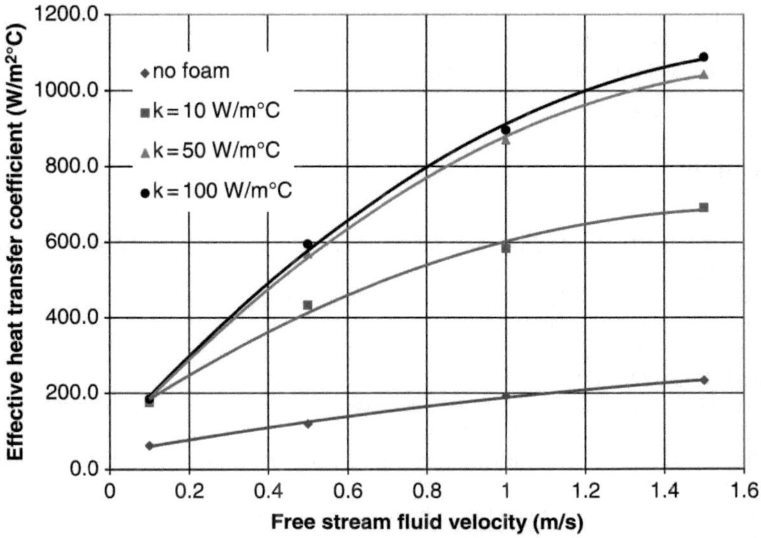

Fig. 15 Effective heat transfer coefficient as a function of Darcy velocity and foam ligament thermal conductivity

8 Summary

The analysis of porous media has been addressed through new techniques for obtaining true three-dimensional images of true foam geometry at the pore level. This allows the development of pore level models that can be modeled without the use of experimental or semi-empirical parameters such as permeability, friction coefficient, and heat transfer coefficient. Instead, the pore level solid model can be used to determine these parameters by applying the fundamental transport equations for a simple channel flow. The SiC model studied in this chapter has been used successfully by the authors to predict the fluid flow, permeability and heat transfer for a heat sink application. It can be concluded that a pore level model using true 3-D images has the potential to determine the thermal conductivity, fluid flow behavior within the foam and the convection due to fluid flow without using experimental results or semi-empirical parameters. Therefore, this technique provides a tool for prediction and design of transport processes in porous media. This is particularly useful for complex microstructures, such as graphitic carbon foam, for which the accuracy of classical volume averaged equations is not well established. Current work on this modeling effort is focused on the design and development of heat sinks and heat exchangers based on flow through high conductivity foams.

Acknowledgments The authors would like to acknowledge the support provided by the Air Force Research Laboratory (AFRL, Dayton, OH), Ultramet Inc. (Pacoima, CA) and use of HyperMesh, Abaqus, and Fluent provided by Ohio Supercomputer Center. Funding was also provided by Ohio

Aerospace Institute (OAI). The authors wish to thank Dr. Benji Maruyama of AFRL, Mr. Matt Wright of Ultramet (Pacoima, CA, USA), the research staff at Graftech (Parma, OH, USA) and Dr. Dan Vrable (TMMT) for helpful discussions.

References

1. Ultramet: http://www.ultramet.com/refractoryopencells_ceramic.html, Website accessed 2/16/1010 (2010)
2. Gibson, L.J., Ashby, M.F.: Cellular Solids: Structure and Properties, 2nd edn. Cambridge University Press, UK (1997)
3. Alam, M.K., Maruyama, B.: Thermal conductivity of graphitic carbon foams. Exp. Heat Transf. **17**(3), 227–241 (2004)
4. Rowe, M.M., Guth, R.A., Merriman, D.J.: Case studies of carbon foam tooling. In: SAMPE 2005, Long Beach, CA, 1–5 May 2005.
5. Gallego, N.C., Klett, J.W.: Carbon foams for thermal management. Carbon **41**(7), 1461–1466 (2003)
6. Druma, A.: Analysis of carbon foams by finite element method. Doctoral dissertation, Ohio University, Athens, OH, USA (2005)
7. Brow, M., Watts, R., Alam, M.K., Koch, R., Lafdi, K.: Characterisation requirements for aerospace thermal management applications. In: Proceedings of SAMPE 2003, Dayton, OH, 2003
8. Chen, C., Kennel, E., Stiller, A., Stansberry, P., Zondlo, J.: Carbon foam derived from various precursors. Carbon **44**(8), 1535–1543 (2006)
9. Alam, M.K., Anghelescu, M.: Analysis of deformation and residual stresses in composites processed on a carbon foam tooling. J. Compos. Mater. **43**(19), 2057–2070 (2009)
10. Spradling, D., Guth, R.: Carbon foams. Adv. Mater. Process **161**(11), 29–31 (2003)
11. Bauer, T.H.: A general analytical approach toward the thermal conductivity of porous media. Int. J. Heat Mass Transf. **36**(17), 4181 (1993)
12. Anghelescu, M.S., Alam, M.K., Maruyama, B.: Evaluation of thermal transport from accurate 3D geometry of carbon foam. In: Proceedings of the SAMPE'09 Conference, Baltimore, MD, 18–21 May 2009
13. Druma, A.M., Alam, M.K., Druma, C.: Analysis of thermal conduction in carbon foams. Int. J. Therm. Sci. **43**(7), 689–695 (2004)
14. Maxwell, J.C.: A Treatise on Electricity and Magnetism. Oxford University Press, New York (1954)
15. Ashby, M.F., Evans, A., Fleck, N.A., Gibson, L.J., Hutchinson, J.W., Wadley, H.N.G.: Metal Foams: A Design Guide. Butterworth Heinemann, Oxford, UK (2000)
16. Klett, J., McMillan, A., Gallego, N., Walls, C.: The role of structure on the thermal properties of graphitic foams. J. Mater. Sci. **39**(11), 3659–3676 (2004)
17. Vafai, K., Tien, C.L.: Boundary and inertia effects on flow and heat transfer in porous media. Int. J. Heat Mass Transf. **24**(2), 195–203 (1981)
18. Calmidi, V.V.: Transport phenomena in high porosity metal foams. Doctoral dissertation, University of Colorado, Boulder, CO, USA (1998)
19. Nield, D.A., Bejan, A.: Convection in Porous Media, 2nd edn. Springer-Verlag, New York (1992)
20. Kaviany, M.: Principles of Heat Transfer in Porous Media. Springer, New York (1995)
21. Hunt, M.L., Tien, C.L.: Effects of thermal dispersion on forced convection in fibrous media. Int. J. Heat Mass Transf. **31**(2), 301–309 (1988)
22. Beavers, G.S., Sparrow, M.: Non-Darcy flow through fibrous porous media. J. Appl. Mech. Trans. ASME **36**(4), 711–714 (1969)
23. Paek, J.W., Kang, B.H., Kim, S.Y., Hyun, J.M.: Effective thermal conductivity and permeability of aluminum foam materials. Int. J. Thermophys. **21**(2), 453–464 (2000)

24. Boomsma, K., Poulikakos, D.: The effect of compression and pore size variations on the liquid flow characteristics in metal foams. J. Fluid. Eng. Trans. ASME **124**(1), 263 (2002)
25. Antohe, B.V., Lage, J.L., Price, D.C., Weber, R.M.: Experimental determination of permeability and inertia coefficients of mechanically compressed aluminum porous matrices. J. Fluid. Eng. Trans. ASME **119**(2), 404–412 (1997)
26. Straatman, A.G., Gallego, N.C., Yu, Q., Betchen, L., Thompson, B.E.: Forced convection heat transfer and hydraulic losses in graphitic foam. J. Heat Transf. Trans. ASME **129**(9), 1237–1245 (2007)
27. Vafai, K., Tien, C.L.: Boundary and inertia effects on convective mass transfer in porous media. Int. J. Heat Mass Transf. **25**(8), 1183–1190 (1982)
28. Calmidi, V.V., Mahajan, R.L.: Forced convection in high porosity metal foams. J. Heat. Transf. Trans. ASME **122**(3), 557–565 (2000)
29. Sihn, S., Roy, A.K.: Modeling and prediction of bulk properties of open-cell carbon foam. J. Mech. Phys. Solids **52**(1), 167–191 (2004)
30. Druma, C., Alam, M.K., Druma, A.M.: Finite element model of thermal transport in carbon foams. J. Sandw. Struct. Mater. **6**(6), 527 (2004)
31. Yu, Q., Thompson, B.E., Straatman, A.G.: A unit cube-based model for heat transfer and fluid flow in porous carbon foam. J. Heat Transf. Trans. ASME **128**(4), 352–360 (2006)
32. Li, K., Gao, X.L., Roy, A.K.: Micromechanical modeling of three-dimensional open-cell foams using the matrix method for spatial frames. Compos. Part B Eng. **36**(3), 249–262 (2005)
33. Krishnan, S., Murthy, J.Y., Garimella, S.V.: Direct simulation of transport in open-cell metal foam. J. Heat Transf. **128**(8), 793 (2006)
34. Karimian, S.A.M., Straatman, A.G.: CFD study of the hydraulic and thermal behavior of spherical-void-phase porous materials. Int. J. Heat Fluid Flow **29**(1), 292–305 (2008)
35. Karimian, S.A.M., Straatman, A.G.: Numerical modeling of multidirectional flow and heat transfer in graphitic foams. J. Heat Transf. **131**(5), 052602 (2009)
36. Anghelescu, M.S.: Thermal and mechanical analysis of carbon foam. Doctoral dissertation, Ohio University, Athens, OH, USA (2008)
37. Maruyama, B., Spowart, J.E., Hooper, D.J., Mullens, H.M., Druma, A.M., Druma, C., Alam, M.K.: A new technique for obtaining three-dimensional structures in pitch-based carbon foams. Scripta Mater. **54**(9), 1709–1713 (2006)
38. Fiedler, T., Solorzano, E., Garcia-Moreno, F., Ochsner, A., Belova, I.V., Murch, G.E.: Lattice Monte Carlo and experimental analyses of the thermal conductivity of random-shaped cellular aluminum. Adv. Eng. Mater. **10**(11), 843–847 (2009)
39. Hugo, J.-M., Brun, B., Topin, F., Vicente, J.: Conjugate heat and mass transfer in metal foams: a numerical study for heat exchanger design. In: Proceedings of the DSL 2009 Conference, Rome, Italy, 24–26 June 2009
40. Amira: http://www.amiravis.com/, Website accessed 2/9/2010 (2010)
41. Bhattacharya, A., Mahajan, L.: Finned metal foam heat sinks for electronics cooling in forced convection. J. Electron. Packag. **124**(3), 155–163 (2002)
42. Boomsma, K., Poulikakos, D., Zwick, F.: Metal foams as compact high performance heat exchangers. Mech. Mater. **35**(12), 1161–1176 (2003)
43. Ansys Fluent Inc.: FLUENT (6.3) (Computer software). Canonsburg, PA, USA (2006)
44. Altair Engineering: Hypermesh (8.0) (Computer software), Troy, MI, USA (2007)
45. Wright, M.: Ultramet, Inc., Pacoima, CA, USA. Personal communication (2010)
46. Anghelescu, M.S., Alam, M.K.: Finite element modeling of forced convection heat transfer in carbon foams. In: Proceedings of 2006 ASME International Mechanical Engineering Congress and Exposition, Chicago, Illinois, 2006
47. Alam, M.K.: Thermal modeling of flow through foam. *Final report to Ultramet*, Sept. 20, 2005 (2005)
48. Druma, A., Alam, M.K., Druma, C.: Surface area and conductivity of open-cell carbon foams. Journal of Minerals & Materials Characterization & Engineering **5**(1), 73–86 (2006)

Part III
Consideration of a Fluid (solid phase [structure] and fluid [in liquid and/or gaseous aggregate state])

Thermal Instabilities in a Fluid Saturated Porous Medium

A. Barletta

Abstract This chapter contains a review of the thermoconvective instabilities that may occur in a fluid saturated porous medium. Reference is made to a horizontal porous layer. The macroscopic description of the fluid flow in a porous medium is outlined. Then, the mass, momentum and energy balance equations for porous media are described and discussed. As a first basic example of convective instabilities, the classical Darcy-Bénard problem is studied in order to highlight the main characteristics of the linear stability analysis. Extensions of the Darcy-Bénard problem based on the Forchheimer model and on the Brinkman model of momentum flow are analysed. Moreover, the effects of either a horizontal or a vertical throughflow in the layer are discussed. The contribution of the effect of viscous dissipation is investigated as a possible cause of convective instabilities.

1 Introduction

Fluid flow in porous media is of paramount importance both for geophysical applications such as filtration of water, hydrocarbons and gases in the soil and for engineering. For instance, one may point out the interest of porous media with reference to the hydrology of aquifers, underground repositories used for sequestering nuclear waste, heat pipes, underground spreading of chemical waste, drainage and irrigation in agriculture, thermal insulation engineering, enhanced recovery of petroleum reservoirs, grain storage, water flow in geothermal reservoirs. Several treatises on the fluid dynamics in porous media have been published. A complete and updated review of the present knowledge in this field, oriented to convection flows, is the textbook by Nield and Bejan [24].

A. Barletta
Dipartimento di Ingegneria Energetica, Nucleare e del Controllo Ambientale (DIENCA), Alma Mater Studiorum, Università di Bologna, Viale Risorgimento 2, 40136 Bologna, Italy
e-mail: antonio.barletta@unibo.it

A. Öchsner and G.E. Murch (eds.), *Heat Transfer in Multi-Phase Materials*,
Adv Struct Mater 2, DOI 10.1007/8611_2010_8,

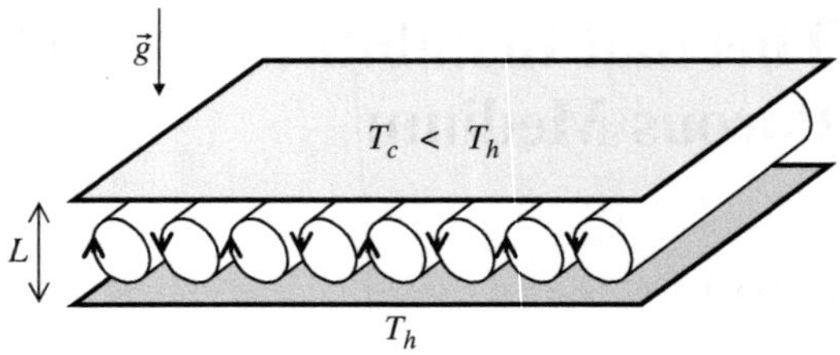

Fig. 1 A sketch of the Rayleigh-Bénard problem

The onset of convective instabilities induced by temperature gradients is a well known phenomenon both for clear fluids and for fluid saturated porous media. A classical reference case is the Rayleigh-Bénard problem of a horizontal fluid layer heated from below [7, 8, 19]. Comprehensive discussions of the Rayleigh-Bénard modes of instability and the study of the critical conditions for the onset of convection cells can be found, for instance, in Normand and Pomeau [25], as well as in the textbooks by Drazin and Reid [12] and by Gebhart et al. [13]. In a simplified formulation of the Rayleigh-Bénard problem, one assumes an infinitely wide horizontal fluid layer with thickness L, bounded by two isothermal planes (see Fig. 1). The fluid is at rest in the undisturbed basic state. The lower boundary plane is kept isothermal at a temperature T_h, while the upper boundary plane is kept isothermal at a temperature $T_c < T_h$. As is well known, convective cells may appear when the difference $T_h - T_c$ becomes greater than a threshold value. In fact, the analysis of the Rayleigh-Bénard problem reveals that the onset of convective rolls occurs when the Rayleigh number,

$$Ra = \frac{g\beta(T_h - T_c)L^3}{\nu\alpha}, \tag{1}$$

exceeds the critical value $Ra_{cr} = 1707.76$ (see, for instance, [13]). Here, g is the modulus of the gravitational acceleration $\vec{g}$, while β, ν and α are respectively the coefficient of isobaric thermal expansion, the kinematic viscosity and the thermal diffusivity of the fluid. The critical value of the Rayleigh number is obtained by a linear stability analysis. This means that a small disturbance of the basic rest state is assumed having the form of a plane wave oriented in any horizontal direction. Since the disturbance has a small amplitude, one keeps only the linear terms in the governing balance equations, while the higher order terms are neglected [12].

In the fluid dynamics of saturated porous media, an issue analogous to the Rayleigh-Bénard problem of clear fluids is the Darcy-Bénard problem, well known also as the Horton-Rogers-Lapwood problem. The former denomination refers to the pioneer of the theory of fluid flow in porous media, Henry Philibert Gaspard Darcy (1803–1858), and to the pioneer of the experiments on convection cells in a fluid layer, Henri Bénard (1874–1939). The latter denomination refers to the forerunners of the investigations on the linear stability analysis of a horizontal fluid saturated porous layer heated from below [14, 18]. In its classical formulation, the Darcy-Bénard problem shares the same geometry, basic state and thermal boundary conditions as the Rayleigh-Bénard problem. The only difference is in the momentum balance being expressed by Darcy's law of porous media instead of the Navier-Stokes equation of clear fluids. The Darcy-Bénard problem is just the simplest of a wide class of convective instability problems in porous media. For previous reviews of the literature on this subject we refer to Nield and Bejan [24], Rees [27] and Tyvand [34]. A remarkable feature of the original Darcy-Bénard problem is that it admits an exact solution. In fact, a fluid saturated porous layer with thickness L has a linearly stable rest state as long as the Darcy-Rayleigh number,

$$R = \frac{g\beta(T_h - T_c)LK}{\nu\tilde{\alpha}}, \tag{2}$$

does not exceed the critical value $4\pi^2$. Here, K is a property of the porous medium called permeability, while $\tilde{\alpha}$ is the average thermal diffusivity of the fluid saturated porous medium.

2 Models of Flow in a Fluid Saturated Porous Medium

In this section, the basic elements for the macroscopic description of fluid flow in a saturated porous medium are discussed. Then, the mathematical models for the expression of the local mass, momentum and energy balance equations are described.

The oldest, the simplest and the most widely employed model of fluid flow in porous media is named after Darcy (see Fig. 2), a French scientist with a strong professional interest in hydraulics. During his life, he was a civil engineer in the city of Dijon in France. He designed and built a pressurized water distribution system in Dijon. A few years before his death, he conducted the experiments that allowed him to formulate what today is well known as *Darcy's law*. His publication *The Public Fountains of the City of Dijon* contains an appendix written in 1856 entitled

Fig. 2 Henry Philibert Gaspard Darcy (1803–1858) [Portrait by F. Perrodin, Bibliothèque Municipale de Dijon]

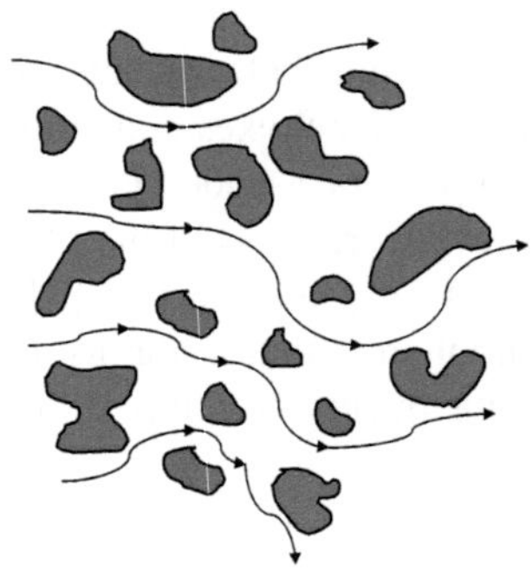

Fig. 3 Flow in a porous medium

Determination of the Laws of Water Flow Through Sand where his law is formulated [11]. A historical survey of Darcy's scientific works and discoveries is given in the paper by Brown [9].

Before formulating Darcy's law, let us review the main features of the macroscopic description of fluid flow in porous media. A porous medium is a solid material with void inner structures saturated by a fluid, liquid or gas. One can think to sand, pebbles or to a metallic foam. One can imagine that the void spaces within the solid are entirely filled by the moving fluid (see Fig. 3).

2.1 Porosity, Volume Averages and the Seepage Velocity

A basic quantity for the description of a porous medium is the ratio between the volume occupied by the fluid (voids) and the total volume including voids and solid. Referring to Fig. 4, one can consider a representative volume V, small on a macroscopic scale even if large on the scale of the single grain, pebble or microchannel that may be present inside the porous medium. If V_f is the void part of V, then let us call *porosity*, φ, the ratio

$$\varphi = \frac{V_f}{V}. \tag{3}$$

The porosity is a dimensionless quantity strictly smaller than unity, whose value can range from $\sim 0.88 \div 0.93$ of fiberglass to $\sim 0.12 \div 0.34$ of bricks. Sand, the material originally considered in Darcy's experiments, has a porosity $\varphi \sim 0.37 \div 0.50$.

The study of convection in porous media is based on the assumption that a fluid saturated porous medium can be described as a continuum. This means that, in the representative volume V of the system, the number of pores is very high. Therefore, one can define a local fluid velocity field as an average value of the local fluid velocity $\vec{u}^*$. There are two possible average values of $\vec{u}^*$ usually introduced: the *intrinsic velocity*, namely

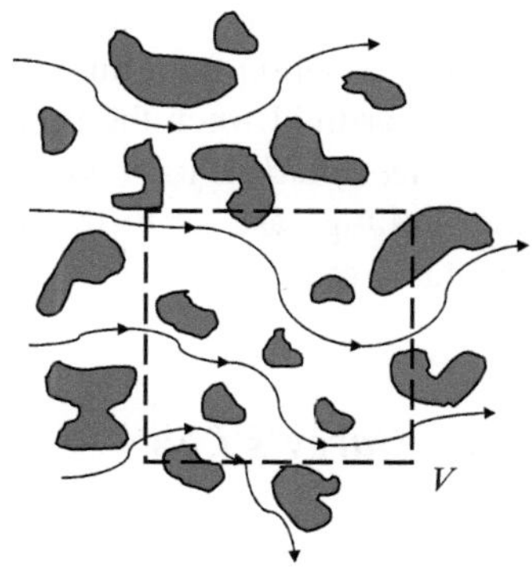

Fig. 4 Representative volume

$$\vec{U} = \frac{1}{V_f} \int_{V_f} \vec{u}^* \, dV, \tag{4}$$

and the *seepage velocity* (also known as *Darcy velocity*), namely

$$\vec{u} = \frac{1}{V} \int_{V} \vec{u}^* \, dV. \tag{5}$$

The intrinsic velocity is defined as an average performed in the void part V_f of the representative volume V. Since $\vec{u}^* = 0$ in the part of V not included in V_f, the two integrals on the right hand sides of (4) and (5) are equal. Then, one can establish a very simple relationship between $\vec{U}$ and $\vec{u}$,

$$\vec{u} = \varphi \vec{U}. \tag{6}$$

This equation is well known as the *Dupuit-Forchheimer relationship*.

The local value of the seepage velocity $\vec{u}$ depends on the shape and the size of the pores as well as on the causes that determine the fluid motion. The relationship between the local value of $\vec{u}$ and the forces acting on the fluid could be deduced by an appropriate local average over the representative volume of the Navier-Stokes momentum balance. However, due to the complexity of the system, in most cases this relationship is postulated through a constitutive equation validated experimentally.

2.2 Mass Balance

By employing a local volume-averaging procedure, the local mass balance equation of a fluid saturated porous medium can be expressed as

$$\varphi \frac{\partial \rho}{\partial t} + \vec{\nabla} \cdot (\rho \vec{u}) = 0, \tag{7}$$

where ρ is the fluid mass density and t is the time. Let us note that the local mass balance equation for a fluid saturated porous medium formally coincides with that of a clear fluid when the flow is stationary, $\partial\rho/\partial t = 0$. Incidentally, the terminology *clear fluids* is used when dealing with fluid saturated porous media, to denote the limiting case when the solid matrix is absent and the fluid occupies all the available space.

2.3 Darcy's Law

The simplest constitutive equation expressing the local seepage velocity is *Darcy's law* (~ 1856), namely

$$\frac{\mu}{K}\,\vec{u} = -\,\vec{\nabla}p + \vec{f}, \tag{8}$$

where K is a property of the system called *permeability*, μ is the dynamic viscosity of the fluid, p is the fluid pressure and $\vec{f}$ is the external body force per unit volume applied to the fluid (in the simplest case, the gravitational body force $\rho\,\vec{g}$).

The ratio behind the assumption given by (8) relies on the observation that a porous medium can be thought of as a network of microscopic ducts where the fluid flows. In the absence of external body forces, the pressure gradient along a duct is proportional to the average fluid velocity in the duct itself, if the flow is laminar. On the other hand, if the flow is highly turbulent (hydraulic regime), the pressure gradient along a duct is proportional to the square of the average fluid velocity in the duct itself. Darcy's law refers to the case of laminar flow within the pores, so that the permeability K is considered as a property of the medium depending on the number of pores per unit area present in a cross-section transverse to the fluid flow, on the shape of the pores and on their size. The units of permeability are m^2. Values range from $\sim 4.8 \times 10^{-15} \div 2.2 \times 10^{-13} m^2$ of a brick to $\sim 1.1 \times 10^{-9} m^2$ of a cigarette.

EXAMPLE – Let us consider a porous medium such that the pores form an ordered array of parallel infinitely-long circular ducts each with a diameter D (see Fig. 5). Let z be the axis parallel to the ducts and let the number of ducts per unit area in a transverse section of the medium be n.

Then, by comparison with the relationship between average velocity u_m^* and pressure drop Δp in this kind of ducts, it is easily verified that $K = n\pi D^4/128$. In fact, it is well known that the average velocity u_m^* of fully developed laminar flow in a circular duct (*Poiseuille flow*) is given by

$$u_m^* = -\frac{D^2}{32\mu}\frac{dp}{dz}.$$

(continued)

One may notice that $u_m^* = U$, i.e. u_m^* coincides with the intrinsic velocity U. Moreover, the porosity is given by $\varphi = n\pi D^2/4$. Then, on account of the Dupuit-Forchheimer relationship, the Darcy velocity u is given by

$$u = \frac{n\pi D^2}{4}\, u_m^* = -\frac{n\pi D^4}{128\,\mu}\,\frac{dp}{dz}.$$

As a consequence of (8), one obtains the expression of K.

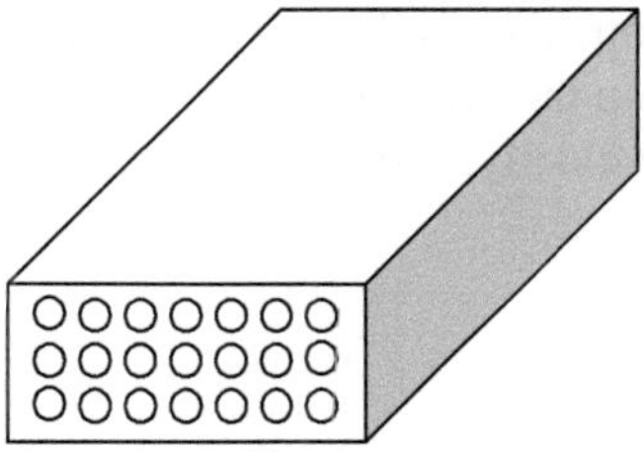

Fig. 5 Drawing of the ordered array of parallel ducts

2.4 Form-Drag Effects: Darcy-Forchheimer's Model

If the hypothesis of laminar fully developed flow in the pores cannot be applied, then proportionality between acting forces and resulting fluid velocity must be released in favour of a gradual transition towards an hydraulic regime where acting forces are proportional to the square of the fluid velocity in each pore. An extended form of (8) has been proposed which accounts for this effect, i.e. *Darcy-Forchheimer's model* ($\sim$ 1901),

$$\frac{\mu}{K}\left(1 + \frac{c_f\sqrt{K}}{\nu}\,|\vec{u}|\right)\vec{u} = -\vec{\nabla}p + \vec{f}, \tag{9}$$

In (9), $\nu = \mu/\rho$ is the kinematic viscosity, $|\vec{u}|$ is the modulus of $\vec{u}$, ρ is the fluid mass density and c_f is a property of the porous medium called *form-drag coefficient*. It is easily verified that c_f is dimensionless. Some authors sustained that the form-drag coefficient c_f is a universal constant, $c_f \cong 0.55$, but later it has been shown that c_f depends on the porous material and one can have, in the case of metal foams, $c_f \cong 0.1$ [24].

Obviously, Darcy-Forchheimer's model includes Darcy's law as a special case, i.e. in the limit $c_f \to 0$. On the other hand, whenever $c_f\,|\vec{u}|\sqrt{K}/\nu \gg 1$, the transition to an hydraulic regime for the fluid flow inside the pores occurs. A widely accepted criterion to establish when Darcy's law must be abandoned in favour of Darcy-Forchheimer's model is constructed with the permeability-based Reynolds number,

$$Re_K = \frac{|\vec{u}|\sqrt{K}}{\nu}. \tag{10}$$

Darcy's law gradually loses its validity when $Re_K \sim 10^2$ or greater. A clever way to apply the criterion is to take $|\vec{u}|$ as the maximum value in the domain.

2.5 Brinkman's Model

A common feature of Darcy's law and of Forchheimer's extension of this law is that they refer to a tight packed solid with a fluid flowing in very small pores. Indeed, this is a circumstance very far from a free flowing fluid. When one applies the curl operator to the left hand sides of (8) and (9), in order to encompass the dependence on the pressure field, one is lead to first order differential equations for the seepage velocity. More precisely, from (8) one obtains

$$\vec{\nabla} \times \left(\frac{\mu}{K}\,\vec{u} - \vec{f}\right) = 0, \tag{11}$$

while from (9) one has

$$\vec{\nabla} \times \left[\frac{\mu}{K}\left(1 + \frac{c_\mathrm{f}\sqrt{K}}{\nu}\,|\vec{u}|\right)\vec{u} - \vec{f}\right] = 0. \tag{12}$$

Being first order, one can complete either (11) or (12) with just one velocity boundary condition on each boundary surface. This boundary condition can be, for instance, impermeability ($\vec{u} \cdot \vec{n} = 0$, where $\vec{n}$ is the unit vector normal to the surface). However, one cannot allow also a no-slip condition on the same surface, as the problem would be over-conditioned. This feature is similar to that arising in perfect clear fluids (Euler's equation). The impossibility to prescribe no-slip conditions at the boundary walls creates a sharp distinction between the Navier-Stokes fluid model and the models of fluid saturated porous media based either on Darcy's law or on Forchheimer's extension of this law.

In some cases a continuous transition from the momentum balance equation of a clear fluid (Navier-Stokes equation) to Darcy's law is considered as realistic. In this direction it has been proposed the so-called *Brinkman's model* ($\sim$ 1948) for fluid flow in a porous medium. This model allows one to prescribe no-slip wall conditions as for a Navier-Stokes clear fluid. According to Brinkman's model (8) must be replaced by

$$\frac{\mu}{K}\,\vec{u} - \tilde{\mu}\nabla^2\vec{u} = -\,\vec{\nabla}p + \vec{f}, \tag{13}$$

the quantity $\tilde{\mu}$ is called *effective viscosity*: it depends on the fluid viscosity μ and on the porosity of the medium where the fluid flows. A commonly employed correlation for the effective viscosity is *Einstein's formula* for dilute suspensions of uncharged spherical colloidal particles, namely

$$\tilde{\mu} = \mu\,[1 + 2.5\,(1 - \varphi)]. \tag{14}$$

If the porosity is equal to 1, one has a clear fluid and (14) implies that $\tilde{\mu} = \mu$. If $\varphi = 1$, (13) reduces to the Navier-Stokes equation without the inertial contribution (negligible acceleration), provided that the limit of infinite permeability is also taken ($K \to \infty$). On the other hand, in the limit of a very small permeability ($K \to 0$), the first term on the left hand side of (13), $\mu\vec{u}/K$, becomes much larger than the second term, $\tilde{\mu}\nabla^2\vec{u}$. Therefore, in the limit $K \to 0$, Brinkman's model reduces to Darcy's law, (8). It must be pointed out that the limit $K \to 0$ yields a singular behavior next to the impermeable boundaries where the no-slip conditions cannot be adjusted anymore.

2.6 The Energy Balance

A local volume-averaging procedure for the fluid and solid phases, similar to that used for the local mass balance (7), yields the local energy balance equation

$$\rho c\left(\sigma\,\frac{\partial T}{\partial t} + \vec{u}\cdot\vec{\nabla}T\right) = \tilde{k}\nabla^2 T + q_g + \Phi, \tag{15}$$

valid under the assumption of constant thermal conductivities of the solid and of the fluid phases. In (15), c is the heat capacity per unit mass and σ is the *heat capacity ratio* defined as

$$\sigma = \frac{\varphi\rho c + (1 - \varphi)\,\rho_s c_s}{\rho c}, \tag{16}$$

while $\tilde{k}$ is the effective thermal conductivity of the fluid saturated porous medium defined as

$$\tilde{k} = \varphi k + (1 - \varphi)\,k_s. \tag{17}$$

The effective thermal diffusivity is then defined as $\tilde{\alpha} = \tilde{k}/(\rho c)$. In (16) and (17), the properties ρ, c and k refer to the fluid, while ρ_s, c_s and k_s refer to the solid matrix.

The term q_g in (15) is the power that may be generated per unit volume in the porous medium by, for instance, Joule heating or chemical reactions. The last term on the right hand side of (15), Φ, is the power per unit volume generated by viscous dissipation.

The expression of Φ depends on the momentum balance model employed. As pointed out in Nield [22], the term Φ can be evaluated according to the general rule

$$\Phi = \vec{F}_d \cdot \vec{u}, \tag{18}$$

where

$$\vec{F}_d = -\vec{\nabla} p + \vec{f} \tag{19}$$

is the drag force. The drag force has an expression that depends on the model adopted:

$$\text{Darcy's law} \quad \rightarrow \quad \vec{F}_d = \frac{\mu}{K}\, \vec{u} \tag{20}$$

$$\text{Darcy-Forchheimer's model} \quad \rightarrow \quad \vec{F}_d = \frac{\mu}{K}\left(1 + \frac{c_{\rm f}\sqrt{K}}{\nu}\, |\vec{u}|\right)\vec{u} \tag{21}$$

$$\text{Brinkman's model} \quad \rightarrow \quad \vec{F}_d = \frac{\mu}{K}\, \vec{u} - \tilde{\mu}\nabla^2\vec{u} \tag{22}$$

Nield's rule expressed by (18) has been the subject of a debate especially with reference to its application in the case of Brinkman's model. Let us refer for simplicity to the case of incompressible flow, $\vec{\nabla} \cdot \vec{u} = 0$. One would expect that, in the limiting case of an infinite permeability $K \rightarrow \infty$, the expression of Φ implied by (18) and (22) is consistent with the expression of the viscous dissipation term for a Navier-Stokes clear fluid, namely

$$\Phi = 2\,\mu\, \mathscr{D}_{ij}\, \mathscr{D}_{ij}, \quad \textit{where} \quad \mathscr{D}_{ij} = \frac{1}{2}\left(\frac{\partial u_i}{\partial x_j} + \frac{\partial u_j}{\partial x_i}\right) \tag{23}$$

is the (i,j) component of the strain tensor and the summation over repeated indices is assumed. On the contrary, in the limit $K \rightarrow \infty$ and $\varphi \rightarrow 1$, (18) and (22) yield

$$\Phi = -\mu\vec{u} \cdot \nabla^2\vec{u}, \tag{24}$$

where $\tilde{\mu} = \mu$ in the limiting case of a clear fluid as it is implied by (14). The difference between the expressions of Φ given in (23) and (24) is apparent as (23) yields an expression containing only first order derivatives of the velocity components, while the right hand side of (24) contains second order derivatives of the velocity components. Moreover, while Φ given by (23) can be only positive or zero, there can be flows such that the right hand side of (24) is negative.

Recently, Al-Hadhrami et al. [1] proposed a different expression of Φ in the case of Brinkman's model, namely

$$\Phi = \frac{\mu}{K}\, \vec{u} \cdot \vec{u} + 2\,\tilde{\mu}\, \mathscr{D}_{ij}\, \mathscr{D}_{ij}. \tag{25}$$

The advantage in the expression of Φ given by (25) is that Φ cannot be negative and that the limiting cases of Darcy's law ($K \to 0$) and of Navier-Stokes clear fluid ($K \to \infty$, $\varphi \to 1$) are correctly recovered. On the other hand, it has been noted that (25) is "an ad hoc formula" generated by the sum of two dissipation terms: one valid in a regime of a very small permeability (Darcy) and the other in a domain of an extremely large permeability (Navier-Stokes) [22].

3 The Darcy-Bénard Problem

As it has been pointed out in Sect. 1, the Darcy-Bénard problem is the study of the conditions for the onset of convective instabilities in a plane porous layer with isothermal impermeable boundaries heated from below. As is well known, convective instabilities are caused by the buoyancy force acting on the fluid. The conceptual scheme for describing buoyant flows is the Oberbeck-Boussinesq approximation [2, 13]. This scheme consists in neglecting the temperature changes of all the fluid properties except for the fluid density. The change of the latter property is considered only with respect to the gravitational body force term $\vec{f} = \rho\,\vec{g}$, where it is assumed to be a linear function of the temperature,

$$\rho = \rho_0[1 - \beta(T - T_0)]. \tag{26}$$

The reference density ρ_0 corresponds to the reference temperature T_0.

3.1 A Horizontal Porous Layer Heated from Below

In analogy with the Rayleigh-Bénard problem, let us consider a horizontal fluid saturated porous layer having thickness L, bounded by two impermeable planes. The lower boundary plane is maintained at temperature T_h, while the upper boundary plane has a uniform temperature $T_c < T_h$. By assuming the validity of the Oberbeck-Boussinesq approximation, as well as of Darcy's law, the following local balance equations hold:

$$\vec{\nabla} \cdot \vec{u} = 0, \tag{27}$$

$$\frac{\mu}{K}\,\vec{u} = -\,\vec{\nabla}P + \rho\beta g(T - T_c)\vec{e}_y, \tag{28}$$

$$\rho c\left(\sigma\,\frac{\partial T}{\partial t} + \vec{u}\cdot\vec{\nabla}T\right) = \tilde{k}\nabla^2 T + \frac{\mu}{K}\,\vec{u}\cdot\vec{u}, \tag{29}$$

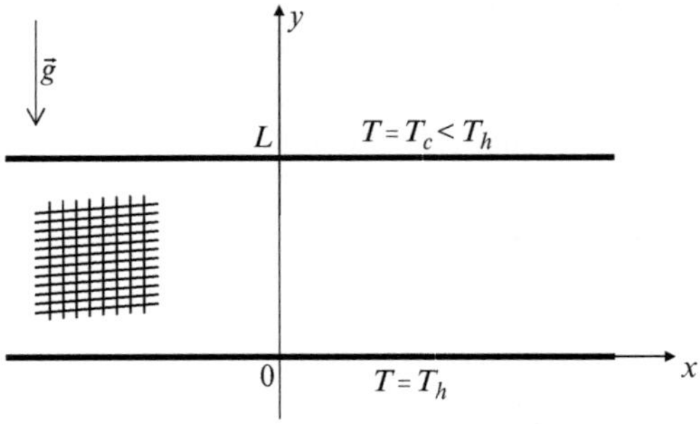

Fig. 6 The Darcy-Bénard problem

where $\vec{e}_y$ is the unit vector along the y-axis. The reference temperature T_0 has been chosen as the temperature of the cold boundary, T_c, and the corresponding reference density has been denoted as ρ for simplicity's sake. In (28), P is the difference between the pressure p and the hydrostatic pressure $-\rho g y$, while $\vec{e}_y$ is the unit vector in the y-direction (see Fig. 6). Let us denote as (u, v, w) the Cartesian components of the vector $\vec{u}$. Then, the boundary conditions are

$$y = 0: \qquad v = 0, \qquad T = T_h, \tag{30}$$

$$y = L: \qquad v = 0, \qquad T = T_c. \tag{31}$$

Equations (27)–(31) can be written in a dimensionless form on defining the following transformation

$$\begin{aligned} &(x, y, z)\frac{1}{L} \to (x, y, z), \qquad t\,\frac{\tilde{\alpha}}{\sigma L^2} \to t, \qquad \vec{u}\,\frac{L}{\tilde{\alpha}} \to \vec{u}, \\ &\frac{T - T_c}{T_h - T_c} \to T, \qquad P\,\frac{K}{\tilde{\alpha}\mu} \to P. \end{aligned} \tag{32}$$

Moreover, let us introduce the Darcy-Rayleigh number, (2), and the Darcy-Brinkman number,

$$Br = \frac{\nu\tilde{\alpha}}{K c (T_h - T_c)}. \tag{33}$$

Then, (27)–(31) are rewritten as

$$\vec{\nabla} \cdot \vec{u} = 0, \tag{34}$$

$$\vec{u} = -\vec{\nabla} P + R T \vec{e}_y, \tag{35}$$

$$\frac{\partial T}{\partial t} + \vec{u} \cdot \vec{\nabla} T = \nabla^2 T + Br\, \vec{u} \cdot \vec{u}, \tag{36}$$

$$y = 0: \qquad v = 0, \qquad T = 1, \tag{37}$$

$$y = 1: \qquad v = 0, \qquad T = 0. \tag{38}$$

In order to carry out the stability analysis of the basic state, let us determine the stationary solution of the governing equations (34)–(38) under the assumption of a rest state, $\vec{u} = 0$,

$$\vec{u}_B = 0, \qquad T_B = 1 - y, \qquad P_B = Ry\left(1 - \frac{y}{2}\right), \tag{39}$$

where the subscript B stands for "basic solution". One may notice that the field P appears in (34)–(38) only through its gradient. Hence, this field can be determined only up to an arbitrary additive constant. In (39), this constant is fixed so that $P_B = 0$ at $y = 0$.

3.2 A Bénard-Like Problem in a Porous Medium

Starting from the basic solution (39), one can define small perturbations of the velocity, temperature and pressure fields,

$$\vec{u} = \vec{u}_B + \varepsilon \vec{U}, \qquad T = T_B + \varepsilon \theta, \qquad P = P_B + \varepsilon \mathscr{P}, \tag{40}$$

where ε is an arbitrarily small perturbation parameter and $\vec{U} = (U, V, W)$.

On substituting (39) and (40) into (34)–(38) and neglecting terms of order ε^2, one obtains

$$\vec{\nabla} \cdot \vec{U} = 0, \tag{41}$$

$$\vec{U} = -\vec{\nabla} \mathscr{P} + R\,\theta\,\vec{e}_y, \tag{42}$$

$$\frac{\partial \theta}{\partial t} - V = \nabla^2 \theta, \tag{43}$$

$$y = 0, 1: \qquad V = 0, \qquad \theta = 0. \tag{44}$$

Equations (41)–(44) reveal an interesting result. The basic solution and the linear disturbances of this solution are not influenced by the effect of viscous dissipation. In fact, the Darcy-Brinkman number Br does not appear either in (39) or in (41)–(44). Obviously, a role of viscous dissipation may arise when a nonlinear disturbance analysis is carried out, i.e. when terms of order ε^2 are taken into account.

3.3 Onset of Convective Instabilities

The solution of (41)–(44) is sought in the form of standing plane waves oriented along an arbitrary direction orthogonal to the y-axis. The basic solution as well as the physics and geometry of the system examined are invariant under rotations around the y-axis. This means that the effect of a wave disturbance oriented along a direction orthogonal to the y-axis is independent of the particular direction selected. Therefore, without any loss of generality, one may choose a plane wave disturbance oriented along the x-direction. This choice implies that the analysis of the disturbances has the form of a 2D problem referred to the (x, y)-plane. On writing (41)–(44) in a 2D form one has

$$\frac{\partial U}{\partial x} + \frac{\partial V}{\partial y} = 0, \tag{45}$$

$$U = -\frac{\partial \mathscr{P}}{\partial x}, \tag{46}$$

$$V = -\frac{\partial \mathscr{P}}{\partial y} + R\,\theta, \tag{47}$$

$$\frac{\partial \theta}{\partial t} - V = \frac{\partial^2 \theta}{\partial x^2} + \frac{\partial^2 \theta}{\partial y^2}, \tag{48}$$

$$y = 0, 1: \qquad V = 0, \qquad \theta = 0. \tag{49}$$

One can easily encompass the dependence on $\mathscr{P}$ in (46) and (47) by differentiating (46) with respect to y, differentiating (47) with respect to x and then subtracting the second resulting equation from the first one,

$$\frac{\partial U}{\partial y} - \frac{\partial V}{\partial x} = -R\,\frac{\partial \theta}{\partial x}, \tag{50}$$

Let us now introduce a dimensionless streamfunction $\Psi(x, y, t)$, such that

$$U = \frac{\partial \Psi}{\partial y}, \qquad V = -\frac{\partial \Psi}{\partial x}. \tag{51}$$

Then, (45) is identically satisfied, while (50) and (48) yield respectively,

$$\frac{\partial^2 \Psi}{\partial x^2} + \frac{\partial^2 \Psi}{\partial y^2} = -R\,\frac{\partial \theta}{\partial x}, \tag{52}$$

$$\frac{\partial \theta}{\partial t} + \frac{\partial \Psi}{\partial x} = \frac{\partial^2 \theta}{\partial x^2} + \frac{\partial^2 \theta}{\partial y^2}. \tag{53}$$

The boundary conditions (49) can be rewritten as

$$y = 0, 1: \qquad \frac{\partial \Psi}{\partial x} = 0, \qquad \theta = 0. \tag{54}$$

Let us seek a solution in the form of a standing wave oriented along the x-axis, namely

$$\Psi = f(y)\, e^{\lambda t} \sin(ax), \qquad \theta = h(y)\, e^{\lambda t} \cos(ax). \tag{55}$$

In (55), a is the dimensionless wave number, while λ is an exponential coefficient. If $\lambda > 0$, the amplitude of the wave increases exponentially in time, thus implying an unstable behaviour of the system. If $\lambda < 0$, the wave undergoes an exponential damping in time, thus implying a stable behaviour of the system. Finally, the case $\lambda = 0$ represents the demarcation between stability and instability. Disturbances such that $\lambda = 0$ are called *neutrally stable* or *marginally stable* as they correspond to a threshold condition for instability.

By substituting (55) into (52)–(54), one transforms a partial differential problem in an ordinary differential problem in the unknown functions $f(y)$ and $h(y)$, namely

$$f''(y) - a^2 f(y) - aR\,h(y) = 0, \tag{56}$$

$$h''(y) - \left(a^2 + \lambda\right) h(y) - a f(y) = 0, \tag{57}$$

$$y = 0, 1: \qquad f = 0, \qquad h = 0, \tag{58}$$

where the primes denote differentiation with respect to y. The ordinary differential problem defined by (56)–(58) is homogeneous. This means that whatever are the prescribed values of the parameters (a, λ, R), (56)–(58) always admit the trivial solution $f = h = 0$. However, for any prescribed pair (a, λ), (56)–(58) may also admit nontrivial solutions for special values of R. These special values of R are called *eigenvalues*. Since one is interested in the nontrivial solutions of (56)–(58), the problem to be solved is an *eigenvalue problem.*

In order to fulfil the boundary conditions (58), $f(y)$ and $h(y)$ can be expressed as

$$f(y) = \sum_{n=1}^{\infty} F_n \sin(n\pi y), \qquad h(y) = \sum_{n=1}^{\infty} H_n \sin(n\pi y). \tag{59}$$

The coefficients F_n and H_n are such that (56) and (57) are satisfied, namely

$$\sum_{n=1}^{\infty} \left[\left(a^2 + n^2\pi^2\right) F_n + aR H_n\right] \sin(n\pi y) = 0, \tag{60}$$

$$\sum_{n=1}^{\infty} \left[(a^2 + \lambda + n^2\pi^2) H_n + a F_n \right] \sin(n\pi y) = 0. \tag{61}$$

Due to the orthogonality relationship between sine functions,

$$\int_0^1 \sin(m\pi y)\ \sin(n\pi y)\, dy = \begin{cases} 0, & m \neq n \\ \dfrac{1}{2}, & m = n, \end{cases} \tag{62}$$

Equations (60) and (61) are fulfilled if and only if all the coefficients of the series vanish. Then, one obtains the algebraic equations

$$(a^2 + n^2\pi^2) F_n + a R H_n = 0, \tag{63}$$

$$(a^2 + \lambda + n^2\pi^2) H_n + a F_n = 0. \tag{64}$$

From (64), one has

$$H_n = -\frac{a}{(a^2 + \lambda + n^2\pi^2)}\, F_n. \tag{65}$$

Then, by substituting (65) in (63) and simplifying F_n, one obtains an expression of the eigenvalue R,

$$R = \frac{(a^2 + n^2\pi^2)\,(a^2 + \lambda + n^2\pi^2)}{a^2}, \qquad n = 1, 2, 3, \ \ldots\ . \tag{66}$$

Let us note that simplifying F_n means that F_n is not vanishing, i.e. that we are seeking a solution (f, h) that is not identically zero.

There are different kinds of instabilities: one for each value of n. The lower one being that for $n = 1$. Unstable roll disturbances ($\lambda > 0$) are such that

$$R > \frac{(a^2 + \pi^2)^2}{a^2}. \tag{67}$$

One can easily see that the right hand side of the inequality (67) is the eigenvalue R corresponding to neutral stability ($\lambda = 0$) for $n = 1$. As a consequence, in the parametric plane (a, R), unstable states correspond to the region above the lower ($n = 1$) neutral stability curve

$$R = \frac{(a^2 + \pi^2)^2}{a^2}, \tag{68}$$

while stable states lie below this curve. One can easily recognize these features in Fig. 7. This figure reveals that the stability of the rest state depends on the wave number a of the roll disturbance. No instability can arise when

$$R < R_{cr} = 4\pi^2 \cong 39.4784, \tag{69}$$

where $R_{cr} = 4\pi^2$ is determined as the minimum of function $R(a)$ defined by (68), i. e. the minimum of the lower ($n = 1$) neutral stability curve. It is easily verified that the minimum $R = R_{cr}$ occurs for

$$a = a_{cr} = \pi. \tag{70}$$

The physical significance of the integer n, labeling the different neutral stability curves represented in Fig. 7, relies in the different kind of instabilities that can be activated at increasing values of the Darcy-Rayleigh number R. The lower neutral stability curve, $n = 1$, defines the first kind of instability to be onset, namely the single row of adjacent convective cells shown in Fig. 8. The second neutral stability curve, $n = 2$, defines a more complicated kind of instability, namely two vertical superposed rows of adjacent convective cells as shown in Fig. 9. One can imagine that, for higher n, new kind of instabilities are activated consisting of an increasing number of vertically superposed rows of adjacent convective cells. As it is easily inferred from (66) by setting $\lambda = 0$, the minimum of each neutral stability curve corresponds to

$$a = n\pi, \qquad R = 4n^2\pi^2. \tag{71}$$

It has been recalled that the critical value of the Rayleigh number equation (1) for the onset of convective cells in Rayleigh-Bénard convection is 1707.76. If one

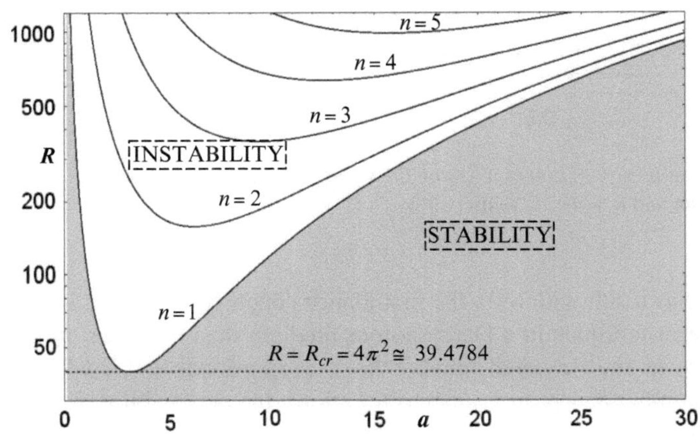

Fig. 7 Stable (*gray* region) and unstable (*white* region) states in the parametric plane (a, R)

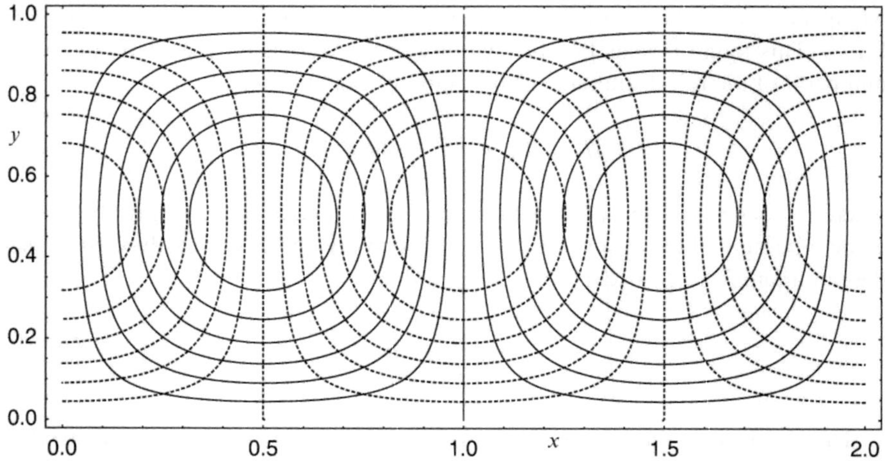

Fig. 8 Streamlines Ψ = constant (*solid lines*) and isotherms θ = constant (*dashed lines*) for $\lambda = 0$, $a = a_{cr} = \pi$ and $R = R_{cr} = 4\pi^2$

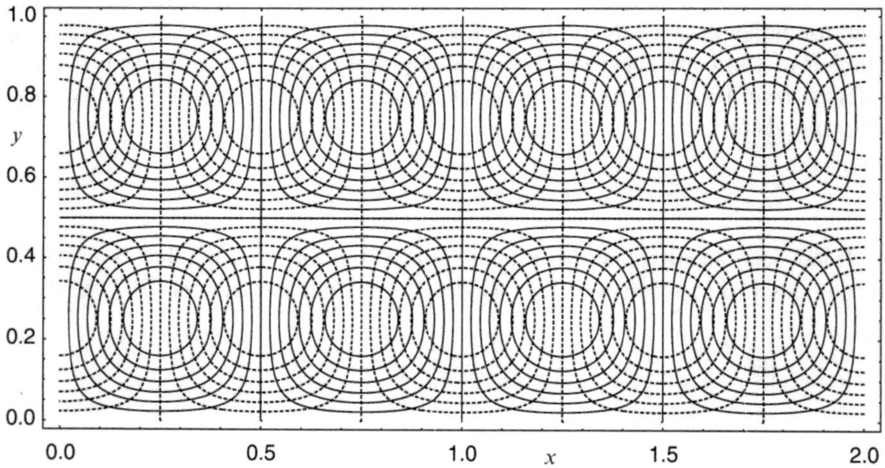

Fig. 9 Streamlines Ψ = constant (*solid lines*) and isotherms θ = constant (*dashed lines*) for $\lambda = 0$, $a = n\pi$ and $R = 4n^2\pi^2$ (with $n = 2$)

compares this result with (69), the first glance conclusion is that it is easier to have convective instabilities in a Darcy porous medium than in a clear fluid. However, this is false as the Rayleigh number Ra is proportional to L^3, while the Darcy-Rayleigh number R is proportional to KL. Since the permeability K is usually very small, it is more common for a clear fluid to have $Ra > 1707.76$, than for a fluid saturated porous medium to have $R > 4\pi^2$.

The classical Darcy-Bénard problem has been recently revisited by Nield and Barletta [23]. In this study, the pressure work has been taken into account in the local energy balance and its effects on the onset conditions for convective instability have been assessed. It has been shown that the pressure work contribution has a stabilizing effect leading to a critical value of R higher than $4\pi^2$.

4 Extensions of the Darcy-Bénard Problem, Beyond Darcy's Law

The discussion of the Darcy-Bénard problem carried out in Sect. 3 is based on the assumption that the local momentum balance equation for the fluid saturated porous medium is correctly expressed by Darcy's law. However, we know that this law may need to be replaced by either the Darcy-Forchheimer model, when the form-drag effects are important, or by the Brinkman model, when the permeability of the porous medium is very large. In this section, the possible changes induced in the analysis of the Darcy-Bénard problem by the use of the Darcy-Forchheimer model and by the use of the Brinkman model are investigated. The absence of any role played by the effect of viscous dissipation in this analysis, proved in Sect. 3 with reference to Darcy's law, holds also for non-Darcy models of momentum transfer. Thus, without any loss of generality, the forthcoming analysis is performed by dropping the viscous dissipation term in the energy balance from the beginning.

4.1 Form-Drag Effects

Let us assume that the Darcy-Bénard problem is defined by (27)–(31) with (28) replaced by

$$\frac{\mu}{K}\left(1+\frac{c_{\mathrm{f}}\sqrt{K}}{\nu}\,|\vec{u}|\right)\vec{u} = -\vec{\nabla}P + \rho\beta g(T-T_c)\vec{e}_y, \tag{72}$$

where (9) has been used. Thus, the dimensionless equations (34)–(38) still hold with (35) replaced by

$$\vec{u}(1+\Lambda\,|\vec{u}|) = -\vec{\nabla}P + RT\vec{e}_y, \tag{73}$$

where the dimensionless parameter Λ is defined as

$$\Lambda = \frac{c_{\mathrm{f}}\,\tilde{\alpha}\sqrt{K}}{\nu L}. \tag{74}$$

One can easily verify that the basic rest state is still described by (39). Then, on introducing the small perturbations of the basic state defined by (40), substituting (39) and (40) into (34), (73), (36)–(38) and finally neglecting terms of order ε^2, one obtains again (41)–(44). This means that the disturbance equations in the linear theory are not affected by the form-drag term. In other words, the stability analysis carried out in Sect. 3 still holds even if the Forchheimer correction of Darcy's law is taken into account. Obviously, the reason of that is the linearisation of the governing equations. Then, the linear stability theory of the Darcy-Bénard problem is not affected by the form-drag effect inasmuch as it is not affected by the effect of viscous dissipation.

4.2 Brinkman's Model Approach to the Onset Conditions

If one uses Brinkman's model, (13), instead of Darcy's law, then (27)–(31) still hold with (28) replaced by

$$\frac{\mu}{K}\,\vec{u} - \tilde{\mu}\nabla^2\vec{u} = -\,\vec{\nabla}P + \rho\,\beta\,g\,(T - T_c)\vec{e}_y. \tag{75}$$

In a dimensionless form, (34)–(38) still hold with (35) replaced by

$$\vec{u} - Da_m\,\nabla^2\vec{u} = -\,\vec{\nabla}P + R\,T\,\vec{e}_y, \tag{76}$$

where Da_m is the modified Darcy number, well-known also as *Brinkman coefficient* [24],

$$Da_m = \frac{\tilde{\mu}}{\mu}\,Da = \frac{\tilde{\mu}}{\mu}\,\frac{K}{L^2}, \tag{77}$$

and $Da = K/L^2$ is the *Darcy number*. One must recognize that the boundary conditions, (38), now include also the no-slip condition $u = w = 0$. Again, the basic rest state is described by (39). Then, on introducing the small perturbations of the basic state defined by (40), substituting (39) and (40) into (34), (76), (36)–(38) and finally neglecting terms of order ε^2, one obtains

$$\vec{\nabla}\cdot\vec{U} = 0, \tag{78}$$

$$\vec{U} - Da_m\,\nabla^2\vec{U} = -\vec{\nabla}\mathscr{P} + R\,\theta\,\vec{e}_y, \tag{79}$$

$$\frac{\partial\theta}{\partial t} - V = \nabla^2\theta, \tag{80}$$

$$y = 0, 1: \qquad U = V = W = 0, \qquad \theta = 0. \tag{81}$$

One can eliminate the dependence on $\mathscr{P}$ by taking the curl of both sides of (79), namely

$$\vec{\nabla} \times \vec{U} - Da_m \, \nabla^2 \left(\vec{\nabla} \times \vec{U} \right) = -R \, \frac{\partial \theta}{\partial z} \, \vec{e}_x + R \, \frac{\partial \theta}{\partial x} \, \vec{e}_z, \tag{82}$$

where $\vec{e}_x$ and $\vec{e}_z$ are the unit vectors along the x- and the z-axes. Let us now assume, as in the analysis of Sect. 3, that the disturbances are 2D and lie in the (x, y)-plane. Then, only the z-component of $\vec{\nabla} \times \vec{U}$ is nonzero and the dimensionless stream-function $\Psi(x, y, t)$ defined by (51) is such that

$$\left(\vec{\nabla} \times \vec{U} \right) \cdot \vec{e}_z = -\nabla^2 \Psi. \tag{83}$$

Then, (78) is identically satisfied, while (82), (80), (81) yield respectively

$$\nabla^2 \Psi - Da_m \, \nabla^2 \left(\nabla^2 \Psi \right) = -R \, \frac{\partial \theta}{\partial x}, \tag{84}$$

$$\frac{\partial \theta}{\partial t} + \frac{\partial \Psi}{\partial x} = \nabla^2 \theta, \tag{85}$$

$$y = 0, 1: \qquad \frac{\partial \Psi}{\partial x} = \frac{\partial \Psi}{\partial y} = 0, \qquad \theta = 0. \tag{86}$$

Then, by assuming plane wave disturbances given by (55), (56)–(58) are now replaced by

$$Da_m \, f''''(y) - \left(2 Da_m a^2 + 1 \right) f''(y) + a^2 \left(Da_m a^2 + 1 \right) f(y) + a R \, h(y) = 0, \tag{87}$$

$$h''(y) - \left(a^2 + \lambda \right) h(y) - a f(y) = 0, \tag{88}$$

$$y = 0, 1: \qquad f = f' = 0, \qquad h = 0. \tag{89}$$

Unlike (56), (87) is a fourth-order differential equation. The eigenvalue problem, (87)–(89), can be solved numerically. The procedure is as follows. One prescribes the input values of (a, λ, Da_m), guesses the eigenvalue R and solves (87) and (88) by, for instance, a Runge-Kutta method under the initial conditions

$$y = 0: \qquad f = f' = 0, \quad f'' = A, \quad f''' = B, \quad h = 0, \quad h' = 1. \tag{90}$$

In (90), A and B are guessed constant values and the additional condition $h' = 1$ is perfectly legitimate in order to fix the scale of the eigenfunctions (f, h), otherwise undetermined. The guessed values (R, A, B) are finally determined through a shooting method by employing the three constraint conditions

$$y = 1: \qquad f = f' = 0, \qquad h = 0. \tag{91}$$

With $\lambda = 0$, this numerical procedure allows one to develop a function $R(a)$ describing the lowest neutral stability curve. As it has been pointed out in Sect. 3, the lowest neutral stability curve is one where the minimum corresponds to the critical conditions for the onset of the first instability mode. More precisely, the critical values (a_{cr}, R_{cr}) are determined numerically by seeking the minimum of $R(a)$. On account of (87)–(89), both function $R(a)$ and the critical values (a_{cr}, R_{cr}) depend parametrically on the modified Darcy number, Da_m.

This study of linear stability was carried out by Walker and Homsy [35] and, more recently, by Rees [28].

The special feature of Brinkman's model is that it represents a bridge from Darcy's law to the Navier-Stokes model of a clear fluid. These limiting cases are recovered as $Da_m \to 0$ (Darcy) and $Da_m \to \infty$ (clear fluid). With reference to the critical conditions for the onset of convective instabilities, it is well known that a Darcy porous medium has a critical Darcy-Rayleigh number, R, equal to $4\pi^2$, while a clear fluid has a critical Rayleigh number, Ra, equal to 1707.67. One may easily verify that

$$\lim_{Da_m \to \infty} \frac{R}{Da_m} = Ra. \tag{92}$$

Therefore, one expects that R_{cr}/Da_m, evaluated from (87)–(89), tends to 1707.67 when $Da_m \to \infty$. This asymptotic behaviour is confirmed by Figs. 10 and 11. These figures display the plot of either R_{cr}/Da_m or R_{cr} versus Da_m compared with the two asymptotes: that for Darcy's law behaviour (dashed line, $R_{cr} = 4\pi^2$) and that for a clear fluid (dashed line, $R_{cr} = 1707.67\,Da_m$).

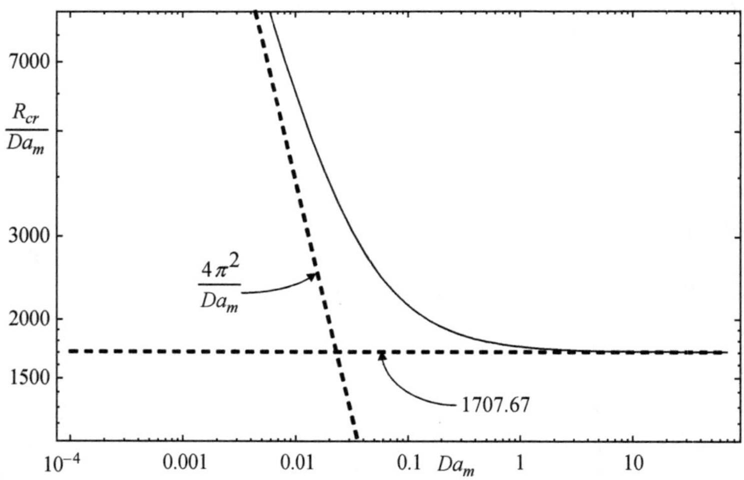

Fig. 10 Critical values of R/Da_m versus Da_m for the Brinkman model formulation of the Darcy-Bénard problem (solid line). The asymptotic cases of Darcy's law ($R_{cr} = 4\pi^2$) and of Navier-Stokes clear fluid ($R_{cr}/Da_m = 1707.67$) are given as dashed lines

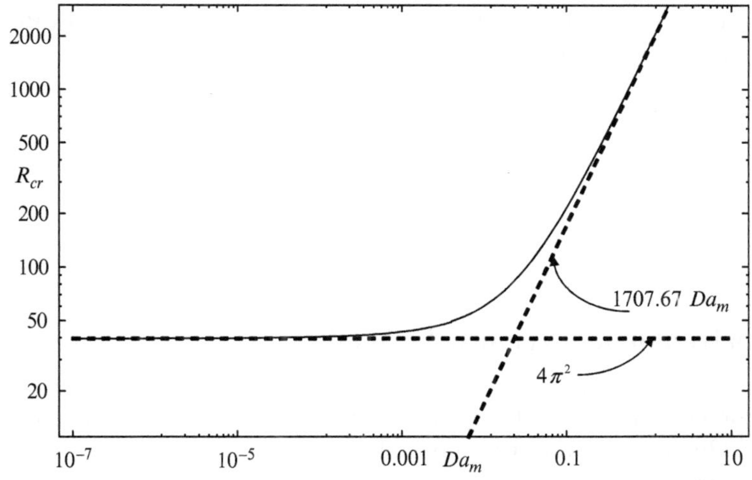

Fig. 11 Critical values of R versus Da_m for the Brinkman model formulation of the Darcy-Bénard problem (solid line). The asymptotic cases of Darcy's law ($R_{cr} = 4\pi^2$) and of Navier-Stokes clear fluid ($R_{cr} = 1707.67\,Da_m$) are given as dashed lines

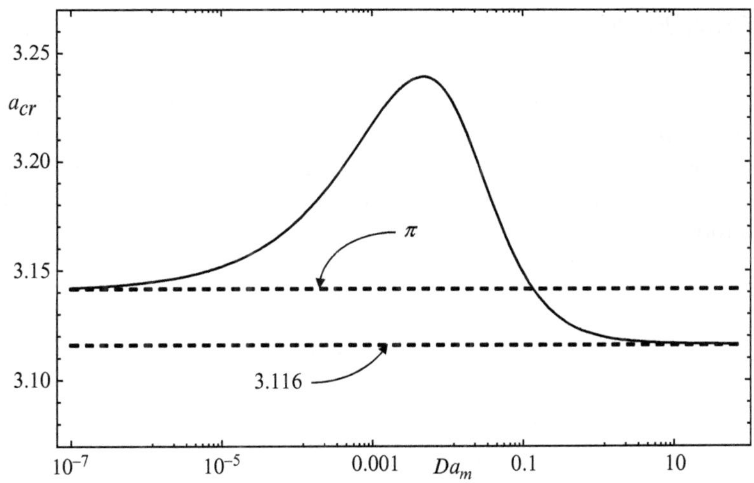

Fig. 12 Critical values of a versus Da_m for the Brinkman model formulation of the Darcy-Bénard problem (solid line). The asymptotic cases of Darcy's law ($a_{cr} = \pi$) and of Navier-Stokes clear fluid ($a_{cr} = 3.116$) are given as dashed lines

Figure 12 shows the plot of a_{cr} versus Da_m. This figure highlights the two asymptotic behaviours: the limit $Da_m \to 0$ corresponding to Darcy's law (dashed line, $a_{cr} = \pi$); the limit $Da_m \to \infty$ of a clear fluid (dashed line, $a_{cr} = 3.116$). An interesting feature is the existence of a maximum of the critical wave number, $a_{cr} = 3.239$, for $Da_m = 4.895 \cdot 10^{-3}$.

Very useful asymptotic expansions of (a_{cr}, R_{cr}) have been obtained by Rees [28] in a range of extremely small values of Da_m,

$$a_{cr} \approx \pi\left(1 + \sqrt{Da_m}\right), \tag{93}$$

$$R_{cr} \approx 4\pi^2\left(1 + 2\sqrt{Da_m} + 28.1337\, Da_m\right). \tag{94}$$

5 Horizontal and Vertical Throughflow

An important variant of the Darcy-Bénard problem arises when a horizontal (Prats problem [26]) or a vertical basic throughflow is assumed instead of the rest state. In the following, it will be shown that the presence of a basic flow modifies the critical conditions for the onset of convective roll instabilities only in the case of a vertical throughflow. The analysis presented in this section is based on Darcy's law, as well as on the assumption of a negligible effect of viscous dissipation.

5.1 Prats Problem

Let us consider the same horizontal fluid layer considered in the study of the Darcy-Bénard problem carried out in Sect. 3. The governing equations and the boundary conditions are again (27)–(31), where we assume as negligible the contribution of viscous dissipation, $\mu \vec{u} \cdot \vec{u}/K$, in the local energy balance, (29). Then, in a dimensionless form, these equations are written as

$$\vec{\nabla} \cdot \vec{u} = 0, \tag{95}$$

$$\vec{u} = -\vec{\nabla} P + R T \vec{e}_y, \tag{96}$$

$$\frac{\partial T}{\partial t} + \vec{u} \cdot \vec{\nabla} T = \nabla^2 T, \tag{97}$$

$$y = 0: \qquad v = 0, \qquad T = 1, \tag{98}$$

$$y = 1: \qquad v = 0, \qquad T = 0, \tag{99}$$

where the definitions, (32), have been used. A stationary solution of the governing equations (95)–(99) exists under the assumption of a horizontal uniform throughflow in the x-direction, $\vec{u} = Pe\, \vec{e}_x$,

$$\vec{u}_B = Pe\, \vec{e}_x, \qquad T_B = 1 - y, \qquad P_B = -Pe\, x + R y\left(1 - \frac{y}{2}\right). \tag{100}$$

Here, Pe is the Péclet number, namely the ratio between the dimensional throughflow velocity in the x-direction and the quantity $\tilde{\alpha}/L$.

One can switch from the rest reference frame to a comoving reference frame defined by

$$\begin{aligned} &\tilde{\vec{x}} = \vec{x} - Pe\, t\, \vec{e}_x, \quad \tilde{t} = t, \quad \tilde{\vec{u}} = \vec{u} - Pe\, \vec{e}_x, \\ &\tilde{P} = P + Pe\, x, \quad \tilde{T} = T. \end{aligned} \tag{101}$$

Here, $\vec{x}$ is the position vector and the tilded quantities are those defined in the comoving frame. Under the transformation defined by (101), the governing equations (95)–(99) are left invariant, while the basic solution (100) is transformed into the basic solution of the Darcy-Bénard problem, (39). This means that the linear stability analysis of the Prats problem yields exactly the same results obtained for the Darcy-Bénard problem. In other words, the critical conditions for the onset of convective roll instabilities are still given by

$$a_{cr} = \pi, \qquad R_{cr} = 4\pi^2. \tag{102}$$

Moreover, the linear stability analysis is independent of the particular direction in the horizontal (x, z)-plane for the propagation of the disturbance plane waves. In fact, in the comoving reference frame, one has an invariance of the governing equations and of the basic solution under arbitrary rotations around the vertical y-axis.

5.2 Effects of a Vertical Throughflow

If the effect of a horizontal uniform throughflow as described in the Prats problem does not lead to novel features with respect to the Darcy-Bénard problem, the effect of a vertical throughflow is a less trivial phenomenon. Classical studies on this subject where carried out by Sutton [32] and Nield [21]. Quite recently, this subject has been revisited by Barletta et al. [5].

The governing equations for this problem are just the same as for the Darcy-Bénard problem and for the Prats problem and, therefore, are given again in a dimensionless form by (95)–(97). On the contrary, the velocity boundary conditions are changed with respect to (98) and (99) as the horizontal boundaries are no more impermeable but subject to a uniform throughflow, namely

$$y = 0: \qquad v = Pe, \qquad T = 1, \tag{103}$$

$$y = 1: \qquad v = Pe, \qquad T = 0. \tag{104}$$

The steady-state basic solution is more complicated than in the Prats problem and is expressed as

$$\vec{u}_B = Pe\,\vec{e}_y, \quad T_B = \frac{e^{Pe} - e^{Pe\,y}}{e^{Pe} - 1}, \quad P_B = \left(\frac{R\,e^{Pe}}{e^{Pe} - 1} - Pe\right)y - \frac{R\,e^{Pe\,y}}{Pe(e^{Pe} - 1)}. \quad (105)$$

The linear disturbance equations for small perturbations defined by means of (40) are now written as

$$\vec{\nabla} \cdot \vec{U} = 0, \quad (106)$$

$$\vec{U} = -\vec{\nabla}\mathscr{P} + R\,\theta\,\vec{e}_y, \quad (107)$$

$$\frac{\partial \theta}{\partial t} - Pe\,V\,F(y) + Pe\,\frac{\partial \theta}{\partial y} = \nabla^2 \theta, \quad (108)$$

$$y = 0, 1: \qquad V = 0, \qquad \theta = 0, \quad (109)$$

where $F(y)$ is a function defined as

$$F(y) = \frac{e^{Pe\,y}}{e^{Pe} - 1}, \quad (110)$$

As for the Darcy-Bénard problem, it is not restrictive seeking two-dimensional solutions of the disturbance equations in the (x, y)-plane. In fact, the system is invariant under rotations around the y-axis, so that every horizontal direction is equivalent. Then, one may set $W = 0$ and the components U and V of the disturbance velocity can be expressed through the streamfunction Ψ defined by (51). Again we assume plane standing wave disturbances defined by (55). If one is interested in the neutral stability condition, $\lambda = 0$, (56)–(58) are now replaced by

$$f''(y) - a^2 f(y) - aR\,h(y) = 0, \quad (111)$$

$$h''(y) - Pe\,h'(y) - a^2 h(y) - a\,Pe\,F(y) f(y) = 0, \quad (112)$$

$$y = 0, 1: \qquad f = 0, \qquad h = 0. \quad (113)$$

Then, (111)–(113) can be solved numerically as an eigenvalue problem by the procedure described in Sect. 4.2. Important features of the numerical solution are that the critical value of the wave number, a_{cr}, and the critical value of the Darcy-Rayleigh number, R_{cr}, depend on the absolute value of Pe, but not on the sign of Pe [5, 32]. This means that the direction of the vertical throughflow, upward or downward, does not influence the onset of the convective instabilities. A table of (a_{cr}, R_{cr}) versus Pe is reported in Table 1. This table reveals that the effect of the vertical throughflow, whatever be its direction, is stabilizing. In fact, R_{cr} is an increasing function of $|Pe|$. The critical conditions for the onset of instabilities obtained for the Darcy-Bénard problem are recovered when the limit $Pe \to 0$ is approached.

Table 1 Vertical throughflow: critical conditions for the onset of convective instabilities [5]

Pe	a_{cr}	R_{cr}
0	π	$4\pi^2$
10^{-3}	3.14159	39.4784
10^{-1}	3.14196	39.4924
1	3.17868	40.8751
2	3.29218	45.0776
5	4.19616	73.4146
8	6.09212	114.833
10	7.59035	143.518
15	11.3830	215.283

6 Viscous Dissipation

The effect of the viscous dissipation can be important in the analysis of the onset of convective instabilities. Therefore, its correct modelization with respect to the local energy balance in the framework of the Oberbeck-Boussinesq approximation of buoyant flows is a significant point. This subject is often underestimated or misunderstood by several authors. A recent analysis of this topic [2] is here outlined first with reference to a clear fluid and then extended to the case of a fluid saturated porous medium.

6.1 The Oberbeck-Boussinesq Approximation and the Local Energy Balance

Let us consider a clear fluid, one whose momentum balance is given by the Navier-Stokes equation. The nature of the Oberbeck-Boussinesq approximation has been already recalled at the beginning of Sect. 3. The assumption is that the fluid properties are considered as constants with the only exception of the density, whose change is taken into account only in the gravitational body force term of the momentum balance. The linear equation of state (26) implies that the density is evaluated at constant pressure and that the temperature changes are very small. Thus, the mass and momentum balance equations are given by

$$\vec{\nabla} \cdot \vec{u} = 0, \tag{114}$$

$$\frac{\partial \vec{u}}{\partial t} + \vec{u} \cdot \vec{\nabla} \vec{u} = -\frac{1}{\rho_0} \vec{\nabla} P - (T - T_0)\beta \vec{g} + \nu \nabla^2 \vec{u}. \tag{115}$$

Equations (114) and (115) must be completed with the energy balance in order to achieve the closure of the problem. In the literature, there is a manifold answer to the question of the proper formulation of the local energy balance. In fact, one may have Chandrasekhar's [10] and White's [36] formulation

$$\rho_0\, c_v \left(\frac{\partial T}{\partial t} + \vec{u} \cdot \vec{\nabla} T \right) = k\, \nabla^2 T + q_g + 2\, \mu\, \mathscr{D}_{ij}\, \mathscr{D}_{ij}, \tag{116}$$

where c_v is the specific heat at constant volume and $\mathscr{D}_{ij}$ is the strain tensor, (23).

The source term in (116), $2\, \mu\, \mathscr{D}_{ij}\, \mathscr{D}_{ij}$, is the thermal power generated by the viscous dissipation.

One may have the enthalpy formulation [20, 33]

$$\rho_0\, c_p \left(\frac{\partial T}{\partial t} + \vec{u} \cdot \vec{\nabla} T \right) = k\, \nabla^2 T + q_g + 2\, \mu\, \mathscr{D}_{ij}\, \mathscr{D}_{ij} + \beta\, T \left(\frac{\partial p}{\partial t} + \vec{u} \cdot \vec{\nabla} p \right), \tag{117}$$

where c_p is the specific heat at constant pressure and the last term on the right hand side is an additional source term: the pressure work acting on the fluid element.

Finally, one may have Landau-Lifshitz's [17], Bejan's [6] and Kundu-Cohen's [16] formulation

$$\rho_0\, c_p \left(\frac{\partial T}{\partial t} + \vec{u} \cdot \vec{\nabla} T \right) = k\, \nabla^2 T + q_g + 2\, \mu\, \mathscr{D}_{ij}\, \mathscr{D}_{ij}. \tag{118}$$

Equations (116), (117) and (118) are different answers to the same question: which is the correct formulation of the local energy balance with respect to the Oberbeck-Boussinesq approximation? An analysis of the different procedures usually followed in the literature in order to answer this question and the common logical pitfalls in these deductions have been recently discussed by Barletta [2]. In this recent paper, a thermodynamic reasoning is presented in order to prove that the correct formulation of the local energy balance with respect to the Oberbeck-Boussinesq approximation is

$$\rho_0\, c \left(\frac{\partial T}{\partial t} + \vec{u} \cdot \vec{\nabla} T \right) = k\, \nabla^2 T + q_g + 2\, \mu\, \mathscr{D}_{ij}\, \mathscr{D}_{ij}. \tag{119}$$

In (119), the heat capacity per unit mass c does not coincide in general either with c_v or with c_p, but it is defined as

$$c = c_p - \frac{p\beta}{\rho_0}. \tag{120}$$

The heat capacity per unit mass c coincides with the specific heat at constant volume, c_v, for a perfect gas and it is definitely well approximated by the specific heat at constant pressure, c_p, for a liquid [2].

6.2 *From the Clear Fluid to the Fluid Saturated Porous Medium*

The analysis presented in Sect. 6.1 is easily extended to a fluid saturated porous medium [2]. In cases of local thermal equilibrium between the solid phase and the fluid phase, the correct formulation of the local energy balance with respect to the Oberbeck-Boussinesq approximation is

$$\rho_0 c \left(\sigma \frac{\partial T}{\partial t} + \vec{u} \cdot \vec{\nabla} T \right) = \tilde{k} \nabla^2 T + q_g + \Phi, \qquad (121)$$

where c is again given by (120).

7 Dissipation-Induced Instabilities

An important feature of the Darcy-Bénard problem and of its variants investigated in the preceding sections is that the convective instabilities are a consequence of the thermal boundary conditions prescribed. In fact, the thermal boundary conditions assigned are such that a vertical temperature gradient is present in the system. When, this gradient is sufficiently intense, the basic solution becomes unstable. There is another important cause of possibly unstable temperature gradients in the basic flow: the effect of viscous dissipation. The viscous dissipation, being a mechanism of internal heat generation due to the fluid friction, may contribute to the instability of the flow in porous media or possibly may be the sole cause of this instability.

7.1 *Viscous Dissipation as the Sole Cause of Convective Instabilities*

In the analysis of the Darcy-Bénard problem presented in Sect. 3, it has been pointed out that the viscous dissipation does not affect the linear stability analysis. The reason is that the basic solution assumed for the Darcy-Bénard problem is such that the velocity field is zero. This feature implies that the effect of viscous dissipation becomes of order ε^2 and can thus be neglected with respect to the linear terms of order ε. Obviously, this argument cannot be invoked if a basic throughflow occurs in the fluid layer. In fact, a basic throughflow may result in a contribution of order ε in the local energy balance due to the effect of viscous dissipation.

Conditions may exist such that no temperature gradient is impressed in the system through the external environment. This circumstance may occur when the external environment is kept at a uniform temperature, or if one or more system boundaries are thermally insulated, i.e. adiabatic. This is not the case of the Darcy-Bénard problem and of its variants discussed in the preceding sections. In fact, the temperature difference between the horizontal boundary planes means exactly an externally induced thermal forcing on the system. On the other hand, no external thermal forcing is present if the boundary planes are kept at the same temperature, or if one or both planes are kept adiabatic. In the absence of thermal forcing caused by the boundary conditions, the viscous dissipation may possibly become the sole cause of thermal gradients within the fluid. As a consequence, viscous dissipation alone may be the cause of convective instabilities. This simple physical fact has been recognized in the papers by Joseph [15], Sukanek et al. [31] and Subrahmaniam et al. [30] with reference to clear fluid flows. Recently, several studies of the dissipation-induced instabilities in fluid saturated porous media have been carried out [3–5, 29]. In the following sections, the dissipation-induced instabilities are discussed for the cases of horizontal and vertical basic throughflow in a plane porous layer.

7.2 Plane Layer with Horizontal Throughflow

Let us consider an horizontal fluid saturated porous layer with infinite width and thickness L. Let us assume that the boundary planes are impermeable: the bottom boundary is adiabatic, while the top boundary is isothermal with temperature T_c (see Fig. 13). The validity of Darcy's law and of the Oberbeck-Boussinesq approximation is assumed.

The governing equations are given by (27)–(29), while the velocity and temperature boundary conditions are expressed as

$$y = 0: \quad v = 0 = \frac{\partial T}{\partial y}, \qquad y = L: \quad v = 0, \quad T = T_c. \tag{122}$$

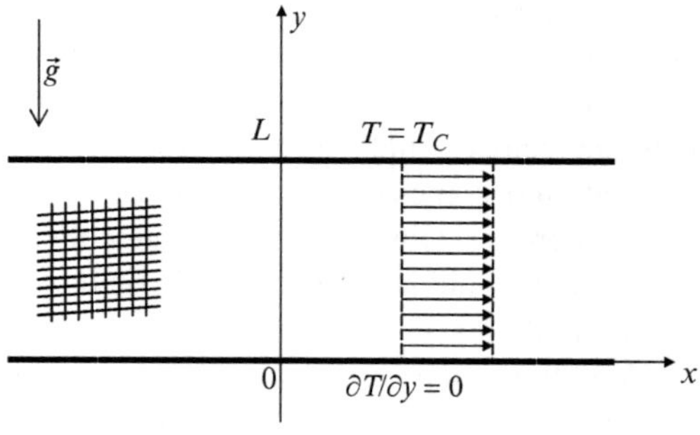

Fig. 13 Boundary conditions considered in the analysis of dissipation-induced instabilities in a porous layer with horizontal basic throughflow [4]

Equations (27)–(29) and (122) can be written in a dimensionless form by the following transformation:

$$(x,y,z)\frac{1}{L} \to (x,y,z), \quad t\,\frac{\tilde{\alpha}}{\sigma L^2} \to t, \quad \vec{u}\,\frac{L}{\tilde{\alpha}} \to \vec{u}, \quad (T - T_c)\frac{Kc}{\nu \tilde{\alpha}} \to T. \tag{123}$$

Then, one obtains the dimensionless equations

$$\vec{\nabla} \cdot \vec{u} = 0, \tag{124}$$

$$\vec{\nabla} \times \vec{u} = Ge\, \vec{\nabla} \times \left(T \vec{e}_y\right), \tag{125}$$

$$\frac{\partial T}{\partial t} + \vec{u} \cdot \vec{\nabla} T = \nabla^2 T + \vec{u} \cdot \vec{u}, \tag{126}$$

$$y = 0: \quad v = 0, \quad \frac{\partial T}{\partial y} = 0, \qquad y = 1: \quad v = 0, \quad T = 0, \tag{127}$$

where

$$Ge = \frac{g \beta L}{c} \tag{128}$$

is the Gebhart number. The thermal boundary conditions are such that no thermal forcing is applied on the system from the external environment. This circumstance is evident since no characteristic temperature difference arises from the thermal boundary conditions.

The basic solution of (124)–(127), which is analysed for stability, is given by

$$u_B = Pe, \quad v_B = 0, \quad w_B = 0, \quad T_B = \frac{Pe^2}{2}(1 - y^2). \tag{129}$$

A vertical temperature gradient appears in the basic state caused only by the frictional heating. This effect produces a bottom boundary temperature, $T_B(0) = Pe^2/2$, higher than the top boundary temperature, $T_B(1) = 0$. Hence, the basic state is possibly unstable to convective rolls for a sufficiently intense viscous dissipation. One expects that instabilities arise when the Péclet number becomes sufficiently high, so that the bottom to top boundary temperature difference, $Pe^2/2$, exceeds the threshold value for convection rolls to take place. This critical value of the Péclet number is a function of the dimensionless parameter Ge and has been determined through a linear stability analysis by Barletta et al. [4].

The analysis has been carried out by assuming two-dimensional disturbances in the form of plane waves travelling in any arbitrary direction in the horizontal (x,z)-plane. Let us note that, in the present problem, the possible horizontal directions are not equivalent, as in the Darcy Bénard problem. In fact, a preferred direction is defined by the basic flow direction, namely the x-axis. Thence, one defines a general *oblique roll* disturbance as a plane wave propagating in a direction lying in the (x,z)-plane and inclined with respect to the x-axis of an arbitrary angle χ. When $\chi = 0$, namely when the plane waves propagate along the x-direction, one speaks of *transverse rolls*. When $\chi = \pi/2$, namely when the plane waves propagate along the z-direction, one speaks of *longitudinal rolls*.

Barletta et al. [4] proved that the basic solution (129) is more unstable to transverse rolls than to any other kind of oblique rolls. The critical values (a_{cr}, Pe_{cr}) for the onset of transverse rolls are given approximately by

$$a_{cr} = 2.4483, \qquad Pe_{cr} = \frac{7.8655}{\sqrt{Ge}}. \tag{130}$$

Equation (130) holds if Ge is sufficiently small. This is not an unphysical restriction since, for most convection problems involving liquids, Ge can hardly be greater than 10^{-6}, unless $L \cong 1\,\text{m}$ or higher. For values of Ge approximately greater than 10^{-3}, (130) yields an overestimate of both a_{cr} and Pe_{cr}.

7.3 Plane Layer with Vertical Throughflow

The vertical throughflow system defined in the studies by Sutton [32] and Nield [21] have been recently extended to the case of non negligible effects of viscous dissipation [5].

The system examined is exactly the same as that described in Sect. 5.2. The only difference is that the local energy balance equation includes the viscous dissipation term, $\mu \vec{u} \cdot \vec{u}/K$. This difference implies that the basic solution is more complicated than (105). In particular, by employing the dimensionless formulation defined through (32), the basic velocity and temperature fields are expressed as

$$\vec{u}_B = Pe\, \vec{e}_y, \quad T_B(y) = \frac{e^{Pe} - e^{Pe\, y}}{e^{Pe} - 1} + \frac{Ge\, Pe}{R} \left(y - \frac{e^{Pe\, y} - 1}{e^{Pe} - 1} \right). \tag{131}$$

The effect of viscous dissipation breaks the symmetry between upward throughflow, $Pe > 0$, and downward throughflow, $Pe < 0$. For downward throughflow, there are two competing effects: the forced convection due to the throughflow, and the viscous dissipation. The former effect causes the cooling of the system due to the fluid input from the upper environment at temperature T_c. The latter effect causes an internal heating of the layer. The competition may yield a perfect balance between these effects when $Ge\,Pe = -R$. In this case, (131) yields the basic linear temperature profile (39), as in the Darcy-Bénard problem.

When the viscous dissipation is taken into account, the sign of Pe affects the critical conditions for the onset of convective instabilities. Increasing values of Ge, namely a more and more intense viscous dissipation, imply a stabilizing effect in the case of downward throughflow and a destabilizing effect in the case of upward throughflow. The destabilizing effect for $Pe > 0$ may be so intense that the critical value of the Darcy-Rayleigh number for the onset of convective rolls, R_{cr}, becomes zero. This phenomenon however may take place only when $Ge > 4.67910$. Values of Ge in this range are definitely unrealistic in practical cases. Then, in the case of vertical throughflow, the effect of viscous dissipation is unlikely to be the sole cause of convective instabilities. This conclusion, drawn by Barletta et al. [5], marks a sharp difference with respect to the case of horizontal throughflow.

8 Concluding Remarks

In this review, some remarkable topics within the wide literature on the convective instabilities of fluid saturated porous media have been outlined. The amount of published papers and book chapters on this subject is really large, so that this review is definitely far from being exhaustive. The aim has been the general description of the basic phenomena behind the thermoconvective roll instabilities in a porous medium. With this in mind, a detailed analysis of the Darcy-Bénard problem and of its most important extensions has been carried out. We mentioned the use of non-Darcy models of momentum transfer and the possible presence of a horizontal or vertical throughflow. An important recent topic is the analysis of the effect of viscous dissipation with respect to the onset conditions of the convective instability. This effect arises when the basic state is a throughflow state. The role played by viscous dissipation in the onset of the instability may be a mere interplay with the effect of the basic thermal gradients induced by the boundary conditions. In some cases, the viscous dissipation may be the sole cause of the instability. The continuing exploration of this possibility is an interesting opportunity for future research.

References

1. Al-Hadhrami, A.K., Elliott, L., Ingham, D.B.: A new model for viscous dissipation in porous media across a range of permeability values. Transp Porous Media **53**, 117–122 (2003)
2. Barletta, A.: Local energy balance, specific heats and the Oberbeck-Boussinesq approximation. Int J Heat Mass Transfer **52**, 5266–5270 (2009)
3. Barletta, A., Rees, D.A.S.: Stability analysis of dual adiabatic flows in a horizontal porous layer. Int J Heat Mass Transfer **52**, 2300–2310 (2009)
4. Barletta, A., Celli, M., Rees, D.A.S.: The onset of convection in a porous layer induced by viscous dissipation: a linear stability analysis. Int J Heat Mass Transfer **52**, 337–344 (2009)
5. Barletta, A., Rossi di Schio, E., Storesletten, L.: Convective roll instabilities of throughflow with viscous dissipation in a horizontal porous layer. Transp Porous Media **81**, 461–477 (2010)
6. Bejan, A.: Heat Transfer. Wiley, New York (1993)

7. Bénard, H.: Tourbillons cellulaires dans une nappe liquide. Revue. Gén. Sci. Pure Appl. **11**, 1261–1271 and **11**, 1309–1328 (1900)
8. Bénard, H.: Les tourbillons cellulaires dans une nappe liquide transportant de la chaleur par convection en régime permanent. Ann Chim Phys **23**, 62–144 (1901)
9. Brown, G.O.: Henry Darcy and the making of a law. Water Resour. Res. **38**(7) (2002). doi:10.1029/2001WR000727
10. Chandrasekhar, S.: Hydrodynamic and Hydromagnetic Stability. Oxford University Press, Oxford (1961)
11. Darcy, H.: Les fontaines publiques de la ville de Dijon. Dalmont, Paris (1856)
12. Drazin, P.G., Reid, W.H.: Hydrodynamic Stability, 2nd edn. Cambridge University Press, Cambridge, UK (2004)
13. Gebhart, B., Jaluria, Y., Mahajan, R.L., Sammakia, B.: Buoyancy-Induced Flows and Transport. Hemisphere, New York (1988)
14. Horton, C.W., Rogers Jr., F.T.: Convection currents in a porous medium. J Appl Phys **16**, 367–370 (1945)
15. Joseph, D.D.: Stability of frictionally-heated flow. Phys Fluids **8**, 2195–2200 (1965)
16. Kundu, P.K., Cohen, I.M.: Fluid Mechanics, 2nd edn. Elsevier, New York (2002)
17. Landau, L.D., Lifshitz, E.M.: Fluid Mechanics, 2nd edn. Pergamon Press, Oxford (1987). Chapter 5
18. Lapwood, E.R.: Convection of a fluid in a porous medium. Proc Camb Philol Soc **44**, 508–521 (1948)
19. Martynenko, O.G., Khramtsov, P.P.: Free-Convective Heat Transfer. Springer, Berlin (2005)
20. Nield, D.A.: Convective instability in porous media with throughflow. AIChE J **33**, 1222–1224 (1987)
21. Nield, D.A.: The modeling of viscous dissipation in a saturated porous medium. ASME J Heat Transfer **129**, 1459–1463 (2007)
22. Nield, D.A., Barletta, A.: The Horton-Rogers-Lapwood problem revisited: the effect of pressure work. Transp Porous Media **77**, 143–158 (2009)
23. Nield, D.A., Bejan, A.: Convection in Porous Media, 3rd edn. Springer, New York (2006)
24. Normand, C., Pomeau, Y.: Convective instability: a physicist's approach. Rev Mod Phys **49**, 581–624 (1977)
25. Prats, M.: The effect of horizontal fluid flow on thermally induced convection currents in porous mediums. J Geophys Res **71**, 4835–4838 (1966)
26. Rayleigh, L.: On convection currents in a horizontal layer of fluid, when the higher temperature is on the under side. Philos Mag **32**, 529–546 (1916)
27. Rees, D.A.S.: The stability of Darcy-Bénard convection. In: Vafai, K. (ed.) Handbook of Porous Media, pp. 521–558. Marcel Dekker, New York (2000)
28. Rees, D.A.S.: The onset of Darcy-Brinkman convection in a porous layer: an asymptotic analysis. Int J Heat Mass Transfer **45**, 2213–2220 (2002)
29. Storesletten, L., Barletta, A.: Linear instability of mixed convection of cold water in a porous layer induced by viscous dissipation. Int J Thermal Sci **48**, 655–664 (2009)
30. Subrahmaniam, N., Johns, L.E., Narayanan, R.: Stability of frictional heating in plane Couette flow at fixed power input. Proc R Soc Lond A **458**, 2561–2569 (2002)
31. Sukanek, P.C., Goldstein, C.A., Laurence, R.L.: The stability of plane Couette flow with viscous heating. J Fluid Mech **57**, 651–670 (1973)
32. Sutton, F.M.: Onset of convection in a porous channel with net through flow. Phys Fluids **13**, 1931–1934 (1970)
33. Turcotte, D.L., Hsui, A.T., Torrance, K.E., Schubert, G.: Influence of viscous dissipation on Bénard convection. J Fluid Mech **64**, 369–374 (1974)
34. Tyvand, P.A.: Onset of Rayleigh-Bénard convection in porous bodies. In: Ingham, D.B., Pop, I. (eds.) Transport Phenomena in Porous Media II, pp. 82–112. Pergamon, London (2002)
35. Walker, K., Homsy, G.M.: Note on convective instabilities in Boussinesq fluids and porous media. ASME J Heat Transfer **99**, 338–339 (1977)
36. White, F.M.: Fluid Mechanics, 4th edn. McGraw-Hill, New York (2002)

New Bio-Inspired Multiphase Thermal Functional Fluid

José L. Lage

Abstract Efforts in harvesting the potential benefits of mimicking the gas exchange in alveolar capillary for channel heat transfer processes has led to a new bio-inspired multiphase thermal functional fluid (MTFF). This MTFF is originally conceived as encapsulated phase-change material particles, with diameter comparable to the channel size, flowing with the cooling liquid. The two main benefits of this new MTFF are not only the phase-change effect of the particles in the heat transfer process, but also the specific geometry of the particle and channel leading to the sweeping of the boundary layer in the channel. This last effect is believed to be responsible for the very high efficiency of the gas exchange taking place in the alveolar capillaries. Preliminary numerical simulation results seem to confirm the benefit of both effects. A groundbreaking experimental apparatus, designed as a pumpless flow loop, uses vortical effects created by a magnetic stirrer to set the liquid and particles of the MTFF in motion, overcoming the settling and clogging difficulties so characteristic of a multiphase fluid flow. Experimental tests, with octadecane paraffin (EPCM) particles or with acrylonitrile butadiene styrene (ABS) plastic particles (with no latent heat capacity), both flowing in water, have been performed and the results compared to results obtained with clear (of particulates) water flow. All tests indicate the advantages of using the MTFF in comparison to clear water, even at relatively low particle concentrations. Moreover, the tests seem to confirm the same behavior found in capillary blood flow, namely the detrimental effect of increasing the particle concentration beyond an optimum concentration, either leading to a reduction in the boundary layer sweeping effect or to an increased competition among particles for the heat transfer. This effort highlights the importance of learning from efficient biological systems.

J.L. Lage
Mechanical Engineering Department, Bobby B. Lyle School of Engineering, Southern Methodist University, Dallas, TX 75275-0337, USA
e-mail: JLL@SMU.EDU

A. Öchsner and G.E. Murch (eds.), *Heat Transfer in Multi-Phase Materials*,
Adv Struct Mater 2, DOI 10.1007/8611_2011_53,

1 Introduction

The quest for increasing the amount of heat transfer in channel flow has been driven by many engineering applications. Electronics, in particular, have dominated the interest of thermal engineers in the past few years with great challenges. The tremendous increase in the power-flux dissipated by electronics, from about 1.0 W/cm^2 by the logic chips manufactured in the mid 1980s to about 100 W/cm^2 by the microprocessors produced in the 2000s [1], led to the increase in the working temperature of the devices. To understand how significant this electronics power-flux strength is, consider the now projected power-flux dissipation in the near future of over 1,000 W/cm^2 and realize this value goes beyond typical values of power-flux found in nuclear reactors (about 100 W/cm^2) and in rocket nozzles (slightly below 1,000 W/cm^2). The increased temperature caused by the increase in power-flux affect negatively the reliability of the electronics. Hence, the cooling problem is essentially a reliability problem. An additional challenge posed by the cooling of electronics is the very small area in which the power-flux is dissipated, requiring small cooling devices with very small channels.

The same main challenging characteristics found in electronics, i.e. to have more heat transferred through a limited real-estate, have been confronted by thermal engineers in many other engineering areas for years. In trying to mitigate these challenges, thermal engineers have focused on creating new, more efficient fluid flow heat exchangers by applying new passive and active design techniques [2]. Passive techniques involve mainly channel shape modification, such as changing the channel curvature, roughing the channel surface, introducing flow disruption elements in the channel, and including constrictions and expansions along the channel. Active techniques, which tend to be more elaborated and costly, include flow pulsation and/or vibration, usually induced by piezoelectric or electric actuators placed along the walls of the channel. The objective of all these techniques has been to create enough mixing in the flow to break the thermal boundary layer that otherwise forms along the channel wall and hinders the heat transfer process. By doing so, the amount of heat transferred between the channel wall and the fluid flowing in it would then increase.

An important additional passive technique for channel heat transfer augmentation is the use of additives in the fluid flow. A dominant sub-group in this realm is the use of phase-change particles (the dispersed phase) to flow together with the flowing fluid (the continuous phase) forming a multiphase heat transfer agent called *latent functionally thermal fluid* (LFTF). Compared with the conventional single phase fluid with no particles, the LFTF provides high energy storage density and small temperature variation during the heat transfer process. The use of LFTFs was shown to yield 1.5–4 times more efficient heat transfer than single phase flow under ideal conditions [3], making this alternative very attractive for several application [3–18].

Recent advances in the production of encapsulated phase change materials, especially micrometer and nanometer scale particles [19–26], have broadened the possible engineering applications and revived interest in LFTFs [27–33]. A very particular characteristic of the recent LFTF thrust is the use of a very large number of small (microscale or nanoscale) particles mixed with the fluid, in a slurry form. The LFTF slurry is advantageous for the particles being very small and, as such, flowing easily with the fluid, facilitating the pumping and minimizing the clogging of the solid particles. Nevertheless, the slurry presents a few disadvantages of its own, including the inefficiency of having particles flowing far away from the channel heat exchange surface (the thermally active particles tend to be restricted to a region adjacent to the heated channel surface), the high pump-power required to circulate the slurry, and the settling of the particles in the recirculation or dead zones formed along the flow loop.

Notwithstanding, the recent interest in LFTF slurry flows brings back interest to the more general *particulate flow*, a class of multiphase flow in which particles (not necessarily of phase change material) flow with a fluid. This flow class is not restricted to channel heat exchangers, but found very often in many engineering processes, such as in chemical reactions (fluidized beds), spray-painting, coating (of solid particles), combustion (fuel-injection), and packaging (of cereals, grains), and in several natural processes as well, for instance in rain fall, river flow (particle sedimentation, erosion), and blood flow (red and white cells flowing with plasma).

The last process, i.e. the biological process of blood flow, is of great interest here, more particularly the alveolar capillary blood flow. The gas exchange process between the alveolar region of the lungs (filled with air) and the blood (containing liquid plasma and red blood cells, RBCs) flowing through an alveolar capillary is one of the most efficient mass transfer processes known. A distinctive characteristic of the alveolar capillary blood flow is the similarity between the diameter of the RBCs and the size of the capillaries (measured as the distance between top and bottom capillary membranes). This particulate multiphase flow is quite different from that obtained with an LFTF slurry: the number of particles (RBCs) in the alveolar capillary flow is much smaller and their size much bigger than in LFTF slurry flows. Nevertheless, the strong analogy between heat and mass transfer, and the particular role played by the RBCs in alveolar gas exchange warranted further examination of this biological mass transfer process.

2 Alveolar Capillary: The Biological Gas Exchanger

Respiration is perhaps one of the most important functions of the human body. Moreover, the gas exchange process in the human lungs is one of the most efficient exchange processes known to humans.

In studying the respiration process in humans, it is essential to identify the gas transport path inside the lungs. This path can be divided into two, namely a gas-path and a liquid-path. The gas-path begins at the nose and mouth, from where the gas flows through the trachea, to then follow through about 23 bifurcations, Fig. 1. The first bifurcation ($z = 1$) is from the trachea to the primary bronchial tubes, which

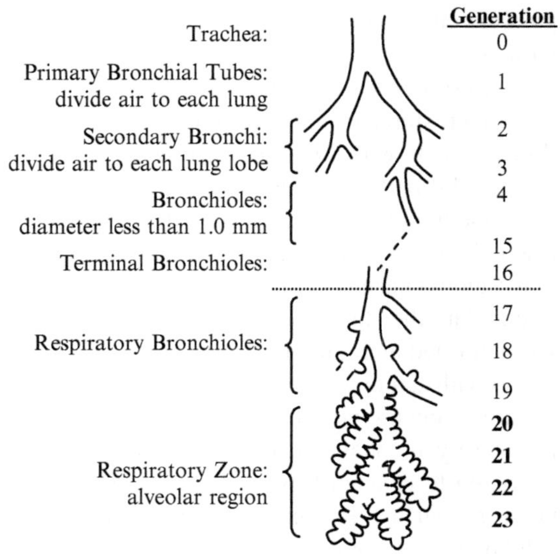

Fig. 1 Lung gas-path: a sequence of 23 bifurcation generations, from the trachea all the away to the alveolar region. Notice the terminal bronchioles, bifurcation generation z = 16, are so named because they mark the end of the *conductive zone*, or the convection dominated zone. Beyond the terminal bronchioles, the gas transport switches to diffusion dominated

take the air to each of the two lungs. The gas-path then bifurcates further, taking the gas to the secondary bronchi, to the bronchioles, and eventually to the terminal bronchioles, at bifurcation generation z = 16.

The terminal bronchioles mark an important transition location inside the gas-path of the lungs. Because the gas-path of the lungs is a dead-end path, and because of the increased flow area provided by the bifurcations (the available flow area increases as $2^{z/3}$ after each bifurcation, where z is the bifurcation generation number past the trachea, where z = 0), the gas speed progressively decreases as it flows into the lungs. Hence, the convection that drives the gas into the lungs through the mouth/nose eventually subsides, becoming negligible at around z = 16, the terminal bronchioles. From there, with the transport being diffusion dominated, the gas still has to go past the respiratory bronchioles, through bifurcations 17–19, to eventually reach the respiratory zone, or the alveolar region located in bifurcations 20 through 23.

It is in the alveolar region that the path changes from a gas-path to a liquid-path. Figure 2a shows a cast of the air-path of the human lungs. Subsequent shots, shown clockwise, details the gas-path as the gas goes deep into the lungs all the way to the alveolar region, Fig. 2d. Once in the alveolus, the gas has nowhere else to go (dead-end gas-path) but to enter the capillary bed.

At this stage, the alveolar gas transport process evolves from diffusion in the gas region of the alveolus to dissolution of the gas into the alveolar membrane. After dissolving into the membrane, the gas then flows by diffusion through the interstitial fluid and the capillary membrane, to eventually reach the liquid-path provided by the blood.

The alveolar membrane, the interstitial fluid and the capillary membrane form a barrier separating the gas-path from the liquid-path of the lung. From the capillary membrane the gas transport process switches again to convection, but this time the

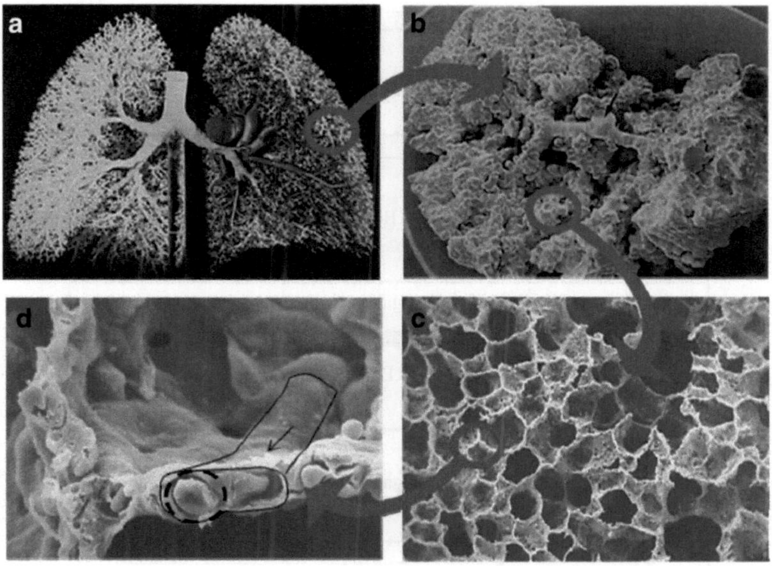

Fig. 2 Scaling of human respiratory system from entire lungs (**a**), to acinus (**b**), to alveolar region (**c**), to alveolar capillary (**d**). The drawn detail in (**d**) highlights the parallel-plate nature of the capillary, as opposed to a circular pipe; the *dashed line* shows a red blood cell. (**c**) and (**d**) are lung micrographs provided by Prof. E. Weibel

convection takes place in the liquid-path provided by the plasma of the blood. Notice that some gases, such as carbon-dioxide, convects away from the alveolar capillary mainly dissolved in the blood plasma. Other gases, such as oxygen and carbon-monoxide, go beyond the plasma, also permeating through the red blood cell membrane into the red blood cell to eventually attach to the hemoglobin. Once in the blood, the gas is then transported out of the alveolus to other parts of the body by convection.

Observe the alveolar respiration process is driven essentially by two potentials, namely the gas concentration in the gas region of the alveolus and the gas concentration inside the blood, either in the plasma or in the red blood cells. Studies of the three-dimensional, unsteady gas diffusion process in the lungs without concern to the blood flow effect, [34–39], have demonstrated the importance of the red cells location vis-à-vis the lung diffusing capacity, which is a measure of the overall lung diffusivity. Perhaps one of the most significant results from these efforts is the observation of the *shielding effect* caused by red blood cells located near the gas source. These cells effectively shield from the gas other cells placed further away from the gas source, making them less effective participants in the overall gas transport process. This same effect might explain why the cells passing through an alveolar capillary do so in a single string fashion – to avoid shielding!

More detail studies of alveolar gas exchange, focusing in the alveolar capillary, were performed recently, [40–42]. These studies were conducted considering the transport of carbon-monoxide, which is the gas choice for measuring the lung diffusing

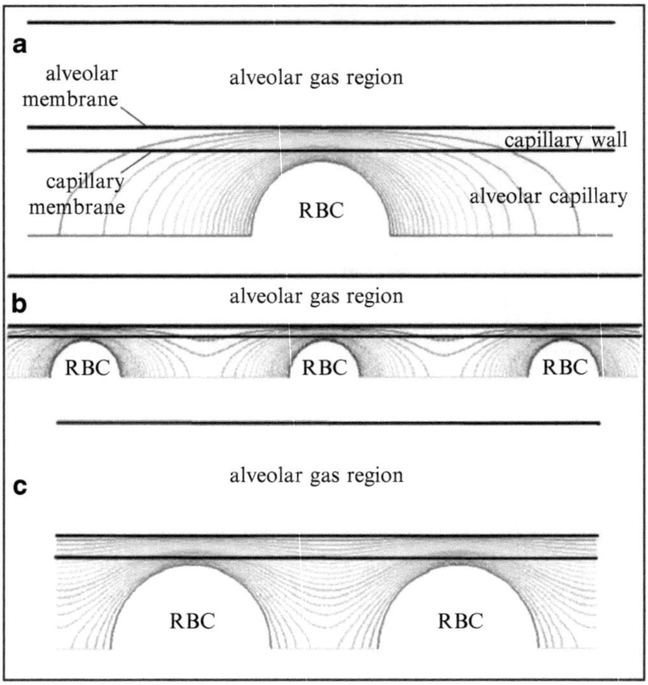

Fig. 3 Numerical simulation results of carbon monoxide diffusion in alveolar capillary, showing the isoconcentration lines in the capillary wall (alveolar membrane, interstitial fluid and capillary membrane) and in the plasma, for one (**a**), three (**b**), and eight (**c**) red blood cells in the capillary. Observe (**c**) has been amplified for better viewing. Only the top half of each capillary is shown

capacity in the laboratory. The same interference between closely spaced red blood cells, leading to the shielding effect, has been identified in the alveolar capillary in the case of pure diffusion – no blood flow, Fig. 3.

The results in Fig. 3, for diffusion of carbon monoxide in an alveolar capillary (only half-capillary is presented for simplicity), show the distribution of the iso-concentration lines from the alveolar membrane, where the gas concentration is unity, to the RBCs, where the gas concentration is zero (observe the RBCs are considered an infinite sink for CO). Observe further the resistance to gas transport is much smaller in the gas region than in the capillary wall or in the blood plasma, making the gas concentration uniform and equal to unity there. The top display, Fig. 3a, shows the concentration distribution for the case of a single RBC in the capillary. Observe how the isolines progress almost radially from the RBCs toward the alveolar gas region in this case, wrapping the RBC completely with ellipses. When more RBCs are placed inside the same capillary, as seen in Fig. 3b with three RBCs for instance, the isolines get flattened in the region in between two RBCs, with a few isolines not wrapping completely around each RBC anymore. This aspect is more pronounced in Fig. 3c, where the amplified view shows two of the eight RBCs placed in the same capillary. The close proximity of the RBCs make them compete for the gas being transferred from the alveolar membrane, that is why the isolines tend to flatten out in between the RBCs.

Figure 4 shows the result of the RBCs competition effect, obtained by [40] and [43], in quantitative fashion in terms of the long diffusing capacity D_L [in μm^3/(s Torr)] versus the number of RBCs in the capillary, N_{RBC}. The curves are distinct for being affected by the value of the specific rate of CO uptake by a single RBC, θ_{CO}, in μm^3/(s TorrRBC), as a result of the competition between CO and O_2 used in the simulations. Observe in Fig. 5 the inclusion of results for the case $\theta_{CO} \rightarrow \infty$.

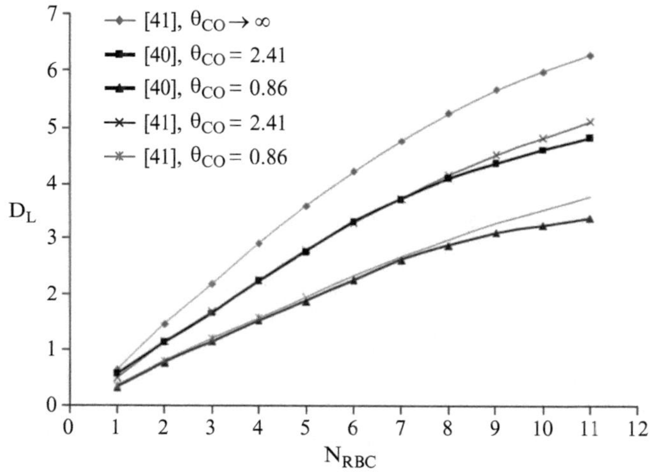

Fig. 4 Pure diffusion (no blood flow) results, from [40] and [41], of lung diffusing capacity D_L versus the number of circular red blood cells inside the capillary N_{RBC}. [θ_{CO} is the specific rate of CO uptake by a single RBC, as a result of the competition between CO and O_2, at different alveolus oxygen tension levels in μm^3/(s TorrRBC) and D_L in μm^3/(s Torr)]

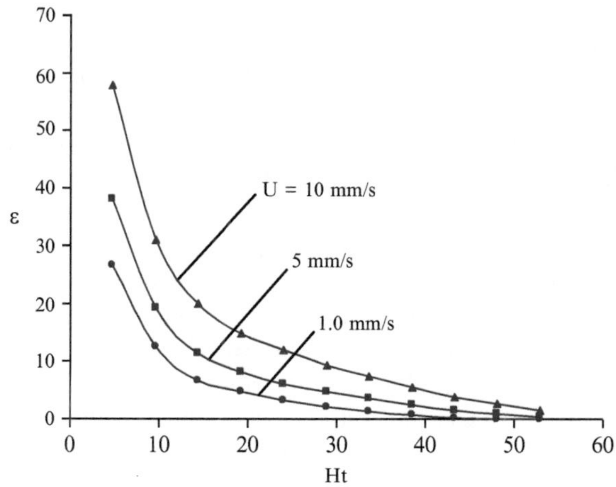

Fig. 5 Percentage relative difference ε between the D_{Ld} for the pure diffusion case (no blood flow, U = 0, with $\theta_{CO} \rightarrow \infty$) and the D_{Lc} for the convection case (with blood flow), versus capillary hematocrit Ht [41]

This limiting case, in which oxygen does not compete with CO, is included to show the consistent trend of the results as θ_{CO} increases to a limit value.

More importantly in Fig. 4, observe how the increase in D_L with the number of RBCs in the capillary is not linear, as one might expect. Rather, the D_L value tends to flatten out when N_{RBC} grows beyond six red cells in the capillary – a result of the competing effect of the red cells for the diffusing gas.

Now, when the blood flow effect is taken into consideration, the results are quite different [41], as shown in Fig. 5. The effect on the lung diffusing capacity in terms of the percentage relative difference ε between the D_{Ld} for the pure diffusion case (no blood flow, $U = 0$, with $\theta_{CO} \to \infty$) and the D_{Lc} for the convection case (with blood flow), namely

$$\varepsilon = \left| \frac{D_{Ld} - D_{Lc}}{D_{Ld}} \right| \times 100 \tag{1}$$

for several percent hematocrit Ht values, point out that the blood convection effect on D_L increases with the blood velocity U, as expected. However, the flow velocity effect decreases with an increase in the hematocrit. Notice the percent blood hematocrit, Ht, defined as the percentage volume occupied by the RBCs in the capillary, ($V_{RBC}N_{RBC}$), divided by the total capillary volume, $V_{capillary}$, is directly related to the number of RBCs in the capillary N_{RBC} as: $Ht = 100(V_{RBC}N_{RBC})/V_{capillary} = (4.8N_{RBC})\%$. Observe that three blood velocities are considered, namely $U = 1$, 5, and 10 mm/s.

The results of Fig. 5 are quite remarkable as they seem to indicate the competing effect of the red blood cells placed close to each other is also present when the blood is flowing, that is, when convection effects are present in the alveolar capillary.

Figure 6, [41], shows for the case of a single circular RBC in the capillary, three sets of isolines from top to bottom, namely for: (a) $U = 1$ mm/s, (b) $U = 5$ mm/s, and (c) $U = 10$ mm/s. Each set shows at the top the isoconcentration lines (same CO partial pressure) and at the bottom the corresponding streamlines, in the membrane and plasma regions.

The streamlines of Fig. 6 are plotted from the point-of-view of an observer traveling with the RBCs. Observable for all speeds are characteristic circulating cells inside the capillary, before and after the RBC, within the plasma region. For the case of $U = 1$ mm/s the blood convection seems to have little effect on the transport of CO to the RBC: observe the symmetry of the isoconcentration lines with respect to a vertical line passing through the RBC center. However, the distortion on the symmetry of the isoconcentration lines when U increases, Fig. 6b and 6c, is evident. Only the plasma flow could be responsible for this distortion. Evident in Fig. 6c is a squeeze of the isoconcentration lines (i.e., the lines become more closely spaced) along the anterior (left) face of the RBC, indicating an enhanced transport of CO, and a slight expansion of the lines (i.e., the lines become spaced far apart) along the posterior (right) face of the RBC.

a

b

c

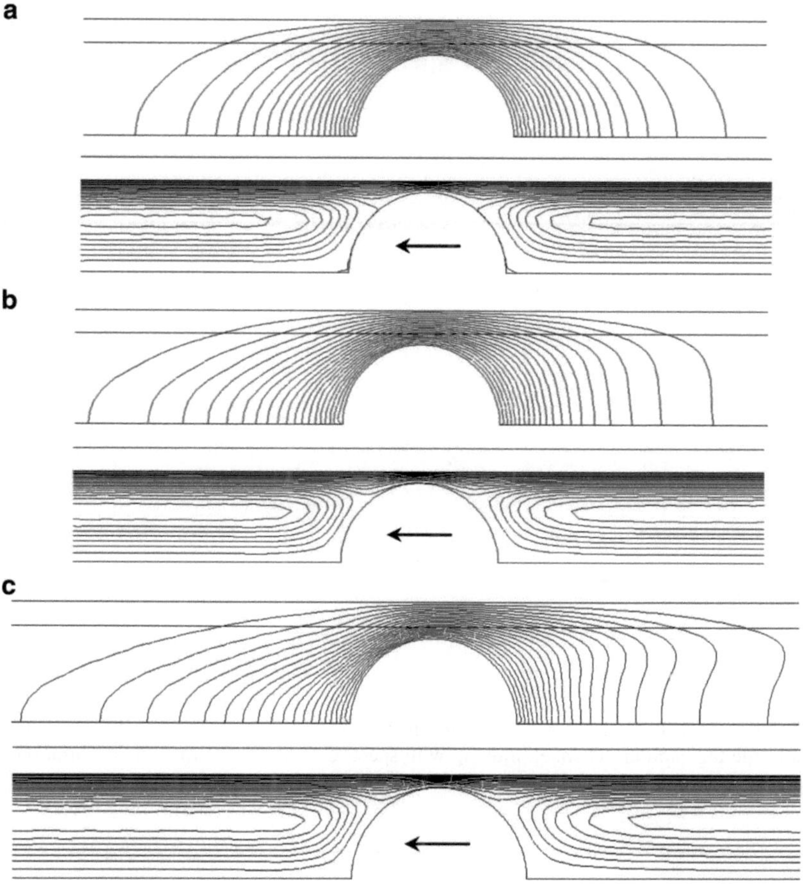

Fig. 6 Convection results for a single RBC in the capillary, for: (**a**) U = 1 mm/s, (**b**) U = 5 mm/s, and (**c**) U = 10 mm/s [41]. Each set shows at the *top* the isoconcentration lines (same CO partial pressure) and at the *bottom* the corresponding streamlines, in the membrane and plasma regions. U is the speed of the RBC in the capillary

Figure 7, [41], depicts results for when eleven RBCs, $N_{RBC} = 11$ or Ht = 52.8%, are placed side by side in the capillary channel, and set to move with speed U = 10 mm/s. The figure shows the resulting isoconcentration lines, Fig. 7a, streamlines, Fig. 7b, and velocity vectors, Fig. 7c. In this case, with a large number of red blood cells in the capillary, the flow circulation in the plasma region is subdued in comparison to the results shown in Fig. 6, by the proximity of the red cells. Consequently, there is little distortion in the isoconcentration lines due to the convection effect, with the effect being restricted to the small region available for plasma flow in between the red cells. Finally, observe the small gap between the red cells and the capillary membrane allow for some plasma to flow through (see the velocity vector plot, Fig. 7c).

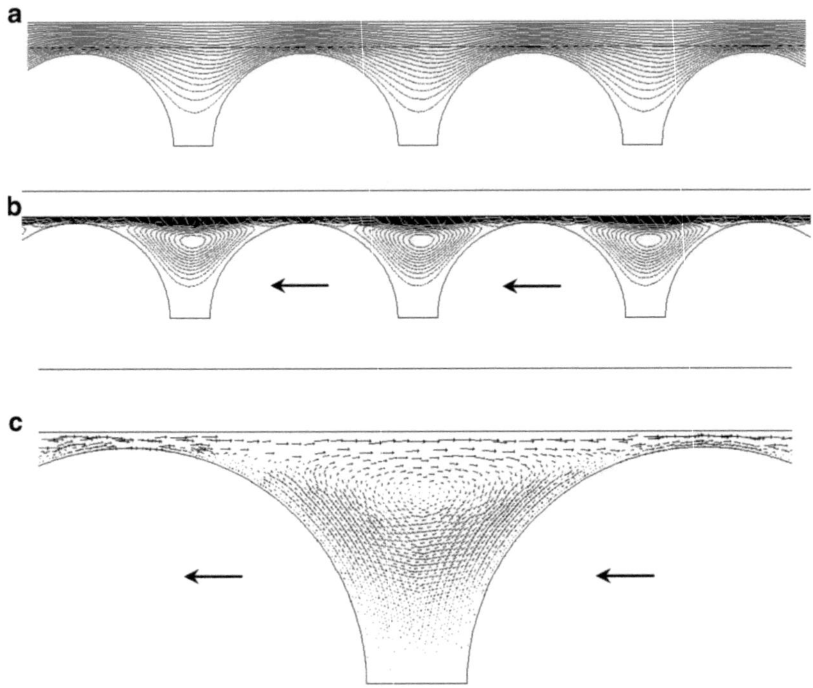

Fig. 7 Numerical simulation results, [41], for the blood convection, with 11 uniformly distributed RBCs inside the capillary channel, moving with speed U = 10 mm/s: (**a**) isoconcentration lines, (**b**) streamlines, (**c**) velocity vector distribution

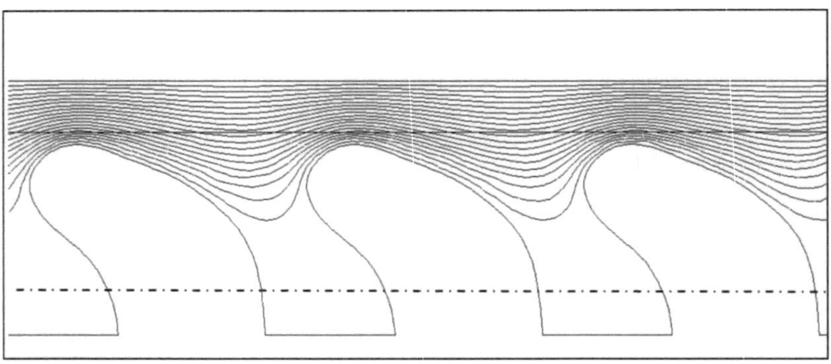

Fig. 8 Alveolar capillary showing three parachute shaped red cells under gas diffusion only (U = 0). Shown also are isoconcentration lines, [43], for Ht = 52.8%

Another aspect considered in studying convection in the alveolar capillary is the effect of the RBC shape. By simulating the more common parachute shape of RBCs, as depicted in Fig. 8, results were obtained by [43] showing the shape effect as compared to spherical RBCs, Fig. 9.

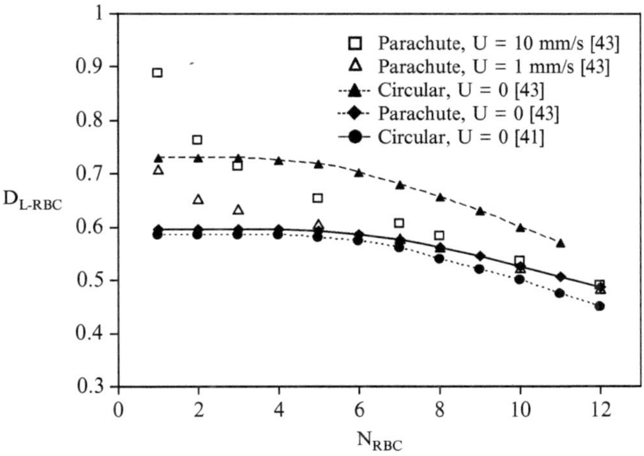

Fig. 9 Numerical simulation results, [41, 43], in terms of lung diffusing capacity per unit of red cell in the capillary, $D_{L\text{-}RBC}$, versus the total number of red cell in the capillary, N_{RBC}, for diffusion ($U = 0$) and blood convection ($U > 0$), with parachute and circular shaped RBCs

The shape effect on the diffusion process is expected because the red cells are participants in the diffusion process (recall the red cell has the low potential for CO diffusion), and by changing their shape the gas exchange surface area changes as well. In the case of the parachute shape RBCs the surface area is slightly larger than the surface area for the circular RBC case. Not so evident is the blood flow effect. Notice how the convection effect gets more pronounced as the number of red cells is small in the capillary. When the N_{RBC} increases too much, not enough space is available in between consecutive red cells for a strong convection effect to develop.

The three-dimensional effect of a flowing parachute-shaped red cell is seen in Fig. 10, from [44], with the top capillary surface and only half the capillary (and red cell) being shown. The sweeping effect of the flowing RBC is remarkable, as is the effect of the trailing edge (extra mixing) also seen in the figure.

Perhaps the most remarkable aspect of the alveolar capillary blood flow is not the diffusion aspect, or the convection effect of blood flow. Rather, the most striking detail is the fact that the capillary allows for a single train of red blood cells to flow through it. That is, the capillary does not accommodate layers of red cells flowing parallel to each other between the top and bottom capillary membranes.

From an engineering designer point-of-view, considering the red cells as the transporters of the gas (in the cases of oxygen and carbon monoxide gases, for instance) the larger the number of red cells inside the capillary, the higher should the efficiency of the process be. For some reason, however, alveolar capillaries are designed to function with a single train of red cells passing through it, Fig. 8. This is

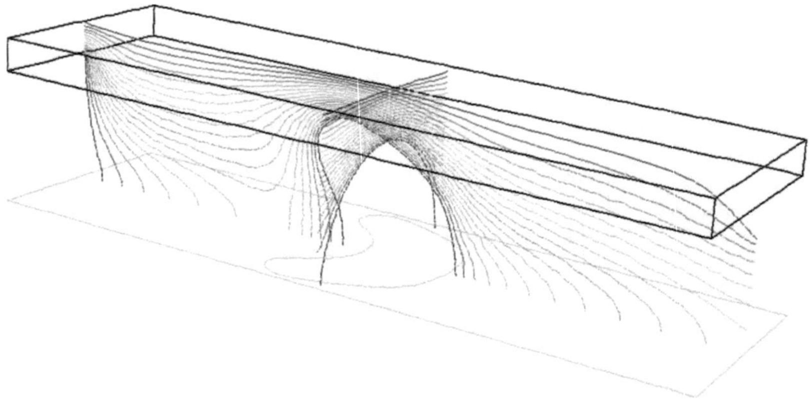

Fig. 10 Sample numerical simulation results of a single parachute-shaped red-blood cell flowing through a capillary. Only the top half of the capillary is shown for simplicity. The *lines* are isoconcentration lines [44]

one of the key factors believed to be responsible for the high gas transfer efficiency of the lungs [42].

True that the alveolar capillaries do not allow for larger white cells (leukocytes), for instance, entering the alveolar capillary when providing a snug fit to the red cells (notice the most common leukocytes, called neutrophils, have diameter about twice the 6–9 μm diameter of RBCs). However, if the capillaries were large enough to allow larger cells to flow through, then the number of red cells would increase as well, and not all cells would be near the capillary membrane. Keep in mind that in liquid flow through a straight pipe, solid particles with diameter less than the pipe diameter would tend to flow along the center of the pipe, a region with the least flow resistance. Then, if several red cells were to fill the capillary, most cells would flow along the center of the capillary, far from the capillary membrane where the gas exchange takes place.

Moreover, if the capillaries were large enough to allow several layers of red cells flowing parallel to each other, the cell in the center of the capillary would be shielded from the capillary membrane by the red cells flowing near the membrane. The resulting gas exchange process, on a per RBC basis, would be less efficient.

In conclusion, the tendency to flow along the center of the capillary and the shielding effect are two important reasons for not allowing more than one single train of red cells to flow through the capillary. Another aspect can be considered as well: the sweeping effect the red cells have along the capillary membrane as they flow snugly through the capillary.

Specifically, it is conceivable that a solid particle with diameter comparable to the channel dimension would act like a broom sweeping along the channel surface, mixing ("breaking") the boundary layer, reducing the transport resistance, and enhancing the convection process. This aspect is considered in more detail in the next section.

3 Mimicking an Alveolar Capillary: The *Sweep* Heat Exchanger

In studying the gas exchange in the alveolar capillary, [41–43] showed the effects of blood flow in the gas exchange. In particular, every time an RBC flows through a capillary, the particle sweeps the capillary wall breaking the boundary layer formed along it, as shown in Fig. 11, from [44].

The figure shows the flowing of a single RBC with blood plasma through a capillary conceived as a parallel-plate channel. Only the top half of the channel is shown for simplicity. The plasma and the RBC enter the capillary with zero concentration of the transferring gas (zero partial pressure P_{CO}), in this case carbon-monoxide, CO, coming from the capillary surface (labeled as *tissue* in the figure). The tissue separates the gas region of the alveolus from the liquid region in the capillary, and it has a uniform gas concentration, equivalent to $P_{CO} = 1$ Torr at the top surface (the alveolar membrane).

The lines in Fig. 11, in the capillary and tissue regions, show isoconcentration lines. Three regions are of importance. The first is the region to the left of the flowing RBC, around the inlet region of the capillary. There the isolines develop by the flowing effect of the plasma alone, and indicate the growth of the mass transfer boundary layer along the capillary. The second region is downstream, to the right of the RBC, where the furthest away isoline extends itself towards the outlet of the capillary. Finally, the third region is the region around the RBC, where the isolines are squeezed against the capillary membrane and the tissue. This squeezing effect characterizes the sweeping of the boundary layer by the flowing RBC. This effect is easy to understand if we consider first the flow of plasma alone (with no RBC) in the capillary.

In a parallel-plate channel, the plasma flow would eventually develop either into a parabolic velocity profile, in the case of laminar flow, or into a more flatten profile, in the case of turbulent flow. It is worth pointing out here that laminar flow predominates within the small dimensions encountered in alveolar capillary. Notwithstanding, the flow induces the convection mass transfer process from the

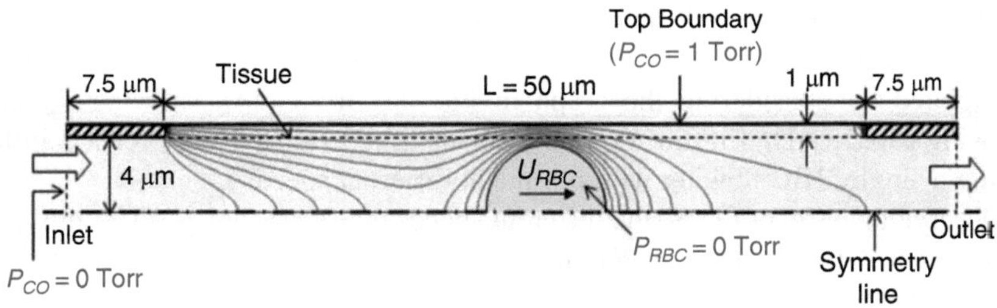

Fig. 11 Numerically obtained isoconcentration CO lines inside an alveolar capillary modeled as a parallel-plate channel (only the top-half of the channel is shown), with a single red blood cell (RBC) flowing through it. The channel top boundary (alveolar membrane) has uniform CO concentration while the flowing fluid – plasma and RBC particle – enters the channel with zero CO concentration [44]

capillary membrane, with a mass transfer boundary layer developing downstream the channel. Eventually the mass transfer process reaches a steady-state, and from this point on the mass transfer distribution is set with a high transfer rate along the inlet section of the capillary, where the mass boundary layer still develops, and a lower transfer rate from where the boundary layer converges at the center of the channel.

Although resulting from a natural and unavoidable process caused by the mass diffusion and convection in the fluid, the boundary layer demarcates a region within the flow of high mass transfer resistance because the potential driving the mass transfer process, which at the entrance of the channel is zero, is now a value closer to the alveolar membrane potential. Hence, the effective potential difference driving the mass transfer decreases as the fluid flows downstream the capillary, hindering the mass transfer process. Ideally, one would like for the fluid potential to remain zero throughout the entire capillary, but this is impossible as the fluid absorbs some gas. However, the concentration gradient across the boundary layer is contrary to an efficient mass transfer process.

Now, when a solid particle, such as an RBC, enters the capillary with a certain speed, it forces an accommodation of the fluid velocity profile because the entire solid particle travels with the same speed. That is, the portion of the particle traveling near the stationary boundaries of the channel, where the fluid originally has low speed, will induce an acceleration of the fluid in this region, while decelerating the fluid in a region near the center of the channel, where the fluid originally flows faster. The net effect is the flattening of the fluid velocity profile, similar to the profile transitioning from laminar to turbulent as the fluid speed increases. This aspect, by itself, would already benefit the heat or mass transfer convection process, similar to the benefit achieved by turbulent flow. However, the solid imposes an additional effect: the mixing of the fluid nearby, inducing the breaking of the mass transfer boundary layer. Keep in mind the acceleration of the fluid by the solid particle does not carry the same pump-power penalty the transition from laminar to turbulent carries because the particle sweeps the channel surface only intermittently, when it pass along it. The pressure-drop penalty of turbulent flow, on the other hand, is continuous in time as the flow is always turbulent along the channel (unless, of course, the flow is made intermittent).

The resulting effect of the particle sweeping the channel surface is better appreciated by considering the graph in Fig. 12, from [44], showing the lung diffusing capacity D_L [in $\mu m^3/(s\ Torr)$] versus time t (in seconds), for a capillary having a single RBC flowing with plasma at different speeds U.

Observe the effect of U on the different shapes the curves have. When the speed is low, $U = 1$ mm/s, the convection process increases when the RBC enters the channel, reaches a plateau as the RBC travels inside the capillary, then it decreases as the RBC leaves the channel. The plateau period is not so long when the RBC speed increases to 5 and 10 mm/s. That is because the RBC flows too fast through the capillary for the process to reach a pseudo steady-state and form a long plateau as in the case of $U = 1$ mm/s. Notice also the cyclic nature of the curves reflecting the entering and exiting of red cells in the capillary. The D_L difference between

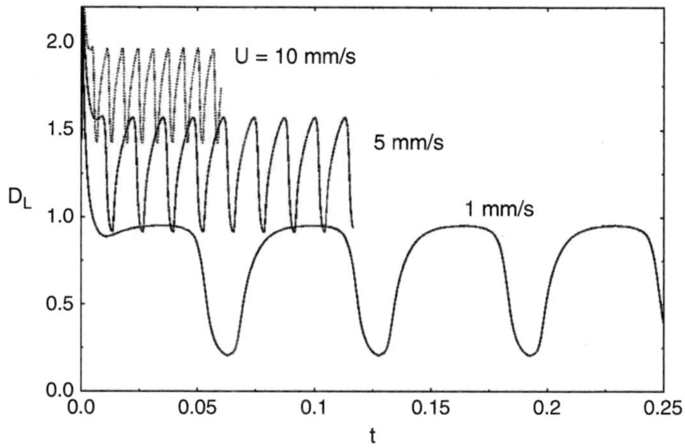

Fig. 12 Effect of a single flowing RBC in and out of an alveolar capillary, in terms of lung diffusing capacity D_L and time, for three distinct RBC speeds U. Notice a new RBC only enters the capillary when the one in the channel leaves it; hence, one, and only one RBC is in the capillary at any given time [44]

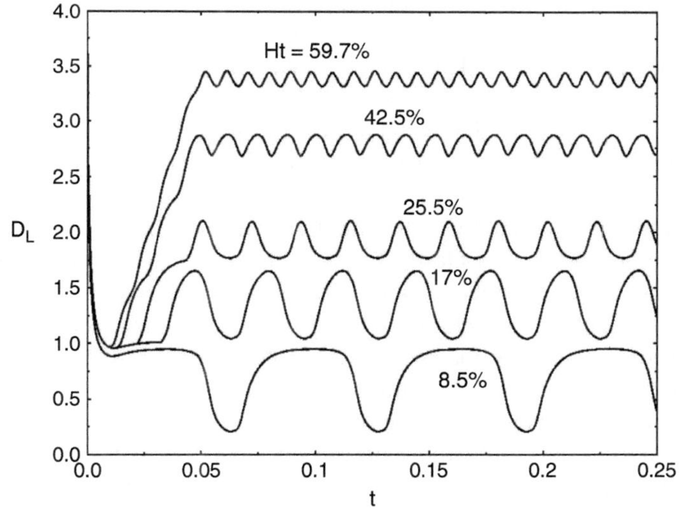

Fig. 13 Effect of increasing the number of RBCs flowing through a capillary, with fixed speed U = 1 mm/s [44]

peaks and valleys represent the RBC flowing effect – if only plasma flows in the capillary, the resulting D_L would be yet lower than the lowest value in the valley of each curve (the simulations are set for the capillary to always have one RBC in it, so the plasma-only convection configuration never sets in completely).

Another interesting similar result is shown in Fig. 13, from [44]. In this case, the results are for the same RBC speed, U = 1 mm/s, but for different numbers of

RBCs flowing with plasma, shown in terms of percent hematocrit Ht (note: Ht $= 8.49N_{RBC}$). As can be seen by the reducing amplitude of the curves, the flow effect is less pronounced when the number of RBCs flowing inside the capillary is large. This observation confirms some of the previous results: too large a number of RBCs inside the capillary leads to direct competition among them for the gas transfer, reducing the overall efficiency of the process. Moreover, when too many RBCs are present, the distortion in the plasma velocity profile will be less because of the small spacing between consecutive RBCs, which tend to dump the convection effect of the RBCs.

From all these studies of gas exchange in an alveolar capillary, a few important observations emerge: (1) the similarity between the dimension of the capillary and of the flowing particle (RBC) seems paramount to an efficient gas transport process for it avoids the shielding effect and it induces a higher flow speed near the surfaces of the channel and the sweeping of the mass/heat boundary layer; (2) the RBC is a participative component of the convection transport, playing the role of an infinite sink for the transporting gas (CO in the studies reviewed here); (3) in the context of (1) and (2), the plasma and the RBCs would form a *multiphase functional fluid* for the mass transfer process in the capillary; (4) a large number of RBCs flowing in the capillary, equivalent to a high hematocrit, is detrimental to the efficiency of the gas transfer process as the RBCs tend to compete with each other for the gas and the accelerating convection effect becomes less evident in this case as well; (5) increasing the speed of the RBCs tend to increase the gas transfer efficiency, with this effect being less dramatic when large numbers of RBCs flow through the capillary.

These observations, together with the strong analogy between mass transfer and heat transfer, can form the backbone of an audacious effort, namely that of building a cold plate similar to an alveolar capillary, [45–48], to be cooled by a new, bio-inspired multiphase thermal functional fluid, such as the blood. Because of the gas-absorbing, infinite-sink role played by the red blood cells in the mass transfer process in the alveolar capillary (in the case of carbon monoxide), a good equivalent particle candidate was determined to be spherical encapsulated phase-change material. The phase-change material would make the particles behave similarly to the red cells in the equivalent heat transfer configuration. During heat transfer, the particles temperature would reach the melting point, during which the temperature tends to remain constant (although in practice the temperature during melting varies within a small range), as if the particles were an infinite sink for heat. True that when all the phase-change material melts, the particles lose the infinite sink characteristic; nevertheless, the phase-change material encapsulated particle seems to be the best alternative in mimicking the red blood cells. Observe that even in the carbon monoxide capillary gas exchange, the red cells would eventually saturate with the gas if left through the capillary long enough.

Preliminary modeling and numerical simulations were performed by [45], considering a square cross section channel mimicking an alveolar capillary, heated uniformly from the top and bottom surfaces, and filled with liquid water flowing with encapsulated phase-change particles, such as the one depicted in Fig. 14. It is interesting that a configuration similar to the one shown in Fig. 14 was considered

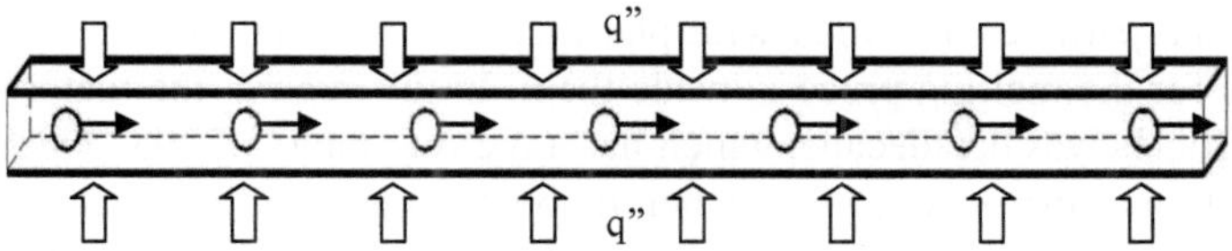

Fig. 14 Sketch of top and bottom uniformly heated square-cross section channel cooled by a bio-inspired multi-phase thermal functional fluid. The *circles* inside the channel are the encapsulated phase-change material particles flowing with the liquid. The front and back surfaces of the channel are assumed adiabatic

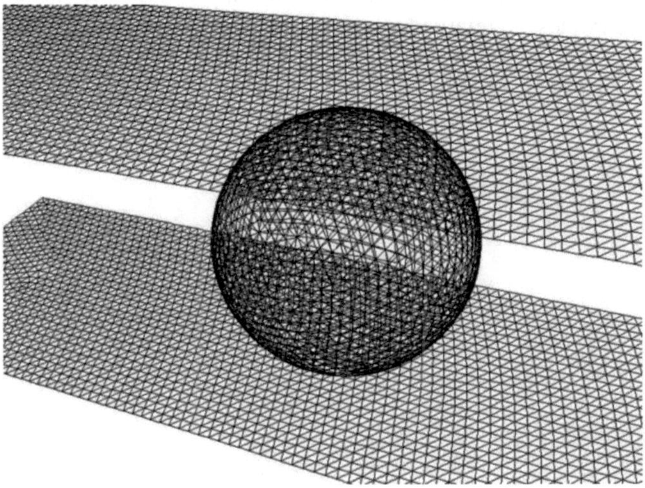

Fig. 15 Sample grid used for numerical simulations [45]

recently, [49–51], for the flow of ice and water through a circular pipe in a train like fashion.

The diameter of the spherical particles was chosen as to allow only one particle to occupy the cross section of the channel. That is, particles could flow in a train line configuration (one after the other, Fig. 14) but not side-by-side. The inlet temperature of the liquid and of the particles was assumed uniform and slightly below the melting point of the phase-change material inside the particles.

The simulations were performed in a frame of reference attached to the particles, with the channel surfaces moving with a constant speed. This choice of frame of reference allowed for the simulations to be performed with a fixed, unstructured grid. Had a frame of reference been attached to the surfaces of the channel instead, the simulation of the moving particles would require a moving mesh, which would make the numerical requirements for the simulations much tougher. Although the choice of reference frame helped alleviate the numerical requirements, it also limited the simulations to the case of an infinitely long channel. Nevertheless, the simulations were very intense, with very fine grid (see Fig. 15) and time-step necessary for capturing the phenomena.

Recall the particles are filled with phase-change material, and the simulations accounted for the melting process inside them. In this regard, the material chosen for the particles was octadecane, which has a melting range of 298–308 K. Hence, the simulations were very realistic when it comes to the melting effect.

The number of particles used in the simulations varied, with N = 3, 6, 11 and 23, or equivalently in terms of volume-fraction (hematocrit equivalent) as $\phi = 3.6\%$, 7.3%, 13.3% and 27.9%, respectively. As the number of particles changes in the channel, so does the flow field around the particles, Fig. 16.

Observe in the top picture, the flow field when N = 3 shows circulation to the right of the particles, but not to the left. This is because the distance between consecutive particles is relatively large with a low number of particles in the

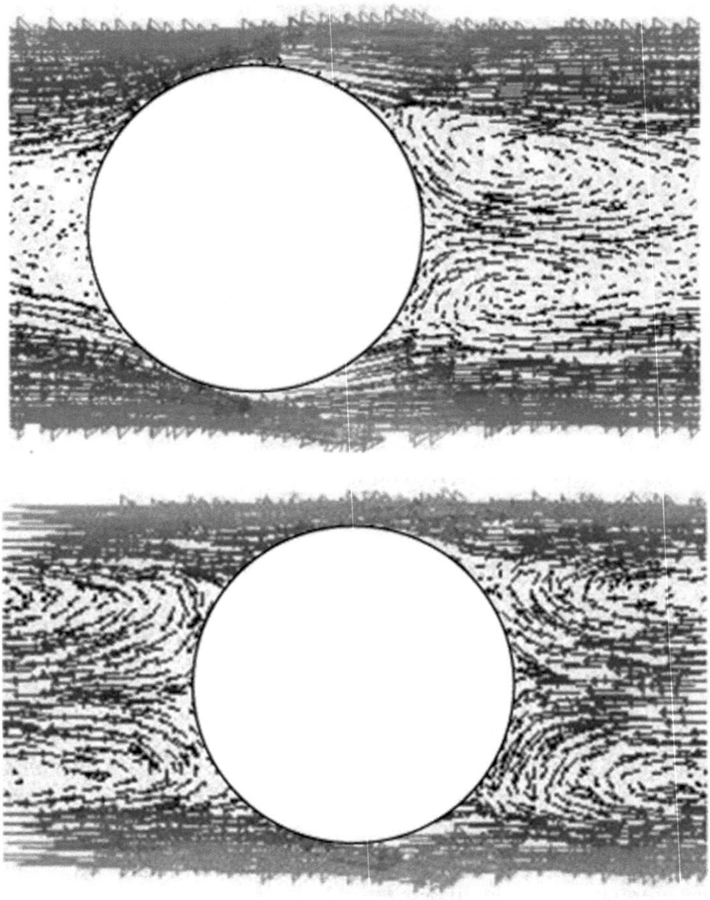

Fig. 16 Flow velocity distribution within the channel with particles, for particle speed equal to 0.03 m/s. The *top* picture is for N = 3 (ϕ = 3.6 %), and the *bottom* is for N = 11 (ϕ = 13.3 %). Only one particle is shown for detailing the flow field [45]

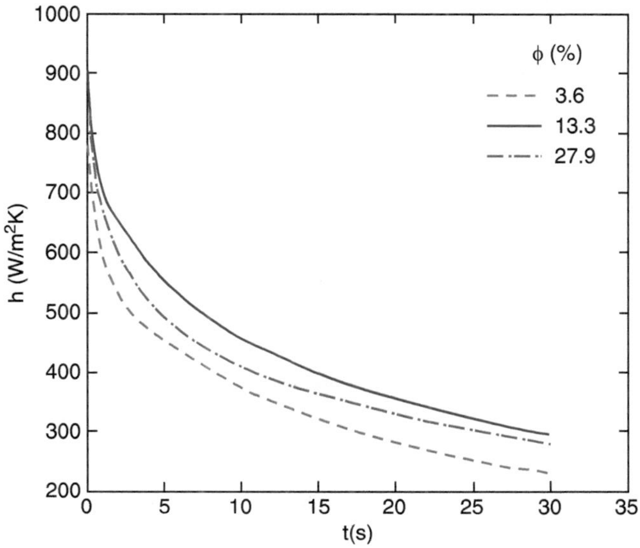

Fig. 17 Effect of particle concentration in the surface-averaged heat transfer coefficient h [45]

channel. However, when the number of particles is large, they get closer together and the circulation becomes evident upstream and downstream of the particles (see bottom picture, for N = 11). This same behavior was observed in the case of the RBCs.

The time-varying surface-average heat transfer coefficient h is affected by the number of particles in the channel, as indicated in Fig. 17. When the particles increase, from N = 3 to 11, to 23, the heat transfer coefficient first increases (from 3 to 11), but then decreases (from 11 to 23). This aspect is quite interesting because it seems to confirm for the heat transfer process with phase-change particles the same phenomenon observed with the mass transfer process in a capillary: if the number of particles in the channel is too large, the process eventually becomes less efficient. This efficiency decrease, shown in Fig. 17 in terms of h, can be due to the competition between the particles for the heat coming from the channel surfaces, and by the change in the flow velocity profile caused by the particles getting closer to each other when N increases (see Fig. 16).

The results obtained through the numerical simulations provided support for another preliminary effort, that of building a cold plate for experimental testing.

4 Preliminary Bio-Inspired MTFF Performance

The design and construction of an experimental apparatus for testing the performance of the bio-inspired multi-phase thermal functional fluid, [48], faced tremendous challenges. A first design was proposed, shown in Fig. 18, as a closed loop, in

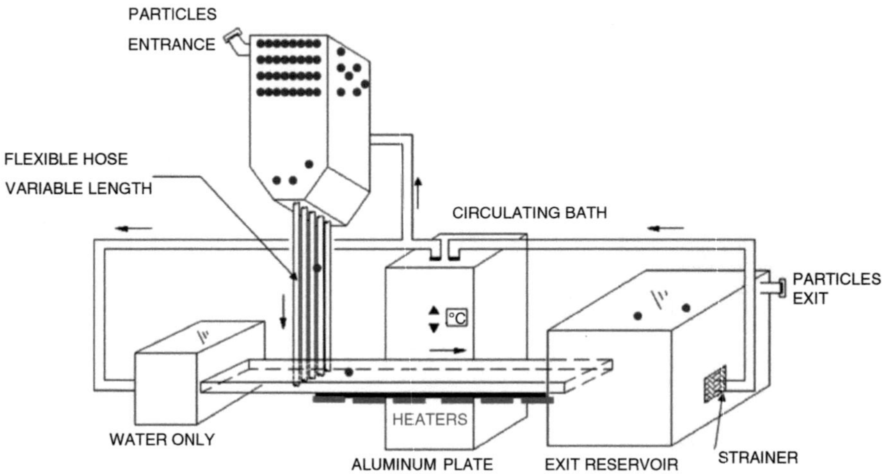

Fig. 18 Flow apparatus for testing new bio-inspired multiphase thermal functional fluid [48]

which the liquid water would flow with the particles through the heated channel, being collected at the exit reservoir. From there, the particles would be separated from the liquid, which would then circulate through a circulating bath (chiller) to be brought back to the required temperature. The required particles would be housed in a separate reservoir, and from there flow down through auxiliary channels connected to the main flow channel leading to the heated channel. This preliminary design did not work mainly because the particles would clog the restriction from the particle reservoir to the auxiliary channels, and stop flowing through it.

An alternative design was tested, Fig. 19, this time with the particles reservoir placed underneath the heated channel, and having straighteners to reduce the clogging problem. This design did not work either, and for the same reason: particles tend to clog near the entrance of the auxiliary channels. It became clear that the flow of the particles was key for the success of any design.

After further observation of the previous designs, it became apparent that the clogging problem would only be resolved if the particles were dispersed in the liquid prior to entering a restriction. An additional problem to be tackled was the need for having the particles circulating with the liquid, and not removed from the flow loop as the initial designs were set up. An ingenious final pumpless design was proposed for eliminating both problems, Fig. 20 [46].

The flow apparatus shown in Fig. 20, [46], has a heated channel (the testing section), heated electrically from the bottom and insulated at the top and sides, with several thermocouples distributed uniformly along the bottom surface for temperature monitoring. From the heated channel, the multiphase thermal functional fluid (MTFF) exits to a reservoir (exit reservoir) having a suction pipe placed flush with the liquid level inside the reservoir. This pipe transfers the fluid back to the inlet reservoir, through a heat exchanger to recondition the fluid and the flowing particles back to their original state.

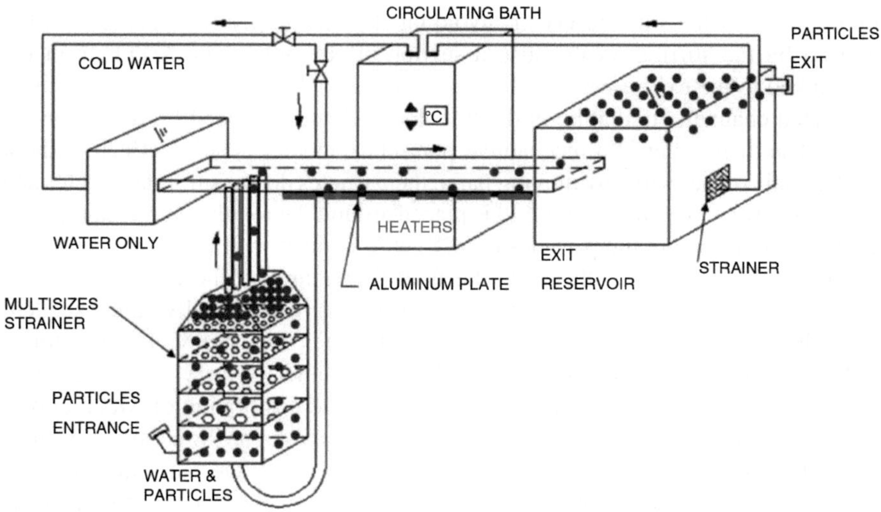

Fig. 19 Alternative flow apparatus for testing new bio-inspired multiphase thermal functional fluid [48]

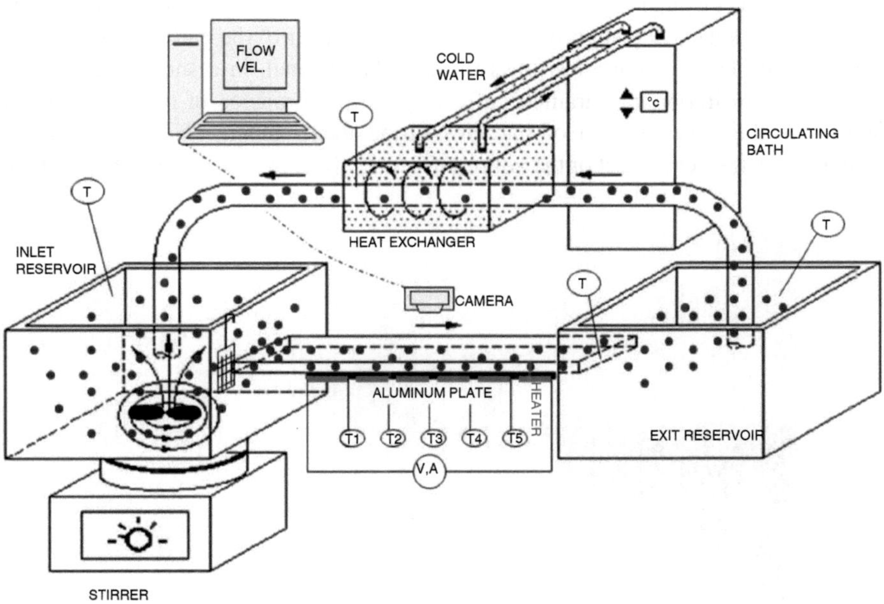

Fig. 20 Final, pumpless flow apparatus for testing new bio-inspired multiphase thermal functional fluid [46]

Once in the inlet reservoir, the particles are stirred by a rotating magnet set in motion by a magnetically coupled stirrer. Notice that the stirring of the MTFF accomplishes several tasks. It not only disperses the particles very uniformly in the liquid while in the inlet reservoir, avoiding the clogging effect, but the vortex induced by the stirring creates also a suction effect pulling the MTFF from the tube linked to the exit reservoir. This last aspect sets the MTFF in motion through the heated channel, circulating in the flow loop without the need for a mechanical pump. The dispensing of a mechanical pump has another positive effect: without mechanical contact with any moving parts, the particles do not suffer an accelerated degrading effect (a course metal grid is placed around the rotating stirrer in the inlet reservoir to make sure the particles do not get in contact with it).

Moreover, the speed with which the fluid flows through the channel is easily controlled by selecting the speed of the stirrer as well. Finally, the excellent dispersion of the particles achieved inside the inlet channel yields an almost uniform feeding of the particles to the heated channel. A very simple metallic grid gate was included at the inlet of the channel, the positioning of which allows for more or less particles to flow through it, for the same flow speed. Hence, different particle concentration inside the channel is possible, all with the same flow speed. (Although the apparatus does not allow for the precise control of the number of particles entering the channel, this can be accomplished with a simple electronic gate that can be installed at the entrance of the channel.) A picture of the final apparatus is shown in Fig. 21 [48].

Observe the channel is built as a parallel-plate channel, with the vertical distance between the plates being slightly larger than the diameter of the particles. Therefore, only a single layer of particles flows through the channel, in a snug fit, as shown in Fig. 22. An important parameter affecting the performance of the heat transfer process is the concentration of particles inside the channel. This concentration, equivalent to the hematocrit or the volume-fraction of the particles inside the channel,

Fig. 21 Final apparatus showing the inlet reservoir (*left*), testing channel (*center*) and outlet reservoir (*right*), with liquid water and microencapsulated phase-change particles, as the bio-inspired multiphase thermal functional fluid, flowing by the exclusive action of the magnetic stirrer placed underneath the inlet reservoir (*left*) [48]

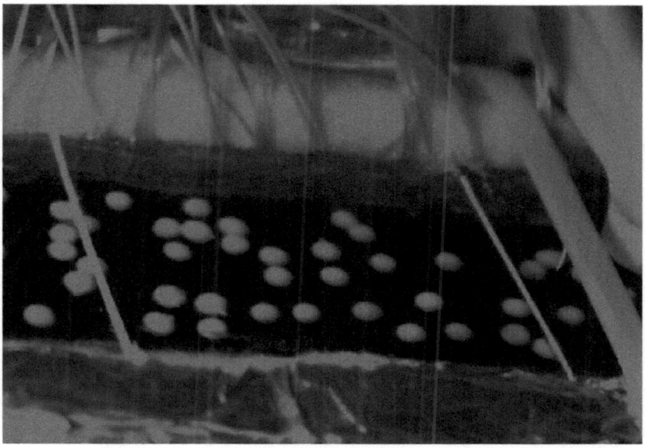

Fig. 22 Flow visualization: top view of testing heated channel. The top channel surface being made of Plexiglas allows for the visualization of the particles (octadecane paraffin, $C_{18}H_{33}$, microencapsulated in a thin melamine shell), flowing with water, and the calculation of the particle density inside the channel [46]

is determined by counting the number of particles inside the channel at any instant in time. Tests have shown this number to be fairly constant (except by the exit and entrance effects) once a flow speed and a position for the inlet gate are chosen.

The experimental results, [46], were obtained in terms of the surface averaged temperature along the heated section of the channel, shown in Fig. 23, for several particle concentration and a fixed flow speed. Observe the flow concentration is relatively low, varying from 1.15% to 4.2%, a result of using a parallel-plate channel. Also noteworthy is the inclusion of results for zero particle concentration, i.e., for flow of clear (of particles) fluid.

Aside from the fact the average surface temperature increases with the increase in the heat flux at the channel surface, as expected, the effect of increasing the particle concentration in the channel is of great interest. Observe the highest channel surface temperatures are obtained for zero concentration (clear water), indicating the worst performance of all tests is when no particles are used. The introduction of particles always lowers the surface temperature, as compared to clear fluid, indicating the benefit of using the bio-inspired MTFF.

What would not be expected perhaps is the effect of increasing the particle concentration: as the particle concentration increases the temperature decreases, but from 1.9% to 2.7% concentration the surface temperature increases, instead of decreasing as one would expect. That is, the benefit of continuously increasing the number of particles flowing with the liquid inside the heated channel seems to have a limit in achieving high heat transfer efficiency. There seems to exist an optimum particle concentration, beyond which the addition of more particles has a lesser and somewhat detrimental effect to the heat transfer process (see how the curves for particle concentration greater than 1.9% are closer together). Even more

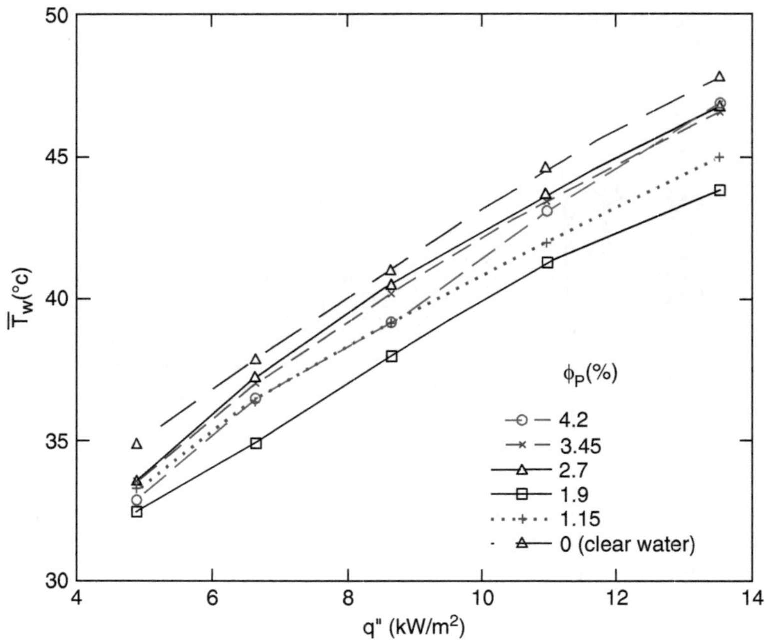

Fig. 23 Surface averaged temperature results versus surface heat flux, for flow speed equal to 0.35 cm/s and several particle concentrations

surprising is that the beneficial effect of adding more particles seems to return as the particle concentration increases again from 2.7 to 3.45 and finally to 4.2%.

Using what was learned in the alveolar capillary, it is conjectured the presence of two effects play a role in the behavior of the curves in Fig. 23. One effect is the reduction in the effectiveness of the extra mixing induced by the particles in the channel when the number of particles increases too much. The other effect is the melting power of the particles: for the same heat flux, a small number of particles could melt completely before exiting the heated channel, whereas a large number of particles might not. This phenomenon would explain the decrease in surface temperature when the particle concentration increases from 2.7 to 3.45 and 4.2%: the large number of particles flowing in the channel allows for more heat to be absorbed as latent heat, decreasing the average surface temperature.

In trying to elucidate further the complex interplay between the phase-change effect and the flow mixing effect, further tests were made with solid particles made of plastic, without phase change capabilities within the temperatures achieved in the testing section [47]. Results shown in Fig. 24, for particle concentration equal to 3%, seem enlightening.

First, the results with plastic particles seem to parallel the results with clear water, but with a higher average heat transfer coefficient. This behavior seems to indicate the sweeping effect of the particles in the flow field (observe the thermal conductivity of the plastic is less than that of the water, so the plastic particles would

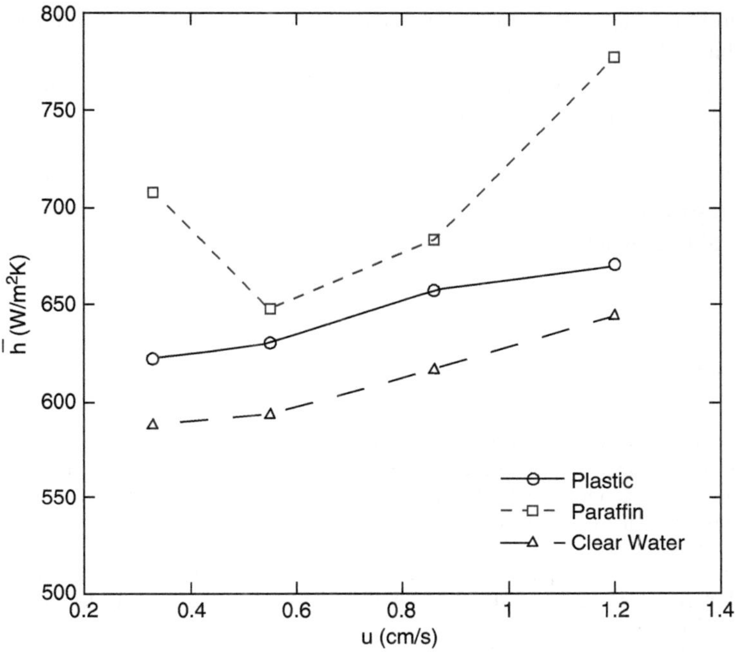

Fig. 24 Surface averaged heat transfer coefficient versus flow speed, for 14 kW/m^2 heat flux, clear water, or 3% concentration plastic or paraffin (phase-change) particles [47]

tend to hinder the heat transfer process). Second, the heat transfer coefficient of the microencapsulated phase-change particles, although higher than that for the plastic particles, does not parallel the heat transfer behavior for clear water. In fact, when the phase-change particles are used, an initial increase in speed leads to a decrease in the heat transfer coefficient, likely because of the smaller latent heat effect as the particles spend less time in the channel. However, as the speed increases further, even though the particles spend less time in the channel the heat transfer coefficient between the liquid and the particles would increase as well (because of the more mixing at higher speed), getting the phase-change effect by the particles to predominate again. Notice the heat transfer coefficient increases, in respect to the value obtained with clear water, by about 20% for the configuration shown in Fig. 24, with about 9% increase estimated to be due to the mixing effect by the particles alone.

5 Summary and Conclusion

A new bio-inspired multiphase thermal functional fluid (MTFF) is proposed for heat transfer in line with the main characteristics found in the alveolar capillaries, namely liquid and particles, with the particles having dimensions similar to the dimensions of the channel through which the exchange takes place.

Preliminary numerical simulations and experimental tests with microencapsulated phase-change material particles flowing with water seem to confirm the suitability of the new MTFF as heat transfer fluid combining the boundary layer sweeping effect and the phase-change effect yielding substantially higher transfer efficiency with minimum pump-power penalty.

Finally, the scalability of the MTFF and the flow system opens up the opportunity for resolving a few of the issues pertaining to the use of nanofluids (or slurries), as to obtain yet much more efficient results.

References

1. Darwish, T., Bayoumi, M.: Trends in low-power VLSI design. In: Chen, W-K. (ed) Electrical Engineering Handbook, vol. 3(5), pp. 263–280. Academic Press, San Diego (2005)
2. Steinke, M.E., Kandlikar, S.G.: Review of single phase heat transfer enhancement techniques for application in microchannels, minichannels and microdevices. Int. J. Heat Technol. **22**(2), 3–11 (2004)
3. Charunyaorn, P., Sengupta, S., Roy, S.K.: Forced convective heat transfer in microencapsulated phase change material slurries: flow in circular ducts. Int. J. Heat Mass Transf. **34**(3), 819–833 (1991)
4. Sohn, C.W., Chen, M.M.: Microconvective thermal conductivity in disperse two-phase mixture as observed in a low velocity Couette flow experiment. J. Heat Transf. **103**, 47–50 (1991)
5. Choi, E., Cho, Y., Lorsch, H.G.: Forced convection heat transfer with phase-change-material slurries: turbulent flow in a circular tube. Int. J. Heat Mass Transf. **37**(2), 207–215 (1993)
6. Goel, M., Roy, S.K., Sengupta, S.: Laminar forced convection heat transfer in microencapsulated phase change material suspensions. Int. J. Heat Mass Transf. **37**(4), 593–604 (1994)
7. Zhang, Y., Faghri, A.: Analysis of forced convection heat transfer in microencapsulated phase change material suspensions. J. Thermophys. Heat Transf. **9**(4), 727–732 (1995)
8. Mulligan, J.C., Colvin, D.P., Bryan, Y.G.: Microencapsulated phasechange material suspensions for heat transfer in spacecraft thermal systems. J. Spacecr. Rockets **33**(2), 278–284 (1996)
9. Yamagishi, Y., Takeuchi, H., Pyatenko, A., Kayukawa, N.: Characteristic of microencapsulated PCM slurry as a heat-transfer fluid. AIChE J **45**(4), 696–707 (1999)
10. Alisetti, L., Roy, S.K.: Forced convection heat transfer to phase change material slurries in circular ducts. J. Thermophys. Heat Transf. **14**(1), 115–118 (2000)
11. Roy, S.K., Avanic, B.L.: Laminar forced convection heat transfer with phase change material suspensions. Int. Commun. Heat Mass Transf. **28**(7), 895–904 (2001)
12. Royon, L., Guiffant, G., Perrot, P.: Forced convection heat transfer in a slurry of phase change material in an agitated tank. Int. Commun. Heat Mass Transf. **27**(8), 1057–1065 (2000)
13. Hu, X.X., Zhang, Y.P.: Theoretical analysis of the convective heat transfer enhancement of latent functionally thermal fluid with isothermal wall. Acta Energiae Solaris Sin. **23**, 626–633 (2002)
14. Hu, X.X., Zhang, Y.P.: Novel insight and numerical analysis of convective heat transfer enhancement with microencapsulated phase change material slurries: laminarflow in a circular tube with constant heat flux. Int. J. Heat Mass Transf. **45**, 3163–3172 (2002)
15. Ayel, V., Lottin, O., Peerhossaini, H.: Rheology, flow behavior and heat transfer of ice slurries: a review of the state of the art. Int. J. Refrig. **26**, 95–107 (2003)

16. Xin, W., Yinping, Z., Xlanxu, H.: Turbulent heat transfer enhancement of microencapsulated phase change material slurries with constant wall heat flux. Enhanced Heat Transf. **11**(1), 13–22 (2003)
17. Zhang, Y.P., Hu, X.X., Wang, X.: Theoretical analysis of convective heat transfer enhancement of microencapsulated phase change material slurries. Heat Mass Transf. **40**, 59–66 (2003)
18. Ho, C.J., Lin, J.F., Chiu, S.Y.: Heat transfer of solid-liquid phase change material suspension in circular pipes: effects of wall conduction. Num. Heat Transf. **45**((A)), 171–190 (2004)
19. Tiarks, F., Landfester, K., Antonietti, M.: Preparation of polymeric nanocapsules by miniemulsion polymerization. Langmuir **17**(3), 908–918 (2001)
20. Momoda, L.A., Phelps, A.C.: Nanometer sized phase change materials for enhanced heat transfer fluid performance. US Patent US 6447692, 2002
21. Cho, J.S., Kwon, A., Cho, C.G.: Microencapsulation of octadecane as a phase-change material by interfacial polymerization in an emulsion system. Colloid Polym. Sci. **280**, 260–266 (2002)
22. Zhang, X.X., Fan, Y.F., Tao, X.M., Yick, K.L.: Fabrication and properties of microcapsules and nanocapsules containing n-octadecane. Mater. Chem. Phys. **88**, 300–307 (2004)
23. Luo, Y.W., Zhou, X.D.: Nanoencapsulation of a hydrophobic compound by a miniemulsion polymerization process. J. Polym. Sci. A Polym. Chem. **42**(9), 2145–2154 (2004)
24. Ozonur, Y., Mazman, M., Paksoy, H.O., Evliya, H.: Microencapsulation of coco fatty acid mixture for thermal energy storage with phase change material. Int. J. Energy Res. **307**, 41–749 (2006)
25. Sarier, N., Onder, E.: The manufacture of microencapsulated phase change materials suitable for the design of thermally enhanced fabrics. Thermochim. Acta **452**, 149–160 (2007)
26. Fang, Y., Kuang, S., Gao, X., Zhang, Z.: Preparation of nanoencapsulated phase change material as latent functionally thermal fluid. J. Phys. D Appl. Phys. **42**, 1–8 (2009)
27. Inaba, H., Kim, M.J., Horibe, A.: Melting heat transfer characteristics of microencapsulated phase change material slurries with plural microencapsules having different diameters. J Heat Transf. **126**(4), 558–565 (2004)
28. Wang, X., Niu, J., Li, Y., Wang, X., Chen, B., Zeng, R., Song, Q., Zhang, Y.: Flow and heat transfer be behaviors of phase change material slurries in a horizontal circular tube. Int. J. Heat Mass Transf. **50**, 2480–2491 (2007)
29. Alvarado, J.L., Marsh, C., Sohn, C., Vilceus, M., Hock, V., Phetteplace, G., Newell, T.: Thermal performance of microencapsulated phase change material slurry in turbulent flow under constant heat flux. Int. J. Heat Mass Transf. **50**, 1938–1952 (2007)
30. Chen, B.J., Wang, X., Zeng, R.L., Zhang, Y.P., Wang, X.C., Niu, J.J., Li, Y., Di, H.F.: An experimental study of convective heat transfer with microencapsulated phase change material suspension: laminar flow in a circular tube under constant heat flux. Exp. Therm. Fluid Sci. **32**, 1638–1646 (2008)
31. Royon, L., Guiffant, G.: Forced convection heat transfer with slurry of phase change material in circular ducts: a phenomenological approach. Energ. Convers. Manag. **49**(5), 928–932 (2008)
32. Wang, X., Niu, J., Li, Y., Zhang, Y., Wang, X., Chen, B., Zeng, R., Song, Q.: Heat transfer of microencapsulated PCM slurry flow in a circular tube. AIChE J **54**, 1110–1120 (2008)
33. Zenga, R., Wanga, X., Chena, B., Zhanga, Y., Niub, J., Wang, X., Dia, H.: Heat transfer characteristics of microencapsulated phase change material slurry in laminar flow under constant heat flux. Appl. Energy **86**, 2661–2670 (2009)
34. Koulich, V.V., Lage, J.L., Hsia, C.C.W., Johnson Jr., R.L.: A porous medium model of alveolar gas diffusion. J. Porous Media **2**, 263–275 (1999)
35. Kulish, V.V., Lage, J.L., Hsia, C.C.W., Johnson Jr., R.L.: Three-dimensional, unsteady simulation of alveolar respiration. ASME J. Biomech. Eng. **124**, 609–616 (2002)
36. Kulish, V.V., Lage, J.L.: Fundamentals of alveolar diffusion: a new modeling approach. AUTOMEDICA Int. J. Bio-Med. Engg. Technol. **20**, 225–268 (2002)

37. Kulish, V.V., Sourin, A.I., Lage, J.L.: Simulation and visualization of gas diffusion in human lungs. J Vis. (The Visualization Society of Japan) **5**, 260–266 (2002)
38. Kulish, V.V., Lage, J.L.: Impact of microscopic solid particles on the alveolar diffusion. In: Kulish, V. (ed.) Human respiration: anatomy and physiology, mathematical modeling, numerical simulation and applications – Advances in Bioengineering Series, vol. 3, pp. 13–22. WIT Press, Southampton (2006)
39. Lage, J.L., Merrikh, A.A., Kulish, V.V.: A porous medium model to investigate the red cell distribution effect on alveolar respiration: numerical simulations to CO diffusion in the alveolar region of the lungs. Emerging Technologies and Techniques in Porous Media, vol. 25, pp. 381–407. Kluwer, Dordrecht (2004)
40. Hsia, C.C.W., Chuong, C.J.C., Johnson Jr., R.L.: Critique of conceptual basis of diffusion capacity estimates: a finite element analysis. J. Appl. Physiol. **79**, 1039–1047 (1995)
41. Merrikh, A.A., Lage, J.L.: Effect of blood flow on gas transport in a pulmonary capillary. ASME J. Biomech. Eng. **127**, 432–439 (2005)
42. Merrikh, A.A., Lage, J.L.: The role of red cell movement on alveolar gas diffusion. Mater. Sci. Eng. Techn. **36**, 497–504 (2005)
43. Merrikh, A.A., Lage, J.L.: Plasma microcirculation in alveolar capillaries: effect of parachute shaped red cells on gas exchange. Int. J. Heat Mass Transf. **51**, 5712–5720 (2008)
44. Merrikh, A.A.: Convection-diffusion analysis of gas transport in a pulmonary capillary. PhD Dissertation, SMU (2004)
45. Hassanipour, F., Lage, J.L.: Numerical simulation of capillary convection with encapsulated phase-change particles. Num. Heat Transf. A **55**, 893–905 (2009)
46. Hassanipour, F., Lage, J.L.: Preliminary experimental study of a bio-inspired, phased-change particle capillary heat exchanger. Int. J. Heat Mass Transf. **53**, 3300–3307 (2010)
47. Hassanipour, F., Lage, J.L.: New bio-inspired, multi-phase forced convection cooling by ABS plastic or encapsulated paraffin beads. ASME J. Heat Transf. **132**, 149–152 (2010)
48. Hassanipour, F.: A particulate flow heat exchanger inspired by gas diffusion in lung capillaries. PhD Dissertation, SMU (2009)
49. Ulusarslan, D., Teke, I.: An experimental investigation of the capsule velocity, concentration rate and the spacing between the capsules for spherical capsule train flow in a horizontal circular pipe. Powder Technol. **159**, 27–34 (2005)
50. Ulusarslan, D., Teke, I.: An experimental determination of pressure drops in the flow of low density spherical capsule train inside horizontal pipes. Exp. Therm. Fluid Sci. **30**, 233–241 (2006)
51. Teke, I., Ulusarslan, D.: Mathematical expression of pressure gradient in the flow of spherical capsules less dense than water. Int. J. Multiph. Flow **33**, 658–674 (2007)

Simulation of Turbulent Combustion in Porous Materials with One- and Two-Energy Equation Models

Marcelo J.S. de Lemos

Abstract The objective of this chapter is to present one- and two-dimensional numerical simulations of combustion of an air/methane mixture in porous materials using a mathematical model that explicitly considers the intra-pore levels of turbulent kinetic energy. Transport equations are written in their time-and-volume-averaged form and a volume-based statistical turbulence model is applied to simulate turbulence generation due to the porous matrix. Four different thermo-mechanical models are compared, namely Laminar, Laminar with Radiation Transport, Turbulent, Turbulent with Radiation Transport. Combustion is modeled via a unique simple closure. Preliminary testing results indicate that a substantially different temperature distribution is obtained depending on the model used. In addition, for high excess air peak gas temperature are reduced and the flame front moves towards the exit of the burner. Also, increasing the inlet flow rate for stoichiometric mixture pushes the flame out of the porous material.

Nomenclature

Latin Characters

A	Pre-exponential factor
c_F	Forchheimer coefficient
c_p	Specific heat
$\mathbf{D} = \left[\nabla\mathbf{u} + (\nabla\mathbf{u})^T\right]/2$	Deformation rate tensor
$\mathbf{D}_{diff}$	Macroscopic diffusion tensor
$\mathbf{D}_{disp}$	Dispersion tensor

M.J.S. de Lemos
Instituto Tecnológico de Aeronáutica - ITA, 12228-900, São José dos Campos, São Paulo Brazil
e-mail: delemos@ita.br

A. Öchsner and G.E. Murch (eds.), *Heat Transfer in Multi-Phase Materials*, Adv Struct Mater 2, DOI 10.1007/8611_2010_9,

$\mathbf{D}_{disp,t}$	Dispersion tensor due to turbulence
$\mathbf{D}_{eff}$	Effective mass transport tensor
f_2	Damping function
f_μ	Damping function
K	Permeability
k_f	Fluid thermal conductivity
k_s	Solid thermal conductivity
$\mathbf{K}_{eff}$	Effective Conductivity tensor
m_ℓ	Mass fraction of species ℓ
Pr	Prandtl number
S_{fu}	Rate of fuel consumption
T	Temperature
$\mathbf{u}$	Microscopic velocity
$\mathbf{u}_D$	Darcy or superficial velocity (volume average of $\mathbf{u}$)

Greek Characters

α	Thermal diffusivity
β_r	Extinction coefficient
ΔV	Representative elementary volume
ΔV_f	Fluid volume inside ΔV
ΔH	Heat of combustion
μ	Dynamic viscosity
ν	Kinematic viscosity
ρ	Density of air/fuel mixture
ϕ	$\phi = {}^{\Delta V_f}/_{\Delta V}$, Porosity
Ψ	Excess air-to-fuel ratio

Special Characters

φ	General variable
$\langle\varphi\rangle^i$	Intrinsic average
$\langle\varphi\rangle^v$	Volume average
${}^i\varphi$	Spatial deviation
ϕ	Time average
${}^i\varphi$	Spatial deviation
$\lvert\varphi\rvert$	Absolute value (Abs)
$\boldsymbol{\phi}$	Vector general variable
$()_{s,f}$	Solid/fluid
$(\ )_{eff}$	Effective value, $\phi\varphi_f + (1-\phi)\varphi_s$
$()_\phi$	Macroscopic value

1 Introduction

Flow with chemical reaction in inert porous media has been extensively investigated due to the many engineering applications and demand for developing high-efficiency power production devices. The growing use of efficient radiant burners can be encountered in the power and process industries and, as such, proper mathematical models of flow, heat and mass transfer in porous media under combustion can benefit the development of such engineering equipment.

Accordingly, the advantages of having a combustion process inside an inert porous matrix are today well documented in the literature [1–8], including recent reviews on combustion of gases [9] and liquids [10] in such burners. Hsu et al. [11] points out some of its benefits including higher burning speed and volumetric energy release rates, higher combustion stability and the ability to burn gases of a low energy content. Driven by this motivation, the effects on porous ceramics inserts have been investigated in Peard et al. [12], among others.

Turbulence modeling of combustion within inert porous media has been conducted by Lim and Matthews [13] on the basis of an extension of the standard k-ε model of Jones and Launder [14]. Work on direct simulation of turbulence in premixed flames, for the case when the porous dimension is of the order of the flame thickness, has also been reported in Sahraoui and Kaviany [15].

Further, non-reactive turbulence flow in porous media has been the subject of several studies [16–18], including many applications such as flow though porous baffles [19], channels with porous inserts [20] and buoyant flows [21]. In such line of work, intra-pore turbulence is accounted for in all transport equations, but only non-reactive flow has been previously investigated in [16–21].

Motivated by the foregoing, this chapter extends previous work on turbulence modeling in porous media to include simulation of reactive flows. Computations are carried out for inert porous material considering one- and two-dimensional turbulent flows with one- and two-energy equation closures. Four different thermo-mechanical models are here compared, namely Laminar Flow, Laminar Flow with Radiation Transport, Turbulent Flow and Turbulent Flow with Radiation Transport. As such, this contribution compares the effects of radiation and turbulence in smoothing temperature distributions within porous burners.

2 Macroscopic Mathematical Model

As mentioned, the thermo-mechanical model here employed is based on the "double-decomposition" concept [16, 17], which has been also described in detail in a book [18]. In that work, transport equations are volume averaged according to the Volume Averaging Theorem [22–24] in addition of using time decomposition of flow variables followed by standard time-averaging procedure for treating turbulence. As the entire equation set is already fully available in the open literature, these equations will be reproduced here and details about their derivations can be obtained in the aforementioned references. Essentially, in all the above-mentioned work the flow variables are decomposed in a volume mean and a deviation (classical porous media analysis) in addition to being also decomposed in a time-mean and a fluctuation (classical turbulent flow treatment).

These final equations in their steady-state form are:

2.1 Continuity Equation

$$\nabla.\rho\,\bar{\mathbf{u}}_D = 0 \tag{1}$$

where, $\bar{\mathbf{u}}_D$ is the average surface velocity (also known as seepage, superficial, filter or Darcy velocity) and ρ is the fluid density. Equation (1) represents the macroscopic continuity equation for the gas.

2.2 Momentum Equation

$$\begin{aligned}\nabla\cdot\left(\rho\frac{\bar{\mathbf{u}}_D\bar{\mathbf{u}}_D}{\phi}\right) &= -\nabla\left(\phi\langle\bar{p}\rangle^i\right)+\mu\nabla^2\bar{\mathbf{u}}_D+\nabla\cdot\left(-\rho\phi\langle\overline{\mathbf{u}'\mathbf{u}'}\rangle^i\right)+\phi\rho\mathbf{g}\\ &\quad-\left[\frac{\mu\phi}{K}\bar{\mathbf{u}}_D+\frac{c_F\phi\rho\,|\bar{\mathbf{u}}_D|\bar{\mathbf{u}}_D}{\sqrt{K}}\right]\end{aligned} \tag{2}$$

where the last two terms in (2), represent the Darcy and Forchheimer contributions. The symbol K is the porous medium permeability, $c_F = 0.55$ is the form drag coefficient, $\langle p\rangle^i$ is the intrinsic (fluid phase averaged) pressure of the fluid, μ represents the fluid viscosity and ϕ is the porosity of the porous medium.

Turbulence is handled via a macroscopic $k-\varepsilon$ model given by:

$$\begin{aligned}\nabla\cdot\left(\rho\,\bar{\mathbf{u}}_D\langle k\rangle^i\right) &= \nabla\cdot\left[\left(\mu+\frac{\mu_{t_\phi}}{\sigma_k}\right)\nabla\left(\phi\langle k\rangle^i\right)\right]\\ &\quad-\rho\langle\overline{\mathbf{u}'\mathbf{u}'}\rangle^i:\nabla\bar{\mathbf{u}}_D+c_k\rho\frac{\phi\,\langle k\rangle^i\,|\bar{\mathbf{u}}_D|}{\sqrt{K}}-\rho\phi\langle\varepsilon\rangle^i\end{aligned} \tag{3}$$

$$\nabla\cdot\left(\rho\,\bar{\mathbf{u}}_D\langle\varepsilon\rangle^i\right)=\nabla\cdot\left[\left(\mu+\frac{\mu_{t_\varphi}}{\sigma_\varepsilon}\right)\nabla\left(\phi\langle\varepsilon\rangle^i\right)\right]+c_1\left(-\rho\langle\overline{\mathbf{u}'\mathbf{u}'}\rangle^i:\nabla\bar{\mathbf{u}}_D\right)\frac{\langle\varepsilon\rangle^i}{\langle k\rangle^i}$$
$$+c_2c_k\rho\frac{\phi\,\langle\varepsilon\rangle^i|\bar{\mathbf{u}}_D|}{\sqrt{K}}-c_2\,\rho\phi\frac{\langle\varepsilon\rangle^{i2}}{\langle k\rangle^i} \tag{4}$$

where

$$-\rho\phi\langle\overline{\mathbf{u}'\mathbf{u}'}\rangle^i=\mu_{t_\phi}2\langle\mathbf{D}\rangle^v-\frac{2}{3}\phi\rho\langle k\rangle^i\mathbf{I} \tag{5}$$

and

$$\mu_{t_\phi}=\rho\,c_\mu\frac{\langle k\rangle^{i2}}{\langle\varepsilon\rangle^i} \tag{6}$$

Details on the derivation of the above equations can be found in [18].

2.3 *Macroscopic Energy Models*

There are two possibilities to handle energy transport across porous burners. In the simplified model, we assume the so-called "Local Thermal Equilibrium Model – LTE", in which the average solid temperature is seen to be numerically equal to the mean gas temperature. When the solid and the gas temperate differ by a considerable amount, the "Local Non-thermal Equilibrium Model – LNTE" applies and distinct energy balances become mandatory for each phase. Transport equations for both models follow below.

2.3.1 One Energy Equation Model (LTE)

Although the Local Thermal Equilibrium hypothesis (LTE) is known to be inappropriate to handle large temperature differences between the solid matrix and the burning gas, this simple mathematical framework may provide insight for investigating the role of the mechanisms of turbulence and radiation, ultimately contributing to the developments of more advanced and sophisticated simulation tools. As an example, one energy equation models have been applied for analyzing combustion in porous media by Mohamad et al. [8] and de Neef et al. [25].

The governing equation for energy transport is:

$$\frac{\partial([\rho\, c_{p_f}\, \phi + \rho_s\, c_{ps}\,(1-\phi)]\langle \overline{T}\rangle^i)}{\partial t} + \nabla\cdot(\rho\, c_{p_f}\cdot \bar{\mathbf{u}}_D \langle \overline{T}\rangle^i) = \nabla\cdot\left\{\mathbf{K}_{eff}\cdot\nabla\langle \overline{T}\rangle^i\right\} + \phi \Delta H\, S_{fu} \tag{7}$$

where, $\langle T\rangle^i$ is the averaged temperature for both the solid and the liquid, $\mathbf{K}_{eff}$, given by:

$$\mathbf{K}_{eff} = \left\{ \overbrace{\phi\, k_f + (1-\phi)\,[k_s}^{conduction} + \overbrace{\frac{16\sigma(\langle \overline{T}\rangle^i)^3}{3\beta_r}]}^{radiation} \right\}\mathbf{I} + \underbrace{\mathbf{K}_{tor}}_{tortuosity} + \underbrace{\mathbf{K}_{disp}}_{dispersion} + \underbrace{\mathbf{K}_t + \mathbf{K}_{disp,t}}_{turbulence} \tag{8}$$

is the effective conductivity tensor, ΔH is the heat of combustion [J/kg], β_r is the extinction coefficient [1/m], σ is the Stephan-Boltzmann constant, S_{fu} is the rate of fuel consumption [kg/m^{3}s] to be commented below and $\mathbf{I}$ is the unit tensor. In (8) all mechanisms contributing to heat transfer within the medium, together with turbulence and radiation, are included in order to compare their effect on temperature distribution.

A steady state form of (7) reads:

$$\nabla\cdot\left(\rho\, c_{p_f}\, \bar{\mathbf{u}}_D \langle \overline{T}\rangle^i\right) = \nabla\cdot\left\{\mathbf{K}_{eff}\cdot\nabla\langle \overline{T}\rangle^i\right\} + \phi \Delta H\, S_{fu} \tag{9}$$

where all additional mechanisms of transfer, as mentioned, are included in $\mathbf{K}_{eff}$.

2.3.2 Two Energy Equation Model (NLTE)

In cases where average temperatures in distinct phases are substantially different, macroscopic energy equations are obtained for both fluid and solid phases by also applying time- and volume- average operators to the instantaneous local equations [26]. As in the flow case, volume integration is performed over a Representative Elementary Volume (REV). After including the heat released due to the combustion reaction, one gets for both phases:

Gas:

$$\nabla \cdot (\rho_f c_{pf} \overline{\mathbf{u}}_D \langle \overline{T_f} \rangle^i) = \nabla \cdot \left\{ \mathbf{K}_{eff,f} \cdot \nabla \langle \overline{T_f} \rangle^i \right\} + h_i a_i \left(\langle \overline{T_s} \rangle^i - \langle \overline{T_f} \rangle^i \right) + \phi \, \Delta H \, S_{fu}, \tag{10}$$

Solid:

$$0 = \nabla \cdot \left\{ \mathbf{K}_{eff,s} \cdot \nabla \langle \overline{T_s} \rangle^i \right\} - h_i a_i \left(\langle \overline{T_s} \rangle^i - \langle \overline{T_f} \rangle^i \right), \tag{11}$$

where, $a_i = A_i/\Delta V$ is the interfacial area per unit volume, h_i is the film coefficient for interfacial transport, $\mathbf{K}_{eff,f}$ and $\mathbf{K}_{eff,s}$ are the effective conductivity tensors for fluid and solid, respectively, given by:

$$\mathbf{K}_{eff,f} = \left\{ \overbrace{\phi \, k_f}^{conduction} \right\} \mathbf{I} + \underbrace{\mathbf{K}_{f,s}}_{local\ conduction} + \underbrace{\mathbf{K}_{disp}}_{dispersion} + \underbrace{\mathbf{K}_t + \mathbf{K}_{disp,t}}_{turbulence} \tag{12}$$

$$\mathbf{K}_{eff,s} = \left\{ \overbrace{(1-\phi)\,[k_s}^{conduction} + \overbrace{\frac{16\sigma(\langle \overline{T} \rangle^i)^3}{3\beta_r}]}^{radiation} \right\} \mathbf{I} + \underbrace{\mathbf{K}_{s,f}}_{local\ conduction} \tag{13}$$

In (12) and (13) all mechanisms contributing to heat transfer within the medium, together with turbulence and radiation, are here also included in order to compare their effect on temperature distribution. Further, such distinct contributions of various mechanisms are the outcome of the application of gradient type diffusion models, in the form (see [26] for details).

Turbulent heat flux:

$$-\left(\rho c_p\right)_f \left(\phi \ \overline{\langle \mathbf{u}' \rangle^i \langle T'_f \rangle^i} \right) = \mathbf{K}_t \cdot \nabla \langle \bar{T}_f \rangle^i . \tag{14}$$

Thermal dispersion:

$$-\left(\rho c_p\right)_f \left(\phi \ \langle {}^i\overline{\mathbf{u}}\, {}^i\overline{T_f} \rangle^i \right) = \mathbf{K}_{disp} \cdot \nabla \langle \bar{T}_f \rangle^i . \tag{15}$$

Turbulent thermal dispersion:

$$-\left(\rho c_p\right)_f \left(\phi \ \langle \overline{{}^i\mathbf{u}'\, {}^iT'_f} \rangle^i \right) = \mathbf{K}_{disp,t} \cdot \nabla \langle \bar{T}_f \rangle^i . \tag{16}$$

Local conduction:

$$\nabla \cdot \left[\frac{1}{\Delta V} \int_{A_i} \mathbf{n}_i \, k_f \overline{T_f} \, dA \right] = \mathbf{K}_{f,s} \cdot \nabla \langle \bar{T}_s \rangle^i$$
$$-\nabla \cdot \left[\frac{1}{\Delta V} \int_{A_i} \mathbf{n}_i \, k_s \overline{T_s} \, dA \right] = \mathbf{K}_{s,f} \cdot \nabla \langle \bar{T}_f \rangle^i \tag{17}$$

The sum of the two terms in (17) gives rise to the tortuosity tensor appearing in (8). For the sake of simplicity, theses terms are here neglected. In (10) and (11) the heat transferred between the two phases was modeled by means of a film coefficient h_i. A numerical correlation for the interfacial convective heat transfer coefficient was proposed by Kuwahara et al. [27] for laminar flow as:

$$\frac{h_i D}{k_f} = \left(1 + \frac{4(1-\phi)}{\phi}\right) + \frac{1}{2}(1-\phi)^{1/2} \mathrm{Re}_D \mathrm{Pr}^{1/3} \quad \text{valid for} \quad 0.2 < \phi < 0.9 \tag{18}$$

For turbulent flow, the following expression was proposed in Saito and de Lemos [26]:

$$\frac{h_i D}{k_f} = 0.08 \left(\frac{Re_D}{\phi}\right)^{0.8} Pr^{1/3}; \text{ for } 1.0x10^4 < \frac{Re_D}{\phi} < 2.0x10^7 \quad \text{valid for} \quad 0.2 < \phi < 0.9, \tag{19}$$

2.4 *Mass Transport*

Transport equation for the fuel reads:

$$\nabla \cdot (\rho \, \bar{\mathbf{u}}_D \langle \bar{m}_{fu} \rangle^i) = \nabla \cdot \rho \, \mathbf{D}_{eff} \cdot \nabla (\phi \langle \bar{m}_{fu} \rangle^i) - \varphi \, S_{fu} \tag{20}$$

where $\langle \bar{m}_{fu} \rangle^i$ is the mass fraction for the fuel. The effective mass transport tensor, $\mathbf{D}_{eff}$, is defined as:

$$\mathbf{D}_{eff} = \underbrace{\mathbf{D}_{disp}}_{dispersion} + \overbrace{\mathbf{D}_{diff}}^{diffusion} + \underbrace{\mathbf{D}_t + \mathbf{D}_{disp,t}}_{turbulence} = \mathbf{D}_{disp} + \frac{1}{\rho}\left(\frac{\mu_\varphi}{Sc_\ell} + \frac{\mu_{t_\varphi}}{Sc_{\ell,t}}\right)\mathbf{I}$$
$$= \mathbf{D}_{disp} + \frac{1}{\rho}\left(\frac{\mu_{\varphi,ef}}{Sc_{\ell,ef}}\right)\mathbf{I} \tag{21}$$

where Sc_ℓ and $Sc_{\ell,t}$ are the laminar and turbulent Schmidt numbers for species ℓ, respectively, and "*ef*" denotes an effective value.

2.5 Simple Chemistry

In this work, for simplicity, the chemical exothermic reaction is assumed to be instantaneous and to occur in a single step, kinetic-controlled, which, for combustion of a mixture air/methane, is given by the chemical reaction [8]:

$$CH_4 + 2(1+\Psi)(O_2 + 3.76N_2) \rightarrow CO_2 + 2H_2O + 2\Psi O_2 + 7.52(1+\Psi)N_2 \tag{22}$$

For N-heptane, a similar equation reads [28]:

$$\begin{aligned} C_7H_{16} + 11(1+\Psi)(O_2 + 3.76N_2) \rightarrow\ & 7CO_2 + 8H_2O + 11\Psi\, O_2 \\ & + 41.36(1+\Psi)N_2 \end{aligned} \tag{23}$$

and for Octane, one has:

$$\begin{aligned} C_8H_{18} + 12.5(1+\Psi)(O_2 + 3.76N_2) \rightarrow\ & 8CO_2 + 9H_2O + 12.5\Psi\, O_2 \\ & + 47(1+\Psi)N_2 \end{aligned} \tag{24}$$

where Ψ is the excess air in the reactant stream at the inlet of the porous foam. For the stoichiometric ratio, $\Psi = 0$. In all of these equations, the reaction is then assumed to be kinetically controlled and occurring infinitely fast. A general expression for them can be derived as:

$$\begin{aligned} C_nH_{2m} + (n + \tfrac{m}{2})(1+\Psi)(O_2 + 3.76N_2) \rightarrow\ & nCO_2 + mH_2O \\ +(n + \tfrac{m}{2})\Psi\, O_2 + (n + \tfrac{m}{2})& 3.76(1+\Psi)N_2 \end{aligned} \tag{25}$$

where the coefficients "n" and "m" can be found in Table 1. Equation (25) is here assumed to hold for the particular examples given in the table.

The rate of fuel consumption over the total volume (gas plus solid) was determined by a one step Arrhenius reaction [29] given by:

$$S_{fu} = \rho^2 A\, \langle \bar{m}_{fu} \rangle^i \langle \bar{m}_{ox} \rangle^i\, \exp[-E/R\, \langle \bar{T} \rangle^i] \tag{26}$$

Table 1 Coefficients in the general combustion equation (25)

Gas	n	m	$(n + m/2)$	$(n + m/2) \times 3.76$
Methane	1	2	2	7.52
N-heptane	7	8	11	41.36
Octane	8	9	12.5	47

where $\langle \bar{m}_{fu} \rangle^i$ and $\langle \bar{m}_{ox} \rangle^i$ are the volume-time averaged mass fractions for the fuel and oxidant, respectively, A is the pre-exponential factor [1×10^{10} m^3/(kg s)] and E is the activation energy [1.4×10^8 J/kmol], where all values used are the ones commonly used in the literature for combustion of methane. The heat generation rate is calculated as:

$$\dot{Q} = \Delta H \, S_{fu} \tag{27}$$

where S_{fu} is computed by (26).

Density ρ in the above equations is determined from the perfect gas equation for a mixture of perfect gases:

$$\rho = \frac{P_o}{RT_f \sum_{1}^{\ell} \frac{m_\ell}{M_\ell}} \tag{28}$$

where P_o is the absolute pressure, R is the universal gas constant [8.134 J/(mol K)] and M_ℓ is the molecular weight of species ℓ.

3 Results and Discussion

For two-dimensional cases using the LTE model, the set of equations above, in Sects 2.2 and 2.3.1, were solved for a two dimensional combustor whereas for the NLTE model, detailed in Sect. 2.3.2, one-dimensional flow was considered.

For one-dimensional cases and the NLTE model, simulations assumed given temperatures (solid and gas) and fuel mass fraction at inlet, $x = 0$. At exit, $x = 12\,\text{cm}$, a zero diffusion condition $\partial(\;)/\partial x = 0$ for all variables was considered. For the solid temperature, a balance between the energy conducted to the exit of the burner and the radiation leaving the porous material to the environment was applied. Further, an initial length of 2 cm was considered to be made of a material that prevents flash back of the flame, which is commonly referred to in the literature as "flame trap" [30]. Ignition, is existing, was then calculated for $x > 2$ cm.

In both 1D and 2D cases, the finite–volume technique was employed to discretize the transport equations. The resulting algebraic equation set was relaxed using the well-known segregated method SIMPLE. Further, the flame front position was the sole outcome of the solution process and no artificial numerical set-up was implemented for holding the flame at some particular location.

As mentioned, two sets of results are reported below, namely those obtained with the LTE hypotheses, for 2D flows in a burner, and additional computations using the NLTE approach, for simple 1D calculation.

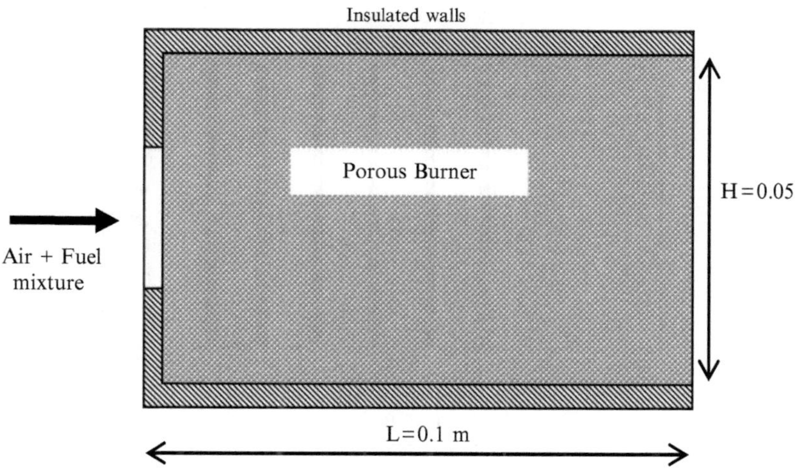

Fig. 1 Two-dimensional combustor model, LTE model

3.1 Two-Dimensinal Flow: LTE Model

For the first set of results, the problem considered consists in having a porous media confined in a channel, through which a mixture of fuel and air enters from the left, as shown in Fig. 1. The fuel/air mixture is injected through an inlet clearance of size less than the burner height, so that flow expansion occurs past the chamber entrance.

Figure 2a shows the effect of inlet velocity U_{in} on the flame front, whose location is here assumed to be related with the heat release rate computed by (26). Figure 2b repeats same simulation of Fig. 2a, using now a higher air-to-fuel ratio, $\Psi = 0.8$. Combustion of a lean mixture reveals that the flame front is more sensitive to the incoming mass flow rate and that the flame front is pushed towards the exit, leading eventually to the opening of the combustion front for higher velocities. As such, unburned gas leaves the chamber for higher mass flow rates.

Finally, Fig. 2c presents the effect of air-to-fuel ratio while keeping the total mass flow rate constant through the combustor. As the mixture becomes leaner, the flame front is pushed towards the right, and the effect seen above, the opening of the combustion front, is observed. Consequently, unburned fuel leaves the chamber. This effect is more pronounced as Ψ increases, a result which is coherent when comparing Fig. 2a, b.

3.2 One-Dimensinal Flow: NLTE Model

For one-dimensional flow using the Non-Thermal Equilibrium model NLTE, the computational grid was generated with a concentration of points close to the

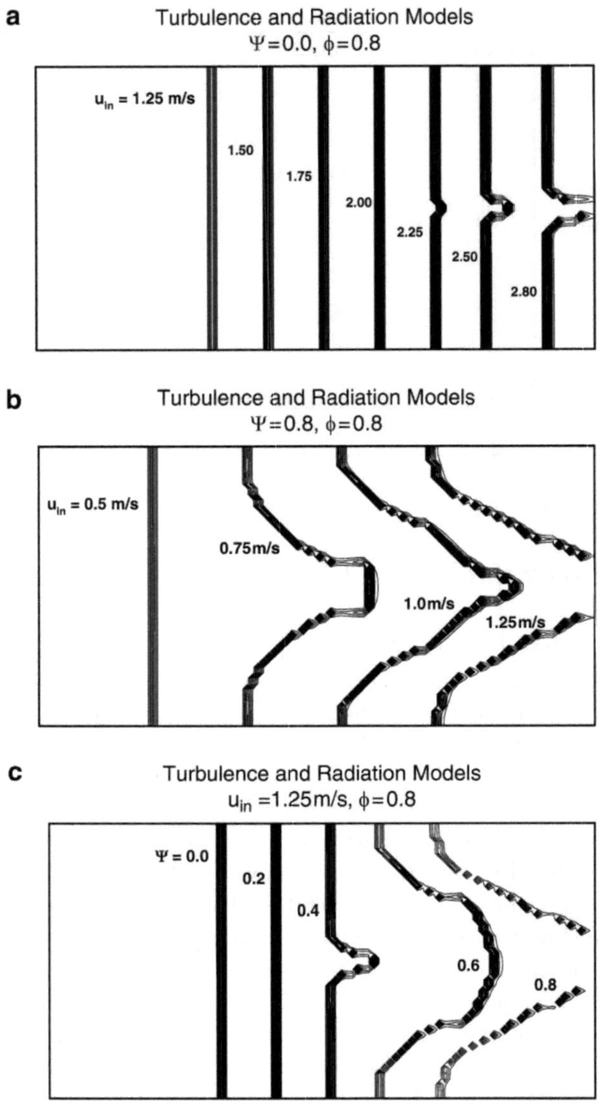

Fig. 2 Flame front location for turbulence and radiation models, $\phi = 0.8$: (**a**) effect of inlet velocity, $\Psi = 0.0$, (**b**) effect of inlet velocity, $\Psi = 0.8$, (**c**) effect of excess air Ψ, $U_{in} = 1.25$ m/s

beginning of the combustion section ($x = 2$ cm), where steep temperature and species gradients were expected to appear. Two grids were employed with 120 and 240 nodes in the x direction, respectively. Figure 3 shows temperature profiles calculated with both mentioned grids and indicates that no detectable differences exists between the two sets of results. For this reason, all simulations in this work make use of the 120 node stretched grid.

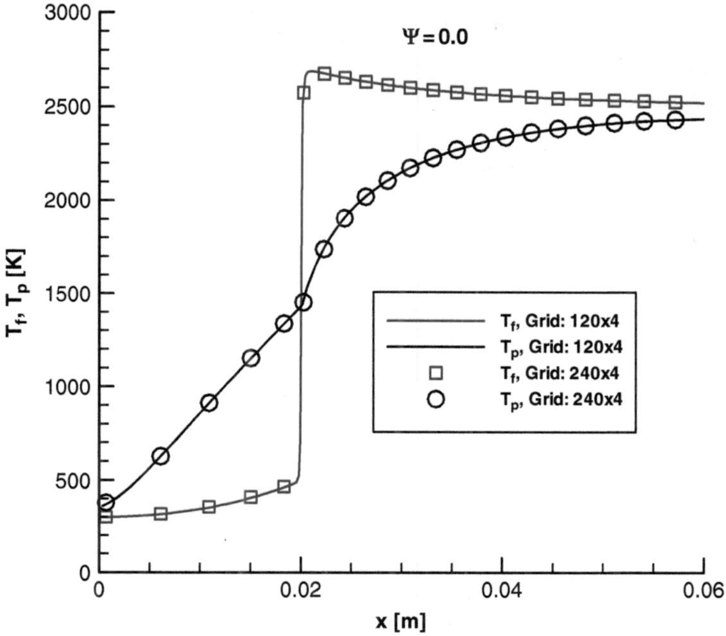

Fig. 3 Grid independence studies for one-dimensional cases, NLTE model

Figure 4a shows the effect of excess air Ψ on the gas temperature, T_f, and solid temperature, T_p. Temperature levels for the stoichiometric case and for $\Psi = 0.8$ gave numerical values close to those from [31]. Likewise, mass fraction behavior of species CH_4, CO_2 (Fig. 4b) and H_2O, O_2 (Fig. 4c) follow closely those reported by [31], for the one-equation simple combustion model here presented. Excess air reduces the final mass fraction of CO_2 and water and raises the amount of oxygen not participating in the combustion reaction. These results are the outcome of the single step reaction (22) that links the consumption and production rates of individual constituents of the mixture.

Figure 5 shows the dependence of temperature levels on inlet velocities U_{in}. As axial flow is increased, one can note a slight reduction of peak values of temperatures, follow by the movement of the flame towards the exit of the burner. Although the movement of the flame front is in accordance with simulations by [32], here a reduction on the maximum values of temperatures was calculated, which is in disagreement with findings in the literature [32] where the temperature rises as the inlet mass flow rate is increased. One possible explanation for this contrary behavior is that there are a number of distinct parameters and assumptions in both calculations sets, here and in [32], spanning from mathematical to numerical modeling hypotheses, which might affect the final results.

Four different thermo-mechanical models are now compared, namely Laminar, Laminar with Radiation Transport, Turbulent, Turbulent with Radiation Transport.

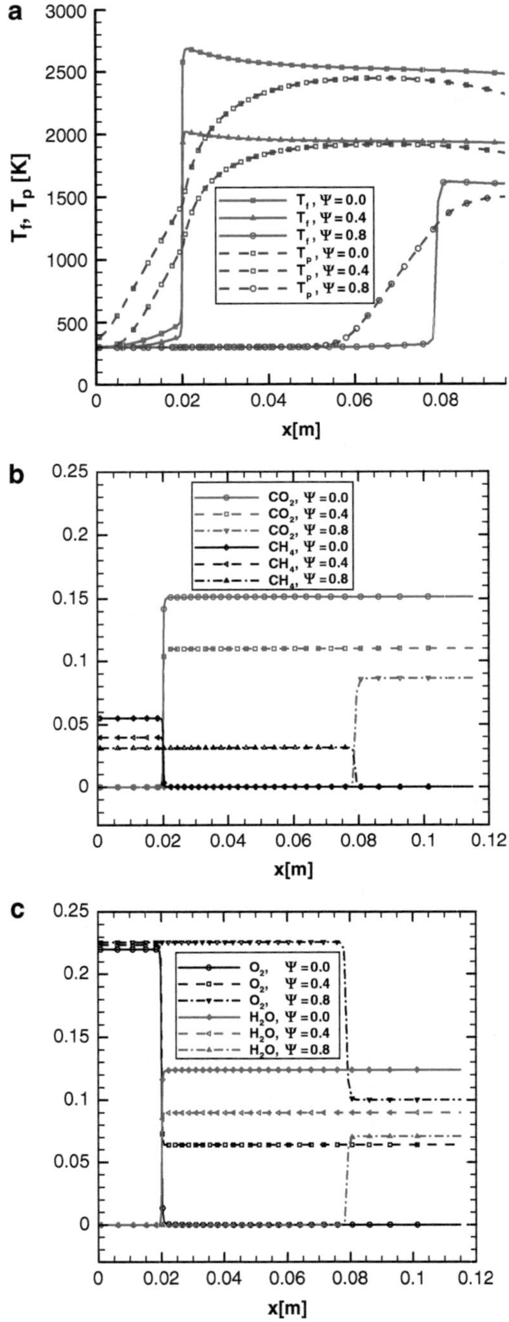

Fig. 4 Effect of excess air Ψ for $U_{in} = 0.1$ m/s on: (**a**) temperature fields, (**b**) fuel and carbon dioxide, (**c**) water and oxygen

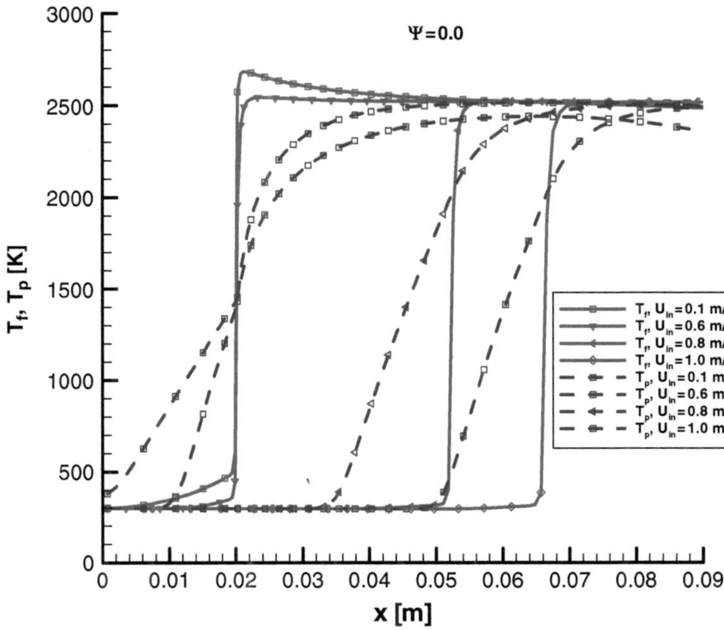

Fig. 5 Effect of inlet gas velocity on temperature fields

Radiation model is included by considering the radiation transport term in the $T_?$ – equation (13). Turbulence modeling is handled by resolving the $k-\varepsilon$ model equations (3) and (4) in addition to solving for the macroscopic turbulent eddy viscosity μ_{t_φ}, equation (6). In all cases, combustion is simulated via a unique simple closure, which is presented by equations (22) and (26).

Numerical simulations obtained with different models are presented in Fig. 6 for two values of U_{in}. Figure 6a shows that for a low value of U_{in}, the flame (solid lines) stabilizes close to the beginning of the burning section ($x = 2$ cm), independently of the mathematical model applied. Solid temperature are influenced by radiation transport, which tends to smooth out temperature differences within the solid matrix, enhancing, as such, the regenerative advantage of porous burners (dashed lines). Regeneration is achieved by preheating the gas prior to the combustion zone. In fact, the use of a turbulence model in conjunction with radiation transport gives the higher temperature peak of the gas temperature at the flame position. Increasing the inlet mass flow rate (Fig. 6b), the flame is pushed towards the burners exit, regardless of the model used. Here also radiation transport substantially affects the solid temperature distribution, but definitive conclusions on the appropriateness of each model can only be reached after careful comparison with experimental measurements. This shall be the subject of future research efforts.

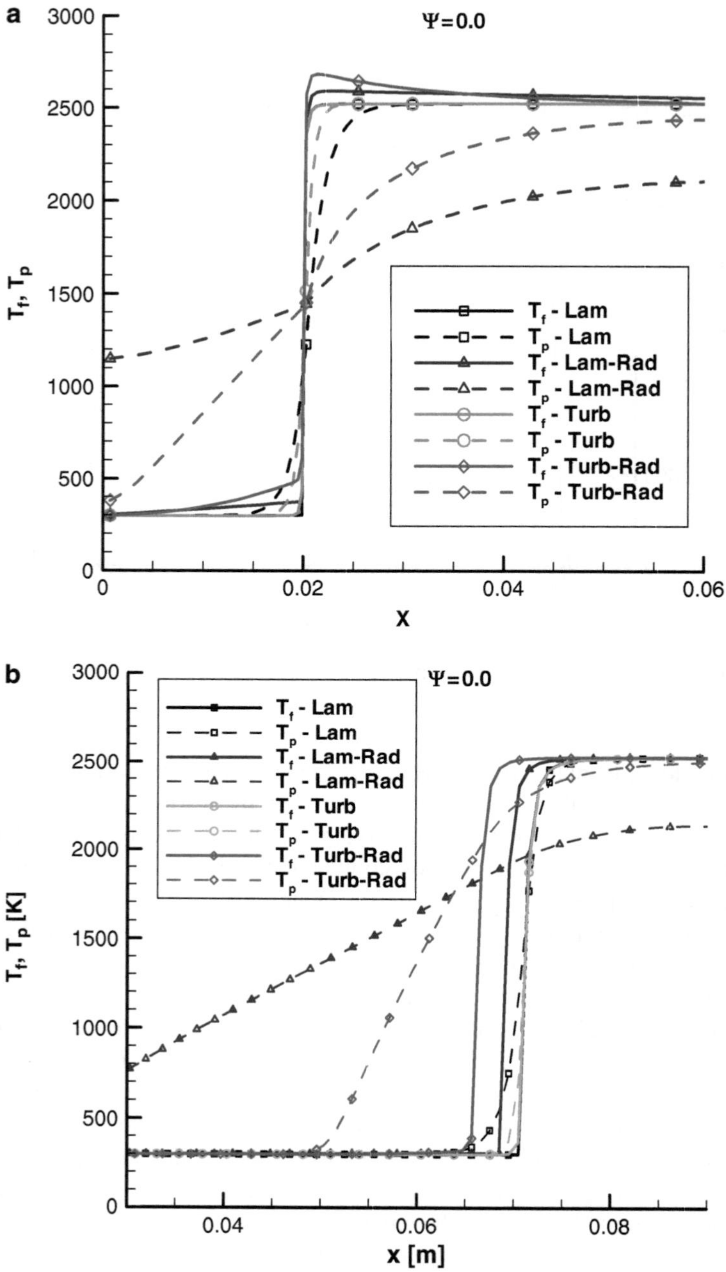

Fig. 6 Comparison of different model solutions: (**a**) $U_{\text{in}} = 0.1$ m/s, (**b**) $U_{\text{in}} = 1.0$ m/s

4 Concluding Remarks

This chapter presented two- and one-dimensional simulations for a mixture of air and methane burning in a porous material. Both LTE and NLTE models were applied. Four different thermo-mechanical models were compared along with a unique simple closure for combustion. Results indicate that a substantially different temperature distribution pattern is obtained depending on the model used. For high excess air or gas velocity, the flame front moves towards the exit of the burner. Results herein motivates further research work on the subject of reactive turbulent flow in porous burners and should be seen as a preliminary step towards reliable simulation of real porous combustors.

Acknowledgments The author is indebted to CNPq, CAPES and FAPESP, Brazil, for their invaluable continuous support during the course of his research work at ITA, Brazil.

References

1. Howell, J.R., Hall, M.J., Ellzey, J.L.: Combustion of hydrocarbon fuels within porous inert media. Progress in Energy and Combustion Sciences **22**(2), 121–145 (1996)
2. Oliveira, A.A.M., Kaviany, M.: Non equilibrium in the transport of heat and reactants in combustion in porous media. Progress in Energy and Combustion Sciences **27**(5), 523–545 (2001)
3. Henneke, M.R., Ellzey, J.L.: Modeling of filtration combustion in a packed bed. Combust. Flame **117**(4), 832–840 (1999)
4. Bouma, P.H., De Goey, L.P.H.: Premixed combustion on ceramic foam burners. Combust. Flame **119**(1–2), 133–143 (1999)
5. Babkin, V.S.: Filtration combustion of gases – present state of affairs and prospects. Pure Appl. Chem. **65**(2), 335–344 (1993)
6. Leonardi, S.A., Viskanta, R., Gore, J.P.: Analytical and experimental study of combustion and heat transfer in submerged flame metal fiber burners/heaters. J. Heat Transf. **125**(1), 118–125 (2003)
7. Lammers, F.A., De Goey, L.P.H.: A numerical study of flash back of laminar premixed flames in ceramic-foam surface burners. Combust. Flame **133**(1–2), 47–61 (2003)
8. Mohamad, A.A., Ramadhyani, S., Viskanta, R.: Modeling of combustion and heat-transfer in a packed-bed with embedded coolant tubes. Int. J. Heat Mass Transf. **37**(8), 1181–1191 (1994)
9. Wood, S., Harries, A.T.: Porous burners for lean-burn applications. Progress in Energy and Combustion Sciences **34**, 667–684 (2008)
10. Abdul, M.M., Abdullah, M.Z., Abu Bakar, M.Z., Mohamad, A.A., Abdullaha, M.K.: A review of investigations on liquid fuel combustion in porous inert media. Progress in Energy and Combustion Sciences **35**, 216–230 (2009)
11. Hsu, P.-F., Howell, J.R., Matthews, R.D.: A numerical investigation of premixed combustion within porous inert media. J. Heat Transf. **115**, 744–750 (1993)
12. Peard, T.E., Peters, J.E., Brewster, B., Buckius, R.O.: Radiative heat transfer augmentation in gas-fired radiant tube burner by porous inserts: effect on insert geometry. Exp. Heat Transf. **6**, 273–286 (1993)
13. Lim, I.G., Matthews, R.D.: Development of a model for turbulent combustion within porous inert media. Transp. Phenom. Therm. Eng., Begell House Inc. Publ., pp. 631–636 (1993)

14. Jones, W.P., Launder, B.E.: The prediction of laminarization with two-equation model of turbulence. Int. J. Heat Mass Transf. **15**, 301–314 (1972)
15. Sahraoui, M., Kaviany, M.: Direct simulation vs time-averaged treatment of adiabatic. Premixed flame in a porous medium. Int. J. Heat Mass Transf. **18**, 2817–2834 (1995)
16. Pedras, M.H.J., de Lemos, M.J.S.: Computation of turbulent flow in porous media using a low-Reynolds k-epsilon model and an infinite array of transversally displaced elliptic rods. Numer. Heat Transf. A Appl. **43**(6), 585–602 (2003)
17. de Lemos, M.J.S.: Turbulent kinetic energy distribution across the interface between a porous medium and a clear region. Int. Commun. Heat Mass Transf. **32**(1–2), 107–115 (2005)
18. de Lemos, M.J.S.: Turbulence in Porous Media: Modeling and Applications. Elsevier, Amsterdam (2006). 384 pgs. ISBN 0-08-044491-1
19. Santos, N.B., de Lemos, M.J.S.: Flow and heat transfer in a parallel-plate channel with porous and solid baffles. Numer. Heat Transf. A Appl. **49**(5), 471–494 (2006)
20. Assato, M., Pedras, M.H.J., de Lemos, M.J.S.: Numerical solution of turbulent channel flow past a backward-facing step with a porous insert using linear and nonlinear k-epsilon models. J. Porous Mater. **8**(1), 13–29 (2005)
21. Braga, E.J., de Lemos, M.J.S.: Turbulent natural convection in a porous square cavity computed with a macroscopic kappa-epsilon model. Int. J. Heat Mass Transf. **47**(26), 5639–5650 (2004)
22. Slattery, J.C.: Flow of viscoelastic fluids through porous media. AIChE J. **13**, 1066–1071 (1967)
23. Whitaker, S.: Advances in theory of fluid motion in porous media. Ind. Eng. Chem. **61**, 14–28 (1969)
24. Gray, W.G., Lee, P.C.Y.: On the theorems for local volume averaging of multiphase system. Int. J. Multiphas. Flow **3**, 333–340 (1977)
25. de Neef, M., Knabner, P., Summ, G.: Numerical bifurcation analysis of premixed combustion in porous inert media. Unpublished (1999). http://citeseer.csail.mit.edu/197085.html.
26. Saito, M.B., de Lemos, M.J.S.: A correlation for interfacial heat transfer coefficient for turbulent flow over an array of square rods. J. Heat Transf. **128**, 444–452 (2006)
27. Kuwahara, F., Shirota, M., Nakayama, A.: A numerical study of interfacial convective heat transfer coefficient in two-energy equation model for convection in porous media. Int. J. Heat Mass Transf. **44**, 1153–1159 (2001)
28. Mohamad, A.A., Viskanta, R., Ramadhyani, S.: Numerical prediction of combustion and heat transfer in a packed bed with embedded coolant tubes. Combust. Sci. Technol. **96**, 387–407 (1994)
29. Kwo, K.K.: Principles of Combustion. Wiley, New York (1986)
30. Trimis, D., Durst, F.: Combustion in a porous medium – advances and applications. Combust. Sci. Technol. **121**, 153–168 (1996)
31. Zhou, X.Y., Pereira, J.C.F.: Comparison of four combustion models for simulating the premixed combustion in inert porous media. Fire Mater. **22**, 187–197 (1998)
32. Mohamad, A.A.: Axial and radial porous burners. In: Ingham, D.B., Mamut, E., Bejan, A., Pop, I. (eds) Chap 24 in Emerging Technologies and Techniques in Porous Media. Springer, Berlin (2004)

MIX
Papier aus verantwortungsvollen Quellen
Paper from responsible sources
FSC® C105338

If you have any concerns about our products,
you can contact us on
ProductSafety@springernature.com

In case Publisher is established outside the EU,
the EU authorized representative is:
Springer Nature Customer Service Center GmbH
Europaplatz 3, 69115 Heidelberg, Germany

Printed by Libri Plureos GmbH
in Hamburg, Germany